Plant Developmental Biology - Biotechnological Perspectives: Volume 2

Eng-Chong Pua • Michael R. Davey
Editors

Plant Developmental Biology - Biotechnological Perspectives

Volume 2

 Springer

Editors

Prof. Dr. Eng-Chong Pua
New Era College
Jalan Bukit
43000 Kajang
Selangor
Malaysia
eng.chong.pua@gmail.com

Dr. Michael R. Davey
Division of Plant and Crop Sciences
School of Biosciences
University of Nottingham
Sutton Bonington
Loughborough LE12 5RD
UK
mike.davey@nottingham.ac.uk

ISBN 978-3-642-04669-8   e-ISBN 978-3-642-04670-4
DOI 10.1007/978-3-642-04670-4
Springer Heidelberg Dordrecht London New York

Library of Congress Control Number: 2009932129

Cover design: WMXDesign GmbH, Heidelberg, Germany

Printed on acid-free paper

Springer is part of Springer Science+Business Media (www.springer.com)

Preface

Many exciting discoveries in recent decades have contributed new knowledge to our understanding of the mechanisms that regulate various stages of plant growth and development. Such information, coupled with advances in cell and molecular biology, is fundamental to crop improvement using biotechnological approaches.

Two volumes constitute the present work. The first, comprising 22 chapters, commences with introductions relating to gene regulatory models for plant development and crop improvement, particularly the use of Arabidopsis as a model plant. These chapters are followed by specific topics that focus on different developmental aspects associated with vegetative and reproductive phases of the life cycle of a plant. Six chapters discuss vegetative growth and development. Their contents consider topics such as shoot branching, bud dormancy and growth, the development of roots, nodules and tubers, and senescence. The reproductive phase of plant development is in 14 chapters that present topics such as floral organ initiation and the regulation of flowering, the development of male and female gametes, pollen germination and tube growth, fertilization, fruit development and ripening, seed development, dormancy, germination, and apomixis. Male sterility and self-incompatibility are also discussed.

Volume 2 has 20 chapters, three of which review recent advances in somatic embryogenesis, microspore embryogenesis and somaclonal variation. Seven of the chapters target plant processes and their regulation, including photosynthate partitioning, seed maturation and seed storage protein biosynthesis, the production and regulation of fatty acids, vitamins, alkaloids and flower pigments, and flower scent. This second book also contains four chapters on hormonal and environmental signaling (amino compounds-containing lipids, auxin, cytokinin, and light) in the regulation of plant development; other topics encompass the molecular genetics of developmental regulation, including RNA silencing, DNA methylation, epigenetics, activation tagging, homologous recombination, and the engineering of synthetic promoters.

These books will serve as key references for advanced students and researchers involved in a range of plant-orientated disciplines, including genetics, cell and molecular biology, functional genomics, and biotechnology.

August 2009 *E-C. Pua and M. R. Davey*

Contents

Part II   Plant Processes and Their Regulation

Contributors

K. Abe Division of Plant Sciences, National Institute of Agrobiological Sciences, 2-1-2 Kannondai, Tsukuba, Ibaraki 305-8602, Japan

A. Berr Institut de Biologie Moléculaire des Plantes (IBMP), Centre National de la Recherche Scientifique (CNRS), Université de Strasbourg (UdS), 12 Rue du Général Zimmer, 67084 Strasbourg Cedex, France

F.C. Botha BSES, P.O. Box 68, Indooroopilly, 4068 Qld, Australia; Institute for Plant Biotechnology, Stellenbosch University, Private Bag X1, Matieland 7602, South Africa

V. Burlat Université François Rabelais de Tours, EA 2106 "Biomolécules et Biotechnologies Végétales", IFR 135 "Imagerie Fonctionnelle", 37200 Tours, France; Surfaces Cellulaires et Signalisation chez les Végétaux, UMR 5546 CNRS - UPS - Université de Toulouse, Pôle de Biotechnologie Végétale, 24 Chemin de Borde-Rouge, BP 42617 Auzeville, 31326 Castanet-Tolosan, France

M.E. Carmody Research School of Biology, Biochemistry and Molecular Biology, The Australian National University, Building 41, Linnaeus Way, Canberra ACT 0200, Australia

C.I. Cazzonelli Research School of Biology, Biochemistry and Molecular Biology, The Australian National University, Building 41, Linnaeus Way, Canberra ACT 0200, Australia

S-K. Chen Australian Research Council Centre of Excellence for Integrative Legume Research, School of Environmental and Life Sciences, The University of Newcastle, University Drive, Callaghan, NSW 2308, Australia

V. Courdavault Université François Rabelais de Tours, EA 2106 "Biomolécules et Biotechnologies Végétales", IFR 135 "Imagerie Fonctionnelle", 37200 Tours, France

S.J. Curtin CSIRO Plant Industry, GPO Box 1600, Canberra, ACT 2601, Australia; School of Wine & Food Science, Charles Sturt University, Wagga Wagga, NSW 2678, Australia; University of Minnesota, St. Paul, MN 55108-6026, USA

K.M. Davies New Zealand Institute for Plant & Food Research Limited, Private Bag 11600, Palmerston North 4442, New Zealand

A. Eamens CSIRO Plant Industry, GPO Box 1600, Canberra, ACT 2601, Australia; School of Molecular and Microbial Biosciences, University of Sydney, Sydney, NSW 2006, Australia

U. Effmert University of Rostock, Institute of Biological Sciences, Albert-Einstein-Str. 3, 18059 Rostock, Germany

M. Endo Division of Plant Sciences, National Institute of Agrobiological Sciences, 2-1-2 Kannondai, Tsukuba, Ibaraki 305-8602, Japan

E.J. Finnegan CSIRO, Plant Industry, GPO Box 1600, Canberra, ACT 2601, Australia

A. Galstyan Centre for Research in Agricultural Genomics (CSIC-IRTA-UAB), c. Jordi Girona 18-26, 08034 Barcelona, Spain

G. Guirimand Université François Rabelais de Tours, EA 2106 "Biomolécules et Biotechnologies Végétales", IFR 135 "Imagerie Fonctionnelle", 37200 Tours, France

N.G. Halford Plant Science Department, Rothamsted Research, Harpenden, Hertfordshire AL5 2JQ, UK

D. Hussain Research School of Biology, Biochemistry and Molecular Biology, The Australian National University, Building 41, Linnaeus Way, Canberra ACT 0200, Australia

T. Kiba RIKEN Plant Science Center, 1-7-22 Suehiro, Tsurumi, Yokohama 230-0045, Japan

S. Kurdyukov Australian Research Council Centre of Excellence for Integrative Legume Research, School of Environmental and Life Sciences, The University of Newcastle, University Drive, Callaghan, NSW 2308, Australia

R. Linacero Departamento de Genética, Facultad de Biología, Universidad Complutense, 28008 Madrid, Spain

J. López-Bucio Instituto de Investigaciones Químico-Biológicas, Universidad Michoacana de San Nicolás de Hidalgo, CP 58030 Morelia, Michoacán, México

F.R. Mantiri Australian Research Council Centre of Excellence for Integrative Legume Research, School of Environmental and Life Sciences, The University of Newcastle, University Drive, Callaghan, NSW 2308, Australia

N. Marsch-Martínez Laboratorio Nacional de Genómica para la Biodiversidad, CINVESTAV-IPN, Campus Guanajuato, C.P. 36821, Irapuato, Gto. México

J.F. Martínez-García Centre for Research in Agricultural Genomics (CSIC-IRTA-UAB), c. Jordi Girona 18-26, 08034 Barcelona, Spain; Institució Catalana de Recerca i Estudis Avançants (ICREA), Passeig Luís Companys 23, 08010-Barcelona, Spain

A. Méndez-Bravo Instituto de Investigaciones Químico-Biológicas, Universidad Michoacana de San Nicolás de Hidalgo, CP 58030 Morelia, Michoacán, México

V. Nachiappan Department of Biochemistry, Bharathidasan University, Tiruchirappalli 620024, India

N. Nisar Research School of Biology, Biochemistry and Molecular Biology, The Australian National University, Building 41, Linnaeus Way, Canberra ACT 0200, Australia

K.E. Nolan Australian Research Council Centre of Excellence for Integrative Legume Research, School of Environmental and Life Sciences, The University of Newcastle, University Drive, Callaghan, NSW 2308, Australia

A. Olmedilla Department of Biochemistry, Cell and Molecular Biology of Plants, Estación Experimental del Zaidín CSIC, Profesor Albareda 1, 18008 Granada, Spain

R. Ortiz-Castro Instituto de Investigaciones Químico-Biológicas, Universidad Michoacana de San Nicolás de Hidalgo, CP 58030 Morelia, Michoacán, México

K. Osakabe Division of Plant Sciences, National Institute of Agrobiological Sciences, 2-1-2 Kannondai, Tsukuba, Ibaraki 305-8602, Japan

A. Pereira Virginia Bioinformatics Institute, Virginia Polytechnic Institute and State University, Washington Street, Virginia Tech, Blacksburg, VA 24061, USA

B. Piechulla University of Rostock, Institute of Biological Sciences, Albert-Einstein-Str. 3, 18059 Rostock, Germany

B.J. Pogson Research School of Biology, Biochemistry and Molecular Biology, The Australian National University, Building 41, Linnaeus Way, Canberra ACT 0200, Australia

R. Rajasekharan Department of Biochemistry, Indian Institute of Science, Bangalore 560012, India; School of Science, Monash University Sunway Campus, 46150 Bandar Sunway, Selangor, Malaysia

R.J. Rose Australian Research Council Centre of Excellence for Integrative Legume Research, School of Environmental and Life Sciences, The University of Newcastle, University Drive, Callaghan, NSW 2308, Australia

H. Sakakibara RIKEN Plant Science Center, 1-7-22 Suehiro, Tsurumi, Yokohama 230-0045, Japan

K.E. Schwinn New Zealand Institute for Plant & Food Research Limited, Private Bag 11600, Palmerston North 4442, New Zealand

M.B. Sheahan Australian Research Council Centre of Excellence for Integrative Legume Research, School of Environmental and Life Sciences, The University of Newcastle, University Drive, Callaghan, NSW 2308, Australia

W.H. Shen Institut de Biologie Moléculaire des Plantes (IBMP), Centre National de la Recherche Scientifique (CNRS), Université de Strasbourg (UdS), 12 rue du Général Zimmer, 67084 Strasbourg cedex, France

N. Sreenivasulu Leibniz Institute of Plant Genetics and Crop Plant Research (IPK), 06466 Gatersleben, Germany

B. St-Pierre Université François Rabelais de Tours, EA 2106 "Biomolécules et Biotechnologies Végétales", IFR 135 "Imagerie Fonctionnelle", 37200 Tours, France

S. Toki Division of Plant Sciences, National Institute of Agrobiological Sciences, 2-1-2 Kannondai, Tsukuba, Ibaraki 305-8602, Japan

A.M. Vázquez Departamento de Genética, Facultad de Biología, Universidad Complutense, 28008 Madrid, Spain

M. Venter Department of Genetics, Stellenbosch University, Private Bag X1, Matieland 7602, South Africa

X-D. Wang Australian Research Council Centre of Excellence for Integrative Legume Research, School of Environmental and Life Sciences, The University of Newcastle, University Drive, Callaghan, NSW 2308, Australia

P.M. Waterhouse CSIRO Plant Industry, GPO Box 1600, Canberra, ACT 2601, Australia; School of Molecular and Microbial Biosciences, University of Sydney, Sydney, NSW 2006, Australia

H. Weber Leibniz Institute of Plant Genetics and Crop Plant Research (IPK), 06466 Gatersleben, Germany

W. Weschke Leibniz Institute of Plant Genetics and Crop Plant Research (IPK), 06466 Gatersleben, Germany

Y. Zhao Section of Cell and Developmental Biology, University of California San Diego, 9500 Gilman Drive, La Jolla, CA 92093-0116, USA

Part I
Cell Differentiation and Development In Vitro

Chapter 1
Developmental Biology of Somatic Embryogenesis

R.J. Rose, F.R. Mantiri, S. Kurdyukov, S-K. Chen, X-D. Wang, K.E. Nolan, and M.B. Sheahan

1.1 Introduction

Somatic embryogenesis (SE) is a remarkable developmental process enabling nonzygotic plant cells, including haploid cells, to form embryos and, ultimately, fertile plants. An expression of totipotency, this asexual process involves dedifferentiation of a nonzygotic cell and subsequent redifferentiation (reprogramming), resulting in the production of all cells characteristic of the mature plant. Although primarily considered in the context of in vitro SE from cultured tissues or cells, SE is also a naturally occurring means of asexual reproduction in some species. For example, somatic embryos form on the succulent leaves of *Kalanchoë* (Garcês et al. 2007) and apomixis, a clonal reproductive process where embryos derive from cells in the ovule wall producing seed genotypically identical to the parent, occurs in a number of evolutionarily divergent plant species (Koltunow and Grossniklaus 2003). This discussion focuses on in vitro SE, but an understanding of this process is applicable to all forms of SE.

SE was first demonstrated 50 years ago by Steward et al. (1958) and Reinert (1958) in carrot. More recently, the emphasis has been on inducing SE in an ever-increasing number of species and cultivars to facilitate genetic transformation. The molecular mechanisms of SE remain poorly understood, with mechanistic studies focused primarily on the hormonology of this developmental process. The classic paradigm is that auxin is initially required to induce SE, while its subsequent withdrawal, or reduction in concentration, drives embryo development (Halperin 1966; Dudits et al. 1991). In many species, cytokinin in addition to auxin is required for SE (Nolan and Rose 1998). These are not the only hormones to consider, but

R.J. Rose, F.R. Mantiri, S. Kurdyukov, S-K. Chen, X-D. Wang, K.E. Nolan,
and M.B. Sheahan
Australian Research Council Centre of Excellence for Integrative Legume Research, School of Environmental and Life Sciences, The University of Newcastle, University Drive, Callaghan, NSW 2308, Australia
e-mail: Ray.Rose@newcastle.edu.au

usually the critical ones. Understanding of how auxin and cytokinin signalling regulates transcription has advanced in recent times (Müller and Sheen 2007; Tan et al. 2007). Application of this understanding to SE, however, is limited and requires an understanding of the interaction of these hormonal signalling pathways with key developmental genes. Increasingly, stress has been recognised as having a critical role in the induction of SE (Touraev et al. 1997; Fehér et al. 2003; Nolan et al. 2006). Thus, hormones and stress collectively induce cell dedifferentiation and initiate an embryogenic program in plants with a responsive genotype (Fehér et al. 2003; Ikeda-Iwai et al. 2003; Rose and Nolan 2006).

SE in vitro takes two forms, indirect and direct, referring respectively to the presence or absence of a phase of callus development (Williams and Maheswaran 1986; Dijak et al. 1986). As indirect SE is more commonly employed, this form is emphasised, and direct SE is also discussed, given that there is likely a continuum in terms of the amount of dedifferentiation and cell proliferation required before somatic embryos are induced.

The ability to produce embryos from nonzygotic cells and to regenerate whole plants is one of the fundamental questions of contemporary biology (Vogel 2005). This chapter summarises current understanding of the cell and molecular biology of SE and discusses the implications SE has to biotechnology.

1.2 Basic Requirements for In Vitro SE

Formation of somatic embryos in a given species typically entails culturing an explant (of appropriate tissue type and genotype) in a basal medium with a suitable hormone regime (Rose 2004). Importantly, excision and culture of the tissue introduces a stress component. A number of studies indicate that stress itself acts as an inducer of SE (Kamada et al. 1993; Touraev et al. 1997; Ikeda-Iwai et al. 2003) and, indeed, is the first event experienced by cells on excision of the explant from the plant tissue or the isolation of a cell before initiating the culture process. The basal medium can modify the amount of SE in a given explant (Ammirato 1983) and is, of course, essential for SE, but is usually not a key regulator. In considering the mechanism of SE, Fig. 1.1 needs to be understood, which will serve as the basis for considering SE in this chapter.

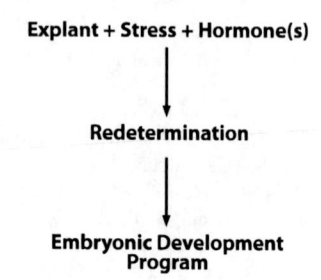

Explant + Stress + Hormone(s)

↓

Redetermination

↓

Embryonic Development Program

Fig. 1.1 The main conceptual components to consider for an understanding of the mechanism of somatic embryogenesis

1.3 Explant and Stem Cell Biology

It may be possible in some circumstances for a somatic embryo to derive from more than a single cell (Williams and Maheswaran 1986). However, it is known from the tracking studies of de Vries and co-workers that an embryo can develop from a single totipotent somatic cell—both in the classic carrot cell suspension system (Schmidt et al. 1997) as well as in other species where somatic embryos are derived from leaf explants (Somleva et al. 2000). These single-cell progenitors of somatic embryos are totipotent stem cells as opposed to the pluripotent cells of plant meristems. Given the developmental programming required for embryo formation and its vascular independence from surrounding cells or callus (Haccius 1978), it seems likely that development from totipotent stem cells is the norm for SE. What is far from clear is which genotype and specific cell type enable SE.

1.3.1 Genotype

It is apparent from many studies, and a serious problem for biotechnology, that only some cultivars within a species are regenerable by SE (Bingham et al. 1975; Ammirato 1983; Vasil 1988). Work by Bingham et al. (1975) in *Medicago sativa* demonstrated the possibility to breed for SE, with an ensuing study by Kielly and Bowley (1992), suggesting that a SE-competent phenotype is dominant and that two genetic loci are involved. However, the identity of the genes residing at these loci is not clear. While genotypic influences may have a genetic basis, another line of enquiry suggests a role for epigenetic phenomena, in particular, the pattern of chromatin condensation. Investigations in plant epigenetics reveal heterochromatin to be marked by methylation of cytosine and histone H3 (Bender 2004). Data supporting epigenetic control of SE are not extensive, but provide support for the control of SE by methylation pattern. Thus, disruption of DNA methylation pattern with 5-azacytidine in *M. truncatula* (Santos and Fevereiro 2002) and carrot (Yamamoto et al. 2005) disrupts SE. Recent studies show the involvement of small interfering RNAs in the condensation of chromatin (Henderson and Jacobsen 2007; Bäurle et al. 2007). The DNA methylation patterns are heritable and can be maintained across mitosis (Henderson and Jacobsen 2007).

In *M. truncatula*, all highly embryogenic genotypes, including Jemalong 2HA (Nolan et al. 1989; Rose et al. 1999), R108 (Hoffman et al. 1997) and M9-10a (Araújo et al. 2004), were obtained after a cycle of tissue culture. These results suggest the regeneration process selects for somatic cells with competence for SE, a heritable trait. Moreover, it suggests the regeneration process may provide a consistent selection for an epigenetic pattern in the cultured somatic cells, as a cycle of tissue culture is always enough to enhance regenerability (Nolan et al. 1989; Hoffman et al. 1997).

In plant development, genes promoting embryogenesis ought to be repressed as nonzygotic cells develop in the plant body. There is evidence that SE could be induced in the Arabidopsis *pickle* mutant (Ogas et al. 1997) by derepressing the embryogenesis pathway. *PICKLE* acts to repress embryonic traits (Henderson et al. 2004) and encodes a CHD3 chromatin-remodelling factor (Ogas et al. 1999). It is possible that activation of the expression of embryogenesis-promoting genes, repressed during development, enables SE. The repression of embryogenesis appears to involve gibberellic acid (GA) metabolism, discussed below in this chapter (Henderson et al. 2004).

1.3.2 Explant Cells

When callus forms from an explant before the appearance of somatic embryos, a question arises as to which cells the embryos derive. Historically, this was investigated and discussed in relation to carrot cell suspension cultures (Halperin 1966). In establishing cell suspension cultures, explants have been taken from various tissues, including hypocotyls (Guzzo et al. 1994; Schmidt et al. 1997), where cell division is initiated in provascular cells that form embryogenic callus that is transferred into liquid cell suspension culture. In the original work on SE by Steward et al. (1958, 1964), explants were obtained from the phloem of the storage root of carrot and then SE induced in liquid culture. What is possibly the case here in the carrot studies is the presence in these tissues of procambial-like cells (involved in the differentiation of vascular tissue) that readily reprogram to a totipotent state. Once the carrot cells are in liquid culture in the presence of auxin, proembryogenic masses (PEMs) form. Further embryonic development does not then occur until removal of auxin. Tracking cells in a newly initiated cell suspension culture via expression of a p*SERK-LUC* fusion reveals that most somatic embryos develop from single cells or small cell clusters originally derived from provascular cells (Schmidt et al. 1997). In established cultures, somatic embryos derive from cells in PEMs.

As plant biotechnology progressed, regeneration via SE was developed for many species using agar-solidified medium. Histological studies often focused on demonstrating that regeneration was via SE, rather than organogenesis (e.g. Sharma and Millam 2004). In the Sharma and Millam (2004) study, meristematic zones from embryogenically induced internode segments were noted and interpreted as PEMs. Our unpublished data with *M. truncatu*la, using both mesophyll protoplasts and leaf explants, provide support for two types of progenitor cells for SE. Callus that derives from leaf explants is permeated by vascular tissue. Most SEs develop from the parenchyma-like callus cells near the epidermis (Fig. 1.2), while others are from the provascular or procambial tissue, mirroring that observed in carrot hypocotyls. There is good supporting evidence for embryos forming at the periphery of callus in a diversity of species such as guinea grass (Lu and Vasil 1985), pea (Loiseau et al. 1998), chickpea (Sagare et al. 1995) and potato (Sharma and Millam 2004). The other common feature with embryogenic callus in a number of studies is the link to vascular tissue (Lu and Vasil 1985, Schwendiman et al. 1988; Schmidt et al. 1997; Somleva et al. 2000), which goes back to the early studies of Steward

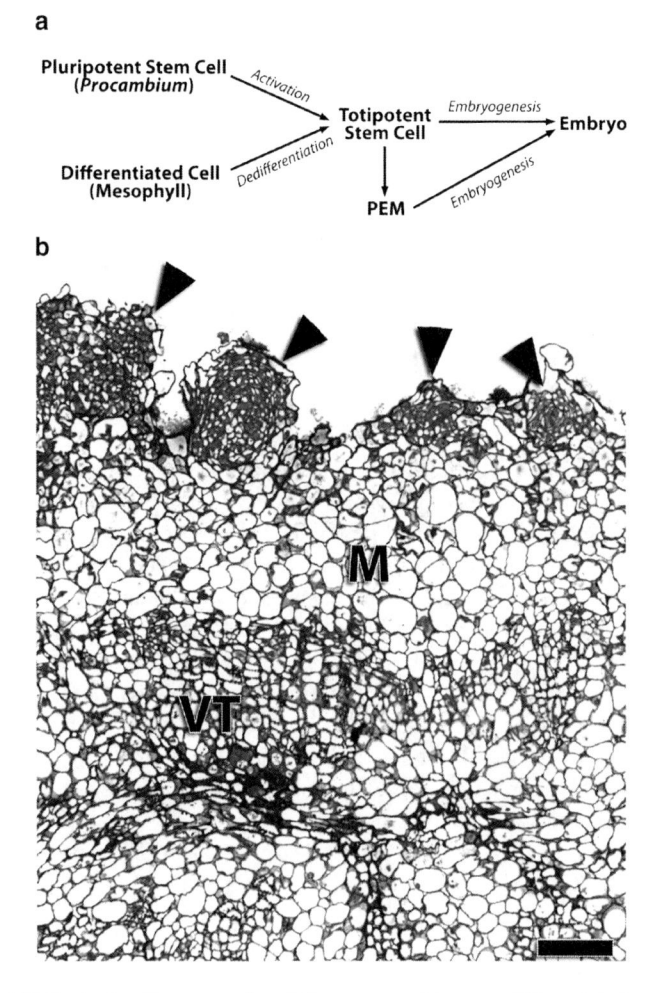

Fig. 1.2 (**a**) Totipotent cells are produced from procambium or differentiated cells, such as mesophyll cells, and initiate embryogenesis, or may divide a number of times to form a PEM before initiating embryogenesis. (**b**) Section through a *Medicago truncatula* leaf explant initiating somatic embryogenesis. The globular-stage somatic embryos have been initiated near the surface of the leaf away from the medium, quite separate from the vascular tissue, and arise from dedifferentiated mesophyll cells. M, dedifferentiating mesophyll cells; VT, vascular tissue; arrowheads, somatic embryos at different stages of development. Bar=100 μm

et al. (1958). Procambial cells are pluripotent stem cells characterised by their capacity to differentiate into cells of the vasculature. This was emphasised in studies of adventitious root formation in *M. truncatula*, where it was clear that cells proliferating from the procambial-like cells of the veins of leaf explants could differentiate into root meristems in response to auxin (Rose et al. 2006). Kwaaitaal and de Vries (2007) subsequently showed that SERK1 (which marks cells that form somatic embryos; see below) expressed in procambium cells and immature vascular cells (also observed in *M. truncatula*) responds to auxin.

The *M. truncatula* protoplast system (Rose and Nolan 1995) where SEs derive from mesophyll cells clearly illustrates that mature plant cells can dedifferentiate and, ultimately, form SEs without any initial vascular tissue being present (Mantiri et al. 2008). Direct SE from mesophyll protoplasts of *M. sativa* has also been demonstrated (Dijak et al. 1986). These data are brought together in a conceptual manner in Fig. 1.2.

1.4 Earliest Event in Embryogenesis—Asymmetric Cell Division

During zygotic embryogenesis, the zygote divides asymmetrically to produce an apical cell, which forms the majority of the embryo, and a basal cell, which forms the suspensor (Scheres and Benfey 1999). The apical cell is smaller and densely cytoplasmic and the basal cell larger and vacuolated (Jenik and Barton 2005). While the basal cell gives rise to the suspensor and tethers the embryo to the ovule, the suspensor cell adjacent to the apical cell (the hypophysis) generates the quiescent centre of the root pole, as well as the root cap and its associated stem cells. The remainder of the root meristem traces back to the apical cell (Jenik and Barton 2005). Therefore, the first cell divisions of the zygote delineate the polarity that determines the root and shoot apical meristems. Tight regulation of each embryonic cell division produces cellular patterning resulting, ultimately, in a heart-stage embryo. Cell division patterning appears to be controlled by auxin fluxes (Kepinski and Leyser 2003; Friml et al. 2003), *WOX* transcription factors (Haecker et al. 2004), cell wall components (Souter and Lindsey 2000; Ingram 2004) and a number of other genes (Chandler et al. 2007).

One protein linked directly to apical–basal pattern formation in the Arabidopsis embryo is GNOM, an ADP-ribosylation factor guanine-nucleotide exchange factor (ARF-GEF) required for correct localisation of, among other proteins, the PIN1 auxin transporter (Geldner et al. 2003). Mutations in *GNOM* can produce seedlings that lack basic polarity and possess unusual shapes (Mayer et al. 1993). This patterning can be related to alteration in the first cell division in the *gnom* mutant where two nearly equal-size cells are produced, rather than the distinctive asymmetric division (Mayer et al. 1993).

1.4.1 Cell Wall in Establishment of Polarity, Division Asymmetry and Cell Fate

One fundamental set of questions that exists in plant biology is how plant cell identity is determined, how cells decide whether to divide or expand, and whether they commit to a specific developmental fate. In the context of SE, specifically, how does a nonzygotic cell become committed to an embryogenic pathway and develop the intricate patterning present in an embryo? Numerous studies, both classical and recent, reveal that plant cell fate can be respecified according to its position relative to other cells (Ingram 2004). In other words, the fate of a cell depends more on its neighbours than its mother cell (Majewska-Sawka and Nothnagel 2000). This implies that cells continually communicate with one another during plant development. Since a rigid cell wall encompasses the plant cell, communication between the cell interior and the environment occurs via the cell wall (Satoh 1998). Indeed, increasing evidence eludes to the fundamental role that cell wall components (both protein and saccharide) have in regulating cell division patterns, cell shape, intercellular attachment, wound/defence-related signalling, programmed cell death (PCD), polarity establishment and determination of cell fate. Arabinogalactan proteins (AGPs), hydroxyproline-rich glycoproteins with complex carbohydrate side-chains comprising approximately 90% of the molecules mass, are believed to be of particular importance in the abovementioned processes (Majewska-Sawka and Nothnagel 2000).

The role of cell wall components in establishing axial–basal polarity (required for division asymmetry) is perhaps best illustrated by early development in *Fucus* zygotes. Here, deposition of sulphated polysaccharides at the base of the cell defines the basal–axial axis leading to the development of thallus (apical) and rhizoid (basal) cell after the first asymmetric division. Interestingly, remnant cell wall in protoplasts isolated from these cells is sufficient to direct the fate of the cells in culture (Souter and Lindsey 2000). Similarly, establishment of the axial-basal polarity in zygotes of higher plants may involve differential distribution of cell wall components. For example, during Brassica embryogenesis, the JIM8 antibody, which recognises a specific AGP epitope, localises only to cells fated to become the suspensor (Pennell et al. 1991; Souter and Lindsey 2000). Thus, polarity can be established before the first asymmetric division through differential localisation of AGPs.

Recently, a hydroxyproline-rich glycoprotein, RSH, has been shown to be essential for normal embryo patterning, with influences observed at the first asymmetric division of the zygote (Hall and Cannon 2002). In rapidly dividing cells of the embryo, RSH localises to the cell plate–cell wall junction, suggesting a role in positioning of the cell plate at cytokinesis. Mutation of *RSH* leads to morphologically defective embryos, with irregularly sized and shaped cells. Similarly, the cellulose synthase-like gene *AtCSLA7*, which encodes a processive β-glycosyltransferase, is critical for normal embryogenesis, causing homozygous *AtCSLA7* mutants to have slow-developing embryos with abnormal cell patterning that arrest at the globular stage (Goubet et al. 2003). Interestingly, mutation in *GNOM* produces a phenotype most similar to *RSH* and *AtCSLA7* mutants, suggesting that RSH and AtCSLA7 are cargo in GNOM-associated vesicles. Whether the role for GNOM in embryogenesis reflects secretion of cell wall components or localisation of auxin transporters is unclear (Ingram 2004).

1.4.2   Division Asymmetry in the Initiation of SE

Somatic embryos have a similar developmental pattern to zygotic embryos, passing through globular, heart, torpedo and cotyledonary stages (Rose 2004). However, somatic cells that form embryos do so in various different environments and tissues, notably without endosperm (Williams and Maheswaran 1986). Thus, are the signalling environment and the division asymmetry necessary for zygotic embryogenesis also required for SE? Evidence for the existence of asymmetric divisions comes from a substantive study involving cell suspension cultures of carrot (McCabe et al. 1997). In these cultures, JIM8+ (state B) cells undergo asymmetric division giving rise to a JIM8+ (state F) cell and a JIM8– (state C) cell. State F cells eventually undergo PCD, while state C cells are competent for SE providing that conditioned media factors from JIM8+ cultures (presumed to be AGPs) are present. Soluble signals from JIM8+ cells that direct state C cells along a SE pathway may be akin to endosperm-derived signals believed to direct development of the zygotic embryo (Ingram 2004), suggesting possible commonalities in the signalling environments of zygotic and somatic embryogenesis.

In *M. sativa* where leaf protoplasts may undergo direct SE, asymmetric cell divisions result in SE, whereas symmetrically dividing cells form callus (Dudits et al. 1991). Asymmetric cell divisions of elongated cells were also observed in the provascular tissue of carrot explants in response to auxin (Guzzo et al. 1994). These asymmetrically divided cells give rise to PEMs, from which embryos subsequently develop in the absence of auxin (presumably following a further asymmetric division). Morphological asymmetry, however, was not observed in *Cichorium* (Blervacq et al. 1995). The evidence for the involvement of asymmetric divisions in SE is best in the case of isolated single cells or protoplasts. The evidence in tissue explants or calli derived from protoplasts, however, is more difficult to obtain. The current authors have observed early asymmetric divisions in somatic embryo formation in calli from isolated protoplasts. In many cases of SE it is probably hard to obtain the clear-cut asymmetric division observed in the zygote, a possible explanation for why many somatic embryos do not progress to plants. Overall, however, there is support for asymmetrical divisions being an integral component of SE (Fehér et al. 2003).

1.4.3   Asymmetric Division and the Suspensor in SE

One of the difficulties in interpreting early development of somatic embryos is that of the suspensor. Are the asymmetric cell divisions of state B cells analogous to the first division of the zygote? Certainly, the JIM8 labelling pattern of divided state B cells is analogous to that observed after the first embryonic division in Brassica (see Sect. 1.4.1) and so, potentially, the cell undergoing PCD after the asymmetric division represents an aberrant suspensor. In the cell tracking study of Schmidt et al.

(1997), which used carrot cell suspension cultures expressing a *SERK*-reporter, a suspensor attached to elongated cells in the culture can be seen in one tracking series but not in another. Because of the unusual developmental context of SE, suspensor development is rather variable, ranging from a large group of cells to absent (Williams and Maheswaran 1986). Thus, the hypophysis–root pole relationship is unlikely to be the same in both zygotic and somatic embryogenesis.

1.5 Stress Component in the Initiation of SE

Stress has been proposed for some time as being an important component of SE (Dudits et al. 1991). The first signal perceived by cells destined to become somatic embryos relates to the wounding that accompanies excision of the tissue or isolation of the cells. In *M. truncatula*, there are many stress-related proteins associated with SE (Imin et al. 2004). A number of these proteins are differentially expressed between regenerators and non-regenerators (Imin et al. 2005).

1.5.1 Reactive Oxygen Species

One of the earliest responses to wounding and, indeed, many other abiotic and biotic stresses is increased production of reactive oxygen species (ROS) at the site of stress, the so-called oxidative burst (Orozco-Cardenas and Ryan 1999). ROS, including superoxide and hydroxyl radicals and hydrogen peroxide (H_2O_2), are produced as byproducts of oxygen metabolism (occurring in organelles, the cytosol, plasma membrane and apoplast) and can cause serious oxidative damage to the cell. However, it is increasingly recognised that ROS are important signalling molecules involved in plant development and environmental perception (Noctor 2006). Furthermore, ROS, particularly H_2O_2, function in lignification and oxidative cross-linking of the cell wall, control of cell growth, destruction (oxidation) of microbial pathogens and PCD/hypersensitive response (Dangl and Jones 2001; Gechev and Hille 2005).

Thus, ROS are produced by cultured explants in response to wounding. Indeed, staining explants with 3,3-diamidinobenzidine (DAB) reveals continual production of H_2O_2 in callus tissue during culture, although importantly not in somatic embryos that form on the callus (N.A. Saeed, M.B. Sheahan and R.J. Rose, unpublished data). Moreover, culturing explants in the presence of diphenyleneiodonium (DPI), an inhibitor of flavoprotein-dependent ROS production, prevents the re-initiation of cell division in *M. truncatula*, Arabidopsis and tobacco leaf explants, indicating a critical role for ROS in explant regeneration (N.A. Saeed, M.B. Sheahan and R.J. Rose, unpublished data). In addition, it is known that oxidative stress causes plants cells to acquire a less differentiated state

(Pasternak et al. 2002) and that oxidative stress enhances auxin-dependent cell cycle reactivation (Pasternak et al. 2005). Although ROS initially promote dedifferentiation and re-initiation of cell division, an inability to control levels of ROS may be a causal factor in recalcitrance of some species to undergo SE. Plants possess an arsenal of small-molecule antioxidants (ascorbate, glutathione and tocopherol) and antioxidant enzymes (including superoxide dismutases, peroxidases and catalase) and, in cultured tobacco protoplasts, up-regulation of the activities of superoxide dismutase, ascorbate peroxidase and glutathione reductase follows the oxidative burst, causing reduced forms of glutathione and ascorbate to predominate. Significantly, this change does not occur in non-totipotent cells and appears necessary for the expression of totipotency (Papadakis et al. 2001).

The nature of downstream events involved in wound-induced ROS signalling remains unclear. However, they may involve direct interaction between ROS and redox-sensitive transcription factors such as NPR1 (Mou et al. 2003), MAPK signalling cascades (Kovtun et al. 2000) or modification of thiol residues on proteins such as protein tyrosine phosphatases or glutathione peroxidases (Gupta and Luan 2003, Miao et al. 2006). ROS signalling may also be sensed via the level of small-molecule antioxidants such as glutathione. Indeed, progression from the G_1 to S phases of the cell cycle requires adequate levels of reduced glutathione that, in turn, may influence the expression or activity of A-type cyclins or the cyclin-dependent kinase inhibitor, CKI, respectively (Vernoux et al. 2000). Undoubtedly, manifestation of stress signalling involves cross-talk between hormonal and ROS signalling pathways and there exists substantial evidence linking auxin, abscisic acid, ethylene, jasmonic and salicylic acid signalling pathways with ROS signalling (Desikan et al. 2005).

Thus, ROS signalling appears critical to cellular reprogramming and re-initiation of the cell cycle. However, ROS also appear to be important for subsequent development of somatic embryos. Hence, H_2O_2 stimulates SE in *Lycium barbarum* (Cui et al. 1999; Kairong et al. 2002), glutathione redox state influences meristem development in *Picea glauca* SE (Stasolla et al. 2004; Belmonte et al. 2005; Belmonte and Stasolla 2007), while regions of somatic embryo formation are marked by high expression levels of an H_2O_2-producing oxalate oxidase in wheat (Caliskan et al. 2004). The effects of ROS on somatic embryo maturation may be mediated by its effects on cell walls that prevent cell expansion and cell–cell separation, thus promoting an 'embryogenically competent' microenvironment (Caliskan et al. 2004; Belmonte and Stasolla 2007).

1.5.2 Stress-Related Hormone Signalling

In our recent study in *M. truncatula* (Mantiri et al. 2008), where SE is induced by auxin and cytokinin, it was shown that the stress-related hormone ethylene is synthesised within the first 24 h of initiating culture. Moreover, inhibitors of ethylene perception or biosynthesis prevent SE and expression of the ethylene response transcription factor (ERF), *MtSERF1*, a member of the AP2/ERF super-family essential for SE (Mantiri et al. 2008). Knockdown of *MtSERF1* expression by RNAi results in almost total suppression of SE. The expression of *MtSERF1* is dependent on not only ethylene but also auxin and cytokinin. Furthermore, the *MtSERF1* promoter contains ethylene and auxin response elements and cytokinin-responsive ARR motifs. These data start to provide a molecular basis for the integration of stress and auxin/cytokinin-dependent signalling in SE.

Another stress-related hormone, abscisic acid (ABA), in certain systems can initiate SE (Nishiwaki et al. 2000). ABA can also promote SE induced by auxin and cytokinin in *M. truncatula* (Nolan and Rose 1998); so, taken together with the ROS and the MtSERF1 data, there may be a signalling web that is likely to be similar to, albeit subtly different from other areas of plant development (Beveridge et al. 2007).

Existing evidence provides strong support for the involvement of stress in SE. As discussed in the next section, auxin is the key player in the SE paradigm and, indeed, 2,4-D, perhaps the most commonly used auxin, has been suggested to promote SE because of its herbicidal properties (Pasternak et al. 2002). Other auxins, however, can substitute for 2,4-D. So, while 2,4-D may add to the stress response, it is more likely to be acting to derepress genes, in line with current evidence for auxin action (Rose and Nolan 2006).

1.6   Hormones and the Initiation of SE

Classically, auxin has been considered the key player in the induction of SE and the formation of somatic embryos. As understanding of the mode of auxin action and the regulation of its transport increases (Friml et al. 2003; Tan et al. 2007), so too does evidence that auxin is the central player in regulation of plant (and somatic embryo) growth and development (Beveridge et al. 2007). The early work of Halperin showed that high concentrations of 2,4-D could produce PEMs from cultured petioles (in a medium containing reduced nitrogen) and somatic embryos were produced when transferred to liquid medium with no or low auxin. It was essential that the auxin concentration be lowered for SE to occur. Though the carrot system has served as the model for SE, there were always some difficulties in studying the activation events in SE, given the proembryogenic nature of the suspension culture (Dudits et al. 1991).

The advent of model plants such as Arabidopsis and *M. truncatula* (Rose and Nolan 2006) has, to a degree, lessened the role of the carrot system in SE studies. Nevertheless, it provides an important perspective to SE (Mordhorst et al. 1997). The work with the *SERK1* gene in carrot ushered in a more concerted effort into the molecular biology of SE. With a luciferase reporter gene for cell tracking, SERK has been shown to express in single cells that developed into somatic embryos and, in carrot, it acted as a marker of cells competent for SE (Schmidt et al. 1997). SERK1, when over-expressed in Arabidopsis, is capable of stimulating SE (Hecht et al. 2001). In *M. truncatula*, the *MtSERK1* gene is rapidly induced by auxin in cultured leaf explants and its expression is associated with both SE (SE requires auxin plus cytokinin in *M. truncatula*) and root morphogenesis. The extensive data now available on SERK (Rose and Nolan 2006) are consistent with an important role for SERK1 in SE competency, although it appears to have a broader role in differentiation as well (Nolan et al. 2003; Kwaaitaal and de Vries 2007).

Investigations into the homeobox gene, *WUSCHEL* (*WUS*), have also served to link SE to hormonal activation of specific developmental genes. *WUS* encodes a transcription factor that regulates the stem cell population in the shoot meristem. Over-expression in Arabidopsis formed embryo-like structures and it was suggested that WUS acted as both a meristem and embryo organiser (Zuo et al. 2002). Other work by Gallois et al. (2004) has shown in Arabidopsis that ectopic expression of WUS in roots could induce shoot organogenesis (low auxin) or SE (high auxin). An important consequence of this work is the concept that genes directing stem cell identity can reprogram somatic cells, with the direction of developmental change dependent on additional cues. There is rapid induction of *WUS* expression in *M. truncatula* (Chen et al. 2009). These data have been interpreted to suggest that genes involved with maintenance of pluripotent stem cells in meristems may have a role in inducing totipotent stem cells in culture. This has been included in the model shown in Fig. 1.3.

1.7 Induction of SE by Over-Expression of Leafy Cotyledon Transcription Factors and Their Relationship to SE Induction and Repression—the GA Connection

A framework is emerging of how stress and hormones can induce the activation of specific genes that can form stem cells capable of initiating the embryogenic pathway. Several transcription factors can induce SE independent of hormones, raising a number of questions in relation to the mechanism of SE. A connection between these different transcription factors is emerging in some cases; for example, the leafy cotyledon (*LEC*) group of transcription factors, required for SE in Arabidopsis, have an interesting connection to GA and ABA. *LEC2* when over-expressed induces both SE and the MADS transcription factor AGL15 (Braybrook et al. 2006). Ectopic expression of AGL15 promotes somatic embryo formation in Arabidopsis (Harding et al. 2003). Further evidence indicates that AGL15 acts by targeting the GA2-oxidase, AtGA2ox6, which catabolises biologically active GA. This decrease in biologically active GA causes increased SE, while addition of GA_3 causes a decrease in SE (Wang et al. 2004). A possible explanation for this effect lies with the *pickle* (*pkl*) mutant of Arabidopsis. The *pkl* mutant, so named because of its abnormal root phenotype, when cultured in the absence of plant hormones generates callus-like growth and forms somatic embryos on roots (Ogas et al. 1997). Wild-type roots do not produce callus or somatic embryos when cultured under the same conditions as the *pkl* roots. The surprising observation is that GA

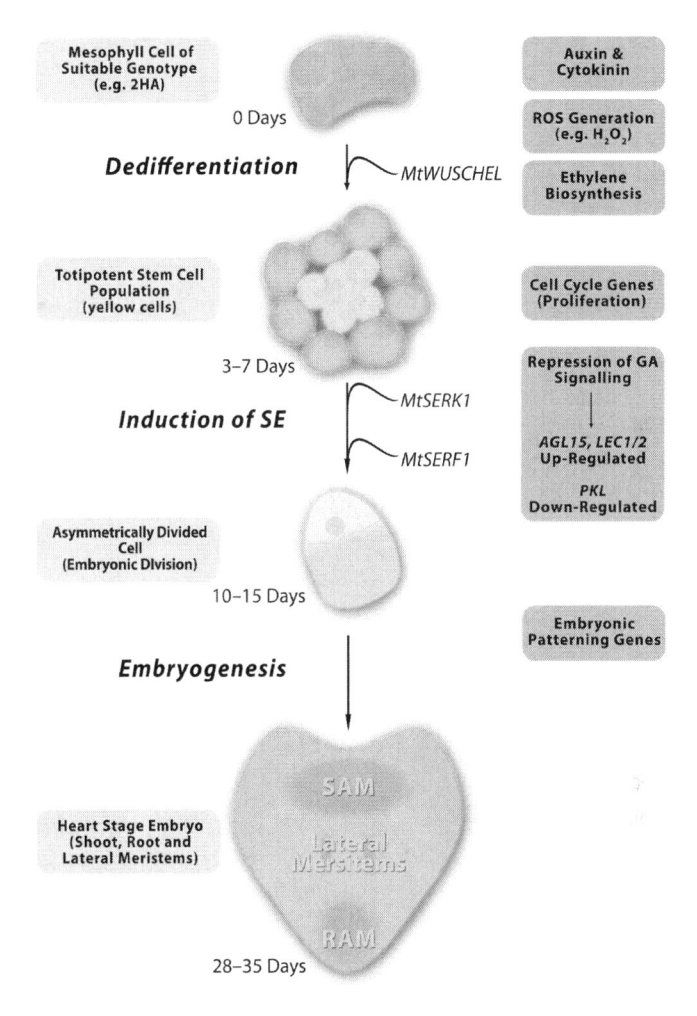

Fig. 1.3 A model for the induction of somatic embryogenesis and the development of the heart-stage embryo from *Medicago truncatula* mesophyll cells, drawing on research from the authors' laboratory and the literature discussed in this review

suppresses the *pkl* phenotype. The *PKL* gene has been cloned (Ogas et al. 1999) and encodes a CHD3 protein linked to chromatin remodelling (component of a histone deacetylase complex). The *PKL* gene also appears to be necessary for the repression of *LEC1* (Ogas et al. 1999; Rider et al. 2003).

Thus, AGL15 and PKL investigations reveal a GA-modulated pathway that represses embryonic identity. It is essential to prevent inappropriate embryo formation as germination and subsequent plant development occurs. This repression is happening in parallel to the GA activation of postembryonic development in Arabidopsis. In principle, this is similar to the situation shown by Haerizadeh et al. (2006), who reported the repression of germ cell-specific genes in somatic cells. Results of GA-related studies imply that it is necessary to ensure GA-repression is derepressed before the SE induction pathway can be initiated.

1.8 ABA, Stress and GA

As discussed above, ABA can stimulate or induce SE in some systems (Nolan and Rose 1998; Nishiwaki et al. 2000) and, as ABA and GA act antagonistically in germination (Henderson et al. 2004), it might be that the ABA effect on SE works through antagonising the GA effect. However, Henderson et al. (2004) found that ABA had only a small effect on the penetrance of the *pkl* phenotype. The authors postulated that PKL might regulate a subset of GA-dependent responses. Gazzarrini et al. (2004) have shown FUSA, a LEC transcription factor required for SE, is up-regulated by ABA and down-regulated by GA. Overall, there is some coherency in these data that emphasise the need to have an SE repression system. The stress-related hormone ABA may well exert part of its action by contributing to overriding this repression.

1.9 Soluble Signals and Cell–Cell Interactions that Promote SE in Suspension Cultures

One observation frequently made in cultured cells, not discussed thus far, is the dependence of SE on cell density. Many compounds, the concentrations of which can vary with cell density, are released into the medium and affect SE. For example, carrot SE does not proceed at high density due to the release and build-up of 4-hydroxyphenol (Kobayashi et al. 2000). Conversely, conditioned medium factors such as phytosulfokine peptides can stimulate cell division (although not SE directly) in cultures of suboptimal density (Matsubayashi and Sakagami 1996).

1.9.1 Secreted Proteins that Influence SE

Some proteins affecting SE progression, such as trypsin inhibitor, are produced by both embryogenic and non-embryogenic carrot cultures, but secreted only by embryogenic cultures, thereby preventing degradation of proteins necessary for progression through SE (Quiroz-Figueroa et al. 2006). The majority of protein secreted into the medium, however, is glycoprotein (Quiroz-Figueroa et al. 2006), most notably AGPs, believed to have important roles in cell–cell interaction and signalling. The importance of AGPs in cell fate determination and in regulating passage through SE is highlighted by tight temporal and spatial regulation of AGP expression in the plant. In carrot cell cultures, a characteristic group of AGPs are secreted into the medium, the composition of which changes with time (Quiroz-Figueroa et al. 2006). These AGPs likely have a structural role. However, that AGP effects (enhancement or reduction in frequency of SE) are exerted at nanomolar concentrations also suggests a signalling role. Interestingly, AGPs can contain GlcNAc and Glc residues, sensitive to cleavage by chitinases (van Hengel et al. 2001). AGPs treated with chitinase produce oligosaccharides, which are more active than AGPs in promoting SE (reviewed in Quiroz-Figueroa et al. 2006). Indeed, addition of EP3 chitinase to temperature-sensitive *ts11* carrot cells is capable of rescuing the defective SE phenotype (Mordhorst et al. 1997). Thus, chitinase-modified AGPs are extracellular matrix molecules able to control plant cell fate (van Hengel et al. 2001). It has been suggested that these signals are indicative of cell–cell communication and, hence, the importance of these molecules.

1.9.2   AGP Signalling in SE: Mechanisms and Interactions Between Signalling Pathways

Experiments in several species indicate that chitinases tend not to be expressed in cells of the embryo proper, but rather surrounding cells, providing evidence for the importance of intercellular communication in SE and reminiscent of the endo-sperm–embryo relationship in zygotic embryogenesis (Wiweger et al. 2003). In *Picea abies* cultures, removal of plant growth regulators up-regulates expression of *Chia4-Pa* chitinase. Chia4-Pa stimulates PCD and promotes the PEM to somatic embryo transition (Wiweger et al. 2003). The presence of JIM13-reactive AGPs, a marker for cells undergoing PCD in maize and zinnia xylem transdifferentiation, on PEMs in *P. abies* provides further evidence to suggest that AGP signalling invokes PCD (Filonova et al. 2000). Thus, AGP signalling likely interacts with an integrated proliferation-PCD control system to regulate SE (Majewska-Sawka and Nothnagel 2000; Filonova et al. 2000).

While soluble signals are undoubtedly important in suspension cultures, this does not necessarily mean that SE from tissue explants depends on the same signals, although clearly the cells that form somatic embryos are part of a stem cell niche dependent on the surrounding cells. It is becoming clear that cell wall-related signalling cascades interact with hormonal and stress-signalling pathways invoked during SE. It is of interest in this regard to note that auxin deprivation (often used to promote PEM to somatic embryo transition) in tobacco BY2 suspension cultures results in a twofold increase in AGP secretion and a fourfold increase in JIM13 reactivity (Winicur et al. 1998). Perturbation of cell walls can signal stress responses. For example, aggregation of AGPs with Yariv reagent produces an expression profile most similar to wounding, and leads to down-regulation of GA signalling and up-regulation of defence-related genes, while inhibition of cellulose synthesis results in increased production of ethylene and jasmonates (Ellis et al. 2002; Guan and Nothnagel 2004; Mashiguchi et al. 2008). Interestingly, AGPs are cross-linked by H_2O_2 and wounding in planta, providing credibility to the hypothesis that stress, hormonal and cell wall signalling interact in SE (Kjellbom et al. 1997). In *Medicago*, ethylene-dependent transcription factors are required for SE induced by auxin plus cytokinin.

1.9.3 Cell–Cell Interaction and Relevance to SE in Suspension Cultures

In carrot and other systems, researchers have noted that the loss of embryogenic competence typically results in a reduction in the size of cell clusters that would normally form somatic embryos (Satoh 1998). Indeed, close observations reveal tighter contact between cells of embryogenic callus than between cells of non-embryogenic callus. Thus, loss of embryogenic competence is associated with a loosening of intercellular attachment, suggesting close attachment is necessary for SE in these culture systems (Satoh 1998). Moreover, in *P. abies*, PEMs readily develop into somatic embryos, whereas single cells (densely cytoplasmic or vacuolated), made by fractionation of crude suspensions, do not (Filonova et al. 2000). This fact demonstrates the obligate requirement for cell–cell interaction in some SE systems. Changes in cell–cell interaction over the course of SE are reflected in, and likely caused by, changes in cell wall composition. For example, pectin comprises a substantially larger proportion of the cell wall in somatic embryos than it does in plantlets regenerated from these embryos (Yeo et al. 1998). Furthermore, hemicelluloses display a transient increase in 5- and 3,5-linked arabinose during embryogenesis (Yeo et al. 1998). Notably, embryogenic cultures have significantly higher levels of neutral sugars like arabinose and xylose than do non-embryogenic cultures (Satoh 1998; Yeo et al. 1998). Neutral sugars are believed to play an important role in intercellular attachment and might act by bridging hemicellulose and pectin molecules (Satoh 1998). One gene product identified as critical for intercellular attachment is C-ESE1 from carrot (Takahata et al. 2004). The *C-ESE1* gene is strongly up-regulated after depletion of 2,4-D from the medium and expressed specifically in the primordial cells of somatic embryos. Co-suppression of *C-ESE1* expression leads to small cell clusters with weak intercellular attachment, thus delaying SE (Takahata et al. 2004). Observation of control and *C-ESE1* co-suppressed cells by scanning electron microscopy reveals smooth cell surfaces of *C-ESE1* co-suppressed cells, which also release large amounts of cell wall polysaccharide into the medium. Thus, C-ESE1 appears to play a role in polysaccharide deposition, in turn important for cell–cell attachment and SE. Notably, control of cell–cell interaction also appears important for SE progression in explant-based systems, such as that of coconut (Verdeil et al. 2001).

1.10 Development Program After SE Induction

This discussion has focused on what is known about the biology of SE induction and the different threads of evidence from a range of culture systems. The tacit assumption is that once SE is induced there is an asymmetric division, albeit with the development fate of the cell that would contribute to the root pole and the suspensor in the zygotic embryo having a more variable history. This latter point is likely one reason for the variable nature of SE. However, once SE initiates, the developmental program is thought to be similar to zygotic embryogenesis, being a primary reason for the adoption of SE as a useful experimental system to understand zygotic embryogenesis. The defining feature of SE, set out in an important review by Zimmerman (1993), is the reprogramming of a somatic cell. Zimmerman's review emphasised that the ability to produce large quantities of embryos at different developmental stages was a distinct advantage in embryogenesis investigations. Zimmerman's validation of SE relied on gene expression data available at that time. Developments since 1993 have revealed the power of genomics, mutants and model organisms, particularly Arabidopsis (Fehér et al. 2003), but also *Medicago* (Rose and Nolan 2006). As pointed out by Fehér et al. (2003), research on the Arabidopsis embryogenesis mutants have provided insights into SE. However, there have been the discoveries of *SERK* and *SERF* genes from SE studies that are also expressed in zygotic embryogenesis. The reality now is that our understanding of embryogenesis benefits from comparisons between zygotic and somatic embryogenesis. While there are likely some gene expression changes that reflect the different environments under which the embryos develop, the accumulated evidence suggests overwhelmingly similar developmental programs (Zimmerman 1993; Dodeman et al. 1997; Mantiri et al. 2008).

1.11 Concluding Remarks and a Model Based on Studies in *Medicago truncatula*

A model that provides a framework for our understanding and future work is shown in Fig. 1.3. While it is based on experience with *M. truncatula*, it draws conceptually from Arabidopsis and the wider literature discussed in this chapter. It is known that ROS are rapidly induced when tissue is excised and placed into culture and that ethylene is essential for SE. It is feasible that both ROS and auxin contribute to the induction of ethylene biosynthesis genes. ROS has been reported to promote auxin-induced ethylene production in mungbean hypocotyls (Song et al. 2007). It is also suggested that WUS is an important early player requiring cytokinins (Chen et al. 2009). WUS, which has been postulated to be an embryo organiser (Zuo et al. 2002), is likely contributing to the production of stem cells, as it does in the shoot meristem in planta. There is some evidence, as discussed above, that auxin-dependent *SERK* expression is indicative of cells primed for SE and commencing to differentiate. *MtSERF1* expression is initiated after *SERK* expression. MtSERF1 is essential for SE. Based on physiological experiments and putative promoter binding sites, there is evidence that *MtSERF1* expression requires cytokinin, auxin, ethylene and WUS (Mantiri et al. 2008). MtSERF1 may interconnect the stress and hormone signalling pathways. Based on the Arabidopsis studies discussed, it is assumed that the SE pathway is derepressed, involving an AGL15-dependent pathway (Zheng et al. 2009) and the genes we have discussed are able to initiate the SE pathway.

1.12 SE and Biotechnology

The immense body of work on SE (more than 400 papers in the past 10 years) has arisen largely in response to the modern era of biotechnology and molecular biology. SE facilitates transformation for many species and, unless the floral dip procedures used in Arabidopsis or other in planta procedures can be developed for other species, SE remains a key component of many transformation protocols (organogenesis is the other regeneration pathway). The SE transformation protocols require the gene to be inserted into the somatic cell, followed by regeneration via SE. The gene insertion is mostly by *Agrobacterium*, but can involve other methods such as microprojectiles.

The cloning of desirable genotypes using SE can be an advantage in high-value plants, such as in forestry, where it is a practicable propagation procedure (Sutton 2002), or in horticulture (Bunn et al. 2007), or with crops like potato (Seabrook and Douglass 2001). Greater knowledge may enable SE to be applied more efficiently and more broadly to encompass more species. Embryo production from nonzygotic cells in anther and isolated microspore culture in the production of doubled haploids have become important in breeding programs (Hosp et al. 2007). As pointed out in the introduction, an understanding of SE will also facilitate an understanding of apomixis. Illuminating how apomixis is induced remains a goal of both fundamental and applied plant research. The ability to produce seed with genotype identical to the parent would provide new plant breeding strategies, particularly with hybrid crops. SE falls under the umbrella of regeneration biology and an understanding of this reprogramming process remains an important biological question (Vogel 2005). Elucidating the mechanisms of somatic embryogenesis will contribute to a fuller understanding of plant development.

Acknowledgements Work in the authors' laboratory has been supported by an ARC Centre of Excellence Grant to the University of Newcastle Node of the ARC Centre of Excellence for Integrative Legume Research, to RJR (Grant CEO348212).

References

Ammirato PV (1983) Embryogenesis. In: Evans DA, Sharp WR, Ammirato PV, Yamada Y (eds) Handbook of plant cell culture, vol 1. Techniques in propagation and breeding. MacMillan, New York, pp 82–123

Araújo SS, Duque ASRLA, Santos DMMF, Fevereiro MPS (2004) An efficient transformation method to regenerate a high number of transgenic plants using a new embryogenic line of *Medicago truncatula* cv. Jemalong. Plant Cell Tiss Organ Cult 78:123–131

Bäurle I, Smith L, Baulcombe DC, Dean C (2007) Widespread role for the flowering-time regulators FCA and FPA in RNA-mediated chromatin silencing. Science 318:109–112

Belmonte MF, Stasolla C (2007) Applications of DL-buthionine-[S,R]-sulfoximine deplete cellular glutathione and improve white spruce (*Picea glauca*) somatic embryo development. Plant Cell Rep 26:517–523

Belmonte MF, Donald G, Reid DM, Yeung EC and Stasolla C (2005) Alterations of the glutathione redox state improve apical meristem structure and somatic embryo quality in white spruce (*Picea glauca*). J Exp Bot 56:2355–2364

Bender J (2004) Chromatin-based silencing mechanisms. Curr Opin Plant Biol 7:521–526

Beveridge CA, Mathesius U, Rose RJ, Gresshoff PM (2007) Common regulatory themes in meristem development and whole plant homeostasis. Curr Opin Plant Biol 10:44–51

Bingham ET, Hurley LV, Kaatz DM, Saunders JW (1975) Breeding alfalfa which regenerates from callus tissue in culture. Crop Sci 15:719–721

Blervacq AS, Dubois T, Dubois J, Vassern J (1995) First divisions of somatic embryogenic cells in *Cichorium hybrid* '474'. Protoplasma 186:163–168

Braybrook SA, Stone SL, Park S, Bui AQ, Le BH, Fischer RL, Goldberg RB, Harada JJ (2006) Genes directly regulated by LEAFY COTYLEDON2 provide insight into the control of embryo maturation and somatic embryogenesis. Proc Natl Acad Sci USA 103:3468–3473

Bunn E, Turner S, Panaia M, Dixon KW (2007) The contribution of *in vitro* technology and cryogenic storage to conservation of indigenous plants. Aust J Bot 55:345–355

Caliskan M, Turet M, Cuming AC (2004) Formation of wheat (*Triticum aestivum* L.) embryogenic callus involves peroxide-generating germin-like oxalate oxidase. Planta 219:132–140

Chandler JW, Cole M, Flier A, Grewe B, Werr W (2007) The AP2 transcription factors DORNRÖSCHEN and DORNRÖSCHEN-LIKE redundantly control *Arabidopsis* embryo patterning via interaction with PHAVOLUTA. Development 134:1653–1662

Chen S-K, Kurdyukov S, Kerestz A, Wang X-D, Gresshoff PM, Rose RJ (2009) The association of homeobox gene expression with stem cell formation and morphogenesis in cultured *Medicago truncatula*. Planta 230:827–840

Cui KR, Xing GS, Liu XM, Xing GM, Wang YF (1999) Effect of hydrogen peroxide on somatic embryogenesis of *Lycium barbarum* L. Plant Sci 146:9–16

Dangl JL, Jones JD (2001) Plant pathogens and integrated defence responses to infection. Nature 411:826–833

Desikan R, Hancock J, Neill S (2005) Reactive oxygen species as signalling molecules. In: Smirnoff N (ed) Antioxidant and reactive oxygen species in plants. Blackwell, Oxford, pp 169–196

Dijak M, Smith DL, Wilson TJ, Brown DC (1986) Stimulation of direct embryogenesis from mesophyll protoplasts of *Medicago sativa*. Plant Cell Rep 5:468–470

Dodeman VL, Ducreux G, Kreis M (1997) Zygotic embryogenesis versus somatic embryogenesis. J Exp Bot 48:1493–1509

Dudits D, Bögre L, Györgyey J (1991) Molecular and cellular approaches to the analysis of plant embryo development from somatic cells *in vitro*. J Cell Sci 99:475–484

Ellis C, Karafyllidis I, Wasternack C, Turner JG (2002) The Arabidopsis mutant *cev1* links cell wall signalling to jasmonate and ethylene responses. Plant Cell 14:1557–1566

Fehér A, Pasternak TP, Dudits D (2003) Transition of somatic plant cells to an embryogenic state. Plant Cell Tiss Organ Cult 74:201–228

Filonova LH, Bozhkov PV, von Arnold S (2000) Developmental pathway of somatic embryogenesis in *Picea abies* as revealed by time-lapse tracking. J Exp Bot 51:249–264

Friml J, Vieten A, Weijers D, Schwarz H, Haqmann T, Offringa R, Jürgens G (2003) Efflux-dependent auxin gradients establish the apical-basal axis of *Arabidopsis*. Nature 426:147–153

Gallois JL, Nora FR, Mizukami Y, Sablowski R (2004) WUSCHEL induces shoot stem cell activity and developmental plasticity in the root meristem. Genes Dev 18:375–380

Garcês HMP, Champagne CEM, Townsley BT, Park S, Malhó R, Pedroso MC, Harada JJ, Sinha NR (2007) Evolution of asexual reproduction in leaves of the genus *Kalanchoë*. Proc Natl Acad Sci USA 104:15578–15583

Gazzarrini S, Tsuchiya Y, Lumba S, Okamoto M, McCourt P (2004) The transcription factor *FUSCA3* controls developmental timing in *Arabidopsis* through the hormones gibberellin and abscisic acid. Dev Cell 7:373–385

Gechev TS, Hille J (2005) Hydrogen peroxide as a signal controlling plant programmed cell death. J Cell Biol 168:17–20

Geldner N, Anders N, Wolters H, Keicher J, Kornberger W, Muller P, Delbarre A, Ueda T, Nakano A, Jürgens G (2003) The *Arabidopsis* GNOM ARF-GEF mediates endosomal recycling, auxin transport, and auxin-dependent plant growth. Cell 112:219–230

Goubet F, Misrahi A, Park SK, Zhang Z, Twell D, Dupree P (2003) AtCSLA7, a cellulose synthase-like putative glycosyltransferase, is important for pollen tube growth and embryogenesis in Arabidopsis. Plant Physiol 131:547–557

Guan Y, Nothnagel EA (2004) Binding of arabinogalactan proteins by Yariv phenylglycoside triggers wound-like responses in Arabidopsis cell cultures. Plant Physiol 135:1346–1366

Gupta R, Luan S (2003) Redox control of protein tyrosine phosphatases and mitogen-activated protein kinases in plants. Plant Physiol 132:1149–1152

Guzzo F, Baldan, B, Mariani P, Lo Schiavo F, Terzi M (1994) Studies on the origin of totipotent cells in explants of *Daucus carota* L. J Exp Bot 45:1427–1432

Haccius B (1978) Question of unicellular origin of non-zygotic embryos in callus cultures. Phytomorphology 28:74–81

Haecker A, Groß-Hardt R, Geiges B, Sarkar A, Breuninger H, Herrmann M, Laux T (2004) Expression dynamics of *WOX* genes mark cell fate decisions during early embryonic patterning in *Arabidopsis thaliana*. Development 131:657–688

Haerizadeh F, Singh MB, Bhalla PL (2006) Transcriptional repression distinguishes somatic from germ cell lineages in a plant. Science 313:496–499

Hall Q, Cannon MC (2002) The cell wall hydroxyproline-rich glycoprotein RSH is essential for normal embryo development in Arabidopsis. Plant Cell 14:1161–1172

Halperin W (1966) Alternative morphogenetic events in cell suspensions. Am J Bot 53:443–453

Harding EW, Tang WN, Nichols KW, Fernandez DE, Perry SE (2003) Expression and maintenance of embryogenic potential is enhanced through constitutive expression of *AGAMOUS-Like 15*. Plant Physiol 133:653–663

Hecht V, Vielle-Calzada J-P, Hartog MV, Schmidt EDL, Boutilier K, Grossniklaus U, de Vries SC (2001) The Arabidopsis *SOMATIC EMBRYOGENESIS RECEPTOR KINASE 1* gene is expressed in developing ovules and embryos and enhances embryogenic competence in culture. Plant Physiol 127:803–816

Henderson IR, Jacobsen SE (2007) Epigenetic inheritance in plants. Nature 447:418–424

Henderson JT, Li H-C, Rider SD, Mordhorst AP, Romero-Severson J, Cheng J-C, Robey J, Sung ZR, de Vries SC, Olgas J (2004) Pickle acts throughout the plant to repress expression of embryonic traits and may play a role in gibberellin-dependent responses. Plant Physiol 134:995–1005

Hoffman B, Trinh TH, Leung J, Kondorosi A, Kondorosi E (1997) A new *Medicago truncatula* line with superior in vitro regeneration, transformation, and symbiotic properties isolated through cell culture selection. Mol Plant-Microbe Interact 10:307–315

Hosp J, de Faria Maraschin S, Touraev A, Boutilier K (2007) Functional genomics of microspore embryogenesis. Euphytica 158:275–285

Ikeda-Iwai M, Umehara M, Satoh S, Kamada H (2003) Stress-induced somatic embryogenesis in vegetative tissues of *Arabidopsis thaliana*. Plant J 34:107–114

Imin N, De Jong F, Mathesius U, van Noorden G, Saeed NA, Wang X-D, Rose RJ, Rolfe BG (2004) Proteome reference maps of *Medicago truncatula* embryogenic cell cultures generated from single protoplasts. Proteomics 4:1883–1896

Imin N, Nizamidin M, Daniher D, Nolan KE, Rose RJ, Rolfe BG (2005) Proteomic analysis of somatic embryogenesis in *Medicago truncatula*. Explant cultures grown under 6-benzylaminopurine and 1-naphthaleneacetic acid treatments. Plant Physiol 137:1250–1260

Ingram GC (2004) Between the sheets: inter-cell-layer communication in plant development. Philos Trans R Soc Lond B Biol Sci 359:891–906

Jenik PD, Barton MK (2005) Surge and destroy: the role of auxin in plant embryogenesis. Development 132:3577–3585

Kairong C, Ji L, Gengmei X, Jianlong L, Lihong W, Yafu W (2002) Effect of hydrogen peroxide on synthesis of proteins during somatic embryogenesis in *Lycium barbarum*. Plant Cell Tiss Organ Cult 68:187–193

Kamada H, Ishikawa K, Saga H, Harada H (1993) Induction of somatic embryogenesis in carrot by osmotic stress. Plant Tiss Cult Lett 10:38–44

Kepinski S, Leyser O (2003) An axis of auxin. Nature 426:132–135

Kielly GA, Bowley SR (1992) Genetic control of somatic embryogenesis in alfalfa. Genome 35:474–477

Kjellbom P, Snogerup L, Stohr C, Reuzeau C, McCabe PF, Pennell RI (1997) Oxidative cross-linking of plasma membrane arabinogalactan proteins. Plant J 12:1189–1196

Kobayashi T, Higashi K, Sasaki K, Asami T, Yoshida S, Kamada H (2000) Purification from conditioned medium and chemical identification of a factor that inhibits somatic embryogenesis in carrot. Plant Cell Physiol 41:268–273

Koltunow AM, Grossniklaus U (2003) Apomixis: a developmental perspective. Annu Rev Plant Biol 54:547–574

Kovtun Y, Chiu WL, Tena G, Sheen J (2000) Functional analysis of oxidative stress-activated mitogen-activated protein kinase cascade in plants. Proc Natl Acad Sci USA 97:2940–2945

Kwaaitaal MACJ, de Vries SC (2007) The *SERK1* gene is expressed in procambium and immature vascular cells. J Exp Bot 58:2887–2896

Loiseau J, Michaux-Ferrière N, Le Denunff Y (1998) Histology of somatic embryogenesis in pea. Plant Physiol Biochem 36:683–687

Lu C-Y, Vasil IK (1985) Histology of somatic embryogenesis in *Panicum maximum* (Guinea grass). Am J Bot 72:1908–1913

Majewska-Sawka A, Nothnagel EA (2000) The multiple roles of arabinogalactan proteins in plant development. Plant Physiol 122:3–10

Mantiri FR, Kurdyukov S, Lohar DP, Sharapova N, Saeed NA, VandenBosch KA, Rose RJ (2008) The transcription factor MtSERF1 of the ERF subfamily identified by transcriptional profiling is required for somatic embryogenesis induced by auxin plus cytokinin in *Medicago truncatula*. Plant Physiol 146:1622–1636

Mashiguchi K, Urakami E, Hasegawa M, Sanmiya K, Matsumoto I, Yamaguchi I, Asami T, Suzuki Y (2008) Defense-related signaling by interaction of arabinogalactan proteins and β-glucosyl Yariv reagent inhibits gibberellin signaling in barley aleurone cells. Plant Cell Physiol 49:178–190

Matsubayashi Y, Sakagami Y (1996) Phytosulfokine, sulfated peptides that induce the proliferation of single mesophyll cells of *Asparagus officinalis* L. Proc Natl Acad Sci USA 93:7623–7627

Mayer U, Büttner G, Jürgens G (1993) Apical-basal pattern formation in the *Arabidopsis* embryo: studies on the role of the *gnom* gene. Development 117:149–162

McCabe PF, Valentine TA, Forsberg LS, Pennell RI (1997) Soluble signals from carrots identified at the cell wall establish a developmental pathway in carrot. Plant Cell 9:2225–2241

Miao Y, Lv D, Wang P, Wang X-C, Chen J, Miao C, Song C-P (2006) An Arabidopsis glutathione peroxidase functions as both a redox transducer and a scavenger in abscisic acid and drought stress responses. Plant Cell 18:2749–2766

Mordhorst AP, Toonen MAJ, de Vries SC (1997) Plant embryogenesis. Crit Rev Plant Sci 16:535–576

Mou Z, Fan W, Dong X (2003) Inducers of plant systemic acquired resistance regulate NPR1 function through redox changes. Cell 113:935–944

Müller B, Sheen J (2007) Advances in cytokinin signaling. Science 318:68–69

Nishiwaki M, Fujino K, Koda Y, Masuda K, Kikuta Y (2000) Somatic embryogenesis induced by the simple application of abscisic acid. Planta 211:756–759

Noctor G (2006) Metabolic signalling in defence and stress: the central roles of soluble redox couples. Plant Cell Environ 29:409–425

Nolan KE, Rose RJ (1998) Plant regeneration from cultured *Medicago truncatula* with particular reference to abscisic acid and light treatments. Aust J Bot 46:151–160

Nolan KE, Rose RJ, Gorst JE (1989) Regeneration of *Medicago truncatula* from tissue culture: increased somatic embryogenesis from regenerated plants. Plant Cell Rep 25:278–281

Nolan KE, Irwanto RR, Rose RJ (2003) Auxin up-regulates *MtSERK1* expression in both *Medicago truncatula* root-forming and embryogenic cultures. Plant Physiol 133:218–230

Nolan KE, Saeed NA, Rose RJ (2006) The stress kinase gene *MtSK1* in *Medicago truncatula* with particular reference to somatic embryogenesis. Plant Cell Rep 25:711–722

Ogas J, Cheng J-C, Sung ZR, Somerville C (1997) Cellular differentiation regulated by gibberellin in the *Arabidopsis thaliana* pickle mutant. Science 277:91–94

Ogas J, Kaufmann S, Henderson J, Somerville C (1999) Pickle is a CHD3 chromatin-remodeling factor that regulates the transition from embryonic to vegetative development in *Arabidopsis*. Proc Natl Acad Sci USA 96:13839–13844

Orozco-Cardenas M, Ryan CA (1999) Hydrogen peroxide is generated systemically in plant leaves by wounding and systemin via the octadecanoid pathway. Proc Natl Acad Sci USA 96:6553–6557

Papadakis AK, Siminis CI, Roubelakis-Angelakis KA (2001) Reduced activity of antioxidant machinery is correlated with suppression of totipotency in plant protoplasts. Plant Physiol 126:434–444

Pasternak TP, Prinsen E, Ayaydin F, Miskolczi P, Potters G, Asard H, Van Onckelen HA, Dudits D, Fehér A (2002) The role of auxin, pH, and stress in the activation of embryogenic cell division in leaf protoplast-derived cells of alfalfa. Plant Physiol 129:1807–1819

Pasternak T, Potters G, Caubergs R, Jansen MAK (2005) Complementary interactions between oxidative stress and auxins control plant growth responses at plant, organ, and cellular level. J Exp Bot 56:1991–2001

Pennell RI, Janniche L, Kjellbom P, Scofield GN, Peart JM, Roberts K (1991) Developmental regulation of a plasma membrane arabinogalactan protein epitope in oilseed rape flowers. Plant Cell 3:1317–1326

Quiroz-Figueroa F, Rojas-Herrera R, Galaz-Avalos R, Loyola-Vargas VC (2006) Embryo production through somatic embryogenesis can be used to study cell differentiation in plants. Plant Cell Tiss Organ Cult 86:285–301

Reinert J (1958) Morphogenese und ihre Kontrolle an Gewebekulturen aus Karotten. Naturwissenschaften 45:344–345

Rider SD, Henderson JT, Jerome RE, Edenberg HJ, Romero-Severson J, Ogas J (2003) Coordinate repression of regulators of embryonic identity by *PICKLE* during germination in *Arabidopsis*. Plant J 35:33–43

Rose RJ (2004) Somatic embryogenesis in plants. In: Goodman RM (ed) Encyclopedia of plant and crop science. Marcel Dekker, New York, pp 1165–1168

Rose RJ, Nolan KE (1995) Regeneration of *Medicago truncatula* from protoplasts isolated from kanamycin-sensitive and kanamycin-resistant plants. Plant Cell Rep 14:349–354

Rose RJ, Nolan KE (2006) Genetic regulation of somatic embryogenesis with particular reference to *Arabidopsis thaliana* and *Medicago truncatula*. In Vitro Cell Dev Biol-Plant 42:473–481

Rose RJ, Nolan KE, Bicego L (1999) The development of the highly regenerable seed line Jemalong 2HA for transformation of *Medicago truncatula* - Implications for regenerability via somatic embryogenesis. J Plant Physiol 155:788–791

Rose RJ, Wang X-D, Nolan KE, Rolfe BG (2006) Root meristems in *Medicago truncatula* tissue culture arise from vascular-derived procambial-like cells in a process regulated by ethylene. J Exp Bot 57:2227–2235

Sagare AP, Suhasini K, Krishnamurthy KV (1995) Histology of somatic embryo initiation and development in chickpea (*Cicer arietinum* L.). Plant Sci 109:87–93

Santos D, Fevereiro P (2002) Loss of DNA methylation affects somatic embryogenesis in *Medicago truncatula*. Plant Cell Tiss Organ Cult 70:155–161

Satoh S (1998) Functions of the cell wall in the interactions of plant cells: analysis using carrot cultured cells. Plant Cell Physiol 39:361–368

Scheres B, Benfey PN (1999) Asymmetric cell division in plants. Annu Rev Plant Physiol Plant Mol Biol 50:505–537

Schmidt EDL, Guzzo F, Toonen MAJ, de Vries SC (1997) A leucine-rich repeat containing receptor-like kinase marks somatic plant cells competent to form embryos. Development 124:2049–2062

Schwendiman J, Pannetier C, Michaux-Ferriere N (1988) Histology of somatic embryogenesis from leaf explants of the oil palm *Elais guineensis*. Ann Bot 62:43–52

Seabrook JEA, Douglass LK (2001) Somatic embryogenesis on various potato tissues from a range of genotypes and ploidy levels. Plant Cell Rep 20:175–182

Sharma SK, Millam S (2004) Somatic embryogenesis in *Solanum tuberosum* L.: a histological examination of key developmental stages. Plant Cell Rep 23:115–119

Somleva MN, Schmidt EDL, de Vries SC (2000) Embryogenic cells in *Dactylis glomerata* L. (Poaceae) explants identified by cell tracking and by SERK expression. Plant Cell Rep 19:718–726

Song YJ, Joo JH, Ryu HY, Lee JS, Bae YS, Nam KH (2007) Reactive oxygen species mediate IAA-induced ethylene production in mungbean (*Vigna radiata* L.) hypocotyls. J Plant Biol 50:18–23

Souter M, Lindsey K (2000) Polarity and signalling in plant embryogenesis. J Exp Bot 51:971–983

Stasolla C, Belmonte MF, van Zyl L, Craig DL, Liu W, Yeung EC, Sederoff RR (2004) The effect of reduced glutathione on morphology and gene expression of white spruce (*Picea glauca*) somatic embryos. J Exp Bot 55:695–709

Steward FC, Mapes O, Smith J (1958) Growth and organised development of cultured cells 1. Growth and division of freely suspended cells. Am J Bot 45:693–703

Steward FC, Mapes MO, Kent AE, Holsten RD (1964) Growth and development of cultured plant cells. Science 143:20–27

Sutton B (2002) Commercial delivery of genetic improvement to conifer plantations using somatic embryogenesis. Ann For Sci 59:657–661

Takahata K, Takeuchi M, Fujita M, Azuma J, Kamada H, Sato F (2004) Isolation of putative glycoprotein gene from early somatic embryo of carrot and its possible involvement in somatic embryo development. Plant Cell Physiol 45:1658–1668

Tan X, Calderon-Villalobos LIA, Sharon M, Zheng C, Robinson CV, Estelle M, Zheng N (2007) Mechanism of auxin perception by the TIRI ubiquitin ligase. Nature 446:640–622

Touraev A, Vicente O, Heberle-Bors E (1997) Initiation of microspore embryogenesis by stress. Trends Plant Sci 2:297–302

van Hengel AJ, Tadesse Z, Immerzeel P, Schols H, van Kammen A, de Vries SC (2001) N-Acetylglucosamine and glucosamine-containing arabinogalactan proteins control somatic embryogenesis. Plant Physiol 125:1880–1890

Vasil IK (1988) Progress in the regeneration and genetic manipulation of cereal crops. Bio/ Technology 6:397–402

Verdeil JL, Hocher V, Huet C, Grosdemange F, Escoute J, Ferriere N, Nicole M (2001) Ultrastructural changes in coconut calli associated with the acquisition of embryogenic competence. Ann Bot 88:9–18

Vernoux T, Wilson RC, Seeley KA, Reichheld J-P, Muroy S, Brown S, Maughan SC, Cobbett CS, Van Montagu M, Inze D, May MJ, Sung ZR (2000) The *ROOT MERISTEMLESS1/CADMIUM SENSITIVE2* gene defines a glutathione-dependent pathway involved in initiation and maintenance of cell division during postembryonic root development. Plant Cell 12:97–110

Vogel G (2005) How does a single somatic cell become a whole plant. Science 309:86

Wang H, Caruso LV, Downie AB, Perry SE (2004) The embryo MADS domain protein AGAMOUS-Like 15 directly regulates expression of a gene encoding an enzyme involved in gibberellin metabolism. Plant Cell 16:1206–1219

Williams EG, Maheswaran G (1986) Somatic embryogenesis: factors influencing coordinated behaviour of cells as an embryogenic group. Ann Bot 57:443–462

Winicur ZM, Feng Zhang G, Andrew Staehelin L (1998) Auxin deprivation induces synchronous golgi differentiation in suspension-cultured tobacco BY-2 cells. Plant Physiol 117:501–513

Wiweger M, Farbos I, Ingouff M, Lagercrantz U, Von Arnold S (2003) Expression of Chia4-Pa chitinase genes during somatic and zygotic embryo development in Norway spruce (*Picea abies*): similarities and differences between gymnosperm and angiosperm class IV chitinases. J Exp Bot 54:2691–2699

Yamamoto N, Kobayashi H, Togashi T, Mori Y, Kikuchi K, Kuriyama K, Tokuji Y (2005) Formation of embryogenic cell clumps from carrot epidermal cells is suppressed by 5-azacytidine, a DNA methylation inhibitor. J Plant Physiol 162:47–54

Yeo U-D, Kohmura H, Nakagawa N, Sakurai N (1998) Quantitative and qualitative changes of cell wall polysaccharides during somatic embryogenesis and plantlet development of asparagus (*Asparagus officinalis* L.). Plant Cell Physiol 39:607–614

Zheng Y, Ren N, Wang H, Stromberg AJ, Perry SE (2009) Global identification of targets of the *Arabidopsis* MADS domain protein AGAMOUS-LIKE 15. Plant Cell (in press). doi:10.1105/tpc.109.068890

Zimmerman JL (1993) Somatic embryogenesis: a model for early development in higher plants. Plant Cell 5:1411–1423

Zuo JR, Niu Q-W, Frugis G, Chua NH (2002) The *WUSCHEL* gene promotes vegetative-to-embryonic transition in *Arabidopsis*. Plant J 30:349–359

Chapter 2
Microspore Embryogenesis

A. Olmedilla

2.1   Introduction

Throughout history, considerable effort has been devoted to improving the quality and yields of crop plants as well as selecting species best adapted to different environmental conditions. Due to the fact that plants are either directly or indirectly the primary source of energy and nourishment for the world's population, plant breeding is the main focus of a great deal of scientific research. Classic breeding approaches are restricted to a certain number of plants that present no problems of self-incompatibility, but even in species such as these it takes a long time to obtain new varieties. To avoid incompatibility barriers and to shorten breeding processes, methods based on tissue culture have been developed and improved to obtain haploid plants. These plants, though infertile, can be used to produce homozygous lines by chromosome doubling. Double haploids are very useful in accelerating the breeding of new cultivars (Forster et al. 2007). This chapter considers the culture of anthers and isolated microspores to obtain haploid plants.

Microspores can be reprogrammed by stress treatment to shift from gameto-phytic development (gametogenesis) towards a sporophytic pathway, a process known as microspore embryogenesis (Fig. 2.1). This developmental process is also known as pollen embryogenesis, androgenesis or, in a more general way, haploid embryogenesis. Haploid embryogenesis refers to embryogenesis induced from both male and female gametic cells. The use of the terms "microspore" or "pollen embryogenesis" depends, amongst other things, upon the structure from which this process is considered to start: from microspores or from pollen (Raghavan 1986; Reynolds 1997; Touraev et al. 1997). Finally, although the

A. Olmedilla

Department of Biochemistry, Cell and Molecular Biology of Plants, Estación Experimental del Zaidín CSIC, Profesor Albareda 1, 18008 Granada, Spain

e-mail: adela.olmedilla@eez.csic.es

E-C. Pua and M.R. Davey (eds.),   27

Plant Developmental Biology – Biotechnological Perspectives: Volume 2,

DOI 10.1007/978-3-642-04670-4_2, © Springer-Verlag Berlin Heidelberg 2010

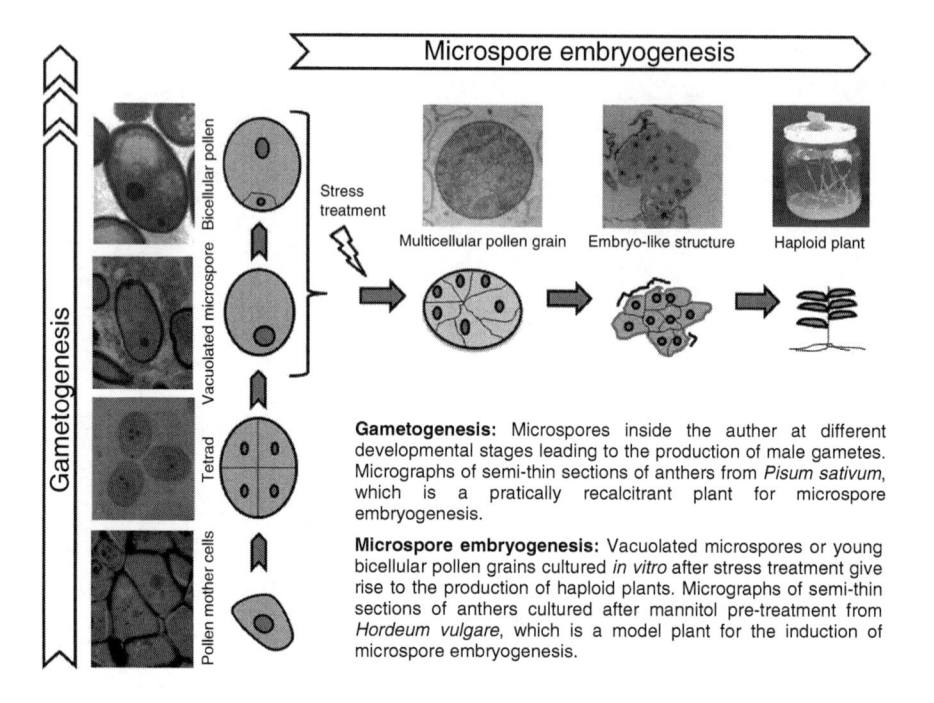

Gametogenesis: Microspores inside the auther at different developmental stages leading to the production of male gametes. Micrographs of semi-thin sections of anthers from *Pisum sativum*, which is a pratically recalcitrant plant for microspore embryogenesis.

Microspore embryogenesis: Vacuolated microspores or young bicellular pollen grains cultured *in vitro* after stress treatment give rise to the production of haploid plants. Micrographs of semi-thin sections of anthers cultured after mannitol pre-treatment from *Hordeum vulgare*, which is a model plant for the induction of microspore embryogenesis.

Fig. 2.1

shorter name "androgenesis" is also frequently used, some authors avoid it because this term was classically used to designate the production of haploid plants by fertilization of an egg cell, the nucleus of which has been eliminated or inactivated, after which the embryo is formed by the division of the male nucleus. Microspore embryogenesis has been induced in a considerable number of plant species and there are some, such as tobacco (*Nicotiana tabacum* L.), oilseed rape (*Brassica napus* L.) and barley (*Hordeum vulgare* L.), which, due to great success in terms of the induction and efficient regeneration of haploid plants, are considered to be model plants. But there are still species of considerable agricultural importance for which it has not been possible until now to establish the correct conditions for the induction of microspore embryogenesis, embryo formation or plant regeneration. This is so with plants of the family Leguminosae, with which only limited success has been achieved (Crosser et al. 2006; Grewal et al. 2009). Another noticeable example of recalcitrance is *Arabidopsis thaliana*, with which the suitable conditions for induction have still not been found.

In vitro technology to induce microspore embryogenesis has contributed significantly to both our fundamental and applied knowledge of the processes involved. The production in vitro of double haploids is a versatile genetic manipulation tool compatible with both classic breeding and new biotechnologies (Maluszynski et al. 2003). Moreover, microspore embryogenesis is very useful for studying the first stages of embryo production as it does not involve the problems of reduced numbers and difficult accessibility associated with zygotic embryos.

2.2 Discovery of the Production of Haploids by Anther Culture

The in vitro culture of anthers was first undertaken by Shimakura (1934) in an attempt to understand the physiology of meiosis. Guha and Maheshwari (1964) cultured anthers of *Datura innoxia* with a similar aim and, in doing so, accidentally stumbled across the formation of small plantlets arising from these anthers after 6 weeks. A basal nutrient medium containing kinetin, coconut milk or grape juice was used to culture the anthers, which after some weeks showed a dead anther wall surrounding a locule, inside which embryo-like structures had formed. In 1966 Guha and Maheshwari, by counting the chromosomes of the plantlets that grew from them, confirmed that these structures originated from microspores. This decisive discovery stimulated a general interest in developing the technique of anther culture to produce haploids in a wide variety of crop plants because of their potential significance in basic and applied genetics and plant breeding. Microspore embryogenesis flourished and was extended not only to other dicotyledonous species but also to species of economic importance, including monocotyledons (Raghavan 1986). Later, Nitsch and Norreel (1973) were able to obtain haploid embryos by culturing microspores isolated from anthers of *D. innoxia* in liquid medium. This was another important achievement in microspore embryogenesis because it represented the beginning of studies into the process without the sporophytic influence of the anther wall.

2.3 Strategies for the Induction of Microspore Embryogenesis

Microspore embryogenesis can be induced either from the anther or from isolated microspores. Anther culture is the most commonly used method for haploid production. Briefly, after pre-treatment of flower buds or spikes (usually cold treatment at 4°C) for embryogenic induction, anthers are extracted under sterile conditions and cultured on a semi-solid medium until the first embryos appear. In many cases, this pre-treatment is combined with another induction treatment applied to the anthers in the culture medium (starvation, osmotic, cold and/or heat shock). Finally, the embryos are transferred to a regeneration medium where the haploid plants are formed.

The culture of isolated microspores also requires the mechanical extraction of microspores when they are not shed spontaneously into the medium from the anther, together with their cleaning and concentration by filtration and differential centrifugation. In anther culture, embryos can be formed via a direct or indirect pathway. In direct embryogenesis, the microspores function as zygotes, dividing successively through different stages similar to those observed in zygotic embryogenesis. Indirect microspore embryogenesis involves the formation of an intermediate callus from which the new plantlet is formed (Maheshwari et al. 1982). Methods have been successfully adjusted for both dicotyledonous species (e.g. *D. innoxia, Hyoscyamus niger, N. tabacum, B. napus* and *Capsicum annuum*) and monocotyledonous ones (e.g. *H. vulgare, Triticum aestivum, Zea mays* and *Oryza sativa*), although these latter were for a long time less responsive to embryogenic induction. Nowadays, embryos are produced mainly from direct microspore embryogenesis because this type of embryogenesis is easier to study and ensures a minimum occurrence of cytogenetic abnormalities.

Although anther culture is less laborious than isolated-microspore culture, it does entail several disadvantages: embryos can be formed from anther tissue and, in cereals, there are greater possibilities of obtaining albino plants and chimeras. The likelihood of introducing culture contaminants is greater with isolated microspores, but in this type of culture the factors playing a role in embryogenesis are better controlled because the anther cells exert no influence. Other advantages also accrue to isolated-microspore culture: the yield of haploids is more readily controlled due to the fact that competition amongst pollen grains can be reduced by optimising the number of microspores per unit volume of culture medium. Transformation and mutation experiments are also easier because microspores are more accessible. Thus, studies into the cellular and molecular evolution of embryos tend to be less complicated (Pretova et al. 1993; Forster et al. 2007).

2.4 Influence of Different Factors in Microspore Embryogenesis

Embryogenesis will occur only when the optimum conditions prevail. The success of embryo induction, maturation and plant regeneration depends upon factors such as genotype, donor-plant physiology, the developmental stage of the pollen, pretreatments and culture conditions.

2.4.1 Genotype

The genome of the donor plant plays a key part in embryogenic induction. Therefore, as in any other tissue-culture technique, it must be considered very early in the protocol. There is no uniformity of response to microspore embryogenesis amongst species of the same family or even amongst cultivars, and it still remains unclear whether the unresponsiveness of certain species is due to their genotype or to unsuitable culture conditions. In the same family, one species, such as *B. napus*, might be a model of microspore embryogenesis whilst another, such as *A. thaliana*, is recalcitrant. Many attempts have been made to determine the level at which genotypic effects influence the successful induction of microspore embryogenesis. Molecular markers have been used to identify the quantitative trait loci (QTL) involved and then determine the magnitude of the genotypic effect upon this process (Forster and Thomas 2005). These studies are still in progress because genetic control of microspore embryogenesis is highly complex, involving genes with additive and dominant effects acting independently during different stages of the process (anther culture response, embryo differentiation, percentage of regenerating haploid plants and the regeneration of green plants) and frequently interacting with environmental conditions.

2.4.2 Donor Plant Physiology

The physiological conditions of donor plants can drastically affect anther development and thus the number of embryogenic microspores induced. In cereals these conditions contribute considerably to the unwanted production of albino plants (Bajaj 1990). Donor plants must be grown under special conditions regarding the intensity, quality and period of light, temperature, water and nutrition so as to yield an optimum percentage of embryogenic microspores (Ferrie et al. 1995). Pest and disease management can also affect embryogenic response. The age of the inflorescence from which anthers are extracted and, in monocotyledons, the position of the spikes on the plant may also influence the response to induction (Maheshwari et al. 1982; Jacquard et al. 2006). The development of annual plants fluctuates throughout the year, thus influencing the competence of isolated microspores for embryogenesis in both mono- and dicotyledonous plants (Jacquard et al. 2006). All of these observations point clearly to the fact that plants of all species growing under unfavourable conditions may well produce altered microspores, which will then respond differently to embryogenic induction.

2.4.3 Stage of Pollen Development

The selection of the most suitable stage of microspore development at which to induce microspore embryogenesis is important for an embryogenic response. In most species, the best initial stages are just before or after the first mitotic division, i.e. vacuolated microspores or young bicellular pollen (Raghavan 1986; Ferrie et al. 1995; Shivanna 2003). To facilitate the selection of the most suitable anthers for culture, bud size, spike characteristics or possibly some other feature of the anther are often correlated to the developmental stage of the pollen. In this way, the stage of development is determined directly by the external appearance of these macroscopic structures.

2.4.4 Pre-treatments

One essential requirement common to practically all species when switching microspore development from the gametophytic to the sporophytic pathway is the application of an external stimulus. The embryogenic potential is usually triggered by stress pre-treatment. Different stress pre-treatments have been used, such as cold or heat shock applied to the plant, flower buds, spikes, anthers or directly to the microspores themselves in culture, starvation by adding non-metabolizable carbohydrates to the culture medium, and osmotic shock, applied to the culture medium in the form of an extra supply of non-metabolizable carbon sources. Combinations of these treatments have also been applied to different parts of the plant to induce embryogenic development (Ferrie et al. 1995; Touraev et al. 1997; Shivanna 2003). In dicotyledonous species such as *D. innoxia* and *H. niger*, for example, cold pre-treatments (4°C) have classically been administered to flower buds (Raghavan 1986). Heat shock (about 32°C) is required to trigger microspore embryogenesis in *B. napus* (Custers et al. 1994), but other types of stress such as the application of gamma irradiation or colchicine (Pechan and Smykal 2001) are also effective. In *N. tabacum*, nitrogen starvation alone (Kyo and Harada 1986) or starvation combined with heat shock (Touraev et al. 1996a) has been applied to the same effect. Cold pre-treatment of spikes enhances the yield of embryos formed after anther culture in several cereals (Ferrie et al. 1995). Studies into *H. vulgare* (Hoekstra et al. 1992; Cistué et al. 1994) found that a combination of cold stress applied to the spikes, followed by osmotic and starvation shock with mannitol in the anther culture medium, increased not only induction rates but also the regeneration of green plants, thus avoiding the problem of albinism that frequently affects cereals. Other authors, however, working with isolated-microspore cultures in *T. aestivum* (Shariatpanahi et al. 2006), suggest that stress conditions may not only provide the trigger for microspore embryogenesis, but also induce recombination events in nuclear genes or chloroplast genomes. This leads to a reduction in plant regeneration and the induction of albino plants or somaclonal variation, and so the workers try to reduce stress in order to control its negative effect in haploid plant production, achieving embryogenesis in microspores without any apparent stress pre-treatment. The trauma caused by the excision and culture of anthers may provide, in some cases, sufficient stress to induce microspore embryogenesis.

2.4.5 Culture Conditions

The successful induction of microspore embryogenesis depends not only upon the factors mentioned above but also upon an adequate combination of culture conditions, amongst which are the composition of the medium (carbohydrate source and concentration, mineral nutrients, osmolarity, pH, gelling matrix and co-culture), the culture environment (temperature and light intensity) and the density of the material in the culture. Culture conditions are important for the nutrition of microspores, embryogenic induction and the regeneration of haploid plants.

2.4.6 Composition of the Medium

In early assays of microspore embryogenesis, hormones were frequently used and different combinations of auxins and/or cytokinins, ethylene, coconut milk, potato or yeast extract were often added to the culture (Raghavan 1986), but in fact with most species hormones are not essential to induce microspore embryogenesis and are frequently responsible for callus formation. Direct embryogenesis is now preferred and the tendency is to simplify the media and avoid hormones.

Sucrose is the most common carbon source in culture media, although the percentage used with different species varies and is generally determined by empirical manipulation of one or a combination of the existing culture media. Basal media such as MS (Murashige and Skoog 1962) or NLN media (Lichter 1982) with slight modifications are commonly used with *Brassica* and other species. With cereals, media such as A2 (Touraev et al. 1996b) and MMS3 (Hu and Kasha 1997) are also used. Nitrogen, phosphorus and calcium sources are also adjusted in these media. As for mineral micronutrients, it has been shown recently that copper sulphate increases the percentage of green plant regeneration in the very responsive winter-barley cultivar Igri (Wojnarowicz et al. 2002). Studies with this cultivar have also shown a marked short- and long-term influence of iron in the culture medium on isolated microspores used for embryogenesis. Quite frequently, the same composition adjusted for anther culture is used in isolated-microspore culture, but it has been shown that iron concentrations adjusted to optimum values for anther culture can be toxic to isolated microspores (Pulido et al. 2005, 2006a).

The osmotic pressure of the medium is another important parameter for both the induction of microspore embryogenesis and green plant regeneration. High concentrations of mannitol are applied to induce embryogenesis in barley and it has also been found that high osmolarity reduces the appearance of albino plants (Cistué et al. 1994; Jacquard et al. 2006).

The pH of the medium is usually about 6 for culture of both anthers and isolated microspores, but some genotypes require moderate adjustments (Ferrie et al. 1995). Anthers are generally cultured on semi-solid media but have also been successfully floated in liquid media enhanced with Ficoll (Cistué et al. 1999). Activated charcoal is often added to the culture medium because it is able to adsorb toxic substances produced during culture, although it is reported also to remove essential compounds and thus to exert a negative effect upon embryogenesis (Gland et al. 1988). One experimental recourse that has been effectively used with wheat to supply the signals needed to induce the development of embryos is that of co-culturing with nurse embryogenic tissues such as ovaries or embryogenic microspores (Hu and Kasha 1997). Light and temperature are also limiting factors. With *Brassica* the situation is very clear: in vitro microspore embryogenesis is induced at 32°C, whilst at 18°C pollen maturation is mimicked (Custers et al. 1994).

Culture density also controls to some extent the effectiveness of embryogenesis as well as the time required for embryo production. In *B. napus*, a plating density of 4×10^4/ml yielded more effective embryogenesis than one of 10×10^4/ml (Huang et al. 1990), whereas with *C. annuum* it has been shown that the number of embryos increases concomitantly with plate density, and that higher densities reduce induction time; above 10×10^4/ml the number of cotiledonary embryos decreases dramatically until 16×10^4/ml, when none are produced at all. The optimum microspore plating for *C. annuum* ($8–10\times10^4$/ml) is therefore greater than for *B. napus* and tends, incidentally, to be species-dependant (Kim et al. 2008).

2.5   Cellular and Molecular Events Associated with Microspore Embryogenesis

Four phases can be distinguished in microspore embryogenic development, namely (1) induction: embryogenic capability is acquired, (2) early embryogenic-division phase: successive divisions inside the exine wall are produced to form multicellular pollen grains, (3) embryo-development phase: multicellular structures are released from the exine and the development of an embryo pattern takes place, and (4) plant regeneration phase: the new plant is formed from the embryo. For these phases to succeed each other successfully, special conditions are required, which have been identified through the empirical modification of factors such as stress, culture medium composition and plant growth ambience.

Cellular and molecular approaches have been applied to improve our knowledge of the mechanisms that control the different phases of the embryogenic process and the identification of key regulators to modify them more precisely (Raghavan 1986, 1997; Sangwan and Sangwan-Norreel 1996; Cordewener et al. 1996; Maraschin et al. 2005a; Hosp et al. 2007). The first two phases are more directly involved in microspore embryogenesis, whilst the second two are common to the various different types of embryogenesis (zygotic, somatic and gametic).

2.5.1   Embryogenic Induction

Since even in the best of cases not all the microspores of an anther respond to embryogenic induction, intensive research has been undertaken to identify microspores with embryogenic capacity to find out at what point this capacity is acquired and to understand more fully the mechanism of transformation of a microspore into an embryogenic cell. Researchers started with light microscopy and were later able to complement their studies with electron microscopy. They focused on the structural changes taking place in microspores in relation to the halting of gametophytic development and its redirection towards the initiation of sporophytic division. Although, as mentioned in Section 2.4.3, the initial pollen developmental stage selected for the induction of microspore embryogenesis may vary according to the species, vacuolated microspores are usually chosen for this purpose. It has been found that induction conditions cause several major changes in these microspores: (1) they are generally enlarged, their central vacuole becoming fragmented and their nuclei moving to occupy a central position, (2) variations in the number of starch granules and lipid bodies are detected in the cytoplasm, and (3) the microspore wall becomes enlarged (Zaki and Dickinson 1990; Hoekstra et al. 1992; Telmer et al. 1995; Sangwan and Sangwan-Norreel 1996; Rodríguez-García et al. 2000; Maraschin et al. 2005a).

The initial pattern of cell division induced by anther or microspore culture varies considerably in different species, but in the earliest species studied the initial division was symmetrical, in contrast to the asymmetric first mitosis produced in normal pollen development, which is why this change in symmetry was taken to be a sign of deviation away from the normal gametophytic development towards the embryogenic programme. Later, it was found that embryos could also be obtained after an initial asymmetrical division and this idea was revised. Different views have since been put forward to explain the possible mechanisms for the induction of microspore embryogenesis, amongst which is one that considers that the embryogenic potential of microspores is predetermined at some early stage of microspore ontogeny (during meiosis) and is expressed later under suitable culture conditions. Another view presumes that microspores are not predetermined, but that they are all potentially embryogenic and this capability is acquired simply by the stress caused in the excision of the anthers and their culture in vitro. These views are still under discussion and the precise mechanisms of induction still remain to be solved (Raghavan 1997; Bonet et al. 1998; Shivanna 2003).

Since the cytoskeleton plays a role in changes in cell division, its configuration has been studied during the induction of embryogenesis in *B. napus* (Zaki and Dickinson 1991; Zarsky et al. 1992; Hause and Hahn 1998). It has been found that microfilaments play no primary role in the induction of embryogenesis, whilst changes in the microtubular cytoskeleton are indeed related to such induction. Cell-cycle transitory arrests as well as de-differentiation signs were found to be linked to the transition of the microspore to the sporophytic pathway in studies carried out with several well-responsive species (Hoekstra et al. 1992; Telmer et al. 1995; Sangwan and Sangwan-Norreel 1996; Maraschin et al. 2005a).

Attempts have been made to find biochemical and molecular markers for the embryogenic process using various strategies (Cordewener et al. 1996; Raghavan 1997; Pechan and Smykal 2001). Recent technological progress has lead to very promising genomic, transcriptomic and metabolomic studies, in which modern researchers are currently engaged (Maraschin et al. 2006; Muñoz-Amatriaín et al. 2006; Malik et al. 2007).

Optimisation of the protocols for embryogenic induction in isolated-microspore culture is crucial to characterize the key genes in the initiation of microspore embryogenesis, which avoids the sporophytic influence of anther tissues and increases the yield of embryogenic induction. To isolate them, differential expression between untreated and stress-treated microspores has been studied (Vrinten et al. 1999) because stress pre-treatment triggers embryogenic development. One of the difficulties in finding these genes is that many of them are similar to those activated at the beginning of zygotic embryogenesis and a number of the genes isolated have turned out to be pollen- or embryo-specific, rather than specific to microspore embryogenesis. Reynolds and Crawford (1996) prepared a cDNA library from the early stages of microspore-derived embryos in *T. aestivum* and analysed it with cDNA probes from pollen at different stages of gametophytic development, thereby isolating a cysteine-labelled metallothionein (*EcMt*) gene. This gene was expressed only in embryogenic microspores, microspore-derived and developing zygotic embryos and was therefore considered to be a possible marker for embryogenic potential. With this approach, different gene members of the heat-shock protein (HSP) family have also been found after heat-shock and starvation pre-treatments used for embryogenic induction in *B. napus* and *N. tabacum* (Zarsky et al. 1995; Smykal and Pechan 2000). Although initially HSPs were considered to be essential for microspore embryogenesis, their role has since been called into doubt because microspore embryogenesis can be induced in *Brassica* with colchicine without affecting the expression of HSPs (Zhao et al. 2003). Furthermore, the specific production of phosphoproteins has been found to be induced in *Nicotiana* after starvation treatment and the phosphorylation of proteins has been associated with the embryogenic dedifferentiation of pollen (Kyo and Harada 1990; Kyo et al. 2003). Vrinten et al. (1999), using a cDNA library of the first stages of microspore embryogenesis, were able to isolate three cDNAs in barley that were not expressed in microspores before treatment: early-culture abundant 1 (*ECA1*), encoding a protein related to AGPs (arabinogalactan proteins), early-culture glutathione S-transferase (*ECGST*), a protein that may play a role in oxidative stress protection, and early-culture lipid-transfer protein (*ECLTP*), a gene encoding a protein with a certain homology with lipid-transfer proteins. Also in *Brassica*, the BABY BOOM gene isolated from embryogenic microspores has been found to mark and stimulate early embryo formation (Boutilier et al. 2002).

The expression of all these genes involved in the acquisition of embryogenic potential has also been studied with modern transcriptomic techniques (Maraschin et al. 2006; Muñoz-Amatriaín et al. 2006; Joosen et al. 2007; Malik et al. 2007; Tsuwamoto et al. 2007). Working with barley, Maraschin et al. (2006) and Muñoz-Amatriaín et al. (2006), using macro- and micro-arrays, analysed the transcripts associated with stress treatment. The former used macro-arrays containing 1,421 expressed sequence tags (ESTs) covering the early stages of barley zygotic embryogenesis to analyse transcripts of vacuolated microspores, embryogenic microspores and bicellular pollen, and observed that the induction of microspore embryogenesis by pre-treatment with mannitol involved the up-regulation of transcripts related to sugar and starch hydrolysis, proteolysis, stress response and the inhibition of programmed cell death and signalling. Further analysis revealed that the induction of genes encoding alcohol dehydrogenase 3, metalloprotease FtsH, cysteine protease 1 precursor, phytepsin precursor (aspartic protease) and a 26S proteasome regulatory subunit was associated with the embryogenic potential of microspores, whereas the induction of transcripts involved in signalling and cytoprotection was associated with stress responses. A study by Muñoz-Amatriaín et al. (2006), using the 22K Barley1 GeneChips and allowing the expression of 22,000 genes to be examined simultaneously, revealed substantial changes in the expression of genes related to central carbon metabolism and stress response accompanying the reprogramming of microspores. The patterns of expression of transcription factors indicated that, after pre-treatment with mannitol, microspores remained undifferentiated but, unexpectedly, no signs of the switch to the embryogenic pathway were found. Other surprising proofs were that the common protection system against stress was not induced and no signs of cell-cycle arrest were found. Despite these new insights, the genes involved in the induction of microspore embryogenesis still remain to be correctly identified.

2.5.2 Early Embryogenic Divisions

There is a certain overlap between the previous phase and this one because, depending upon the species involved, microspore embryogenesis can be induced at either the microspore or immature-pollen stage and, in the latter case, the first mitotic division has taken place before culture. Nevertheless, whenever the first division takes place cytological studies show that the early patterns of embryogenic division produced inside the exine wall are heterogeneous and complex. The first cell division after the induction of embryogenesis in microspores can be either symmetric or asymmetric and, from that division, several possibilities have also been described. Thus, (1) after symmetric division, either the identical cells continue to divide to form the haploid embryo, or their nuclei first fuse and then the new diploid cell gives rise, after successive division, to a diploid embryo, and (2) from asymmetric division, a pollen grain is formed by a generative cell inside the vegetative cytoplasm, similar to the pollen grain obtained during in vivo gametogenesis. These cells may also evolve following different patterns. In most cases, the vegetative cell divides consecutively to form the embryo whilst the generative cell degenerates, although sometimes the generative cell continues to proliferate and the vegetative cell degenerates. Another pattern observed occurs when the two nuclei fuse to form a new central nucleus which then divides to form a diploid embryo. Finally, on some occasions, both cells divide to form the new embryo. Interestingly, these different embryogenic routes may vary within the same species according to whether anther or isolated microspores are used for induction, the type of stress applied and/or the culture medium used (Sunderland 1974; Raghavan 1986; Zaki and Dickinson 1991; Pretova et al. 1993; Ilic-Grubor et al. 1998).

The names of the structures produced after the successive divisions of microspores are not unanimously agreed upon and terms such as "embryoid", "embryogenic structures", "embryo-like structures" and "multicellular pollen grains" can be found in the literature. The main problem involved in choosing suitable terminology is that it is impossible to be sure whether the structures observed after successive microspore divisions will produce embryos or simply degenerate. Cell-tracking technologies have been developed to follow the fate of microspores throughout the embryogenic process (Hause and Hahn 1998; Maraschin et al. 2004), thereby helping to characterize the structures capable of producing embryos. Multicellular structures formed after microspore culture in maize and barley have been characterized using ultrastructural, cytochemical and immunocytochemical techniques (Magnard et al. 2000; Pulido et al. 2002; Testillano et al. 2002). All these studies support previous reports describing the accumulation of endoplasmic reticulum and an increased build-up of plasmodesmata between cells (both signs of embryogenesis induction), whilst in non-embryogenic structures an accumulation of starch and/or lipids has been detected (Zaki and Dickinson 1990; Telmer et al. 1995).

It is known that a number of divisions occur inside the microspore wall, and once structures similar to globular embryos are formed the wall disintegrates and differentiation of the protodermis of the embryo begins. Consequently, the microspore wall plays an important role in this process. It is made up initially of an external exine and inner intine but, immediately after the induction of embryogenesis, this structure changes. In *B. napus* a thick fibrillar wall forms beneath the intine (Zaki and Dickinson 1990) and, in *H. vulgare* and *T. aestivum*, the intine thickens (Bonet and Olmedilla 2000; Pulido et al. 2005). Cellular and molecular evidence for programmed cell death has also been shown in the transition from multicellular structures to globular embryos once the exine has been ruptured in *H. vulgare* (Maraschin et al. 2005b).

2.5.3 Development of Embryo Pattern

The pattern of the formation of the embryogenic structures generated once the pollen wall is broken is similar to that of the zygotic or somatic embryos (Mordhorst et al. 1997). Structural and molecular similarities have been found between zygotic and microspore embryogenesis development (Sangwan and Sangwan-Norreel 1996; Ilic-Grubor et al. 1998; Boutilier et al. 2002; Joosen et al. 2007). It was claimed initially that one of the main differences was that microspore-derived embryos generally lacked a suspensor, but recently Custers et al. (2001) have shown that under special culture conditions it is possible to produce a large number of embryos with suspensors in *B. napus*. Other slight differences, such as the number and size of cells that constitute the embryo, are also clear in comparing the microspore-derived embryo with the zygotic embryo at any determined stage (Sangwan and Sangwan-Norreel 1996).

Gene products expressed during zygotic embryogenesis have been used as probes to study the development of microspore-derived embryos and, via this strategy, one of the first markers found was the 12S storage glycoprotein (Crouch 1982). Using a similar approach, Boutilier et al. (1994) found that napines (seed-storage proteins in *Brassica*) were expressed very early in microspore embryogenesis, coinciding with embryogenic induction, and could be used as markers of these early phases. In the same way, hordeins were found to be expressed in microspore-derived embryos in barley (Pulido et al. 2006b). Also in *Brassica*, the BABY BOOM gene (Boutilier et al. 2002) was found to mark and stimulate early embryo formation.

2.5.4 Plant Formation and Diploidization

Regeneration of the whole plant from the embryos obtained after microspore embryogenesis is another challenge in the production of double haploid plants. Once the embryo pattern is established not all the proembryos are able to regenerate into plants and the culture medium plays an important part in the regeneration rate (Cegielska-Taras et al. 2002). Diploidization is necessary because haploid plants are sterile and although spontaneous diploids may appear because of the fusion of nuclei, which can occur during the initial stages of microspore embryogenesis, a chromosome doubling protocol using an anti-mitotic chemical such as colchicine is commonly required if there is a low level of spontaneous doubling of haploid material (Barnabás et al. 2001). As colchicine is highly toxic to human, several attempts have been made to reduce its concentration or to find alternative methods with other chemicals for chromosome doubling, but colchicine still proves to be a useful method for inducing diploidization (Soriano et al. 2007).

Albino plants are very often produced during the regeneration of microspore-derived plants in cereals. Although different factors have been found to affect the degree of albinism, such as genotype, the state of donor plants and the microspore stage, the precise cause of albinism is as yet unknown and more research is needed to avoid this limiting factor in double haploid production (Touraev et al. 2001).

2.6 Conclusions

Microspore embryogenesis is one of the few biotechnologies that has already found its way into breeding practice, speeding up the production of double haploids. Efficient culture protocols have been obtained on the basis of almost completely trial-and-error empirical research in plant-tissue culture in a restricted number of species. Nevertheless, from the beginning of the discovery of this phenomenon, it has been obvious that to find reliable procedures that work whatever the genotype there is an imperative need to understand this developmental process at both cellular and molecular levels. Despite the considerable effort invested in this direction, the multifaceted mechanisms that lead microspores to divide and evolve to form an embryo while they were initially programmed to form a mature pollen grain are as yet far from understood. It is to be hoped that the advent of new transcriptomic, proteomic and metabolomic studies will help us to obtain a more detailed picture of this fascinating process.

Acknowledgements The author thanks Salvatore Pelliccione, Irene Serrano and Rosa Luque for their scientific contribution and technical help, and Dr. J. Trout for revising the English text.

References

Bajaj YPS (1990) In vitro production of haploids and their use in cell genetics and plant breeding. In: Bajaj YPS (ed) Biotechnology Agriculture and Forestry, vol 12. Haploids in crop improvement I. Springer, Berlin Heidelberg, pp 3–44

Barnabás B, Szakács É, Karsai I, Bed Z (2001) In vitro androgenesis of wheat: from fundamentals to practical application. Euphytica 119:211–216

Bonet FJ, Olmedilla A (2000) Structural changes during early embryogenesis in wheat pollen. Protoplasma 211:94–102

Bonet FJ, Azbaid L, Olmedilla A (1998) Pollen embryogenesis: atavism or totipotency? Protoplasma 202:115–121

Boutilier KA, Gines MJ, DeMoor JM, Huang B, Baszczynski CL, Iyer VN, Miki BL (1994) Expression of the BnmNAP subfamily of napin genes coincides with the induction of *Brassica* microspore embryogenesis. Plant Mol Biol 26:1711–1723

Boutilier K, Offringa R, Sharma VK, Kieft H, Ouellet T, Zhang L, Hattori J, Liu CM, van Lammeren AAM, Miki BLA, Custers JBM, van Lookeren Campage MM (2002) Ectopic expression of BABY BOOM triggers a conversion from vegetative to embryonic growth. Plant Cell 14:1737–1749

Cegielska-Taras T, Tykarska T, Szala L, Kuras M, Krzymanski J (2002) Direct plant development from microspore-derived embryos of winter oilseed rape *Brassica napus* L. ssp. *oleifera* (DC.) Metzger. Euphytica 124:341–347

Cistué L, Ramos A, Castillo AM, Romagosa I (1994) Production of large number of doubled haploids from barley anthers pretreated with high concentrations of mannitol. Plant Cell Rep 13:709–712

Cistué L, Ramos A, Castillo AM (1999) Influence of anther pretreatment and culture medium composition on the production of barley doubled haploids from model and low responding cultivars. Plant Cell Tiss Organ Cult 42:163–169

Cordewener JHG, Custers JBM, Dons HJM, Vanlookeren Campagne MM (1996) Cytological and biochemical aspects of in vitro androgenesis in higher plants. In: Jain SM, Sopory SK, Veilleux RF (eds) In vitro haploid production in higher plants. Kluwer, Dordrecht, pp 111–124

Crosser JS, Lülsdorf MM, Davies PA, Clarke HJ, Bayliss KL, Mallikarjuna N, Siddique KHM (2006) Toward doubled-haploid production in the Fabaceae: progress, constraints and opportunities. Crit Rev Plant Sci 25:139–157

Crouch ML (1982) Non-zygotic embryos of *Brassica napus* L. contain embryo-specific storage proteins. Planta 156:520–524

Custers JBM, Cordewener JHG, Nollen Y, Dons HJM, Van Lookeren Campagne MM (1994) Temperature controls both gametophytic and sporophytic development in microspore cultures of *Brassica napus*. Plant Cell Rep 13:267–271

Custers JBM, Cordewener JHG, Fiers MA, Maassen BTH, van Lookeren Campagne MM, Liu CM (2001) Androgenesis in *Brassica*: a model system to study the initiation of plant embryogenesis. In: Bhojwani SS, Soh WY (eds) Current trends in the embryology of angiosperms. Kluwers, Dordrecht, pp 451–470

Ferrie AMR, Palmer CE, Keller WA (1995) Haploid embryogenesis. In: Thorpe TA (ed) In vitro embryogenesis in plants. Kluwer, Dordrecht, pp 309–344

Forster BP, Thomas WTB (2005) Doubled haploids in genetics and plant breeding. Plant Breed Rev 25:57–88

Forster BP, Heberle-Bors E, Kasha KJ, Touraev A (2007) The resurgence of haploids in higher plants. Trends Plant Sci 12:368–375

Gland A, Lichter R, Schweiger HG (1988) Genetic and exogenous factors affecting embryogenesis in isolated microspore cultures of B. napus L. J Plant Physiol 132:613–617

Grewal RK, Lulsdorf M, Crosser J, Ochatt S, Vandenberg A, Warkentin TD (2009) Doubled-haploid production in chickpea (Cicer arietinum L.): role of stress treatments. Plant Cell Rep 28:1289–1299

Guha S, Maheshwari SC (1964) In vitro production of embryos from anthers of Datura. Nature 204:497

Guha S, Maheshwari SC (1966) Cell division and differentiation of embryoids in the pollen grains of Datura in vitro. Nature 212:97–98

Hause G, Hahn H (1998) Cytological characterization of multicellular structures in microspore cultures of Brassica napus L. Bot Acta 111:204–211

Hoekstra S, van Zijderveld MH, Louwerse JD, Heidekamp F, van der Mark F (1992) Anther and microspore culture of Hordeum vulgare L. cv Igri. Plant Sci 86:89–96

Hosp J, Maraschin SF, Touraev A, Boutilier K (2007) Functional genomics of microspore embryogenesis. Euphytica 158:275–285

Hu TC, Kasha KJ (1997) Improvement of isolated microspore culture of wheat (Triticum aestivum L.) through ovary co-culture. Plant Cell Rep 16:520–525

Huang B, Bird S, Kemble R, Simmonds D, Keller W, Miki B (1990) Effects of culture density, conditioned medium and feeder cultures on microspore embryogenesis in Brassica napus L. cv. Topas. Plant Cell Rep 8:594–597

Ilic-Grubor K, Attree SM, Fowke LC (1998) Comparative morphological study of zygotic and microspore-derived embryos of Brassica napus L. as revealed by scanning electron microscopy. Ann Bot 82:157–165

Jacquard C, Asakaviciute R, Hamalian AM, Sangwan RS, Devaux P, Clement C (2006) Barley anther culture: effects of annual cycle and spike position on microspore embryogenesis and albinism. Plant Cell Rep 25:375–381

Joosen R, Cordewener J, Supena EDJ, Vost O, Lammers M, Maliepaard C, Zeilmaker T, Miki B, America T, Custer J, Boutelier K (2007) Combined transcriptome and proteome analysis identifies pathways and markers associated with the establishment of Brassica napus microspore-derived embryo development. Plant Physiol 144:155–172

Kim M, Jang IC, Kim JA, Park EJ, Yoon M, Lee Y (2008) Embryogenesis and plant regeneration of hot pepper (Capsicum annuum L.) through isolated microspore culture. Plant Cell Rep 27:425–434

Kyo M, Harada H (1986) Control of the developmental pathway of tobacco pollen in vitro. Planta 168:427–432

Kyo M, Harada H (1990) Specific phosphoproteins in the initial period of tobacco pollen embryogenesis. Planta 182:58–63

Kyo M, Hattori S, Yamaji N, Pechan P, Yuasa Y, Fukui H (2003) Cloning and characterization of cDNAs associated with the embryogenic dedifferentiation of tobacco immature pollen grains. Plant Sci 164:1057–1066

Lichter R (1982) Induction of haploid plants from isolated pollen of Brassica napus. Z Pflanzenphysiol 105:427–434

Magnard JL, Le Deunff E, Domenech J, Rogowsky PM, Testillano PS, Rougier M, Risueño MC, Vergne P, Dumas C (2000) Genes normally expressed in the endosperm are expressed at early stages of microspore embryogenesis in maize. Plant Mol Biol 44:559–574

Maheshwari SC, Raschid A, Tyagi AK (1982) Haploids from pollen grains - Retrospect and prospect. Am J Bot 69:865–879

Malik MR, Wang F, Dirpaul JM, Zhou N, Polowick PL, Ferri AMR, Krochko JE (2007) Transcript profiling and identification of molecular markers for early microspore embryogenesis in *Brassica napus*. Plant Physiol 144:134–154

Maluszynski M, Kasha KJ, Szarejko I (2003) Published doubled haploid protocols in plant species. In: Maluszynski M, Kasha KJ, Forster BP, Szarejko I (eds) Doubled haploid production in crop plants, a manual. Kluwer, Dordrecht, pp 309–335

Maraschin SF, Vennik M, Lamers GEM, Spaink HP, Wang M (2004) Time-lapse tracking of barley androgenesis reveals position-determined cell death within pro-embryos. Planta 220:531–540

Maraschin SF, De Priester W, Spaink HP, Wang M (2005a) Androgenic switch: an example of plant embryogenesis from the male gametophyte perspective. J Exp Bot 417:1711–1726

Maraschin SF, Gaussand G, Pulido A, Olmedilla A, Lamers GEM, Korthout H, Spaink HP, Wang M (2005b) Programmed cell death is involved in the transition from multicellular structures to globular embryos in barley androgenesis. Planta 221:459–470

Maraschin SF, Caspers M, Potokina E, Wülfert F, Graner A, Spaink HP, Wang M (2006) cDNA array analysis of stressed induced gene expression in barley androgenesis. Physiol Plant 127:535–550

Mordhorst AP, Toonen MAJ, de Vries SC (1997) Plant embryogenesis. Crit Rev Plant Sci 16:535–576

Muñoz-Amatriaín M, Svensson JT, Castillo AM, Cistué L, Close TJ, Vallés MP (2006) Transcriptome analysis of barley anthers: effect of mannitol treatment on microspore embryogenesis. Physiol Plant 127:551–560

Murashige T, Skoog F (1962) A revised medium for rapid growth and bioassay with tobacco tissue cultures. Physiol Plant 15:473–497

Nitsch C, Norreel B (1973) Effet d'un choc thermique sur le pouvoir embryogène du pollen de *Datura* culturé dans l'anthère ou isolé de l'anthère. C R Acad Sci Paris 276:303–306

Pechan PM, Smykal P (2001) Androgenesis: affecting the fate of the male gametophyte. Physiol Plant 111:1–8

Pretova A, De Ruijter NCA, Van Lammeren AAM, Schel JHN (1993) Structural observations during androgenic microspore culture of the 4cl genotype of *Zea mays* L. Euphytica 65:61–69

Pulido A, Castillo A, Vallés MP, Olmedilla A (2002) In search of molecular markers for androgenesis. Biologia 57:29–36

Pulido A, Bakos F, Castillo A, Vallés MP, Barnabas B, Olmedilla A (2005) Cytological and ultrastructural changes induced in anther and isolated-microspore cultures in barley: Fe deposits in isolated-microspore cultures. J Struct Biol 149:170–181

Pulido A, Bakos F, Castillo A, Vallés MP, Barnabas B, Olmedilla A (2006a) Influence of Fe concentration in the medium of multicellular pollen grains and haploid plants induced by mannitol pretreatment in barley (*Hordeum vulgare* L.). Protoplasma 228:101–106

Pulido A, Hernando A, Bakos F, Méndez E, Devic M, Barnabás B, Olmedilla A (2006b) Hordeins are expressed in microspore-derived embryos and also during male gametophytic development and very early stages of seed development. J Exp Bot 57:2837–2846

Raghavan V (1986) Pollen embryogenesis. In: Barlow PW, Green PB, Wylie CC (eds) Embryogenesis in angiosperms: a developmental and experimental study. Cambridge University Press, New York, pp 152–189

Raghavan V (1997) Embryogenic development of pollen grains. In: Raghavan V (ed) Molecular embryology of flowering plants. Cambridge University Press, Cambridge, pp 500–523

Reynolds TL (1997) Pollen embryogenesis. Plant Mol Biol 33:1–10

Reynolds TL, Crawford RL (1996) Changes in abundance of an abscisic acid-responsive, early cysteine-labeled metallothionein transcript during pollen embryogenesis in breed wheat (*Triticum aestivum*). Plant Mol Biol 32:823–829

Rodríguez-García MI, Olmedilla A, Alché JD (2000) The contributions and limitations of micros-copy in studying the mechanisms of pollen embryogenesis. In: Bohanec B (ed) Biotechnological approaches for utilization of gametic cells. COST 824, European Communities, Belgium, pp 253–259

Sangwan RS, Sangwan-Norreel BS (1996) Cytological and biochemical aspects of in vitro andro-genesis in higher plants. In: Jain SM, Sopory SK, Veilleux RE (eds) In vitro production of higher plants. Kluwer, Dordrecht, pp 95–109

Shariatpanahi ME, Belogradova K, Hessamvaziri L, Heberle-Bors E, Touraev A (2006) Efficient embryogenesis and regeneration in freshly isolated and cultured wheat (Triticum aestivum L.) microspores without stress pre-treatment. Plant Cell Rep 25:1294–1299

Shimakura K (1934) The capability of continuing divisions of Tradescantia pollen mother-cells in saccharose solution. Cytologia 5:363–372

Shivanna KR (2003) Induction of haploids from pollen grains. In: Enfield NH (ed) Pollen biology and biotechnology. Science Publishers, New Hampshire, pp 219–230

Smykal P, Pechan PM (2000) Stress, as assessed by the appearance of sHsp transcripts, is required but not sufficient to initiate androgenesis. Physiol Plant 110:135–143

Soriano M, Cistué L, Vallés MP, Castillo AM (2007) Effects of colchicine on anther and microspore culture of bread wheat (Triticum aestivum L.). Plant Cell Tiss Organ Cult 91:225–234

Sunderland N (1974) Anther culture as means of haploid induction. In: Kasha KJ (ed) Haploids in higher plants: advances and potential. University of Guelph, Guelph, Ontario, pp 91–122

Telmer CA, Newcomb W, Simmonds DH (1995) Cellular changes during heat shock induction and embryo development of cultured microspores of Brassica napus cv. Topas. Protoplasma 185:106–112

Testillano PS, Ramírez C, Domenech J, Coronado MJ, Vergne P, Matthys-Rochon E, Risueño MC (2002) Young microspore-derived maize embryos show two domains with defined features also present in zygotic embryogenesis. Int J Dev Biol 46:1035–1047

Touraev A, Pfosser M, Vicente O, Heberle-Bors E (1996a) Stress as the major signal controlling the developmental fate of tobacco microspores: towards a unified model of induction of microspore/pollen embryogenesis. Planta 200:144–152

Touraev A, Indrianto A, Wratschko I, Vicente O, Heberle-Bors E (1996b) Efficient microspore embryogenesis in wheat (Triticum aestivum L.) induced by starvation at high temperature. Sex Plant Reprod 9:209–215

Touraev A, Vicente O, Heberle-Bors E (1997) Initiation of microspore embryogenesis by stress. Plant Sci 2:297–302

Touraev A, Pfosser M, Heberle-Bors E (2001) The microspore: a haploid multipurpose cell. Adv Bot Res 35:53–109

Tsuwamoto R, Fukuoka H, Takahata Y (2007) Identification and characterization of genes expressed in early embryogenesis from microspores of Brassica napus. Planta 225:641–652

Vrinten PL, Nakamura T, Kasha K (1999) Characterization of cDNAs expressed in the early stages of microspore embryogenesis in barley (Hordeum vulgare L.). Plant Mol Biol 41:455–463

Wojnarowicz G, Jacquard C, Devaux P, Sanqwan RS, Clement C (2002) Influence of copper sulfate on anther culture in barley (Hordeum vulgare). Plant Sci 162:843–847

Zaki MAM, Dickinson HG (1990) Structural changes during the first divisions of embryos resulting from anther and free microspore culture in Brassica napus. Protoplasma 156:149–162

Zaki MAM, Dickinson HG (1991) Microspore-derived embryos in Brassica: The significance of division symmetry in pollen mitosis I to embryogenic development. Sex Plant Reprod 4:48–55

Zarsky V, Garrido D, Rihova L, Tupy J, Vicente O, Heberle-Bors E (1992) Derepression of the cell cycle by starvation is involved in the induction of tobacco pollen embryogenesis. Sex Plant Reprod 5:189–194

Zarsky V, Garrido D, Eller N, Tupy J, Vicente O, Schoffl F, Heberle-Bors E (1995) The expression of a small heat-shock gene is activated during induction of tobacco pollen embryogenesis by starvation. Plant Cell Environ 18:139–147

Zhao JP, Newcomb W, Simmonds D (2003) Heat-shock proteins 70 kDa and 19 kDa are not required for induction embryogenesis of *Brassica napus* L. cv. Topas. Plant Cell Physiol 44:1417–1421

Chapter 3
Stress and Somaclonal Variation

A.M. Vázquez and R. Linacero

3.1 Introduction

Plant species have developed in the course of evolution through a series of mechanisms which optimize their relationships with the environments in which they live. As plants are sessile organisms, they cannot avoid extreme stress situations which can be deleterious. Although the word stress is well known, it has proven to be a very elusive concept, and the term is used in scientific literature in many different ways and contexts (Qureshi et al. 2007). Thus, stress can result from changes in abiotic factors, such as mechanical damage, climatic factors and chemical or physical modifications in the environment, as well as biotic factors such as pathogen attack. In some cases, stress in plants has been defined as any change under growth conditions which disrupt metabolic homeostasis, requiring an adjustment of metabolic pathways in a process usually referred to as acclimation (Shulaev et al. 2008). The concept of acclimation is broad, and induced responses work at different levels according to the nature and duration of stress conditions.

In the immediate short-term responses, cells possess a repertoire of genes which can be activated or inactivated according to their needs. This type of response involves the modification of gene activity. However, under extreme stressful conditions or long-term responses, cells may react by undergoing a more drastic modification, the so-called genomic reprogramming. These two types of responses are discussed below.

A.M. Vázquez and R. Linacero
Departamento de Genética, Facultad de Biología, Universidad Complutense, 28008 Madrid, Spain
e-mail: anavaz@bio.ucm.es

E-C. Pua and M.R. Davey (eds.),
Plant Developmental Biology – Biotechnological Perspectives: Volume 2,
DOI 10.1007/978-3-642-04670-4_3, © Springer-Verlag Berlin Heidelberg 2010

3.2 Stress Responses in Plants

3.2.1 Short-Term Responses

It has been reported that plants respond to biotic and abiotic stresses by upregulating the expression of a large number of pathogenesis-related (PR) genes. The expression of PR genes, which was first known by its rapid induction in response to pathogenic infection, seems to play a role in the plant defence mechanism. Evidence from several lines of study shows that PR genes play a role in some physiological processes during plant growth and development, and that their expression is regulated by factors including abiotic stresses and plant hormones (Seo et al. 2008). To date, new genomic methodologies have been employed and more genes associated with stress have been identified (Aarts and Fiers 2003; Bohnert et al. 2006). Microarray analyses examining transcriptional reprogramming in *Arabidopsis thaliana* induced by several pathogens, messenger molecules and elicitors revealed that hundreds of genes, in addition to PR genes, exhibited differential expression after activation of the defence program (reviewed in Eulgen 2005). As a result, the pattern of protein accumulation is changed in stressed cells, both qualitatively and quantitatively (reviewed in Qureshi et al. 2007). Also, the importance of small RNAs as response modulators in plants under stress conditions has been emphasized recently (Phillips et al. 2007; Sunkar et al. 2007).

Although the response of plants to stress varies with species, common mechanisms seem to be implicated even under different stress conditions. Results from recent studies indicate that signalling pathways used by different defence systems converge and target overlapping gene sets. Cheong et al. (2002) reported interactions between wounding and other signals, including biotic and abiotic stress factors and plant hormones. It has also been observed that different stresses are implicated in the regulation of expression of different but overlapping series of genes, and plant signalling pathways consisting of complex networks (Swindell 2006; Fujita et al. 2006; Walley et al. 2007).

Gene expression is controlled at different levels which operate through changes in DNA methylation, histone modification and chromatin remodelling. As indicated above, stress can modify gene expression. It has been reported that stress can alter the pattern of gene methylation; thus, those genes activated under stress conditions showed a modification in their methylation status. Choi and Sano (2007) reported that some genes were selectively demethylated and, subsequently, transcribed under stress conditions. They proposed that environmental responses of plants are partially mediated through active alteration of the DNA methylation status. In other cases, the modifications were distributed along the genome but not randomly, involving precise sites. This can be illustrated by changes in methylation patterns induced by water stress in pea (*Pisum sativum*; Labra et al. 2002) or by heavy metals (Ni^{2+}, Cd^{2+} and Cr^{6+}) in white clover (*Trifolium repens*; Aina et al. 2004), which involved specific DNA sequences.

3.2.2   Long-Term Responses

Stress is not only the motor of modification of gene expression but it can also generate genome modifications, referred to as "genomic stress", through processes not very well known. In establishing a relation between stress and genomic changes, McClintock (1984) argued that "some responses to stress are especially significant for illustrating how a genome could modify itself when confronted with unfamiliar conditions". These genomic changes, sometimes referred to as an adaptive process of mutation, can provide the cells with a new genetic variability which can enhance adaptability in the new situation. As a result, plants can "prepare" to overcome adverse situations which can compromise cellular life, even with the possible occurrence of undesirable mutations. It is possible that adaptive mutations in response to stress exist in plants, and this may reflect an inducible mechanism which generates genetic variability. These mutation events may provide the cells with a new genetic repertoire with which they can adapt and survive under transient, or permanent, new conditions.

Several types of genome changes induced by stress have been reported. These include mobile elements (ME) activation and rearrangements, recombination and point mutations. ME activation was described by McClintock (1984). There has been increasing evidence showing that different types of stress can activate both transposons and retrotransposons. In retrotransposons, transcription and subsequent transposition are detectable under some stressing conditions (Grandbastien 1998; Kumar and Bennetzen 1999). Transcription of the retroelements is not always followed by the new copy insertions, and only in a few cases has the completed activity of retrotransposons been reported (Feschotte et al. 2002; Casacuberta and Santiago 2003). However, as the study of the appearance of new insertions is quite complex, probably on many occasions we cannot prove their existence even when they occur.

Several subfamilies of *Tnt1*, a copia retrotransposon of tobacco (*Nicotiana tabacum*) transcribed under stress, are associated with plant defence reactions and can be induced by salicylic acid or 2,4-dichlorophenoxyacetic acid (2,4-D; Pouteau et al. 1994; Beguiristain et al. 2001). In this species, expression of *Tto1*, another copia retroelement, has been shown to be induced in leaves by wounding and the exogenous supply of methyl jasmonate (Takeda et al. 1998). In addition, transcriptional activation of type II ME can also be induced by stress. *Rim2*, a *CACTA-like* element of rice (*Oryza sativa*), has been shown to be transcriptionally activated in response to pathogen infection (He et al. 2000). Xu et al. (2004) reported that the transposon *Jittery*, distantly related to *Mutator*, was activated in a maize (*Zea mays*) inbreed line and excised, but without reinsertion, after infection with the barley stripe mosaic virus (BSMV).

Transposon activation and mobilisation in response to abiotic stress have also been reported. In alfalfa (*Medicago sativa*), cold-induced transcriptional activation of a retrotransposon, the *MCIRE* element, was observed (Ivashuta et al. 2002). The expression of this ME was associated with genotypic variation in cold acclimation.

In the Tam element of *Antirrhinum majus*, the transposition was controlled by temperature (Hashida et al. 2003). The element transposed rarely at 25°C, but this frequency increased 1,000-fold at 15°C. Reactivation has also been reported of inactive *Mutator* lines of maize induced with gamma irradiation (Walbot 1988). The miniature inverted-repeat transposable element (MITE) *mPing* of rice transposes actively in gamma ray-irradiated plants (Nakazaki et al. 2003). This MITE, together with one of its putative transposase encoding partners, the Pong element, has been shown to mobilise efficiently in somatic cells of intact rice plants, derived from germinating seeds subjected to high hydrostatic pressure, or in plants from seeds treated with high hydrostatic pressure (Lin et al. 2006).

An increase in somatic recombination is another response of plants under stress conditions. Transgenic plants with selectable markers or reporter genes carrying different overlapping deletions have been used to detect somatic recombination which has been expressed by the restored function of the mutated transgenes due to different recombination events. Abiotic stress seems to increase somatic recombination. In *Arabidopsis*, low doses of X-rays (1.25 Gy) enhanced the relative recombination frequency to approximately twice the spontaneous value, while exposure to an elevated temperature (50°C) resulted in a 6.5-fold increase in the frequency (Lebel et al. 1993). Recombination frequencies were also enhanced twofold by NaCl (Puchta et al. 1995), or several fold by UV-light (Puchta et al. 1995; Ries et al. 2000; Molinier et al. 2006). Biotic stresses also increased the rate of somatic recombination. Infection by pathogens has been shown to stimulate somatic recombination in Arabidopsis, and a similar stimulatory effect has been observed when plant pathogen defence mechanisms are activated (Lucht et al. 2002), or when plants are treated with xylanase, a general fungal elicitor (Molinier et al. 2005), or flagellin, a bacterially derived elicitor (Molinier et al. 2006). In tobacco, plants infected with either tobacco mosaic virus (TMV) or oilseed rape mosaic virus (ORMV) exhibited a threefold increase in homologous recombination frequency in both infected and non-infected leaves (Kovalchuk et al. 2003). Moreover, when leaves of those plants which did not contain the virus were grafted onto healthy, non-infected plants, a recombination-inducing signal was detected. The recombination frequency was 2.3 times higher when plants were grafted with a "signal-carrying" leaf, compared with those grafted with a leaf from a mock-inoculated plant.

Alteration of particular DNA sequences has also been associated with temperature. Ceccarelli et al. (2002) reported that some repeated sequences were more represented in the genome of *Festuca arundinacea* seedlings grown at 30°C compared to seedlings raised at 10°C, whereas other sequences were more represented in the genome of seedlings grown at the lower temperature.

One of the most striking environmental-induced and heritable changes is the genotrophs of flax (*Linum usitatissimum*). After the plants were grown for one generation under the induced environment (e.g. different fertilizer combination or temperature regimes), stable and genetically altered plants were obtained, referred to as genotrophs (Cullis 1986, 2005). Some phenotype alterations were observed, and the extreme genotrophs were characterized by a marked difference in nuclear

DNA content. Other observed alterations include changes in the number of genes coding for the large and 5S ribosomal RNAs and in other repetitive sequence families. Oh and Cullis (2003) described RAPDs (randomly amplified polymorphic DNA) which were derived from all parts of the genome, including highly repetitive, middle-repetitive and low-copy-number sequences (Cullis et al. 1999), and their changes, such as deletions and/or rearrangements of both low-copy-number and highly repetitive sequences. Chen et al. (2005) reported a novel 5.8-kb element, designated as *Linum Insertion Sequence 1* (*LIS-1*), which appeared in the genome of five independent flax lines at the same position when subjected to the same environmental conditions.

Interspecific hybridization and allopolyploid formation have been considered as one of the most stressful situations for genome stability. Madlung and Comai (2004) reported phenotypic modification, widespread changes in DNA methylation, gene silencing, ME activation and loss of sequences, amongst other changes, in the hybrids or newly obtained allopolyploids.

3.2.3 Modifications Induced by Stress Could be Inheritable

In the course of the plant's life, different patterns of gene expression are established and epigenetic marks appear. For a long time, it has been accepted that this epigenetic information would be erased between generations and epigenetic changes would not be transmitted through the germ line. However, results from several lines of study show that at least some of these modifications can be inherited (Grant-Downton and Dickinson 2005; Bond and Finnegan 2007; Henderson and Jacobsen 2007).

As discussed above, plants react to stress through short- or long-term responses. There is evidence showing that some of these modifications can be inherited. Short-term responses act throughout modification of gene activity and, consequently, through chromatin modifications at different levels. One assumption is that an environmental stimulus can induce heritable chromatin modification as an adaptive response (Jablonka and Lamb 2002). As a result, changes in chromatin can play a role in passing on the "memory" of being exposed to environmental stress (Bond and Finnegan 2007). Bruce et al. (2007) considered that plants may possess a mechanism for storing information from previous exposures to stressful situations. As a result, plants have the capacity for some form of "memory" which has been referred to as "stress imprint". This stress imprint has been defined as a genetic or biochemical modification of a plant which occurs after stress exposure, and which causes future responses to future stresses to be different. Thus, exposure to a stressful agent can activate a gene or set of genes but, instead of reverting to the transcriptionally silent state once the stimulus is removed, an epigenetic mark can perhaps be left, keeping the region in a "permissive" state, thereby facilitating quicker and more potent responses to subsequent attacks.

Some stress imprint effects in plants have even been shown to be perpetuated into the following generations. The results of several studies have indicated the importance of DNA methylation in the maintenance of the stress memory throughout generations. Vaillant and Paszkowski (2007) reported that CG methylation plays a key role in the transgenerational inheritance of epigenetic states of transcriptional activity. Furthermore, Mathieu et al. (2007) proposed that the immediate, non-heritable stress responses might be associated with alteration of non-CG methylation patterns, while long-term, heritable adaptation of plant populations to a changing environment would require modulation of mCG patterns. Akimoto et al. (2007) investigated the inheritance of phenotypic changes directly correlated with the methylation status of some genes. DNA methylation was reduced by treating rice seeds with 5-azadeoxycytidine, and the progenies were cultivated in the field for 10 years. One line showed a clear marker phenotype of dwarfism, which was stably inherited by the progeny over nine generations. It was also reported that in other lines demethylation activated a disease-resistance gene, thereby conferring a disease-resistant trait. Both hypomethylation and acquired phenotype were stably inherited in the progeny.

As discussed above, the rate of somatic recombination in plants increased in response to stress. Molinier et al. (2006) suggested that the transfer of epigenetic information between generations provided a memory of environmental stresses experienced in earlier generations. This "memory" of exposure to ultraviolet radiation and flagellin, a bacterially derived elicitor, resulted in increased somatic homologous recombination, which was inheritable in successive generations. Boyko et al. (2007) analysed the progeny of TMV-sensitive tobacco plants, the parental lines of which were infected with TMV. It was found that these plants had a higher frequency of rearrangements in the loci containing R-genes—for pathogen recognition—and alterations in global genome and R-gene loci-specific methylation. In those cases where ME change their position or are amplified, the variations promoted are inheritable, although the rate of reversion in some transposons can be high. Also the genomic changes which take place during hybridization and polyploidization are inheritable. In flax, the extensive phenotypic and genomic variations which occurred in the genothrops were also observed in the subsequent generations and were stably inherited, irrespective of the conditions under which they were grown (Cullis 2005).

3.3 Tissue Culture Imposes a Stress to Cultivated In Vitro Cells

Cultured cells are exposed to a complex process of overlapping stresses which may trigger the expression of totipotency. The process of totipotency may require major modifications of gene expression, as the status of differentiated cells needs to be altered and redirected to another developmental pathway. Under certain conditions, these cells follow a precise morphogenic pathway, and cell divisions progress according to a pattern defined in time and space, resulting in the formation of a new organ or an embryo.

Plant tissue culture procedures impose a range of stresses on cultured cells, as cells are removed from their normal location in the whole plant and placed in a particular culture medium in order to induce in most cases a morphogenic response. Most of the stresses which we have already indicated are exerted on cultured cells. Frequently, wounding is the first shock imposed on the cells because different organs are cut and the fragments used as explants. Normally as a result of wounding, the cells of the explant divide, mainly those contiguous to the wounded cells. Generally, these cells are in a non-dividing state and they must re-enter into the cell cycle and in most cases experience a high rate of division. These phenomena have been frequently compared with the behaviour of tumoral cells, particularly in cases where a callus was formed. The culture medium usually includes different compounds and plant hormones which may impose stress on cultured cells and tissues.

Other than manipulation of the medium composition, a wide range of modifications have been investigated to develop the methodology enabling one to obtain the desired response in a particular plant species. The beneficial, or crucial, role of stress treatments has been well documented in the induction of somatic embryogenesis (SE). The addition of exogenous 2,4-D is one of the key steps in the induction of SE and, in particular, in the acquisition of embryogenic competence in many plant cells cultured in vitro. However, the role of 2,4-D is controversial, because it has been considered to function as a stress substance rather than as a phytohormone, triggering the acquisition of embryogenic competence (Kikuchi et al. 2006). In addition to 2,4-D, abscisic acid (ABA) may act as a stress-causing agent for inducing SE in carrot (Kikuchi et al. 2006). The important role of ABA signalling in the induction of SE has been confirmed by a mutant study in which ABA-hypersensitive and ABA-insensitive mutants of A. thaliana exhibited decreased SE potential (Gaj et al. 2006). Results of other studies also support the implication of stress in SE induction. These include osmotic shock with sucrose or sodium chloride (Wetherell 1984; Kiyosue et al. 1989) or application of heavy metal ions (Cd^{2+}, Ni^{2+}, Cu^{2+} and Co^{2+}) or high temperature (37°C; Kiyosue et al. 1990; Kamada et al. 1994). Another class of stress (e.g. auxin starvation) can also effectively induce SE in several plant species (Fehér et al. 2003; Quiroz-Figueroa et al. 2006). Interestingly, some genes involved in SE, such as *OsSERK1* in rice, have been shown to be activated by the rice blast fungus and inducible by defence signalling molecules, such as salicylic acid, jasmonic acid and ABA (Hu et al. 2005). These results suggest a possible connection between SE and stress response pathways.

Stress has also been found to play an important role in androgenesis, using anther or microspore culture. Different types of stress, including heat shock, osmotic treatment and starvation, are capable of triggering the switch of microspore development from the gametophytic pathway towards the sporophytic route. Other types of stress, such as nitrogen starvation, auxin, chemicals, gamma irradiation or cold treatments, have been shown to trigger androgenesis at lower rates (Maraschin et al. 2005; Hosp et al. 2007).

Other approaches of plant tissue culture are also a source of stress for the cells. For example, protoplast isolation is a stress-inducing technique, especially during enzymatic cell wall degradation, which resembles the infective process of pathogens. In protoplast culture, cells experience variable osmotic pressure and, occasionally, electrical shock when electrical treatment is applied to stimulate protoplast growth. Somatic hybridization by protoplast fusion represents an additional stress: the genomic stress caused by the coexistence of two different genomes as well as two different cytoplasms in the same cell (Davey et al. 2005). During cell cryopreservation, cells are subject not only to a progressive decrease in temperature, but also to the presence of cryoprotective agents (Harding 2004). Additionally, transformation techniques impose another stress on cells already stressed under in vitro conditions (Filipecki and Malepszy 2006). Apart from the genome alterations which could be caused by the transgene insertions, the transformation process itself is a source of new stress. Cell injury caused by wounding is promoted in the case of transformation mediated by particle bombardment, or the transformed cells may also be stressed by co-cultivation of cells with *Agrobacterium tumefaciens*, when this bacterium is used as vehicle for transgene transfer. These bombarded/inoculated cells are subjected to further stress by growth on media containing antibiotics or herbicides for transgenic plant selection.

3.4 Cultured Cells and Regenerated Plants Show Variations

It has been recognized that various tissue culture techniques impose stress on cultured cells and tissues. However, it is not clear whether the modifications or changes which occur under stress conditions discussed above also operate in cultured cells or regenerated plants. Modifications of gene expression in cells grown in vitro are expected, and a reprogramming of gene activity may be induced during the process of cell dedifferentiation and the acquisition of a new developmental program, leading to organ differentiation or SE. Che et al. (2002) studied the modification of the activity of 8,000 genes during the process of shoot regeneration in Arabidopsis, using oligonucleotide array analysis. They describe the global program of gene expression during this process and the changes in the expression of genes involved in cytokinin and auxin signalling. The modification of gene expression during the induction and development of SE has been examined. In soybean (*Glycine max* L.), Thibaud-Nissen et al. (2003), using a 9,280-cDNA clone array, studied the transcriptional reprogramming occurring during the time necessary for the development of globular embryos. They observed changes in mRNA abundance of genes characteristic of oxidative stress, cell division, storage proteins, and the genes involved in the synthesis of gibberellic acid. Macroarrays and suppression subtractive hybridization (SSH) was performed to generate transcripts highly enriched in SE-related genes in cotton (*Gossypium hirsutum* L.; Zeng et al. 2006). The cDNA collection was composed of a broad repertoire of SE genes encoding proteins which are involved in the initial and subsequent developmental stages of somatic embryos. Sequencing clones from c-DNAs libraries and applying reverse Northern analysis, Singla et al. (2007) showed that SE in wheat (*Triticum aestivum*) follows a unique developmental pathway regulated by temporal and spatial patterns of gene expression. Analysis of data of select genes suggested that the induction phase of somatic embryogenesis is accompanied by the expression of genes which might also be involved in zygotic embryogenesis. The proteomic changes of the stress-responsive pathways have also been investigated. In *Medicago truncatula* more than 60% of the differentially expressed protein spots had very different patterns of gene expression for a wild-type line and another line which had a 500-fold greater capacity to regenerate plants by SE (Imin et al. 2005). In *Cyclamen persicum* Mill, a comparison of the proteomes of zygotic and somatic embryos was performed and selected proteins were identified following mass spectrometry (Winkelmann et al. 2006). These analyses revealed a great similarity between both proteomes.

To study changes in gene expression during the induction of microspore embryogenesis, several techniques have been used, including macroarray and other methodologies such as suppression subtractive hybridization or metabolic profiling, in conjunction with gas chromatography/mass spectrometry (Hosp et al. 2006, 2007). Joosen et al. (2007) characterized and compared the transcriptome and proteome of rapeseed (*Brassica napus*) microspore-derived embryos differentiated and developed from the few-celled to the globular/heart stage.

Although changes in gene expression have been detected, processes such as genome modifications similar to those recorded in long-term stress responses are also induced in cultured cells. Plant tissue culture is commonly used to clonally propagate a particular genotype. Thus, all the cells, organs or plants regenerated from a given genotype by tissue culture are genetically identical to the plant from which they derived. However, often plants with a phenotype different from the expected one have been obtained. Larkin and Scowcroft (1981) described the phenomenon as a novel source of variability, considering it as a process of mutation. It was referred to as somaclonal variation (SV) which, since then, has been applied in many cases to any kind of phenotypic change observed in regenerated plants and/or cells in culture. Several factors may also be involved in phenotypic change of the regenerated plants. Bouman and De Klerk (2001) observed that plants produced vegetatively in tissue culture were different from the donor plants because of physiological or genetic changes, loss of pathogens and/or loss of chimeric structure, although in our opinion the last two possibilities may not occur frequently.

Different types of changes have been described according to the method used to assess the variation. Many phenotypic modifications have been reported. Equally, cytogenetic abnormalities have been found among tissue culture regenerants. Modifications of the DNA sequence have been described in the cultured cells or the regenerated plants (Karp 1991; Vázquez 2001). Furthermore, various regenerated plants and their progeny have shown variation in their DNA methylation patterns (Kaeppler et al. 2000; Jain 2001).

3.4.1   Heritable Changes Versus Non-Heritable Changes

Phenotypic changes, both in quantitative and qualitative traits, have been described above. In many cases, phenotypic variation of regenerated plants is transient and is not transmitted to the progeny. It has been reported that other modifications which are permanent in the regenerated plant are also not detectable in the progeny. In the case where the variation is transmitted to the descendants in a Mendelian fashion, it is assumed to be caused by a mutation. Although recessive mutations have been observed frequently, a high rate of dominant mutations has also been reported. Cytogenetic abnormalities, including ploidy changes and chromosome rearrangements, have been observed in cultured cells or regenerants. In some cases, the regenerated plants exhibit a chromosome mosaic which is frequently non-heritable. In contrast, some regenerants are solid mutants, and cytogenetic modification can be transmitted to the next generations (Lee and Phillips 1987; Karp 1991). The occurrence of erroneous mitosis and, consequently, modification of the normal caryotype in a particular organ could be heritable (Linacero and Vázquez 1992a).

Variation at the DNA level has been studied most extensively and is inheritable per se. Modification of the DNA sequence, such as point mutations, increase or decrease in the length of repetitive sequences or movements of ME have been described in cultured cells or regenerated plants. Although it can be assumed that a large number of point mutations occur, based on the high number of published works in which the phenotypic alteration corresponded to a single gene mutant, only in a few cases has this possibility been confirmed. One of the confirmed cases is a tissue culture-derived mutant of the *Adh1* gene in maize (Dennis et al. 1987). In rice, point mutations have been observed in 20-year-old cell lines at the 3-kb *EPSPs–RPS20* region, in which different nucleotide substitutions have been characterized by a high frequency of transition mutations of A/T to G/C (Noro et al. 2007). Furthermore, point mutations can cause DNA modifications resulting in different DNA banding patterns, as shown by RFLPs (restriction fragment length polymorphism) or other molecular techniques used to assess variability based on the polymerase chain reaction (PCR). However, it is only in a few cases that the implication of point mutations in the modification of these patterns of bands has been experimentally proved (Li et al. 2007). Although it is not clear which factors cause these variations, the presence of 2,4-D in the culture medium has been shown to play an important role in the occurrence of point mutations (Filkowski et al. 2003). Modification of the length of repetitive DNA sequences has been reported. Simple sequence repeats (SSRs) or microsatellite variation have been detected in some plant species (Palombi and Damiano 2002; Wilhelm et al. 2005; Burg et al. 2007).

Also, the implication of ME in SV has been reported. In tobacco, some retro-transposons, such as *Tnt1A*, were highly activated in protoplasts (Grandbastien 1998), while *Tto1* could be activated during callus formation, as well as in cultured cell lines (Hirochika 1993). In both cases, new copies of retrotransposons were generated. Three of the rice retrotransposons (*Tos10*, *Tos17* and *Tos19*) have also been activated under tissue culture conditions (Hirochika et al. 1996; Hirochika 1997). The copy number of *Tos17* increased during prolonged culture periods and, in all of the plants regenerated, including transgenic plants, 5–30 new transposed *Tos17* copies were detected (Hirochika et al. 1996; Hirochika 1997). Interestingly, the transcription of *Karma*, a LINE retrotransposon of rice, has also been activated in cultured cells, but no increase in the copy number was detected in cultured cells or in the regenerated plants (Komatsu et al. 2003). However, an increase in the copy number of *Karma* has been observed in plants of the R1 and later generations.

Maize plants regenerated from cell culture contained newly activated transposons such as *Activator* (*Ac*; Peschke et al. 1987; Brettell and Dennis 1991) and *Suppressor-mutator* (*Spm*; Peschke and Phillips 1991). Also, the *Mutator* transposon of maize could be activated and mobilised during culture and, moreover, plants regenerated from these cultures transmitted the *Mutator* activity to their progeny (Planckaert and Walbot 1989). Alves et al. (2005) reported a fold-back transposon of rye (*Secale cereale*), the *RYS1* element, which could be activated under in vitro conditions and which changed its position in the genome. In rice, the MITE *mPing* was activated in anther culture excised efficiently from original sites and reinserted into new loci (Kikuchi et al. 2003). This MITE and its related *Pong* transposon (Jiang et al. 2003) could also be transposed in an indica rice cell culture line. The *ZmTPAPong-like* element was also active in maize tissue culture, and the plants regenerated from callus showed new insertions of the element (Barret et al. 2006).

Somatic recombination has been reported previously (Lörz and Scowcroft 1983). Auxin (e.g. 2,4-D) has been shown to increase the frequency of somatic recombination (Filkowski et al. 2003).

3.4.2 Genetic Versus Epigenetic Changes

As discussed above, epigenetic changes occur when plants are under stress. Epigenetic changes have also been considered to act as a promoter to the variation observed in tissue culture (Kaeppler et al. 2000; Valledor et al. 2007). This may involve both heritable and non-heritable changes. Recent studies show changes in the DNA methylation under tissue culture conditions, although other epigenetic modifications, such as histone acetylation, have also been investigated (Law and Suttle 2005). Phenotypic changes related to modification of DNA methylation have been reported. Joyce and Cassells (2002) observed a relationship between the potato (*Solanum tuberosum*) micro-plants with most juvenile leaf shapes and the highest frequency of DNA methylation.

Sometimes, some regenerated plants which looked abnormal could recover the normal phenotype, or the abnormal trait was not inherited by its progeny. The phenomenon may be explained if the affected gene was methylated and, consequently, its expression repressed. In contrast, loss of methylation may account for reversion to the normal phenotype. In most cases, methylation does not persist in the next generation and, accordingly, phenotype abnormality is not observed in the progeny. These types of variation, even if they are not inheritable, are important in cases where the primary regenerants are the end product, as in the micropropagation of forest trees (Valledor et al. 2007).

Recent studies have focused on the methylation state of regenerated plants in order to ascertain whether their patterns of methylation differ from those of the donor plant (Xu et al. 2004; Li et al. 2007; Smýkal et al. 2007), or between the in vivo- and in vitro-produced plants (Peraza-Echeverria et al. 2001). However, work aiming to discern whether epigenetic alterations are related to genetic alterations has been limited. In *Codonopsis lanceolata*, a significant correlation has been observed between genetic variations detected by RAPD and ISSR (inter-simple sequence repeats), and epigenetic changes revealed by MSAP (methylation-sensitive amplified polymorphism; Guo et al. 2007). In *Hordeum brevisubulatum*, correlation analysis of the genetic and epigenetic instabilities shows that there is a significant correlation between MSAP and SSAP (sequence-specific amplification polymorphism), whereas the correlation between MSAP and AFLP (amplified fragment length polymorphism) is not statistically significant (Li et al. 2007). Results of these studies indicate that modification of the methylation state of DNA can cause variation in the DNA sequence.

As DNA methylation may be inheritable (cf. above), it is speculated that some of the theoretical "mutants" may possess a permanent and stable methylation. Some epigenetic changes induced by tissue culture can also be inherited. Meins and Thomas (2003) reported in tobacco the heritable change of a cytokinin habituation phenotype, habituated leaf (Hl), which was inherited meiotically as a dominant trait, but which exhibited a high rate of reversion in the successive sexual generations.

The modification of DNA methylation may be associated with some mutations in which a connection exists between genetic and epigenetic variation. It has been suggested that the presence of methylations could trigger mechanisms similar to the RIP (repeat-induced point mutation) known to operate in fungi, or the methylation-induced premeiotically (MIP) type mechanism which could induce point mutations as well as possibly small amplifications or deletions (Phillips et al. 1994). Gonzalgo and Jones (1997) reported that methylation of cytosines at CpG dinucleotides has mutational consequences on mammalian genomes, because methylated cytosines represent hotspots for spontaneous base substitutions through the generation of transition mutations via deamination-driven events.

There may be a direct relationship between DNA methylation and ME activation. ME can be methylated under normal conditions, and their reactivation is correlated with reduced methylation (Bender 2004). Under tissue culture conditions, ME can be demethylated and become active and, as a result, mobilisation of these elements in the genome may lead to insertional mutations (Hirochika et al. 1996). In most cases, mainly in that of retrotransposons, the methylation is recovered in the regenerated plants and the elements become inactive. However, the rice Karma is not subject to de novo methylation, and retrotransposition persists through several generations (Komatsu et al. 2003). Also, activation of the rice retrotransposon *Tos17* by tissue culture may cause heritable alterations in cytosine-methylation patterns of the flanking genomic regions (Liu et al. 2004).

3.4.3 Variation Promoted by Tissue Culture is Not Randomly Distributed in the Genome

Although changes in DNA or genes are believed theoretically to be distributed randomly in the genome, the presence of loci with a high rate of mutation promoted by in vitro conditions has been reported (Xie et al. 1995). In the family Poaceae, variation in the chlorophyll content has been commonly observed in plants regenerated from microspores or immature embryos. It has been reported that the albino appearance in plants derived from anther culture is controlled by the nuclear genes in barley (*Hordeum vulgare*; Larsen et al. 1991) and wheat (Zhou and Konzak 1992). In rye, a mutation has been shown to occur at the same locus in plants regenerated from different cell lines (Linacero and Vázquez 1992b). Furthermore, some frequent phenotypic changes, such as abnormal leaf morphology, so-called potato leaf, have been detected in tomato (*Lycopersicon esculentum*; Gavazzi et al. 1987).

Some ME activated by in vitro culture have preferential insertion sites such as the retrotransposon *Tos17* of rice (Miyao et al. 2003) or the *RYS1* transposon of rye (Alves et al. 2005). The use of molecular markers to analyse the stability of the in vitro-derived plants also revealed that some loci are highly variable under in vitro conditions. Linacero et al. (2000) employed RAPD analyses to show that some DNA sequences in rye plants were more variable under tissue culture conditions; the same sequence varied in several plants obtained from different cell lines. Similar results have been reported in garlic (*Allium sativum*; Al-Zahim et al. 1999), Arabidopsis (Polanco and Ruiz 2002) and banana (*Musa* spp.; Oh et al. 2007). These loci appeared to be hotspots for variation arising during culture.

Other phenotypic modifications probably resulting from epigenetic changes have also been observed in regenerated plants. These include the occurrence of juvenile plants and alteration in floral morphology. Bouman and De Klerk (2001) studied the extent of variation in *Begonia* plants regenerated under various culture conditions. They observed the occurrence of 2.5% juvenile plants regenerated via an intermediate callus phase and suggested that this aberration could be related with changes in the DNA methylation status. Furthermore, the mantled phenotype commonly observed in oil palm (*Elaeis guineensis*) regenerated from tissue culture (Corley et al. 1986) may be transmitted by genetic crossing. However, the reversion from a mantled to a wild-type phenotype for some, but not all, clones has been observed under field conditions, indicating an epigenetic origin. It has been proposed this variation is associated with differential genomic DNA methylation (Jaligot et al. 2000). Changes in genomic methylation under tissue culture conditions appear to occur non-randomly. A novel MSAP fragment in barley, which is absent from the donor plant, can be detected in 90–100% of regenerants (Bednarek et al. 2007). In all the hop (*Humulus lupulus*) callus-derived plants, the same fragments— 13.37% of the total MSAP variation—were present (Peredo et al. 2006).

3.4.4 Are New Alleles Originated by In Vitro Stress Already Present in Other Plants of In Vivo Populations?

It has been reported that some plants regenerated from a cultivar of soybean generated new alleles which were present in other cultivars (Roth et al. 1989). In rye the situation is similar and some loci, which have a high rate of mutation, originate changes which were already present in other individuals in the in vivo populations (Alves et al. 2005). In *Rauwolfia* species, RFLP was employed to analyse 18S-25S and 5S rRNA genes to investigate the pattern of rDNA variability which occurred as a result of species evolution in nature, and which was induced by tissue culture (Andreev et al. 2005). The changes in rRNA genes induced by tissue culture affect the gene regions which show interspecies polymorphism. The same sequences are involved in variation, arising either from natural evolution or tissue culture conditions, suggesting that similar mechanisms operate in both cases. Thus, tissue culture conditions appear to increase the rate of mutation which occurs in nature, and some of those changes are induced by the stress exerted by the environment.

3.5 Concluding Remarks

It is clear that environmental stress in plants is remarkably similar to the stress of plant cells grown under tissue culture conditions. Stress responses in plant cells are similar in both cases, in terms of transient or stable modification in gene expressions or genome alteration. Similarities could also be established because the variation promoted by tissue culture, at least in some cases, was not randomly scattered in the genome, and the modifications promoted by stress were distributed along the genome but involved precise sites. These findings suggest that the same mechanisms for variability may be operating under both natural and tissue culture conditions.

Acknowledgements This work on SV is supported by the MEC of Spain (AGL 2006-14249-C02-01) and the SANTANDER-COMPLUTENSE grant PR41/06-14923.

References

Aarts MGM, Fiers MWEJ (2003) What drives plant stress genes? Trends Plant Sci 8:99–102
Aina R, Sgorbati S, Santagostino A, Labra M, Ghiani A, Citterio S (2004) Specific hypomethylation of DNA is induced by heavy metals in white clover and industrial hemp. Physiol Plant 121:472–480
Akimoto K, Katakami H, Kim H-J, Ogawa E, Sano CM, Wada Y, Sano H (2007) Epigenetic inheritance in rice plants. Ann Bot 100:205–217

Alves E, Ballesteros I, Linacero R, Vázquez AM (2005) *RYS1*, a foldback transposon is activated by tissue culturetissue culture and shows preferential insertion points into the rye genome. Theor Appl Genet 111:431–436

Al-Zahim MA, Ford-lloyd BV, Newbury HJ (1999) Detection of somaclonal variation in garlic (*Allium sativum* L.) using RAPD and cytological analysis. Plant Cell Rep 18:473–477

Andreev IO, Spiridonova KV, Solovyan VT, Kunakh VA (2005) Variability of ribosomal RNA genes in *Rauwolfia* species: parallelism between culture-induced rearrangements and interspecies polymorphism. Cell Biol Int 29:21–27

Barret P, Brinkman M, Beckert M (2006) A sequence related to rice Pong transposable element displays transcriptional activation by in vitro culture and reveals somaclonal variations in maize. Genome 49:1399–1407

Bednarek PT, Orłowska R, Koebner RMD, Zimny J (2007) Quantification of the tissue-culture induced variation in barley (*Hordeum vulgare* L.). BMC Plant Biol 7:10

Beguiristain T, Grandbastien MA, Puigdomenech P, Casacuberta JM (2001) Three Tnt1 subfamilies show different stress-associated patterns of expression in tobacco. Consequences for retrotransposon control and evolution in plants. Plant Physiol 127:212–221

Bender J (2004) DNA methylation and epigenetics. Annu Rev Plant Biol 55:41–68

Bohnert HJ, Gong Q, Li P, Ma S (2006) Unraveling abiotic stress tolerance mechanisms – getting genomics going. Curr Opin Plant Biol 9:180–188

Bond DM, Finnegan EJ (2007) Passing the message on: inheritance of epigenetic traits. Trends Plant Sci 12:211–216

Bouman H, De Klerk G-J (2001) Measurement of the extent of somaclonal variation in begonia plants regenerated under various conditions. Comparison of three assays. Theor Appl Genet 102:111–117

Boyko A, Kathiria P, Zemp FJ, Yao Y, Pogribny I, Kovalchuk I (2007) Transgenerational changes in the genome stability and methylation in pathogen-infected plants. Nucleic Acids Res 35:1714–1725

Brettell RIS, Dennis ES (1991) Reactivation of a silent Ac following tissue culture is associated with heritable alterations in its methylation pattern. Mol Gen Genet 229:365–372

Bruce TJA, Matthes MC, Napier JA, Pickett JA (2007) Stressful "memories" of plants: Evidence and possible mechanisms. Plant Sci 173:603–608

Burg K, Helmersson A, Bozhkov P, von Arnold S (2007) Developmental and genetic variation in nuclear microsatellite stability during somatic embryogenesis in pine. J Exp Bot 58:687–698

Casacuberta JM, Santiago N (2003) Plant LTR-retrotransposons and MITEs: control of transposition and impact on the evolution of plant genes and genomes. Gene 311:1–11

Ceccarelli M, Esposto MC, Roscini C, Sarri V, Frediani M, Gelati MT, Cavallini A, Giordani T, Pellegrino RM, Cionini PG (2002) Genome plasticity in *Festuca arundinacea*: direct response to temperature changes by redundancy modulation of interspersed DNA repeats. Theor Appl Genet 104:901–907

Che P, Gingerich DJ, Lall S, Howell SH (2002) Global and hormone-induced gene expression changes during shoot development in *Arabidopsis*. Plant Cell 14:2771–2785

Chen Y, Schneeberger RG, Cullis CA (2005) A site-specific insertion sequence in flax genotrophs induced by environment. New Phytol 167:171–180

Cheong YH, Chang H-S, Gupta R, Wang X, Zhu T, Luan S (2002) Transcriptional profiling reveals novel interactions between wounding, pathogen, abiotic stress, and hormonal responses in Arabidopsis. Plant Physiol 129:661–677

Choi C-S, Sano H (2007) Abiotic-stress induces demethylation and transcriptional activation of a gene encoding a glycerolphosphodiesterase-like protein in tobacco plants. Mol Genet Genomics 277:589–600

Corley RHV, Lee CH, Law IH, Wong CY (1986) Abnormal flower development in oil palm clones. Planter 62:233–240

Cullis CA (1986) Phenotypic consequences of environmentally induced changes in plant DNA. Trends Genet 2:307–309

Cullis CA (2005) Mechanisms and control of rapid genomic changes in flax. Ann Bot 95:201–206

Cullis CA, Swami S, Song Y (1999) RAPD polymorphisms detected among the flax genotrophs. Plant Mol Biol 41:795–800

Davey MR, Anthony P, Power JB, Lowe KC (2005) Plant protoplasts: status and biotechnological perspectives. Biotechnol Adv 23:131–171

Dennis ES, Brettell RIS, Peacock WJ (1987) A tissue culture induced Adh1 null mutant of maize results from a single base change. Mol Gen Genet 210:181–183

Eulgen T (2005) Regulation of the Arabidopsis defense transcriptome. Trends Plant Sci 10:71–78

Fehér A, Pasternak TP, Dudits D (2003) Transition of somatic plant cells to an embryogenic state. Plant Cell Tiss Organ Cult 74:201–228

Feschotte C, Jiang N, Wessler SR (2002) Plant transposable elements: where genetics meets genomics. Nature Rev Genet 3:329–341

Filipecki M, Malepszy S (2006) Unintended consequences of plant transformation: a molecular insight. J Appl Genet 47:277–286

Filkowski J, Besplug J, Burke P, Kovalchuk I, Kovalchuk O (2003) Genotoxicity of 2,4-D and dicamba revealed by transgenic *Arabidopsis thaliana* plants harbouring recombination and point mutation markers. Mutation Res 542:23–32

Fujita M, Fujita Y, Noutoshi Y, Takahashi F, Narusaka Y, Yamaguchi-Shinozaki K, Shinozaki K (2006) Crosstalk between abiotic and biotic stress responses: a current view from the points of convergence in the stress signalling networks. Curr Opin Plant Biol 9:436–442

Gaj MD, Trojanowska A, Ujczak A, Medrek M, Koziol A, Garbaciak B (2006) Hormone-response mutants of *Arabidopsis thaliana* (L.) Heynh. impaired in somatic embryogenesis. Plant Growth Regul 49:183–197

Gavazzi G, Tonelli C, Todesco G, Arreghini E, Raffaldi F, Vecchio F, Barbuzzi G, Biasini MG, Sala F (1987) Somaclonal variation versus chemically induced mutagenesis in tomato (*Lycopersicon esculentum* L.). Theor Appl Genet 74:733–738

Gonzalgo ML, Jones PA (1997) Mutagenic and epigenetic effects of DNA methylation. Mutation Res 386:107–118

Grandbastien M-A (1998) Activation of plant retrotransposons under stress conditions. Trends Plant Sci 3:181–189

Grant-Downton RT, Dickinson HG (2005) Epigenetics and its implications for plant biology. 1. The epigenetic network in plants. Ann Bot 96:1143–1164

Guo WL, Wu R, Zhang YF, Liu XM, Wang HY, Gong L, Zhang ZH, Liu B (2007) Tissue culture-induced locus-specific alteration in DNA methylation and its correlation with genetic variation in *Codonopsis lanceolata* Benth. et Hook. f. Plant Cell Rep 26:1297–1307

Harding K (2004) Genetic integrity of cryopreserved plant cells: a review. Cryo Lett 25:3–22

Hashida S-N, Kitamura K, Mikami T, Kishima Y (2003) Temperature shift coordinately changes the activity and the methylation state of transposon Tam3 in *Antirrhinum majus*. Plant Physiol 132:1207–1216

He ZH, Dong HT, Dong JX, Li DB, Ronald PC (2000) The rice *Rim2* transcript accumulates in response to *Magnaporthe grisea* and its predicted protein product shares similarity with TNP2-like proteins encoded by CACTA transposons. Mol Gen Genet 64:2–10

Henderson IR, Jacobsen SE (2007) Epigenetic inheritance in plants. Nature 447:418–424

Hirochika H (1993) Activation of tobacco retrotransposons during tissue culturetissue culture. EMBO J 12:2521–2528

Hirochika H (1997) Retrotransposons of rice: their regulation and use for genome analysis. Plant Mol Biol 35:231–240

Hirochika H, Sugimoto K, Otsuki Y, Kanda M (1996) Retrotransposons of rice involved in mutations induced by tissue culture. Proc Natl Acad Sci USA 93:7783–7788

Hosp J, Tashpulatov A, Roessner U, Barsova E, Katholnigg H, Steinborn R, Melikant B, Lukyanov S, Heberle-Bors E, Touraev A (2006) Transcriptional and metabolic profiles of stress-induced, embryogenic tobacco microspores. Plant Mol Biol 63:137–149

Hosp J, de Faria Maraschin S, Touraev A, Boutilier K (2007) Functional genomics of microspore embryogenesis. Euphytica 158:275–285

Hu H, Xiong L, Yang Y (2005) Rice *SERK1* gene positively regulates somatic embryogenesis of cultured cell and host defense response against fungal infection. Planta 222:107–117

Imin N, Nizamidin M, Daniher D, Nolan KE, Rose RJ, Rolfe BG (2005) Proteomic analysis of somatic embryogenesis in *Medicago truncatula*. Explant cultures grown under 6-benzylaminopurine and 1-naphthaleneacetic acid treatments. Plant Physiol 137:1250–1260

Ivashuta S, Naumkina M, Gau M, Uchiyama K, Isobe S, Mizukami Y, Shimamoto Y (2002) Genotype-dependent transcriptional activation of novel repetitive elements during cold acclimation of alfalfa (*Medicago sativa*). Plant J 31:615–627

Jablonka E, Lamb MJ (2002) The changing concept of epigenetics. Ann NY Acad Sci 981:82–96

Jain SM (2001) Tissue culture-derived variation in crop improvement. Euphytica 118:153–166

Jaligot E, Rival A, Beulé T, Dussert S, Verdeil J-L (2000) Somaclonal variation in oil palm (*Elaeis guineensis* Jacq.): the DNA methylation hypothesis. Plant Cell Rep 19:684–690

Jiang N, Bao Z, Zhang X, Hirochika H, Eddy SR, McCouchk SR, Wessler SR (2003) An active DNA transposon family in rice. Nature 421:163–167

Joosen R, Cordewener J, Darmo Jaya Supena E, Vorst O, Lammers M, Maliepaard C, Zeilmaker T, Miki B, America T, Custers J, Boutilier K (2007) Combined transcriptome and proteome analysis identifies pathways and markers associated with the establishment of rapeseed microspore-derived embryo development. Plant Physiol 144:155–172

Joyce SM, Cassells AC (2002) Variation in potato microplant morphology in vitro and DNA methylation. Plant Cell Tiss Organ Cult 70:125–137

Kaeppler SM, Kaeppler HF, Rhee Y (2000) Epigenetic aspects of somaclonal variation in plants. Plant Mol Biol 43:179–188

Kamada H, Tachikawa Y, Saitou T, Harada H (1994) Heat stresses induction of carrot embryogenesis. Plant Tissue Cult Lett 11(3):229–232

Karp A (1991) On the current understanding of somaclonal variation. Oxford Surv Plant Mol Cell Biol 17:1–58

Kikuchi K, Terauchi K, Wada M, Hirano HY (2003) The plant MITE mPing is mobilized in anther culture. Nature 421:167–170

Kikuchi A, Sanuki N, Higashi K, Koshiba T, Kamada H (2006) Abscisic acid and stress treatment are essential for the acquisition of embryogenic competence by carrot somatic cells. Planta 223:637–645

Kiyosue T, Kamada H, Harada H (1989) Induction of somatic embryogenesis by salt stress in carrot. Plant Tissue Cult Lett 6:162–164

Kiyosue T, Takano K, Kamada H, Harada H (1990) Induction of somatic embryogenesis in carrot by heavy metal ions. Can J Bot 68:2301–2303

Komatsu M, Shimamoto K, Kyozuka J (2003) Two-Step regulation and continuous retrotransposition of the rice LINE-type retrotransposon *Karma*. Plant Cell 15:1934–1944

Kovalchuk I, Kovalchuk O, Kalck V, Boyko V, Filkowski J, Heinlein M, Hohn B (2003) Pathogen-induced systemic plant signal triggers DNA rearrangements. Nature 423:760–762

Kumar A, Bennetzen JL (1999) Plant retrotransposons. Annu Rev Genet 33:479–532

Labra M, Ghiani A, Citterio S, Sgorbati S, Sala F, Vannini C, Ruffini-Castiglioni M, Bracale M (2002) Analysis of cytosine methylation pattern in response to water deficit in pea root tips. Plant Biol 4:694–699

Larkin PJ, Scowcroft WR (1981) Somaclonal variation. Novel source of variability from cell cultures for plant improvement. Theor Appl Genet 60:197–214

Larsen ET, Tuvesson IKD, Andersen SB (1991) Nuclear genes affecting percentage of green plants in barley (*Hordeum vulgare* L.) anther culture. Theor Appl Genet 82:417–420

Law RD, Suttle JC (2005) Chromatin remodeling in plant cell culture: patterns of DNA methylation and histone H3 and H4 acetylation vary during growth of asynchronous potato cell suspensions. Plant Physiol Biochem 43:527–534

Lebel EG, Masson J, Bogucki A, Paszkowski J (1993) Stress-induced intrachromosomal recombination in plant somatic cells. Proc Natl Acad Sci USA 90:422–426

Lee M, Phillips RL (1987) Genetic variants in progeny of regenerated plants. Genome 29:834–838

Li X, Yu X, Wang N, Feng Q, Dong Z, Liu L, Shen J, Liu B (2007) Genetic and epigenetic instabilities induced by tissue culturetissue culture in wild barley (*Hordeum brevisubulatum* (Trin.) Link). Plant Cell Tiss Organ Cult 90:153–168

Lin X, Long L, Shan X, Zhang S, Shen S, Liu B (2006) *In planta* mobilization of *mPing* and its putative autonomous element *Pong* in rice by hydrostatic pressurization. J Exp Bot 57:2313–2323

Linacero R, Vázquez AM (1992a) Cytogenetic variation in rye regenerated plants and their progeny. Genome 35:428–430

Linacero R, Vázquez AM (1992b) Genetic analysis of chlorophyll-deficient somaclonal variants in rye. Genome 35:981–984

Linacero R, Freitas Alves E, Vázquez AM (2000) Hot spots of DNA instability revealed through the study of somaclonal variation in rye. Theor Appl Genet 100:506–511

Liu ZL, Han FP, Tan M, Shan XH, Dong YZ, Wang XZ, Fedak G, Hao S, Liu B (2004) Activation of a rice endogenous retrotransposon *Tos17*. Theor Appl Genet 109:200–209

Lörz H, Scowcroft WR (1983) Variability among plants and their progeny regenerated from protoplasts of *Su/su* heterozygotes of *Nicotiana tabacum*. Theor Appl Genet 66:67–75

Lucht JM, Mauch-Mani B, Steiner H-Y, Metraux J-P, Ryals J, Hohn B (2002) Pathogen stress increases somatic recombination frequency in *Arabidopsis*. Nature Genet 30:311–314

Madlung A, Comai L (2004) The effect of stress on genome regulation and structure. Ann Bot 94:481–495

Maraschin SF, de Priester W, Spaink HP, Wang M (2005) Androgenic switch: an example of plant embryogenesis from the male gametophyte perspective. J Exp Bot 56:1711–1726

Mathieu O, Reinders J, Caikovski M, Smathajitt C, Paszkowski1 J (2007) Transgenerational stability of the *Arabidopsis* epigenome is coordinated by CG methylation. Cell 130:851–862

McClintock B (1984) The significance of responses of the genome to challenge. Science 226:792–801

Meins F Jr, Thomas M (2003) Meiotic transmission of epigenetic changes in the cell-division factor requirement of plant cells. Development 130:6201–6208

Miyao A, Tanaka K, Murata K, Sawaki H, Takeda S, Abe K, Shinozuka Y, Onosato K, Hirochika H (2003) Target site specificity of the *Tos17* retrotransposon shows a preference for insertion within genes and against insertion in retrotransposon-rich regions of the genome. Plant Cell 15:1771–1780

Molinier J, Oakeley EJ, Niederhauser O, Kovalchuk I, Hohn B (2005) Dynamic response of plant genome to ultraviolet radiation and other genotoxic stresses. Mutation Res 571:235–247

Molinier J, Ries G, Zipfel C, Hohn B (2006) Transgeneration memory of stress in plants. Nature 442:1046–1049

Nakazaki T, Okumoto Y, Horibata A, Yamahira S, Teraishi M, Nishida H, Inoue H, Tanisaka T (2003) Mobilization of a transposon in the rice genome. Nature 421:170–172

Noro Y, Takano-Shimizu T, Syono K, Kishima Y, Sano Y (2007) Genetic variations in rice in vitro cultures at the *EPSPs–RPS20* region. Theor Appl Genet 114:705–711

Oh TJ, Cullis CA (2003) Labile DNA sequences in flax identified by combined sample representational difference analysis (csRDA). Plant Mol Biol 52:527–536

Oh TJ, Cullis MA, Kunert K, Engelborghs I, Swennen R, Cullis CA (2007) Genomic changes associated with somaclonal variation in banana (*Musa* spp.). Physiol Plant 129:766–774

Palombi MA, Damiano C (2002) Comparison between RAPD and SSR molecular markers in detecting genetic variation in kiwifruit (*Actinidia deliciosa* A. Chev). Plant Cell Rep 20:1061–1066

Peraza-Echeverria S, Herrera-Valencia VA, James-Kay A (2001) Detection of DNA methylation changes in micropropagated banana plants using methylation-sensitive amplification polymorphism (MSAP). Plant Sci 161:359–367

Peredo EL, Revilla MA, Arroyo-García R (2006) Assessment of genetic and epigenetic variation in hop plants regenerated from sequential subcultures of organogenic calli. J Plant Physiol 163:1071–1079

Peschke VM, Phillips RL (1991) Activation of the maize transposable element *Suppressor-mutator* (*Spm*) in tissue culture. Theor Appl Genet 81:90–97

Peschke VM, Phillips RL, Gengenbach BG (1987) Discovery of transposable element activity among progeny of tissue culture-derived maize plants. Science 238:804–807

Phillips RL, Kaeppler SM, Olhoft P (1994) Genetic instability of plant tissue cultures: breakdown of normal controls. Proc Natl Acad Sci USA 91:5222–5226

Phillips JR, Dalmay T, Bartels D (2007) The role of small RNAs in abiotic stress. FEBS Lett 581:3592–3597

Planckaert F, Walbot V (1989) Molecular and genetic characterization of *Mu* transposable elements in *Zea mays*: behavior in callus culture and regenerated plants. Genetics 123:567–578

Polanco C, Ruiz ML (2002) AFLP analysis of somaclonal variationsomaclonal variation in *Arabidopsis thaliana* regenerated plants. Plant Sci 162:817–824

Pouteau S, Grandbastien MA, Boccara M (1994) Microbial elicitors of plant defence responses activate transcription of a retrotransposon. Plant J 5:535–542

Puchta H, Swoboda P, Hohn B (1995) Induction of intrachromosomal homologous recombination in whole plants. Plant J 7:203–210

Quiroz-Figueroa FR, Rojas-Herrera R, Galaz-Avalos RM, Loyola-Vargas VM (2006) Embryo production through somatic embryogenesis can be used to study cell differentiation in plants. Plant Cell Tiss Organ Cult 86:285–301

Qureshi MI, Qadir S, Zolla L (2007) Proteomics-based dissection of stress-responsive pathways in plants. J Plant Physiol 164:1239–1260

Ries G, Heller W, Puchta H, Sandermann H, Seidlitz HK, Hohn B (2000) Elevated UV-B radiation reduces genome stability in plants. Nature 406:98–101

Roth EJ, Frazier BL, Apuya NR, Lark KG (1989) Genetic variation in an inbreed plant: variation in tissue cultures of soybean (*Glycine max* (L.) Merrill). Genetics 121:359–368

Seo PJ, Lee A-K, Xiang F, Park C-M (2008) Molecular and functional profiling of *Arabidopsis* pathogenesis-related genes: insights into their roles in salt response of seed germination. Plant Cell Physiol 49:334–344

Shulaev V, Cortes D, Miller G, Mittler R (2008) Metabolomics for plant stress response. Physiol Plant 132:199–208

Singla B, Tyagi AK, Khurana JP, Khurana P (2007) Analysis of expression profile of selected genes expressed during auxin-induced somatic embryogenesis in leaf base system of wheat (*Triticum aestivum*) and their possible interactions. Plant Mol Biol 65:677–692

Smýkal PL, Valledor L, Rodríguez R, Griga M (2007) Assessment of genetic and epigenetic stability in long-term in vitro shoot culture of pea (*Pisum sativum* L.). Plant Cell Rep 26:1985–1998

Sunkar R, Chinnusamy V, Zhu J, Zhu J-K (2007) Small RNAs as big players in plant abiotic stress responses and nutrient deprivation. Trends Plant Sci 12:301–309

Swindell WR (2006) The association among gene expression responses to nine abiotic stress treatments in *Arabidopsis thaliana*. Genetics 174:1811–1824

Takeda S, Sugimoto K, Otsuki H, Hirochika H (1998) Transcriptional activation of the tobacco retrotransposon Tto1 by wounding and methyl jasmonate. Plant Mol Biol 36:365–376

Thibaud-Nissen F, Shealy RT, Khanna A, Vodkin LO (2003) Clustering of microarray data reveals transcript patterns associated with somatic embryogenesis in soybean. Plant Physiol 132:118–136

Vaillant I, Paszkowski J (2007) Role of histone and DNA methylation in gene regulation. Curr Opin Plant Biol 10:528–533

Valledor L, Hasbún R, Meijón M, Rodríguez JL, Santamaría E, Viejo M, Berdasco M, Feito I, Fraga MF, Cañal MJ, Rodríguez R (2007) Involvement of DNA methylation in tree development and micropropagation. Plant Cell Tiss Organ Cult 91:75–86

Vázquez AM (2001) Insight into somaclonal variation. Plant Biosyst 135:57–62

Walbot V (1988) Reactivation of the mutator transposable element system following gamma radiation of seed. Mol Gen Genet 212:259–264

Walley JW, Coughlan S, Hudson ME, Covington MF, Kaspi R, Banu G, Harmer SL, Dehesh K (2007) Mechanical stress induces biotic and abiotic stress responses via a novel *cis*-element. PLoS Genet 3:e172. doi:10.1371/journal.pgen.0030172

Wetherell DF (1984) Enhanced adventive embryogenesis resulting from plasmolysis of cultured wild carrot cells. Plant Cell Tissue Organ Cult 5:221–227

Wilhelm E, Hristoforoglu K, Fluch S, Burg K (2005) Detection of microsatellite instability during somatic embryogenesis of oak (*Quercus robur* L.). Plant Cell Rep 23:790–795

Winkelmann T, Heintz D, Van Dorsselaer A, Serek M, Braun HP (2006) Proteomic analyses of somatic and zygotic embryos of *Cyclamen persicum* Mill. reveal new insights into seed and germination physiology. Planta 224:508–519

Xie QJ, Oard JH, Rush MC (1995) Genetic analysis of a purple-red hull rice mutation derived from tissue culture. J Hered 86:154–156

Xu ZN, Yan XH, Maurais S, Fu HH, O'Brien DG, Mottinger J, Dooner HK (2004) Jittery, a mutator distant relative with a paradoxical mobile behavior: excision without reinsertion. Plant Cell 16:1105–1114

Zeng F, Zhang X, Zhu L, Tu L, Guo X, Nie Y (2006) Isolation and characterization of genes associated to cotton somatic embryogenesis by suppression subtractive hybridization and macroarray. Plant Mol Biol 60:167–183

Zhou H, Konzak CF (1992) Genetic control of green plant regeneration from anther culture of wheat. Genome 35:957–961

Part II
Plant Processes and Their Regulation

Chapter 4
Photosynthate Partitioning

N.G. Halford

4.1 Introduction

Photosynthesis occurs predominantly in a plant's mature leaves. The end product is sucrose, a disaccharide comprising a glucose unit and a fructose unit that is uniquely important to plants as the molecule used for transport of carbon throughout the plant. The sucrose that accumulates in the leaf has a number of possible fates: it can be broken down and used in glycolysis to provide energy, it can be used for growth and development of the leaf itself, it can be stored in the chloroplasts as starch, or it can be exported to other parts of the plant. There is, therefore, a decision-making process that determines how much of the assimilated carbon is used for each purpose, and this process is called photosynthate or carbon partitioning. The broader processes of partitioning nitrogen, sulphur and other minerals, as well as carbon, are called assimilate partitioning.

As far as photosynthate is concerned, the mature leaves are regarded as source organs, while organs such as roots, developing leaves, seeds, tubers and fruits that either do not photosynthesise or are not self-sufficient in carbon fixation are termed sink organs. The partitioning of photosynthate between different organs is a key determinant of yield because the harvested organs of all of the major crops are photosynthate sinks. Once the sucrose arrives at a sink organ it can accumulate, as it does in sugar beet roots, for example, or be broken down to hexoses and hexose phosphates that are used to provide energy and building blocks for growth, or be stored in the form of storage proteins, storage carbohydrates, fats and oils. In other words, the carbon is partitioned again (Fig. 4.1) and this partitioning process is a key determinant of end use quality, processing properties and the yield of specific compounds.

N.G. Halford
Plant Science Department, Rothamsted Research, Harpenden, Hertfordshire, AL5 2JQ, UK
e-mail: nigel.halford@bbsrc.ac.uk

E-C. Pua and M.R. Davey (eds.),
Plant Developmental Biology – Biotechnological Perspectives: Volume 2,
DOI 10.1007/978-3-642-04670-4_4, © Springer-Verlag Berlin Heidelberg 2010

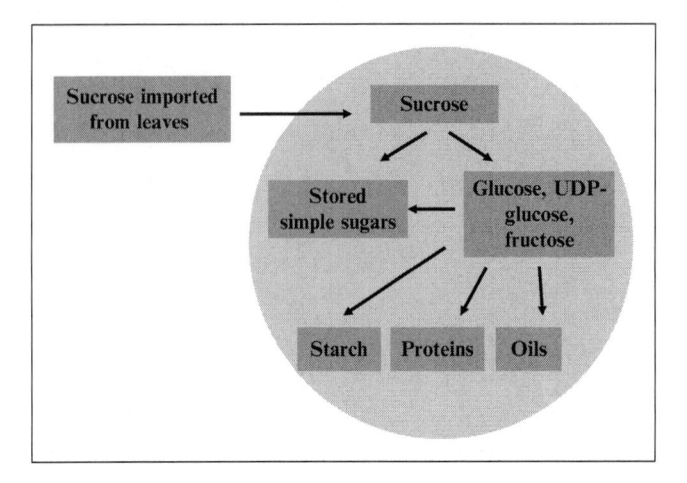

Fig. 4.1 Carbon is transported from source leaves to sink organs as sucrose and partitioned between different storage compounds

Wild plants store starch, oils and proteins in seeds and tubers to support an emergent young plant after germination or sprouting, while in fruits these compounds serve to attract animals that eat the fruit and disseminate the seeds. Cultivated crops are effectively storing resources to provide us with food and many have been bred to accumulate much more than would be required otherwise; assimilate partitioning is perhaps the parameter by which crop species differ most from their wild relatives. For example, approximately 80 % of a cultivated potato plant's dry weight is contained in its tubers, ten times the amount in the tubers of its wild relatives (Inoue and Tanaka 1978). Similarly, modern bread-making wheat produces seeds that are much larger than is required simply to support the growth of a seedling, and much larger than the seeds produced by its wild relatives (Fig. 4.2).

The control of carbon partitioning is closely integrated with metabolic regulation and therefore the survival, structure and biochemistry as well as the productivity of plants. Understanding metabolic regulation provides the basis for rational modification of carbon partitioning. Two decades ago, metabolic regulation was still viewed as a function of the intrinsic properties of enzymes and the levels of different metabolites in a pathway. Now it is known that the control of metabolism involves the coordinated regulation of genes and enzymes at the level of transcription, translation, post-translational modification and protein turnover. Metabolic signalling pathways also crosstalk with other pathways, including hormone and stress response pathways, cell cycle control mechanisms, and nitrogen and sulphur response systems. Metabolic regulation and, therefore, carbon partitioning influence and are influenced by many factors. Understanding how these influences are integrated and relate to the development, survival and productivity of plants is a major challenge for plant biologists.

Fig. 4.2 Grain of hexaploid bread wheat, *Triticum aestivum*, compared with grain from a wild relative, *Triticum urartu*. Cultivated crops such as wheat store more starch and protein in their grain than do their wild relatives. Bar = 1 cm

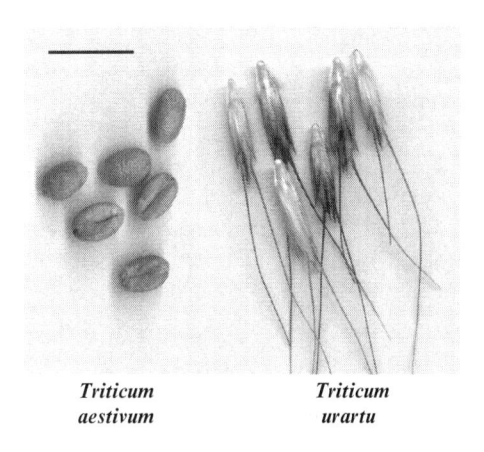

Triticum aestivum   *Triticum urartu*

4.2 Source and Sink

The control exerted by photosynthate production in the source leaves on carbon partitioning in the whole plant can be substantial, accounting for up to 80% of the total source to sink carbon flux regulation, at least in potato (Sweetlove and Hill 2000). The critical question to understanding this control is which enzymatic steps are limiting. Much research during the 1990s established that increasing the activity of many so-called key enzymes would not improve photosynthesis because of the sharing of control amongst enzymes and because feedback regulation of photosynthesis by metabolite signalling could neutralise any improvement (Paul et al. 2001). For example, many studies showed that the benefits of high atmospheric CO_2 concentrations could be short-lived because of such feedback mechanisms (van Oosten and Besford 1996). This emphasises the importance of metabolite signalling in the regulation of photosynthesis and carbon metabolism in the context of the whole plant. Indeed, the manipulation of signalling processes looks to be a promising way forward in the improvement of complex physiological processes such as these, and there is already an example of this in the increase in photosynthetic capacity in transgenic plants over-expressing the gene encoding trehalose phosphate synthase (TPS; Paul et al. 2001). In fact, this works both ways. Modelling studies have indicated that optimising carbon partitioning in crop plants could improve their performance as atmospheric carbon dioxide increases in the coming decades (Zhu et al. 2007).

4.3 Sugars as Signalling Molecules

One of the ways in which photosynthesis and metabolism are regulated is through the action of sugars as signalling molecules, causing feedback regulation of photosynthesis and influencing the expression of genes and the activity of enzymes in many metabolic pathways. Almost two decades ago, Sheen (1990) used photosynthetic gene promoter/reporter gene fusions to show that seven maize photosynthetic genes were repressed by glucose or sucrose in a maize protoplast system. Further evidence of feedback control of photosynthetic gene expression was obtained in experiments where a yeast invertase was expressed in tobacco and Arabidopsis (von Schaewen et al. 1990) or tomato (Dickinson et al. 1991). This resulted in the accumulation of hexoses in the leaves of the transgenic plants, which caused feedback inhibition of photosynthesis. Other genes that are required for photosynthesis and are regulated by sugars include those encoding the ribulose 1,5-bisphosphate carboxylase/oxygenase (RuBisCo) small subunit, chlorophyll a/b binding protein (CAB) and thylakoid ATPase delta subunit (Harter et al. 1993; Krapp et al. 1993; reviewed in Sheen 1994; Halford and Paul 2003).

Other glucose-repressed genes include those encoding isocitrate lyase, malate synthase and α-amylase (Yu et al. 1991; Graham et al. 1994), while the expression of a gene encoding β-amylase has been shown to be induced by sugars in rosette leaves of Arabidopsis (Mita et al. 1995). Another enzyme of carbohydrate metabolism, sucrose synthase, responds to sucrose but is unaffected by glucose (Salanoubat and Belliard 1989; Sowokinos and Varns 1992). The reaction catalysed by sucrose synthase requires uridine diphosphate (UDP) and produces UDP-glucose and fructose (under physiological conditions, the equilibrium of this reaction very much favours cleavage over synthesis, despite the name of the enzyme). Sucrose synthase activity is closely correlated with starch accumulation in potato and maize (Choury and Nelson 1976; Zrenner et al. 1995). Plants contain a second enzyme that cleaves sucrose, invertase, which catalyses a reaction that produces glucose and fructose, principally for flux into glycolysis (Trethewey et al. 1998). Therefore, the relative activities of sucrose synthase and invertase in sink tissues are the key determinant of carbon partitioning between storage and other uses (Fig. 4.3). It also plays a role in the control of development, triggering the initiation of the storage phase of, for example, legume seed development (Weber et al. 1996).

4.4   Key Metabolic Regulators

4.4.1   SNF1-Related Protein Kinase 1 (SnRK1)

The sucrose synthase gene that is expressed in potato tubers is *SUS4* (Fu and Park 1995). Under normal conditions, it is expressed only in tubers but its expression can be induced in leaves in response to incubation with high concentrations of sucrose (Fu and Park 1995); its expression is not affected at all by glucose. Purcell et al. (1998) reported that expression of *SUS4* required the activity of a regulatory protein kinase called sucrose nonfermenting-1-related protein kinase 1 (SnRK1). This protein kinase was shown subsequently to be required for the redox modulation

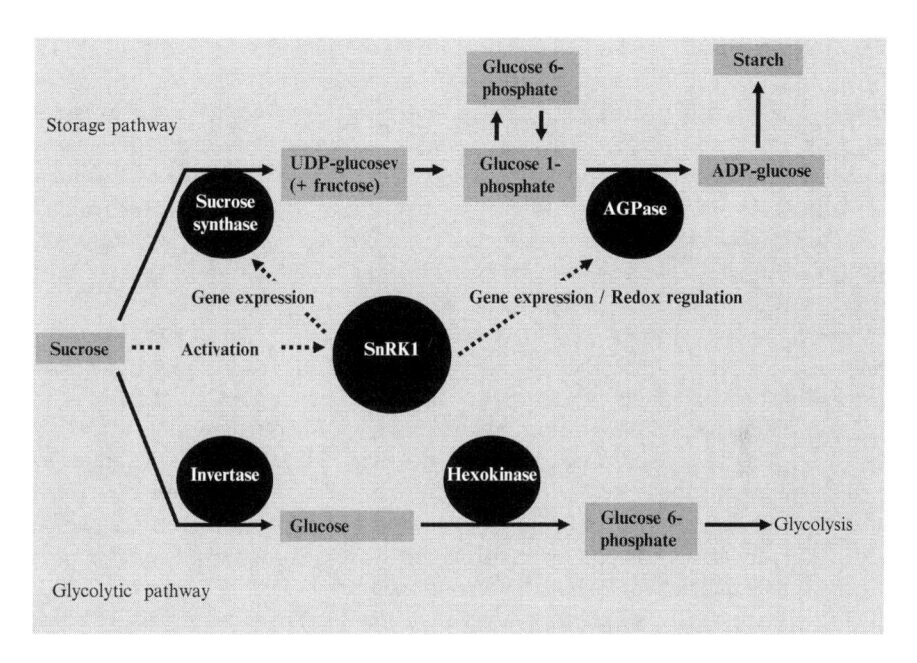

Fig. 4.3 The relative activities of sucrose synthase and invertase in potato tubers determine how much carbon enters the storage pathway for starch biosynthesis and how much enters the glycolytic pathway. The starch biosynthesis pathway is controlled by the metabolic regulator, SnRK1 (McKibbin et al. 2006)

of ADP-glucose pyrophosphorylase (AGPase), another enzyme in the starch bio-synthetic pathway (Tiessen et al. 2002, 2003), also in response to sucrose. Over-expression of SnRK1 resulted in an increase in the expression of both SuSy and AGPase, leading to high starch and low free sugar concentrations (McKibbin et al. 2006; Fig. 4.3).

SnRK1 takes its name from sucrose nonfermenting-1 (SNF1), its homologue in budding yeast (*Saccharomyces cereviseae*). SNF1 is at the heart of a glucose repression signalling pathway but does not appear to be involved in sucrose signalling. Indeed, yeast does not appear to sense sucrose, and glucose repression is the overriding mechanism for controlling carbon metabolism in that organism (reviewed in Dickinson 1999; Halford 2006). In that regard, the yeast and plant systems have diverged, although SnRK1 is required for expression of α-amylase, a glucose-repressible gene, in plants (Laurie et al. 2003; Lu et al. 2007) and has been implicated in controlling the expression of a wide range of other genes in transcriptomic-based studies (Radchuk et al. 2006; Baena-González et al. 2007). SnRK1 has also been shown to regulate another sugar-repressible promoter, the asparagine synthetase (ASN1) gene promoter of Arabidopsis (Baena-González et al. 2007).

There is also a homologous regulatory protein kinase in animals, called adenosine monophosphate (AMP)-activated protein kinase (AMPK). As its name indicates, it is activated allosterically by AMP, the concentrations of which respond very sensitively to the metabolic status of the cell and other stresses. Note that SNF1 and SnRK1 are not activated by AMP in the same way, despite the fact that SNF1, SnRK1 and AMPK have similar functions, form similar complexes and recognise the same target site (Halford and Hardie 1998). They are also regulated by phosphorylation of the same regulatory threonine in the so-called T-loop and similarities in the signalling systems extend to the upstream kinases that are responsible for this phosphorylation (Hey et al. 2007). They are not reviewed further here but are described more fully by Halford and Hardie (1998), Halford and Paul (2003) or Halford (2006).

Purcell and co-workers (1998) produced transgenic potato plants expressing antisense *SnRK1* leaf-specifically under the control of a *ST-LS1* promoter and showed that sucrose induction of sucrose synthase gene expression did not occur in leaves excised from these plants. Subsequent experiments showed that the yield of tubers from these plants was much greater than that from wild-type plants, or plants expressing antisense *SnRK1* under the control of a patatin promoter, when the plants were grown under low light conditions (Fig. 4.4a). Furthermore, leaves expressing antisense *SnRK1* behaved differently to leaves from control plants when excised and incubated for 24 h in the dark on sucrose medium. While control leaves mobilized starch, leaves expressing antisense *SnRK1* appeared to be unable to do so (Fig. 4.4b).

The increase in tuber yield under low light conditions in these plants is consistent with the model in which control of source to sink carbon flux in potato is largely vested in the reactions of the leaf (80% or so under a range of experimental conditions; Sweetlove and Hill 2000). Furthermore, it suggests a mechanistic basis for the model in which SnRK1 is a key regulator of source to sink carbon flux.

Acclimation to carbon supply may occur at two levels. One is the fine control of metabolism through post-translational mechanisms, such as protein phosphorylation, mainly in response to shorter-term changes in carbon supply and demand, and coarse control involving changes in the expression of genes encoding key enzymes. Two hypotheses could be proposed. In the first hypothesis, decreased activity of SnRK1 in leaves would result in more sucrose phosphate synthase (SPS) remaining in an active, dephosphorylated state in the low light conditions, where SPS would normally be inactivated in response to low carbon availability. This would result in more sucrose synthesis, while decreased expression of sucrose synthase would result in less sucrose degradation. More sucrose would be available for export to sinks, leading to an increase in tuber yield. However, carbon supply from photosynthesis varies not only with irradiance but also with other changes in the environment, and demand for carbon for use in growth is also affected by environmental variables, particularly availability of nutrients and temperature. In the second hypothesis, SnRK1 would be required for adaptation to conditions associated with low carbon supply (exactly analogous to the role of SNF1 in yeast) and/ or high carbon demand. This would be expected to be most apparent in plants adapting to new conditions that require a switch in carbon allocation strategy from storage of carbon to mobilization of carbon reserves (exactly analogous to yeast switching between carbon sources).

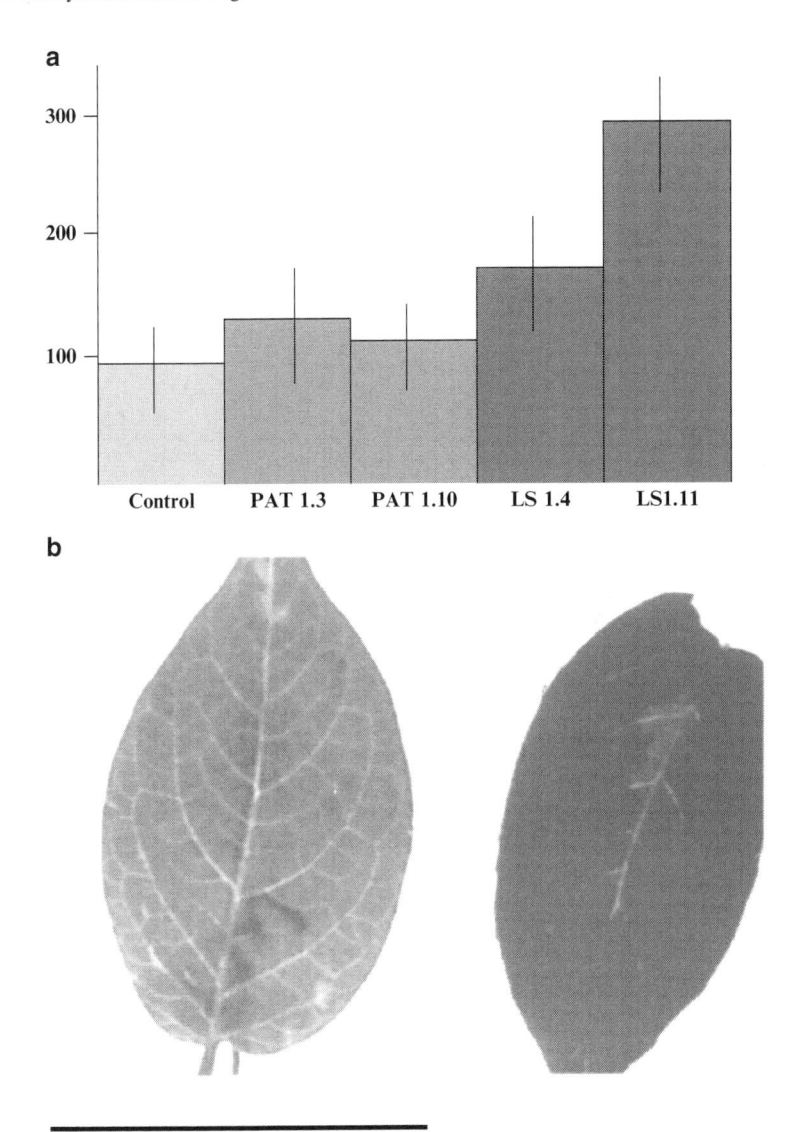

Fig. 4.4 a Yield (g fresh weight) of tubers from control potato plants and potato plants expressing antisense *SnRK1* in the tubers (PAT lines) and leaves (LS lines). The plants were grown under low light conditions (150 µmol m^{-2} s^{-1}), 16 h day, 20°C day temp., 15°C night temp. Mean and standard deviation are shown for ten plants from each line. **b** Iodine staining of leaves from wild type (*left*) and antisense *SnRK1* potato plants (*right*) after excision and culture in the dark for 24 h on medium containing 250 mM sucrose. Bar = 5 cm

Whichever hypothesis is correct, manipulating SnRK1 activity in leaves to increase yield of storage organs would seem to be an obvious target for biotechnologists. However, expression of the transgene in this and subsequent experiments using a different leaf-specific promoter was not maintained; in other words, the transgene appeared to be silenced (unpublished data). Transgene silencing has not been reported in the literature for potato, but there are anecdotal reports of it from industrial sources. Therefore, irrespective of yield gains in the primary transgenics, genetic modification of *SnRK1* gene expression in the leaves of potato plants does not appear to be appropriate for use in the production of a commercial potato variety.

4.4.2   Hexokinase

Another protein that has been implicated in sugar sensing and signalling is hexokinase, an enzyme that catalyses the conversion of glucose to glucose 6-phosphate (G6P), the first stage in glycolysis. Hexokinase has been associated particularly with linking the carbohydrate status of leaves to photosynthetic gene expression as a means of feedback control (Jang and Sheen 1997; Jang et al. 1997). Its proposed role as a sugar sensor and signalling molecule has been controversial (Halford et al. 1999) and its metabolic function means that any attempt to alter its activity in transgenic or mutant plants or with inhibitors inevitably causes changes in metabolism that could be misinterpreted as evidence of a signalling role. The use of targeted mutagenesis of one of the Arabidopsis hexokinase genes, *HXK1*, has been used to produce enzymes that retained signalling but not catalytic activities (Moore et al. 2003). However, the mechanism by which hexokinase signals remains unknown nearly two decades after hexokinase was first proposed as a signalling factor. Consequently, it is not possible to measure its signalling activity.

4.4.3   The Trehalose Pathway

Another signalling molecule that has received increasing attention over the last few years is trehalose 6-phosphate (T6P). T6P is formed from G6P and UDP-glucose through the action of the enzyme TPS and can be converted to trehalose (a disaccharide of two glucose units) through the action of trehalose phosphate phosphatase (TPP). Trehalose itself rarely accumulates in plants other than some 'resurrection plants' that can withstand extreme desiccation. This is because it is broken down by a third enzyme, trehalase. In fungi, both the enzyme TPS and the metabolite T6P have an inhibitory effect on hexokinase and regulate flux into glycolysis, preventing the potentially lethal accumulation of glycolytic intermediates (Bonini et al. 2000; Noubhani et al. 2000). TPS and T6P do not inhibit hexokinase in plants. Nevertheless, there is clear evidence that the trehalose pathway is an essential component of metabolic signalling in plants (Goddijn and Smeekens 1998; Eastmond et al. 2002; Schluepmann et al. 2003, 2004). For example, transgenic plants over-expressing TPP and trehalose phosphate hydrolase

(TPH) have been shown to have low T6P content and cannot grow on sugar-containing media. This is because they accumulate large pools of metabolites and ATP concentrations become depleted, while feeding seedlings with trehalose stimulates starch accumulation (Wingler et al. 2000; Fritzius et al. 2001).

The effect of trehalose on starch metabolism may be brought about through activation of the enzymes sucrose synthase and AGPase (Muller et al. 1998; Kolbe et al. 2005). Both are key enzymes in the starch biosynthetic pathway. These enzymes are also regulated by SnRK1, suggesting that trehalose pathway signalling interacts with the SnRK1 system. Recently, it has been shown that SnRK1 can phosphorylate and inactivate some TPS isozymes from Arabidopsis (Harthill et al. 2006), adding considerable support to this hypothesis.

4.5 Applications in Biotechnology

Understanding the control of assimilate partitioning and how to manipulate it underpins many targets for plant biotechnologists (Fig. 4.5). An important target with broad applications is the manipulation of the carbon/nitrogen balance. In potato tubers and most cereal seeds, this is effectively the ratio of starch to protein, although all cereal seeds also contain some oil. Farmers producing barley for malting, for example, grow varieties that favour the accumulation of starch over protein in the seed and cultivate the crop under low nitrogen conditions to keep seed protein levels low. Similarly, high-starch and low-protein varieties are grown to produce the 6.1 million tonnes of starch from wheat, maize and potato that is used by food and non-food industries each year in Europe. In contrast, farmers producing barley, wheat and maize for animal feed need varieties that favour protein production. The ability to manipulate the C:N balance to suit a particular end-user requirement has long been a target for biotechnologists, but it has become more important than ever recently, because cereal starch has become a favoured raw material in the production of ethanol for use as a biofuel. Indeed, much of the maize (corn) 'surplus' in the United States that, up to a few years ago, would be sold on the world food markets is now being used to make ethanol for fuel. There is a demand for varieties that are not only high in starch but also high in starch that is readily fermentable. How sustainable this industry is may be open to conjecture. The economics are generally unconvincing and the ethics of using an important food crop to produce fuel are being questioned. However, there is a clear benefit in the reduction of carbon emissions and, therefore, mitigating climate change and increasing the use of biofuels has been adopted as a policy by many governments. Nevertheless, changes brought about by policy rather than market forces are always vulnerable.

There is an overlap in many cases between traits that are desirable for a biofuel crop and those that are required for food crops. High-starch lines of potato, for example, are suitable not only for industrial starch production (potentially including starch for fermentation into ethanol for fuel, although this is not currently

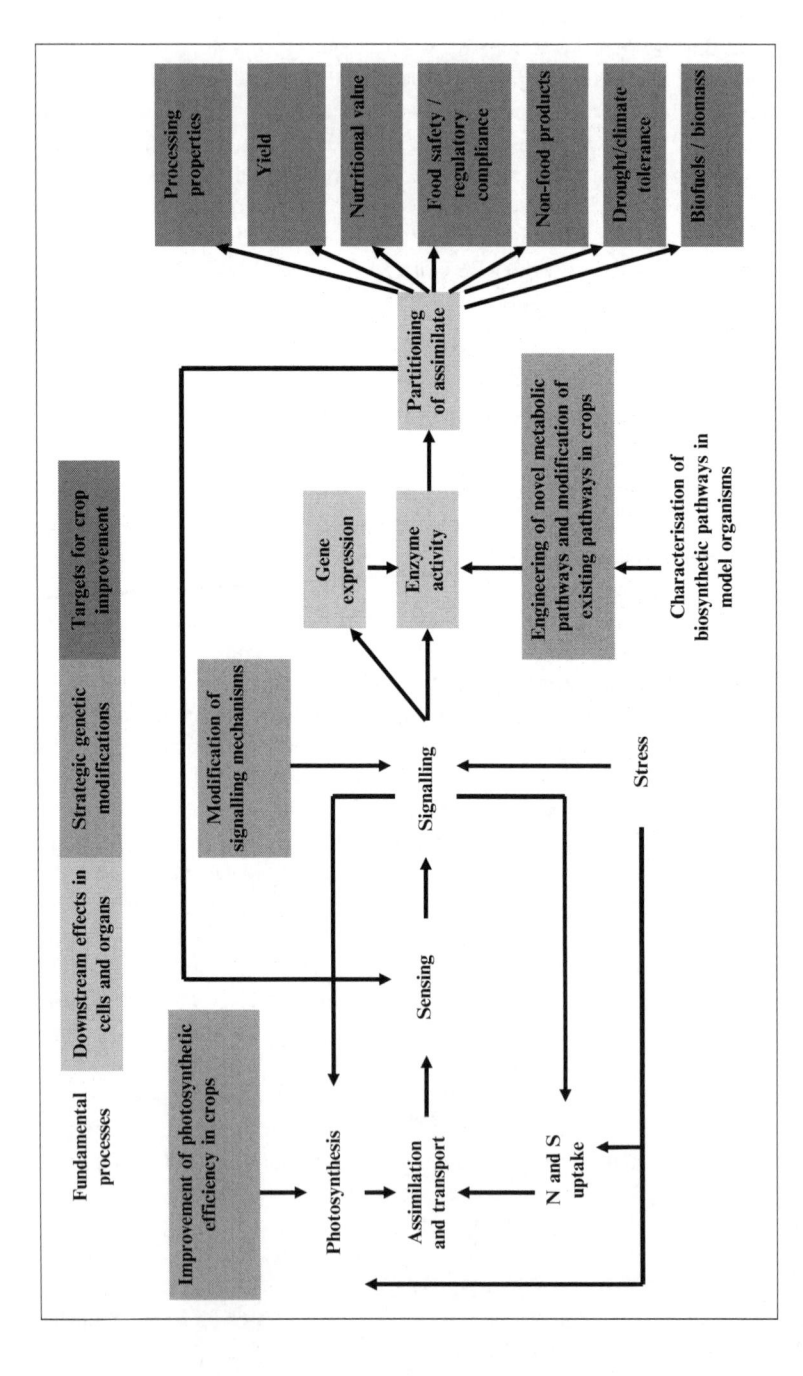

Fig. 4.5 Schematic illustrating how the regulation of metabolism and partitioning is key to achieving many of the strategic objectives being set by government and industry

economically viable), but also for the production of French fries and crisps. High-starch varieties contain less water, and thereby reduce processing costs because the water is replaced by oil during cooking and the oil costs money. It also increases the fat content of the product. Favouring starch production also reduces the concentrations of free sugars. These cause blackening and can react with asparagine during baking, frying or roasting to produce acrylamide, a nerve poison and likely carcinogen (reviewed in Halford et al. 2007; Muttucumaru et al. 2008).

Free sugars are desirable in some crops, the obvious examples being sugar beet and sugar cane. The enzymes targeted by breeders and biotechnologists in this case are those that make sucrose, i.e. SPS and sucrose phosphate phosphatase (SPP), and those that break it down, i.e. sucrose synthase and invertase. Once again, a trait that has been a target for many years for food crops has received new impetus as the biofuel market has become established. Sugar cane producers, in particular, who have been shut out of the European market by import tariffs and the effect of subsidies given to Europe's sugar beet farmers, are now seeing their profit margins increasing at last, and biotechnology companies are again turning their attention to a crop in which previously it may not have been worth investing.

Carbon is also partitioned into free amino acids and proteins and this process also affects the quality of the product. In this case, nitrogen and sulphur nutrition, in particular, also have a major effect, altering the C:N balance and the proportion of storage proteins compared with other proteins (Giese and Hopp 1984; Duffus and Cochrane 1992). Both of these parameters affect the suitability of cereal grain for processes such as bread-making, the production of noodles, starch for industrial uses (including biofuels) and animal feed, as well as malting.

Sulphur starvation affects storage protein synthesis in two ways. Firstly, it results in decreased total storage protein synthesis and increased free amino acids accumulating within the grain (Shewry et al. 1983, 2001). Secondly, it causes an increase in sulphur-poor proteins compared with sulphur-rich proteins as a proportion of the total (Kirkman et al. 1982; Shewry et al. 1983). S-poor proteins also accumulate preferentially under conditions of high nitrogen availability, showing that nitrogen and sulphur interact in the regulation of storage protein gene expression. In wheat and barley, the major free amino acid to accumulate is asparagine, and this again has potentially severe consequences for acrylamide formation during baking or processing into breakfast cereal. Wheat that is sulphur-starved can accumulate up to 30-fold higher free asparagine in the grain than wheat with an adequate supply of sulphur (Muttucumaru et al. 2006; Halford et al. 2007). Acrylamide concentrations in heated flour derived from such grain can reach several thousand parts per billion. The enzymes involved in asparagine synthesis and breakdown are therefore likely to become important targets for breeders and biotechnologists.

Protein content and quality are also the key parameters in the production of animal feed. This is shown graphically in the case of UK barley, the market for which in animal feed has almost disappeared over the last few decades in the face of competition from imported soybean. Barley, like most cereals, contains relatively little lysine in its grain. Although biotechnologists and breeders have been seeking ways of improving cereal lysine content for many years, breakthroughs have been made only recently and have been applied in maize, for which there is now an established biotechnology market, but not in barley. The improvement has been made by the genetic modification of maize to express an unregulated dihydrodipicolinate synthase (DHDPS) enzyme from the CORDAPA gene of the bacterium Corynebacterium glutamicum (Huang et al. 2005).

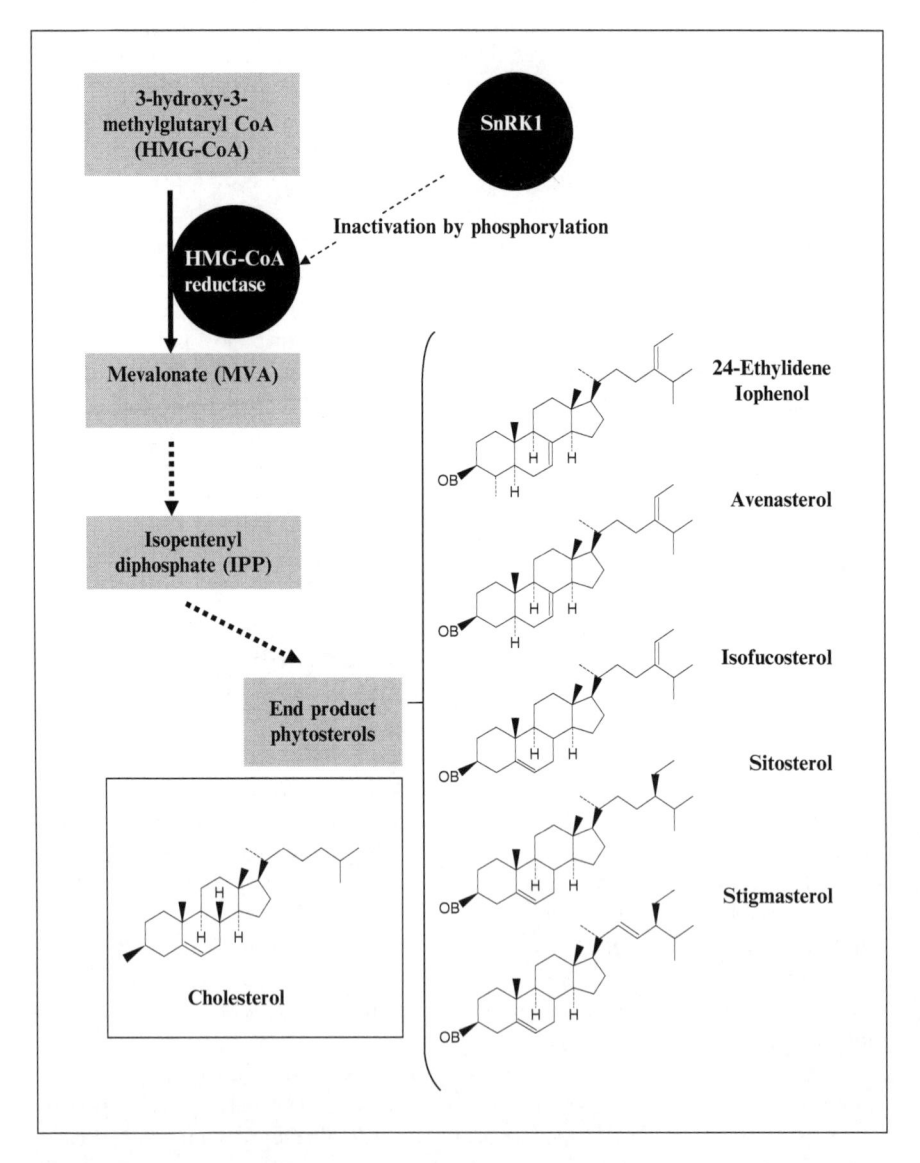

Fig. 4.6 Schematic illustrating the role of SnRK1 in controlling flux into the mevalonic acid pathway for sterol biosynthesis and the similarity in structure of plant sterols and cholesterol

It could be argued that this is metabolic engineering rather than manipulation of partitioning, but any metabolic engineering intervention inevitably changes the partitioning of assimilate in that it alters metabolic pathways to induce more carbon flux to a chosen target. This may seem like stretching the point, but biotechnologists ignore the regulation of partitioning and metabolism at their peril when planning a metabolic engineering project. Indeed, increasing lysine biosynthesis was achieved only by using an unregulated form of the key enzyme in the pathway. Another excellent example of this is the manipulation of the mevalonic acid (MVA) pathway to increase phytosterol yield. The phytosterols are attractive targets for biotechnologists because the predominant naturally occurring phytosterols are structurally related to cholesterol (Fig. 4.6). These compounds are known as 4-desmethylsterols because of the absence of methylation at the fourth carbon in the chain and include β-sitosterol, campesterol and stigmasterol, and they competitively inhibit the uptake of cholesterol from the small intestine in humans. They are purified from plant sources, such as soybean, and added to margarines and other products designed to reduce serum cholesterol levels. Increasing their levels in oilseeds would make them cheaper to purify. Alternatively, oils with enhanced phytosterol levels could be used to make these products without supplementation. The key regulatory step in the pathway is the NADPH-dependent reduction of 3-hydroxy-3-methylglutaryl-coenzyme A (HMG-CoA) to MVA, catalysed by the enzyme HMG-CoA reductase. Several attempts have been made to increase seed sterol biosynthesis by over-expressing this enzyme. However, the best results have been obtained by using deregulated forms of the enzyme: for example, a 3.2-fold increase in seed sterol levels was achieved in tobacco (as a model) by expressing a truncated rubber tree (*Hevea brasiliensis*) HMG-CoA reductase lacking the N-terminal regulatory domain (Harker et al. 2003), while uncoupling HMG-CoA reductase from regulation by SnRK1 through the mutagenesis of a C-terminal SnRK1 target site increased seed sterol content approximately 2.5-fold (Hey et al. 2006).

4.6 Concluding Remarks

Like all 'second generation' targets (in other words those that affect output traits rather than inputs of herbicide or insecticide), the manipulation of photosynthate partitioning has yet to find a market in biotech crops. However, that may be about to change, with high-lysine corn ready for launch and a demand established for high-starch crops for biofuels. Therefore, it may well be that the next decade sees biotech crops with modified partitioning break through on to the market. Furthermore, the possibility that manipulating partitioning could make crops more efficient under high carbon dioxide conditions ensures that this area of research and development will remain a key one in crop biotechnology.

Acknowledgements Rothamsted Research receives grant-aided support from the Biotechnology and Biological Sciences Research Council of the United Kingdom.

References

Baena-González E, Rolland F, Thevelein JM, Sheen J (2007) A central integrator of transcription networks in plant stress and energy signalling. Nature 448:938–942

Bonini BM, Van Vaeck C, Larsson C, Gustafsson L, Ma P, Winderickx J, Van Dijck P, Thevelein JM (2000) Expression of *Escherichia coli otsA* in a *Saccharomyces tps1* mutant restores growth and fermentation with glucose and control of glucose influx into glycolysis. Biochem J 350:261–268

Choury PS, Nelson OE (1976) The enzymatic deficiency conditioned by the *Shrunken-1* mutations in maize. Biochem Genet 14:1041–1055

Dickinson JR (1999) Carbon metabolism. In: Dickinson JR, Schweizer M (eds) The metabolism and molecular physiology of *Saccharomyces cerevisiae*. Taylor and Francis, London, pp 23–55

Dickinson CC, Altabella T, Chrispeels MJ (1991) Slow growth phenotype of transgenic tomato expressing apoplastic invertase. Plant Physiol 95:420–425

Duffus CM, Cochrane MP (1992) Grain structure and composition. In: Shewry PR (ed) Barley. Genetics, biochemistry, molecular biology and biotechnology. CAB International, Wallingford, pp 291–317

Eastmond PJ, van Dijken AJ, Spielman M, Kerr A, Tissier AF, Dickinson HG, Jones JD, Smeekens SC, Graham IA (2002) Trehalose-6-phosphate 1, which catalyses the first step in trehalose synthesis, is essential for *Arabidopsis* embryo development. Plant J 29:225–235

Fritzius T, Aeschbacher R, Wiemken A, Wingler A (2001) Induction of *APL3* expression by trehalose complements the starch-deficient *Arabidopsis* mutant *adg2-1* lacking ApL1, the large subunit of ADP-glucose pyrophosphorylase. Plant Physiol 126:883–889

Fu H, Park WD (1995) Sink- and vascular-associated sucrose synthase functions are encoded by different gene classes in potato. Plant Cell 7:1369–1385

Giese H, Hopp HE (1984) Influence of nitrogen nutrition on the amount of hordein, protein Z and β-amylase messenger RNA in developing endosperms of barley. Carlsberg Res Comm 49:365–383

Goddijn OJ, Smeekens SCM (1998) Sensing trehalose biosynthesis in plants. Plant J 14:143–146

Graham IA, Denby KJ, Leaver CJ (1994) Carbon catabolite repression regulates glyoxylate cycle gene expression in cucumber. Plant Cell 6:761–772

Halford NG (2006) Regulation of carbon and amino acid metabolism: roles of sucrose nonfermenting-1-related protein kinase-1 and general control nonderepressible-2-related protein kinase. Adv Bot Res Inc Adv Plant Pathol 43:93–142

Halford NG, Hardie DG (1998) SNF1-related protein kinases: global regulators of carbon metabolism in plants? Plant Mol Biol 37:735–748

Halford NG, Paul MJ (2003) Carbon metabolite sensing and signalling. Plant Biotechnol J 1:381–398

Halford NG, Purcell PC, Hardie DG (1999) Is hexokinase really a sugar sensor in plants? Trends Plant Sci 4:117–120

Halford NG, Muttucumaru N, Curtis TY, Parry MAJ (2007) Genetic and agronomic approaches to decreasing acrylamide precursors in crop plants. Food Add Contam SI 24:26–36

Harker M, Holmberg M, Clayton JC, Gibbard CL, Wallace AD, Rawlins S, Hellyer SA, Lanot A, Safford R (2003) Enhancement of seed phytosterol levels by expression of an N-terminal truncated *Hevea brasiliensis* (rubber tree) 3-hydroxy-3-methylglutaryl-CoA reductase. Plant Biotechnol J 1:113–121

Harter K, Talke-Messerer C, Barz W, Schäfer E (1993) Light and sucrose-dependent gene expression in photomixotrophic cell suspension cultures and protoplasts of rape. Plant J 4:507–516

Harthill JE, Meek SE, Morrice N, Peggie MW, Borch J, Wong BH, MacKintosh C (2006) Phosphorylation and 14-3-3 binding of Arabidopsis trehalose-phosphate synthase 5 in response to 2-deoxyglucose. Plant J 47:211–223

Hey SJ, Powers SJ, Beale M, Hawkins ND, Ward J, Halford NG (2006) Enhanced seed phytosterol accumulation through expression of a modified HMG-CoA reductase. Plant Biotechnol J 4:219–229

Hey S, Mayerhofer H, Halford NG, Dickinson JR (2007) DNA sequences from Arabidopsis which encode protein kinases and function as upstream regulators of Snf1 in yeast. J Biol Chem 282:10472–10479

Huang S, Kruger DE, Frizzi A, D'Ordine RL, Florida CA, Adams WR, Brown WE, Luethy MH (2005) High-lysine corn produced by the combination of enhanced lysine biosynthesis and reduced zein accumulation. Plant Biotechnol J 8:555–569

Inoue H, Tanaka A (1978) Comparison of source and sink potentials between wild and cultivated potatoes. J Sci Soil Manage Japan 49:321–327

Jang JC, Sheen J (1997) Sugar sensing in higher plants. Trends Plant Sci 2:208–214

Jang J-C, Leon P, Zhou L, Sheen J (1997) Hexokinase as a sugar sensor in higher plants. Plant Cell 9:5–19

Kirkman MA, Shewry PR, Miflin BJ (1982) The effect of nitrogen nutrition on the lysine content and protein composition of barley seeds. J Sci Food Agric 33:115–127

Kolbe A, Tiessen A, Schluepmann H, Paul M, Ulrich S, Geigenberger P (2005) Trehalose 6-phosphate regulates starch synthesis via posttranslational redox activation of ADP-glucose pyrophosphorylase. Proc Natl Acad Sci USA 102:11118–11123

Krapp A, Hofmann B, Schäfer C, Stitt M (1993) Regulation of the expression of rbcS and other photosynthetic genes by carbohydrates: a mechanism for the 'sink' regulation of photosynthesis? Plant J 3:817–828

Laurie S, McKibbin RS, Halford NG (2003) Antisense SNF1-related (SnRK1) protein kinase gene represses transient activity of an α-amylase (α-Amy2) gene promoter in cultured wheat embryos. J Exp Bot 54:739–747

Lu C-A, Lin C-C, Lee K-W, Chen J-L, Ho S-L, Huang L-F, Hsing Y-I, Yu S-M (2007) The SnRK1A protein kinase plays a key role in sugar signalling during germination and seedling growth of rice. Plant Cell 19:2484–2499

McKibbin RS, Muttucumaru N, Paul MJ, Powers SJ, Burrell MM, Coates S, Purcell PC, Tiessen A, Geigenberger P, Halford NG (2006) Production of high starch, low glucose potatoes through over-expression of the metabolic regulator, SnRK1. Plant Biotechnol J 4:409–418

Mita S, Suzuki-Fujii K, Nakamura K (1995) Sugar-inducible expression of a gene for β-amylase in Arabidopsis. Plant Physiol 107:895–904

Moore B, Zhou L, Rolland F, Hall Q, Cheng W-H, Liu Y-X, Hwang I, Jones T, Sheen J (2003) Role of the Arabidopsis glucose sensor HXK1 in nutrient, light and hormonal signalling. Science 300:332–336

Muller J, Boller T, Wiemken A (1998) Trehalose affects sucrose synthase and invertase activities in soybean (Glycine max L. Merr.) roots. J Plant Physiol 153:255–257

Muttucumaru N, Halford NG, Elmore JS, Dodson AT, Parry M, Shewry PR, Mottram DS (2006) The formation of high levels of acrylamide during the processing of flour derived from sulfate-deprived wheat. J Agric Food Chem 54:8951–8955

Muttucumaru N, Elmore JS, Curtis T, Mottram DS, Parry MAJ, Halford NG (2008) Reducing acrylamide precursors in raw materials derived from wheat and potato. J Agric Food Chem 56:6167–6172

Noubhani A, Bunoust O, Rigoulet M, Thevelein JM (2000) Reconstitution of ethanolic fermentation in permeabilised spheroplasts of wild type and trehalose 6-phosphate synthase mutants of the yeast Saccharomyces cerevisiae. Eur J Biochem 267:4566–4576

Paul MJ, Pellny T, Goddijn O (2001) Enhancing photosynthesis with sugar signals. Trends Plant Sci 6:197–200

Purcell PC, Smith AM, Halford NG (1998) Antisense expression of a sucrose nonfermenting-1-related protein kinase sequence in potato results in decreased expression of sucrose synthase in tubers and loss of sucrose-inducibility of sucrose synthase transcripts in leaves. Plant J 14:195–202

Radchuk R, Radchuk V, Weschke W, Borisjuk L, Weber H (2006) Repressing the expression of the sucrose nonfermenting-1-related protein kinase gene in pea embryo causes pleiotropic defects of maturation similar to an abscisic acid-insensitive phenotype. Plant Physiol 140:263–278

Salanoubat M, Belliard G (1989) The steady-state level of potato sucrose synthase mRNA is dependant on wounding, anaerobiosis and sucrose concentration. Gene 84:181–185

Schluepmann H, Pellny T, van Dijken A, Smeekens S, Paul MJ (2003) Trehalose 6-phosphate is indispensable for carbohydrate utilisation and growth in *Arabidopsis thaliana*. Proc Natl Acad Sci USA 100:6849–6854

Schluepmann H, van Dijken A, Aghdasi M, Wobbes B, Paul M, Smeekens S (2004) Trehalose-mediated growth inhibition of Arabidopsis seedlings is due to trehalose-6-phosphate accumulation. Plant Physiol 135:879–890

Sheen J (1990) Metabolic repression of transcription in higher plants. Plant Cell 2:1027–1038

Sheen J (1994) Feedback control of gene expression. Photosynth Res 39:427–438

Shewry PR, Franklin J, Parmar S, Smith SJ, Miflin BJ (1983) The effects of sulfur starvation on the amino acid and protein compositions of barley grain. J Cereal Sci 1:21–31

Shewry PR, Tatham AS, Halford NG (2001) Nutritional control of storage protein synthesis in developing grain of wheat and barley. Plant Growth Regul 34:105–111

Sowokinos JR, Varns JL (1992) Induction of sucrose synthase in potato tissue culture: effect of carbon source and metabolic regulators on sink strength. J Plant Physiol 139:672–679

Sweetlove LJ, Hill SA (2000) Source metabolism dominates the control of source to sink carbon flux in tuberising potato plants throughout the diurnal cycle and under a range of environmental conditions. Plant Cell Environ 23:523–529

Tiessen A, Hendriks JHM, Stitt M, Branscheid A, Gibon Y, Farre EM, Geigenberger P (2002) Starch synthesis in potato tubers is regulated by post-translational redox modification of ADP-glucose pyrophosphorylase: a novel regulatory mechanism linking starch synthesis to the sucrose supply. Plant Cell 14:2191–2213

Tiessen A, Prescha K, Branscheid A, Palacios N, McKibbin R, Halford NG, Geigenberger P (2003) Evidence that SNF1-related kinase and hexokinase are involved in separate sugar-signalling pathways modulating post-translational redox activation of ADP-glucose pyrophosphorylase in potato tubers. Plant J 35:490–500

Trethewey RN, Geigenberger P, Riedel K, Hajirezaei M-R, Sonnewald U, Stitt M, Riesmeier JW, Willmitzer L (1998) Combined expression of glucokinase and invertase in potato tubers leads to a dramatic reduction in starch accumulation and a stimulation of glycolysis. Plant J 15:109–118

van Oosten J-J, Besford RT (1996) Acclimation of photosynthesis to elevated CO_2 through feedback regulation of gene expression: climate of opinion. Photosynth Res 48:353–365

von Schaewen A, Stitt M, Sonnewald U, Willmitzer L (1990) Expression of a yeast-derived invertase in the cell wall of tobacco and arabidopsis plants leads to accumulation of carbohydrate and inhibition of photosynthesis and strongly influences growth and phenotype of transgenic tobacco plants. EMBO J 9:3033–3044

Weber H, Buchner P, Borisjuk L, Wobus U (1996) Sucrose metabolism during cotyledon development of *Vicia faba* L. is controlled by the concerted action of both sucrose phosphate synthase and sucrose synthase: expression patterns, metabolic regulation and implications for seed development. Plant J 9:841–850

Wingler A, Fritzius T, Wiemken A, Boller T, Aeschbacher RA (2000) Trehalose induces the ADP-glucose pyrophosphorylase gene, *ApL3*, and starch synthesis in *Arabidopsis*. Plant Physiol 124:105–114

Yu S-M, Kuo Y-H, Sheu G, Sheu Y-J, Liu L-F (1991) Metabolic derepression of α-amylase gene expression in suspension-cultured cells of rice. J Biol Chem 266:21131–21137

Zhu X-G, de Sturler E, Long SP (2007) Optimizing the distribution of resources between enzymes of carbon metabolism can dramatically increase photosynthetic rate: a numerical simulation using an evolutionary algorithm. Plant Physiol 145:513–526

Zrenner R, Salanoubat M, Willmitzer L, Sonnewald U (1995) Evidence of the crucial role of sucrose synthase for sink strength using transgenic potato plants (*Solanum tuberosum* L.). Plant J 7:97–107

Chapter 5
Molecular Physiology of Seed Maturation and Seed Storage Protein Biosynthesis

H. Weber, N. Sreenivasulu, and W. Weschke

5.1 Introduction

Seeds synthesize proteins, starch and lipids in different amounts and represent the most important source of feed and food for human and animal nutrition. Legume seeds especially are rich in protein, with contents in the range of 20 to 40% of seed dry weight (Müntz 1989). Cereal grain protein content is lower, with an average of about 10 to 12% of dry weight but, for nutrition, cereals provide about three times the amount derived from legume seeds. In addition to their nutritional importance, the amount and quality of cereal seed proteins also influence food processing, in particular bread making quality (Wiesner 2007) and suitability for brewing. Therefore, seed proteins in legumes and cereals became a major topic of research towards the understanding of their structures, and the mechanisms which control storage protein synthesis, accumulation and content in seeds (Müntz 1989; Casey et al. 1993; Kermode and Bewley 1999; Shewry and Halford 2002). Increasing the protein content and/or improving the seed composition of crop seeds is a desirable aim (Wang et al. 2003) and basic research can provide valuable prerequisites for future improvement of agricultural traits. This review summarizes and discusses recent approaches in order to understand seed protein synthesis in the context of seed maturation, with the main emphasis on legume and cereal crop seeds. Most of the basic knowledge of the genetic background, the cellular differentiation and the regulatory network of signal transduction of seed maturation and storage protein formation comes from *Arabidopsis* (Finkelstein et al. 2002; Gutierrez et al. 2007). Although it can be assumed that fundamental regulatory mechanisms are comparable in seeds of different species, too much

H. Weber, N. Sreenivasulu, and W. Weschke
Leibniz Institute of Plant Genetics and Crop Plant Research (IPK), 06466 Gatersleben, Germany
e-mail: weber@ipk-gatersleben.de

E-C. Pua and M.R. Davey (eds.),
Plant Developmental Biology – Biotechnological Perspectives: Volume 2,
DOI 10.1007/978-3-642-04670-4_5, © Springer-Verlag Berlin Heidelberg 2010

generalization should be avoided, especially in cases of possible improvement strategies because, in contrast to *Arabidopsis*, crop plants have been selected for high seed yield and characterised by high metabolic activity and fluxes in their seeds.

5.2 Seed Maturation

The biosynthesis of seed storage proteins is subject to tissue-specific and developmental regulation. Significant amounts accumulate only in the storage parenchyma of embryos or in the cereal endosperm during mid- and late seed maturation. The process is integrated into the seed maturation program. Therefore, a profound understanding of storage protein formation requires knowledge of the regulatory network of seed maturation, which has been studied in some detail in *Arabidopsis* (reviewed in Finkelstein et al. 2002; Gutierrez et al. 2007), legumes (Wobus and Weber 1999; Weber et al. 1997a, 1998a, 2005) and, based on transcriptional profiling, in the barley endosperm (Sreenivasulu et al. 2004, 2006). Seed maturation is controlled by a complex transcriptional network. Its spatial and temporal regulation requires the concerted action of several signalling pathways integrating information from genetic programs, from hormonal signals with greatest importance of abscisic acid (ABA), as well as from metabolic signals, with a role for sucrose and nitrogen (Wobus and Weber 1999; Weber et al. 2005; Sreenivasulu et al. 2006). Seed formation can be divided into two phases. During the initial prestorage phase, which is characterised by maternal control, organogenesis and morphogenesis occur on the basis of high cell division activity. Maturation and storage activities are initiated during the phase transition. When cell divisions cease, the embryo enters the maturation phase and differentiates from a proliferating tissue into a highly specialized storage organ. The maturation phase is characterised by cell expansion and storage activity. Physiological maturity is the stage of maximal dry weight accumulation and of the highest degree of differentiation before desiccation starts (see Fig. 5.1 for a general scheme of legume seed development). From a developmental point of view, maturation interrupts seedling growth and development. Mutations which affect ABA biosynthesis/function/sugar signalling shift seed development towards germination. For example, immuno-modulation of ABA in tobacco seeds reduced ABA amounts and caused a developmental switch from maturation to germination programs characterised by chloroplast formation and reduced storage oil and proteins (Phillips et al. 1997). The seed phenotype of the ABA-immuno-modulated tobacco plants was similar to that of *Arabidopsis aba/abi3* double mutants (Koornneef et al. 1989). The results show that from a developmental point of view seed maturation is not a necessary process, but is a dispensable process initiated by the sugar-ABA signalling network.

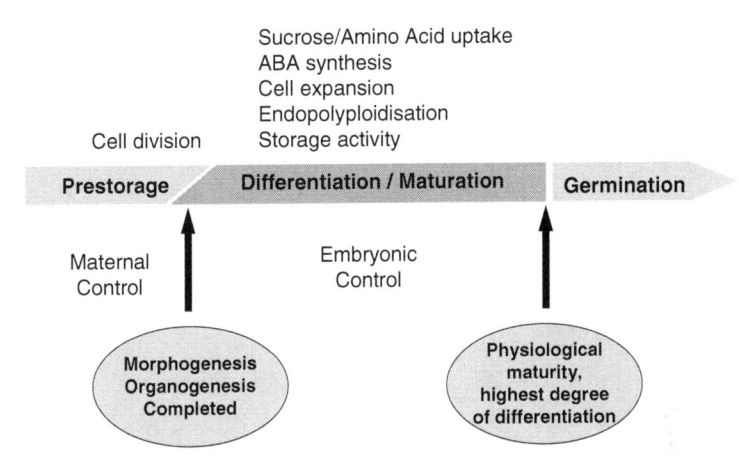

Fig. 5.1 A general scheme of legume seed development (see text for details)

5.3 Sucrose as a Maturation Signal

Sucrose has a dual function as a transport and nutrient sugar and as a signal molecule triggering storage-associated processes (Smeekens 2000; Koch 2004). Increasing sucrose concentration in *Vicia* and pea cotyledons, as well as the barley endosperm at the onset of maturation, is mediated by sucrose transporter activity within newly established epidermal/endospermal transfer cells (Weber et al. 1997b; Tegeder et al. 1999; Weschke et al. 2000). It marks the switch from maternal to filial control of seed growth. In *Vicia* and barley, sucrose induces storage-associated gene expression and at the transcriptional level up-regulates enzymes like sucrose synthase and ADP-glucose pyrophosphorylase (AGP; Heim et al. 1993; Weber et al. 1998b; Weschke et al. 2000). In vitro sucrose feeding disrupts the meristematic state, induces cell expansion and endopolyploidization (Weber et al. 1996) and promotes cotyledonary storage activity at the transcript level (Ambrose et al. 1987; Corke et al. 1990). In transgenic *Vicia* embryos over-expressing a yeast-derived invertase (Weber et al. 1998b), transcript levels of AGP and sucrose synthase and starch accumulation are correlated with sucrose, but not with hexose concentrations, indicating that seed storage product accumulation is subject to sugar signalling and is dependent on sucrose and a sucrose synthase-mediated pathway. Indirect evidence shows that in *Arabidopsis* sucrose can induce a number of storage-related genes, among them oleosins, 2S and 12S globulins as well as Lec and Fus-like transcription factors (TFs; Tsukagoshi et al. 2007). Thus, sucrose initiates maturation and signals the transition of accumulating tissues into the storage mode.

Sucrose also has metabolic effects which may be different from signalling. The cleavage by sucrose synthase, the major mobilizing enzyme, is readily reversible (Geigenberger and Stitt 1993) and inhibited by free hexoses (Ross and Davies 1992;

Weber et al. 1996). The high *Km* of sucrose synthase for sucrose in *Vicia* and pea seeds (Ross and Davies 1992) indicates that flux through sucrose synthase depends on high sucrose concentrations and removal of cleavage products, and can be down-regulated or even reversed at high hexose concentrations (Heim et al. 1993; Weber et al. 1996). The data indicate that sucrose potentially acts on transcriptional and post-transcriptional/metabolic levels (Huber and Hardin 2004) and is a key player within the regulatory network controlling seed maturation. Trehalose signalling may also be involved in the regulation of sucrose conversion into storage products as part of a network regulating seed maturation (Gomez et al. 2006).

5.4 Synthesis and Deposition of Storage Proteins in Crop Seeds

Seed storage proteins are synthesized and deposited in a highly regulated manner. They can be stored for years before becoming mobilized by degradation and being utilized by the germinating seedling. Seed protein usually contains high amounts of amides and arginine resulting in a high N to C ratio, which is consistent with its nitrogen storage function. So far, no other biological activity or function has been detected (Müntz 1998). However, globulins (e.g. legumins and vicilins), which are present in all investigated seeds, reveal structural domains common with a germin-like ancestor involved in cell responses to desiccation, dehydration and osmotic stress (Shutov et al. 2003).

Seed storage proteins are classified according to their solubility as albumins, globulins, prolamins and glutelins. In dicotyledon seeds, the major storage proteins consist of globulins from which the vicilins and legumins are structurally closely related with similar evolutionary roots (Shutov et al. 2003). *Vicia faba* and pea seeds contain globulin storage proteins, hexameric legumins and trimeric vicilins/convicilins, which account for the majority of seed protein. The remainder consists of albumins, including lectins, lipoxygenases, proteinase inhibitors, late embryogenesis abundant proteins and many other soluble proteins (Casey et al. 1993). In cereals, the majority of prolamins and glutelins belong to one class of related seed proteins which have common evolutionary roots (Shewry and Tatham 1990).

Storage proteins are formed at the cytoplasmic side of the rough endoplasmatic reticulum (ER) from where they are translocated into the storage compartment (Müntz 1989). From biosynthesis to deposition, the polypeptides remain protected against uncontrolled breakdown by their sequestration from the cytoplasm into specialized membrane-bound protein bodies and by structural features, preventing cleavage by proteinases which are simultaneously present in the same compartment (Shutov et al. 2003).

N-terminal signal peptides are responsible for their co-translational transport from the cytoplasmic side of the rough-ER into the endomembrane system. After biosynthesis, the polypeptide precursors are translocated through the ER membrane into the ER lumen and enter the secretory pathway. Removal of the signal peptide is followed by oligomerization and further transfer through the secretory pathway. Before leaving the ER, the precursors can be modified by glycosylation, protein disulfide isomerase-catalysed disulfide bridging and chaperone-aided folding. On their way from the ER to the vacuole, the storage protein precursors pass the Golgi and are sorted from secretory proteins by positive targeting signals. Protein bodies are generated within the endomembrane system of the ER or the vacuoles. The origins of the protein bodies and the mechanisms which determine the pathway of storage protein trafficking and deposition are still incompletely understood (Shewry and Halford 2002). The transport is under cytoplasmic control with gene transcription as a major control level for the differential protein supply to vacuoles (Müntz 2007). The precise sorting mechanisms are not fully understood but physical aggregation within the Golgi appears to be important, leading to the formation of electron-dense aggregates which form the contents of dense vesicles (Kermode and Bewley 1999). After transfer into the protein storage vesicles, the seed storage proteins are further processed in order to get effectively deposited (Müntz 1998).

5.5 Storage Proteins in Cereals

5.5.1 Storage Proteins in the Different Grain Parts

The cereal endosperm typically contains over 70% of starch, but it also comprises significant amounts of protein with an average of about 10 to 12% of dry weight (Shewry and Halford 2002). During seed maturation, cereal endosperm storage proteins accumulate in starchy endosperm, sub-aleuron and aleuron as well as in the embryo, and are mobilized during germination to provide nutrients during seedling establishment. Storage proteins in cereal seeds are dominated by two groups, prolamins and globulins. Prolamins are restricted to grasses. With the exception of oats and rice, the major endosperm storage proteins of all cereal grains are prolamins (Shewry et al. 1995). Based on their amino acid sequence, prolamins are classified into three groups (Miflin et al. 1983), namely the sulphur-rich, sulphur-poor and high molecular weight (HMW) prolamins. In barley, prolamins are called hordeins. They are classified into four groups of polypeptides, so-called B, C, D and γ-hordeins. The sulphur-rich prolamins of barley are represented by B hordein and γ-hordein, the sulphur-poor prolamins comprise C hordeins, whereas D hordeins are related to the HMW subunits of polymeric glutenin of wheat (Halford and Shewry 2007). B and C hordeins account for 70 to 80% of the total prolamin amount, while D and γ-hordeins are only minor components (Kreis and Shewry 1992).

Barley and wheat grains show a considerable variation in the protein content between outer and inner layers of the starchy endosperm. In general, sub-aleuron cells are rich in protein but contain little starch. In barley, the central part of the starchy endosperm was enriched in D hordein and the peripheral starchy endosperm in polymeric B hordeins (Millet et al. 1991). In a similar approach, Shewry et al. (1996) found that the sub-aleuron layer of barley grains was enriched in B and C hordeins, whereas the HMW D hordein accumulated in the sub-aleuron and inner starchy endosperm. Localization of the different tissue types of a developing barley caryopsis is shown in Fig. 5.2.

The major storage protein type present in most dicotyledon seeds is an 11-12S globulin called "legumin". Proteins related to this type form the major storage protein fraction in oats and rice (Casey 1999). A legumin-type protein called triticin is also present in wheat (Singh et al. 1991). Although 11S globulin proteins are not yet characterised in barley seeds, proteins related to the 7S globulins ("vicilins") of legumes and other dicotyledonous plants were found in the embryo and/or aleuron layer of not only barley but also wheat and oats (Burgess and Shewry 1986; Heck et al. 1993). However, the globulins in these tissues have limited impact on grain protein content and quality, especially in small-grained cereals like barley and wheat, where the embryo/aleuron fractions account for only 10% of the grain dry weight (Shewry and Halford 2002).

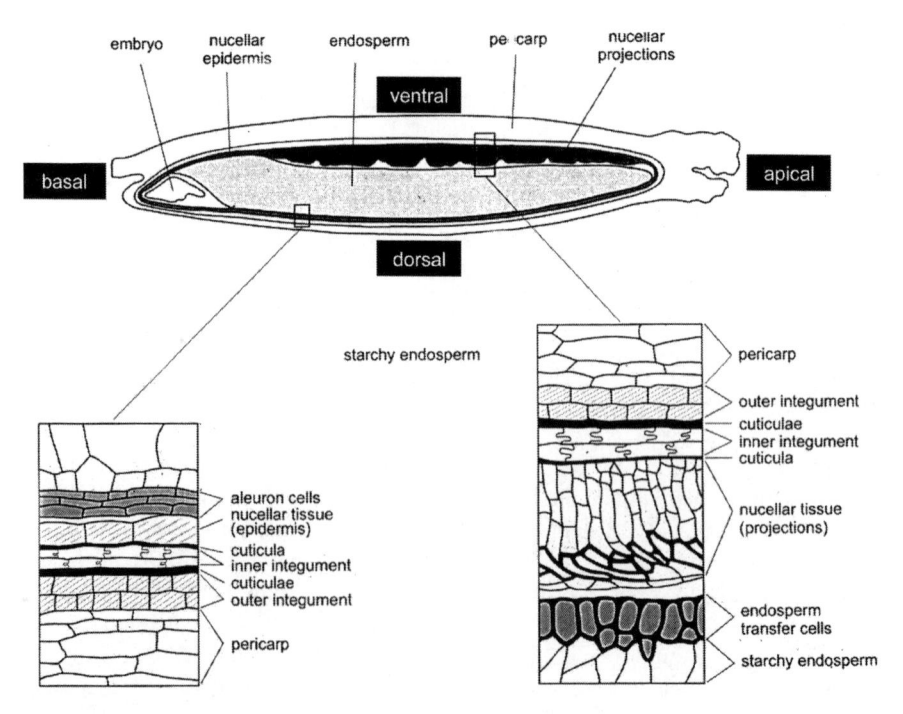

Fig. 5.2 Schematic drawing of a median longitudinal section of a developing barley caryopsis (*above*). The two figures below show details of the cellular organization of the maternal–filial transition at the dorsal (*left*) and ventral caryopsis side (*right*)

5.5.2 Transcriptional Regulation of Arabidopsis Seed Maturation

Transcriptional regulation of *Arabidopsis* seed maturation has been studied extensively and may represent a general model for dicotyledon seeds. A network of master regulators involved includes LEAFY COTYLEDON 1-like (LEC1), belonging to the HAP3 family, and CCAAT-binding factors and the B3 domain proteins LEC2, FUSCA3 (FUS3) and ABSCISIC ACID-INSENSITIVE 3 (ABI3). These transcription factors (TFs) interact directly with regulatory sequences in the promoters of maturation-related genes (reviewed in Vicente-Carbajosa and Carbonero 2005; Gutierrez et al. 2007). LEC1-like proteins act early in embryogenesis and initiate B3 transcription factor expression associated with the transition from embryo morphogenesis to maturation and desiccation tolerance. Seed storage protein gene expression is controlled indirectly by LEC1 through the regulation of ABI3 and FUS3 expression (Kagaya et al. 2005). ABI3 initiates maturation-related, ABA-regulated gene expression and embryo dormancy during mid-to-late embryo development. LEC2, FUS3 and ABI3 activate genes involved in accumulation of storage protein (and lipid reserves). Activation of other downstream genes is mediated by specific binding of the B3 domain to the Sph/RY *cis*-elements. ABI3 mediates ABA-regulated gene expression through interaction with basic Leu zipper TFs which bind ABA response elements (Finkelstein et al. 2002). FUS3/LEC2 represses gibberellin biosynthesis during seed development, indicating that the ratio of ABA to gibberellic acids (GA) is important (Gazzarrini et al. 2004; Gutierrez et al. 2007).

Analyzing *ABI3*, *FUS3* and *LEC2* expression in single, double and triple mutants revealed multiple regulatory links as well as local and redundant control (To et al. 2006). During seed maturation, each regulator activates a set of target genes in an overlapping manner, thereby indirectly regulating seed maturation via different secondary TFs, which activate their own transcriptional programs. For example, *GLABRA2* affects seed oil content (Shen et al. 2006) and *WRINKLED1*, encoding an APETALA2-ethylene-responsive element-binding protein, up-regulates glycolytic gene expression required for the fatty acid biosynthesis (Cernac and Benning 2004). Future studies to discover novel downstream regulators will allow understanding of the regulation of specific metabolic pathways in seeds.

5.5.3 Transcriptional Regulation of Arabidopsis Seed Maturation—a Model Also for Cereal Seeds?

Several aspects of maturation-related gene expression programmes are common in both embryo and endosperm of monocotyledonous and dicotyledonous species. Specific dicotyledon albumins and monocotyledon prolamins can be evolutionarily traced to a common ancestor. Gene promoter *cis*-acting motives and interacting transcription factors are functionally exchangeable, indicating a remarkable conservation in the machinery responsible for maturation. However, comparisons between the transcriptional networks regulating storage protein gene expression in Arabidopsis and cereals are rather scarce.

Functional analysis of promoters of endosperm-specific storage protein genes in barley and other cereals revealed conserved *cis*-motifs. Most important is the tripartite endosperm box with the GCN4 motif, the prolamin box and the AACA/ TA motif (Vicente-Carbajosa and Carbonero 2005). TFs of the opaque type, the Dof- (BPBF) and R2R3Myb-type (GAMYB) respectively bind to these motifs in a synergistic manner. These TFs regulate barley seed storage protein gene expression (Diaz et al. 2005). Results from maize show that *opaque2* encodes a basic leucine zipper (*bZIP*) TF which binds to the promoters of 22-kDa zeins and activates their expression (Schmidt et al. 1990). Although conserved RY-elements are also present in monocotyledon promoters, there was little evidence for a direct regulation by a B3-type TF. The maize *VP1* gene (the monocotyledon orthologue of *ABI3*) is weakly expressed in the starchy endosperm, but highly expressed in aleurone and embryo participating in the ABA-dependent late-embryogenesis-associated (LEA) gene activation. On the contrary, transcripts of the recently identified barley HvFUS-3-gene accumulate in the endosperm and embryo of developing seeds, and reach a peak at the mid-maturation phase (Moreno-Risueno et al. 2008). Using HvFUS-3 for complementation of the loss-of-function mutant *fus3* of Arabidopsis resulted in restored transcription under the control of the At2S3 promoter. Furthermore, in transient expression assays, HvFUS3 is able to transactivate *Hor2* and *ltr1* promoters which contain intact RY boxes. In the yeast two-hybrid system, the HvFUS3 N-terminal region interacts with the TF BLZ2. These results suggest a FUSCA3-mediated mechanism of control of storage protein gene expression which seems to be highly conserved among various angiosperm species (Moreno-Risueno et al. 2008). However, there is no evidence yet for a role of LEC genes in monocotyledon seed maturation.

5.5.4 Unravelling Transcriptional Regulation by Co-Expression Analysis

Transcriptional profiling of *Arabidopsis* and barley seeds (Ruuska et al. 2002; Sreenivasulu et al. 2004, 2006) revealed common patterns of gene expression during development and identified co-regulated TFs which potentially control specific pathways. More than 2,000 TF-genes can be expected in legumes (Udvardi et al. 2007), and about 1,200 TF-genes have been identified from the barley EST collection (Zhang et al. 2004). Of these TFs, 214 were differentially expressed during barley seed development (Sreenivasulu et al. 2006).

Clusters of genes co-expressed with hordeins in the endosperm and globulins in the embryo of barley seeds contain coding sequences associated with related biosynthetic pathways and potential regulatory genes (Fig. 5.3). Eight chromatin

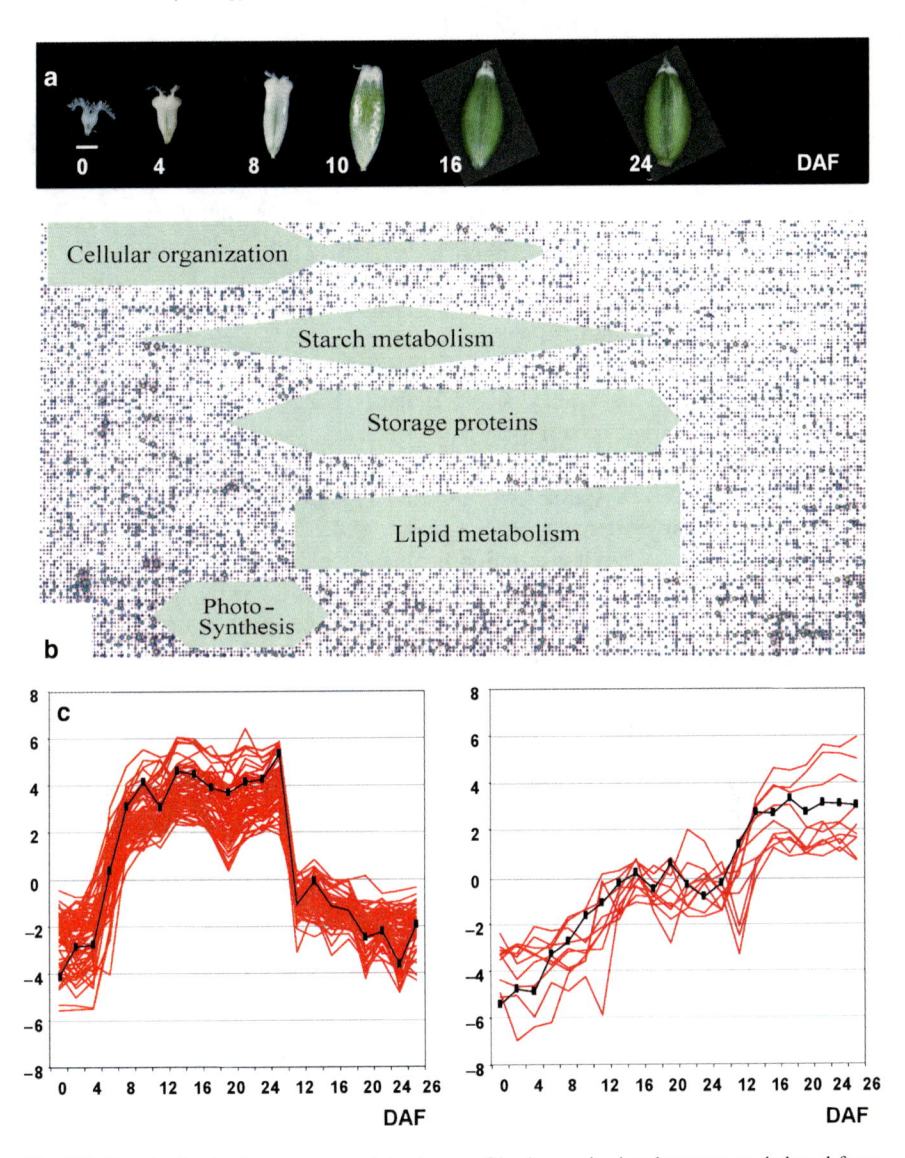

Fig. 5.3 Developing barley caryopses (**a**), phases of barley grain development as deduced from transcript profiling (**b**) and clusters of co-expressed genes in the developing endosperm and embryo of barley (**c**). Regulatory genes and protease inhibitors expressed in the endosperm together with the hordein B1 gene and other globulins are given at the *left-hand side* of **c**. At the *right-hand side*, co-expression of transcriptional regulators and globulin/triacly glycerol genes is shown in the developing embryo. The expression profiles of hordein B1- and 7S globulin genes are given by *black lines*; x-axis, developmental scale (DAF, days after flowering). Gene expression levels (y-axis) are given as $\log_2$ transformed values

remodelling factors are expressed together with hordein genes, protease inhibitors, two DOF, two bHLH, three NAC and 16 unknown transcription factors. It has been shown that chromatin structure interacts with the transcriptional machinery to regulate phaseolin gene expression (Li et al. 2001). Diaz et al. (2005) reported that two DOF transcription factors (SAD and BPBF) activate B1 storage protein genes during the maturation phase. Accordingly, the same two DOF-family members are expressed together with hordein transcripts. Enrichment of prolamin-box *cis* elements was observed in the upstream sequences of rice D, B1 and B3 hordein genes (Sreenivasulu et al. 2006). These *cis*-elements bind TFs of the DOF, bZIP and R2R3MYB classes, including BPBF and SAD proteins (Vicente-Carbajosa et al. 1997, 1998; Mena et al. 1998; Oñate et al. 1999; Diaz et al. 2002, 2005). These results indicate a complex network of regulation necessary for transcriptional activation of these storage protein genes.

In the developing barley embryo, regulators co-expressed with 7S globulins and triacyl glycerol (TAG) biosynthesis genes encode ABF TFs of the bZIP class, a dehydration responsive element binding factor (DREB) from the AP2/EREBP family and two unclassified transcription factors (Sreenivasulu et al. 2006). ABA-responsive ABF as well as ABA-independent DREB TFs are active during embryo maturation. These regulators seem to be involved in the control of globulin and lipid accumulation as well as the expression of late-embryogenesis-associated (LEA) genes.

5.5.5 DNA Methylation and Storage Protein Gene Expression in the Barley Endosperm

Regulation of hordein gene expression is based presumably not only on the action of TF complexes. CpG methylation detected in the promoters of B-hordein and barley prolamin-box binding factor (BPBF) genes during the pre-storage phase may participate synergistically in transcriptional activation. During the pre-storage phase, CpG methylation is accompanied by complete transcriptional suppression of the respective genes. With proceeding endosperm development, decreasing gene expression activities of two methyltransferases (HvMET1 and HvCMT1) correlate with changes of the HpaII digestion pattern and activation of hordein B and BPBF mRNA expression (Radchuk et al. 2005). Furthermore, the *lys-3a* mutant of barley was used to depict existing tight connections between methylation and transcriptional activity of B- and C-hordein, but not of D-hordeins genes. A persistent hypermethylation was found in the promoters of B- and C-type hordeins of the *lys-3a* mutant, resulting in severe reduction of their transcript levels in this mutant (Sørensen 1992; Sørensen et al. 1996). In summary, these findings suggest a coordinated action of transcription factor complexes and methylation/demethylation events for full transcriptional activation of barley hordein genes.

5.6   Metabolic Control of Seed Storage Protein Synthesis

5.6.1   Nitrogen Availability and Signalling

Figure 5.4 shows a scheme of assimilate uptake and primary metabolism during legume seed maturation. Storage protein synthesis is regulated largely by nitrogen metabolites. Storage protein accumulation in pea depends strongly on nitrogen availability in the seed (Golombek et al. 2001; Miranda et al. 2001; Salon et al. 2001). In soybean embryos, asparagine concentrations in high-protein lines are tightly regulated and may act as a metabolic signal of the seed nitrogen status (Hernández-Sebastià et al. 2005). In barley, the endosperm-specific synthesis of hordeins is under nutritional control and is dependent on nitrogen availability. When barley is grown under high nitrogen, hordein genes are transcribed preferentially. The C-hordein promoter exhibits a specific response to amino acids and ammonium. The GCN4 motif ATGA(C/G)TCAT is the dominating *cis*-acting element, but synergistic interaction with other neighbouring endosperm motifs is

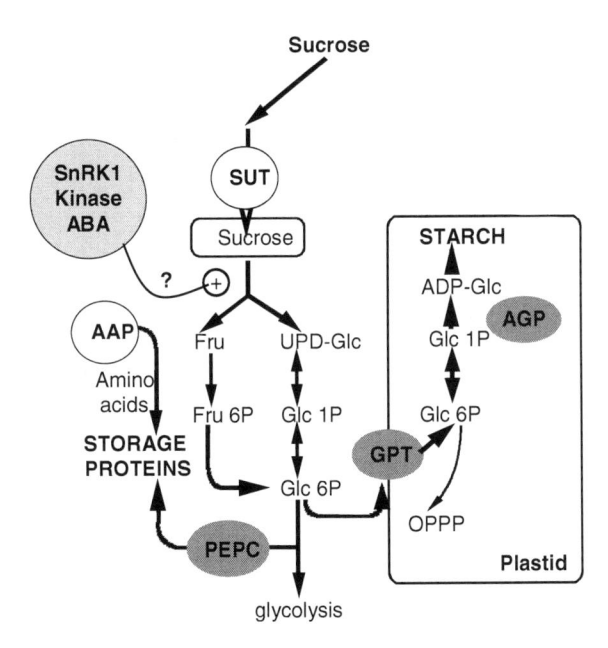

Fig. 5.4 Schematic overview of assimilate uptake and primary storage pathway in maturing legume seeds. Potential targets for the manipulation of metabolic pathways are highlighted by *circles*. Sucrose and amino acids are taken up by transporters for sucrose and amino acids (SUT, AAP). Phospho*enol*caboxylase (PEPC) supply organic acids for amino acid biosynthesis. ADP-glucose pyrophosphorylase (AGP) is a key enzyme of starch production and plastidic glucose-6-P translocator (GPT) import carbon into heterotrophic plastids. Sucrose-non-fermenting-kinase (SnRK1) and abscisic acid (ABA) are involved in the mobilization of sucrose and other aspects of seed maturation (see text for details)

required for a positive regulation by N. Low N levels convert the GCN4 box into a negative motif (Müller and Knudsen 1993). Applying N to maize kernels in vitro increases sucrose synthase, aldolase, phosphoglucomutase, alanine and aspartate aminotransferases and acetolactate synthase activities. These N-induced changes were correlated with enhanced zein-N contents, indicating that yield increase by N fertilization is partly due to altered kernel metabolism in response to N (Singletary et al. 1990).

5.6.1.1 Nitrogen Transport into Seeds

The ability of the embryo to attract and import amino acids can be passive due to high demand, or to active uptake via membrane-localized transporters. An 11S and 7S globulin null mutant of soybean strongly accumulates amino acids in the seeds (Takahashi et al. 2003), suggesting efficient uptake of amino acids independent of globulin biosynthesis. Similarly, high seed-protein content in soybean is determined by the capacity of the embryo to take up nitrogen compounds and to synthesize storage proteins (Hernández-Sebastià et al. 2005).

Glutamine and/or asparagine are translocated through the phloem (Miflin and Lea 1977) and are symplastically unloaded into the seed coat where they are metabolized and reconstructed. Mainly glutamine, alanine and threonine are released from pea seed coats (Lanfermeijer et al. 1992). At maturation, asparagine is also unloaded (Rochat and Boutin 1991). Efflux from pea seed coats is passive and influx into pea embryos is partially passive, especially during early stages (DeJong et al. 1997). Later, a saturable system attributed to H^+-amino acid cotransport becomes important and is inducible by nitrogen starvation, indicating the control by assimilate availability (Bennett and Spanswick 1983). Plant amino acid transporters of the amino acid permease (AAP) subfamily are integral membrane proteins catalysing H^+-coupled amino acid uptake into seeds (Tegeder et al. 2000; Miranda et al. 2001; Okumoto et al. 2002). AtAAP1 in *Arabidopsis* is expressed seed-specifically and may control storage protein synthesis (Hirner et al. 1998). In pea, two AAP-isoforms, PsAAP1 and PsAAP2, are expressed in seed coats, epidermal transfer cells and storage parenchyma cells and in vegetative organs (Tegeder et al. 2000). The *V. faba* AAP1 is expressed during mid-cotyledon development with the beginning of storage protein gene transcription, and is detectable until late cotyledon development (Miranda et al. 2001).

5.6.1.2 Overexpression of an Amino Acid Transporter in Legume Seeds

Two questions arise, namely (1) is active amino acid transport into the seed rate-limiting for storage protein accumulation and (2) what are the physiological and biochemical consequences of increased seed sink strength for nitrogen? To answer these questions, the well-characterised VfAAP1 (Miranda et al. 2001) has been expressed ectopically in maturing embryos of *Vicia narbonensis* and pea (Rolletschek et al. 2005; Weigelt et al. 2008). The transgenic AAP-seeds have been shown to increase the sink strength for amino acids. At maturity, starch is nearly unchanged, but the total nitrogen content increased by 10 to 25%, owing to higher globulins, vicilins and legumins, but not albumins. These results indicate that increased seed nitrogen stimulates globulin synthesis. Thus, legume seed protein synthesis is nitrogen-limited and amino acid transport activity into storage parenchyma cells of the cotyledons may be rate-limiting.

Although it is difficult to predict the relevance of AAP overexpression in terms of agricultural productivity, it has been shown to increase the seed protein content and stimulate growth of vegetative and seed tissues (Götz et al. 2007). Furthermore, the cytokinin (CK) concentrations in the AAP-narbon bean seeds are markedly higher, indicating a possible role of CKs in N-dependent growth stimulation. Pleiotropic effects, such as stimulation of vegetative growth, can mask an increased seed N-accumulation resulting in little changes in grain nitrogen yield. The findings of other studies also show compensation among seed mass, seed number and seed growth rate, as well as compensation by altered seed fill duration and overall plant growth; as a result, there is no benefit in yield or harvest index (reviewed in Sinclair 1998). Therefore, it is necessary to consider possible whole-plant effects whenever improved yield is envisaged, even if some transgenes are expressed specifically in the seed (Götz et al. 2007).

In pea, transgenic lines overexpressing VfAAP1 in seeds exhibited a phenotype similar to those of AAP1-narbon beans (Rolletschek et al. 2005). These lines were tested in field trials at the Leibniz Institute of Plant Genetics and Crop Plant Research in Gatersleben, Germany. The seed nitrogen content was shown to increase by 8%, owing to 15% higher globulins, whereas albumins were 8% lower. Furthermore, the concentrations of sucrose, starch and lipids decreased by 5 to 10%. There was no difference in the seed number per plant but the individual seed weight was lower by about 6% (Weigelt et al. 2008). The results from field-grown plants differ from those from the phyto-chamber, most probably because of different environmental conditions and higher N supply for the chamber-grown plants. Thus, N uptake and utilization in seeds is complex and highly interrelated with the environment as well as conditions at the whole-plant level, and involves many aspects of growth and development (Triboi and Triboi-Blondel 2002; Barneix 2007; Götz et al. 2007). The results further revealed that by using the transgenic approach to increase amino acid uptake into seeds, it is difficult to break the long-known negative correlation between yield and grain protein content. One explanation (Götz et al. 2007) is that high N supply increases cytokinin which stimulates vegetative growth and leads to a lower harvest index and may prevent senescence/proteolysis/remobilization (see Barneix 2007). An alternative explanation is the well-known negative correlation between carbon costs which have to be spent per gram seed and yield (Sinclair 1998). Given that a protein-storing seed must invest more carbon for the synthesis of a gram seed than a starch-storing seed, it follows that seed protein production is energetically more expensive than that of starch (Vertregt and de Vries 1987). Such a negative correlation between yield (seed weight) and protein content is a well-known phenomenon in practice (Sinclair 1998).

5.6.2   Carbon Availability

5.6.2.1   Response to Increased Nitrogen to Carbon Status

Because legume seeds mainly take up amides, additional carbon is required from glycolysis and the tricarbonic-acid cycle to fuel adequate amino acid synthesis. Ammonia-nitrogen released by de-amination of amides is used to build up other amino acids. The carbon key acceptors are oxalacetate and α-ketoglutarate, for the synthesis of the aspartate and glutamate family members respectively. Increased uptake of amino acids by AAP-overexpression may increase carbon demand and potentially leads to source limitation of carbon. This is in accordance with lower starch and sucrose concentrations in mature AAP1-pea seeds and is further supported by transcriptional profiling, where amino acid synthesis is only moderately stimulated compared to the phospho*enol*pyruvate carboxylase (PEPC)-overexpressing seeds (Radchuk et al. 2007), or to AGP-repressed seeds. However, transcriptional changes give evidence for the hypothesis that increasing the N to C ratio, due to AAP1-overexpression, leads to sugar limitation and, subsequently, to compensating effects to improve the carbon state and to reconstitute the N to C ratio. Starch breakdown (starch phosphorylase, glucan-water-dikinase) and amino acid degradation is initiated in mitochondria (γ-amino-butyric-acid (GABA)-shunt enzymes, isovaleryl-CoA-dehydrogenase; Weigelt et al. 2008). Thus, the carbon to nitrogen ratio of the seed is tightly regulated and any deviation caused, for example, by increased amino acid uptake gives rise to compensation. The expression studies also demonstrate the involvement of mitochondrial metabolism to adjust C to N ratio and amino acid metabolism in maturing pea seeds. The observed lower seed weight of the field-grown AAP1-seeds can be explained by carbon limitation of the source.

5.6.2.2   Overexpression of Phospho*enol*pyruvate Carboxylase in Narbon Beans

Carbon acceptors for seed-specific amino acid biosynthesis are supplied via the anaplerotic reaction of phospho*enol*pyruvate carboxylase (PEPC) which, in legume seeds, represents the major CO_2-fixing activity. The enzyme re-fixes HCO_3^- from respiration in a reaction catalysing the conversion of PEP to oxalacetate. PEPC is highly regulated with allosteric properties subject to antagonistic effects of metabolites such as malate and glucose-6-P. Phosphorylation modulates activity with respect to feedback inhibition by malate (Golombek et al. 1999). Correlative evidence points to a rate-limiting role of PEPC in *V. faba* and soybean storage protein synthesis (Flinn 1985; Smith et al. 1989; Sugimoto et al. 1989; Golombek et al. 2001).

Because of strong post-translational regulation, transgenic narbon beans were generated using a PEPC-isoform from *Corynebacterium*. Transgenic seeds show a higher ^{14}C-CO_2 uptake and an incorporation of labelled carbon into proteins. The changed metabolite profiles of maturing cotyledons indicate a shift of metabolic fluxes from sugars/starch into organic and free amino acids, consistent with an increased carbon flow through the anaplerotic pathway. Transgenic seeds accumulate up to 20% more protein per gram (Rolletschek et al. 2004) and partitioning reveals higher uptake of carbon and nitrogen. These lines represent models with increased sink strength and improved nutrient status. Transcriptional profiling of embryos demonstrates a general up-regulation of seed metabolism in terms of protein storage and processing, amino acid metabolism, primary metabolism and transport, and mitochondrial activity (Radchuk et al. 2007). Apparently, activated organic acid production leads to a wide-range activation of nitrogen metabolism, including the machinery of storage protein synthesis, amino acid synthesis, protein processing and deposition, translational activity and methylation cycle.

α-Keto-glutarate and/or oxalacetate may provide signals for the coordinate up-regulation of amino acid biosynthesis in seeds, mechanisms which are well described for bacteria (Galvez et al. 1999; Ninfa and Jiang 2005), whereas evidence in seeds is lower. Transcriptional analysis further revealed activation of stress tolerance genes, indicating partial overlap between nutrient, stress and ABA signals.

5.6.2.3 Repression of ADP-Glucose Pyrophosphorylase in Narbon Beans

Specific down-regulation of one storage product causes a compensatory change of another. Transgenic *V. narbonensis* and pea seeds with repressed ADP-glucose pyrophosphorylase (AGP; Weber et al. 2000; Rolletschek et al. 2002; Weigelt et al. 2009) have moderately lower starch contents but, during maturation, accumulate more sugars which lead to high water uptake and to a wrinkled mature seed phenotype. Total seed nitrogen, globulins and albumins as well as lipid contents are higher. The globulin fractions of storage proteins have a lower legumin to vicilin ratio. This apparently reflects greater sensitivity of legumin synthesis towards high sucrose concentrations, already observed in wrinkled pea mutants (Turner et al. 1990). The higher seed protein content may be due to increased carbon availability. However, other pleiotropic effects are possible. Enhanced levels of soluble compounds in the seed can cause a compensatory water influx, which represents a driving force for increased amino acid uptake. This itself may partly stimulate storage protein biosynthesis, indicating that amino acid uptake into seeds may be dependent on water flow. Elevated protein concentrations could also be explained by higher cytoplasmic volume, which increases the compartment in which protein synthesis takes place and provides more space for protein body accumulation (Weber et al. 2000). This is in accordance with a proportionally larger increase of albumins.

5.6.2.4 Repression of ADP-Glucose Pyrophosphorylase in Pea

Pea seeds with repressed APG show a similar phenotype in terms of reduction of starch, stimulation of protein storage, accumulation of soluble sugars and wrinkled seed phenotype. Transcriptional profiling on these pea seeds (Weigelt et al. 2009) shows that transcriptional changes are in accordance with the hypothesis that increased sugar concentrations stimulate glycolysis and tri-carbonic-acid-cycle (TCA) pathways, obviously leading to additional supply of carbon precursors for amino acid biosynthesis. Genes related to storage protein synthesis (mainly vicilins), as well as amino acid biosynthesis in plastids and mitochondria were transcriptionally up-regulated, reflected by increased concentrations of free amino acids and storage protein. Similar effects have been observed in narbon seeds with increased anaplerotic carbon flux, which also stimulated amino acid biosynthesis (Rolletschek et al. 2005; Radchuk et al. 2007). Therefore, under normal conditions, amino acid synthesis in legume seeds is potentially carbon-limited. Deregulation of starch biosynthesis in the AGP-repressed pea seeds leads to other indirect changes at the gene expression level coming from sugar accumulation, increased mitochondrial metabolism or osmotic stress (severely wrinkled phenotype)-like senescence-related processes, reactive oxygen and oxylipin production (Weigelt et al. 2009). This is in accordance with deregulated genes associated to biotic and abiotic stress. Decreasing carbon flux through the starch pathway can stimulate storage protein synthesis or raise pleiotropic, mainly stress-related effects coming from increased sugar accumulation and metabolism in AGP-repressed seeds.

5.7 Outlook

Seed storage proteins represent an important source for feed and food and improving protein content and seed composition is therefore a highly desirable aim. The synthesis of seed storage proteins is tightly connected with seed development. As many failed metabolic engineering strategies have demonstrated, any successful improvement is likely to require profound understanding of seed metabolism and its regulation. In the future, we need to further analyze the regulatory networks of metabolic and hormonal signals in wild-type seeds and altered models as they relate to the genetic program and the external signals coordinating seed storage protein production. Especially for ABA metabolism, it is necessary to analyze regulation of key enzymes and to study mechanisms involved in nitrogen-dependent signalling. In future, the discovery of novel downstream regulators will allow understanding of the regulation of specific metabolic pathways involved in seed storage synthesis. An important task is to identify those transcription factors which regulate specific pathways for amino acid synthesis and storage product synthesis (Uauy et al. 2006). Much useful information can be gathered from seed metabolic models with altered seed metabolism. In order to obtain a more comprehensive view with respect to seed storage proteins and yield improvement, it is important to study seed phenotypes also at the whole-plant level, rather than to concentrate only on seeds as isolated systems (Götz et al. 2007). Metabolic changes are often complex and affect development and cell structures. Thus, increased efforts are required to understand the intimate association of metabolism with development, at the seed but also at the whole-plant level. It is also necessary to test these models under different environmental conditions—for example, various stresses.

In order to better understand the complexity of phenotypes and metabolic changes, it is necessary to refine our analytical methods to achieve the most comprehensive picture. Seed metabolism, as well as storage protein synthesis, involves different compartments, plastids, mitochondria and cytosol. Therefore, a precise subcellular analysis of seed biochemistry—for example, by non-aqueous fractionation—will be of great assistance.

Acknowledgements The authors are grateful to all the colleagues from IPK Gatersleben who contributed to the results presented here. Thanks to Ulrich Wobus for critically reading the manuscript and for discussions and continuous support. The help of Ursula Tiemann and Karin Lipfert for figure artwork is gratefully acknowledged. This work was supported by the European Union (Integrated project GRAIN LEGUMES), the Bundesministerium für Bildung und Forschung (BMBF, Innoplanta; GABI, The German Plant Genome Research Program), the Deutsche Forschungsgemeinschaft (DFG) and IPK-funds.

References

Ambrose MJ, Wang TL, Cook SK, Hedley CL (1987) An analysis of seed development in *Pisum sativum* L. IV. Cotyledon cell population in vitro and in vivo. J Exp Bot 38:1909–1920

Barneix AJ (2007) Physiology and biochemistry of source-regulated protein accumulation in the wheat grain. J Plant Physiol 164:581–590

Bennett AB, Spanswick RM (1983) Derepression of amino acid-H$^+$ cotransport in developing soybean embryos. Plant Physiol 72:781–786

Burgess SR, Shewry PR (1986) Identification of homologous globulins from embryos of wheat, barley, rye and oats. J Exp Bot 37:1863–1871

Casey R (1999) Distribution of some properties of globulins. In Shewry PR, Casey R (eds) Seed proteins. Kluwer, Dordrecht, pp 4159–4169

Casey R, Domoney C, Smith AM (1993) Biochemistry and molecular biology of seed proteins. In: Casey R, Davies RD (eds) Peas: genetics, molecular biology and biotechnology. CAB International, Wallingford, pp 121–163

Cernac A, Benning C (2004) *WRINKLED1* encodes an AP2/EREB domain protein involved in the control of storage compound biosynthesis in *Arabidopsis*. Plant J 40:575–585

Corke FMK, Hedley CL, Wang TL (1990) An analysis of seed development in *Pisum sativum* XI. Cellular development and the position of storage protein in immature embryos grown *in vivo* and *in vitro*. Protoplasma 155:127–135

DeJong A, Koerselman-Kooij JW, Schuurmans JAMJ, Borstlap AC (1997) The mechanism of amino acid efflux from seed coats of developing seeds as revealed by uptake experiments. Plant Physiol 114:731–736

Diaz I, Vicente-Carbajosa J, Abraham Z, Martínez M, Isabel-LaMoneda I, Carbonero P (2002) The GAMYB protein from barley interacts with the DOF transcription factor BPBF and activates endosperm-specific genes during seed development. Plant J 29:453–464

Diaz I, Martinez M, Isabel-LaMoneda I, Rubio-Somoza I, Carbonero P (2005) The DOF protein, SAD, interacts with GAMYB in plant nuclei and activates transcription of endosperm-specific genes during barley seed development. Plant J 42:652–662

Finkelstein RR, Gampala SS, Rock CD (2002) Abscisic acid signalling in seeds and seedlings. Plant Cell suppl 14:S15–S45

Flinn AM (1985) Carbon dioxide fixation in developing seeds. In: Hebblethwaite PD, Heath MC, Dawkins TCK (eds) The pea crop: a basis for improvement. Butterworths, London, pp 349–358

Galvez S, Lancien M, Hodges M (1999) Are isocitrate dehydrogenases and 2-oxoglutarate involved in the regulation of glutamate synthesis? Trends Plant Sci 4:484–490

Gazzarrini S, Tsuchiya Y, Lumba S, Okamoto M, McCourt P (2004) The transcription factor FUSCA3 controls developmental timing in *Arabidopsis* through the hormones gibberellin and abscisic acid abscisic acid. Dev Cell 7:373–385

Geigenberger P, Stitt M (1993) Sucrose synthase catalyses a readily reversible reaction in vivo in developing potato tubers and other plant tissues. Planta 189:329–339

Golombek S, Heim U, Horstmann C, Wobus U, Weber H (1999) PEP-carboxylase in developing seeds of *Vicia faba*. Gene expression and metabolic regulation. Planta 208:66–72

Golombek S, Rolletschek H, Wobus U, Weber H (2001) Control of storage protein accumulation during legume seed development. J Plant Physiol 158:457–464

Gomez LD, Baud S, Gilday A, Li Y, Graham IA (2006) Delayed embryo development in the *Arabidopsis* trehalose-6-phosphate synthase 1 mutant is associated with altered cell wall structure, decreased cell division and starch accumulation. Plant J 46:69–84

Götz KP, Staroske N, Radchuk R, Emery RJN, Wutzke KD, Herzog H, Weber H (2007) Uptake and allocation of carbon and nitrogen in *Vicia narbonensis* plants with increased seed sink strength achieved by seed-specific expression of an amino acid permease. J Exp Bot 58:3183–3195

Gutierrez L, Van Wuytswinkel O, Castelain M, Bellini C (2007) Combined networks regulating seed maturation. Trends Plant Sci 12:294–300

Halford NG, Shewry PR (2007) The structure and expression of cereal protein genes. In: Olsen OA (ed) Plant Cell Monographs, vol 8. Endosperm. Developmental and Molecular Biology. Springer, Berlin Heidelberg, pp 195–218

Heck GR, Chamberlain AK, Ho DTH (1993) Barley embryo globulin 1 gene, *Beg1*: characterization of cDNA chromosome mapping and regulation of expression. Mol Gen Genet 239: 209–218

Heim U, Weber H, Bäumlein H, Wobus U (1993) A sucrose-synthase gene of *V. faba* L.: Expression pattern in developing seeds in relation to starch synthesis and metabolic regulation. Planta 191:394–401

Hernández-Sebastià CF, Marsolais C, Saravitz D, Israel RE, Dewey S, Huber SC (2005) Free amino acid profiles suggest a possible role for asparagine in the control of storage-product accumulation in developing seeds of low- and high-protein soybean lines. J Exp Bot 56:1951–1963

Hirner B, Fischer WN, Rentsch D, Kwart M, Frommer WB (1998) Developmental control of H^+/ amino acid permease gene expression during seed development of *Arabidopsis*. Plant J 14:535–544

Huber SC, Hardin SC (2004) Numerous posttranslational modifications provide opportunities for the intricate regulation of metabolic enzymes at multiple levels. Curr Opin Plant Biol 7: 318–322

Kagaya Y, Toyoshima R, Okuda R, Usui H, Yamamoto A, Hattori T (2005) LEAFY COTYLEDON1 controls seed storage protein genes through its regulation of FUSCA3 and ABSCISIC ACID INSENSITIVE3. Plant Cell Physiol 46:399–406

Kermode AR, Bewley JD (1999) Synthesis, processing and deposition of seed proteins: the pathway of protein synthesis and deposition of the cell. In: Shewry PR, Casey R (eds) Seed proteins. Kluwer, Dordrecht, pp 807–841

Koch K (2004) Sucrose metabolism: regulatory mechanisms and pivotal roles in sugar sensing and plant development. Curr Opin Plant Biol 7:235–246

Koornneef M, Hanhart CJ, Hilhorst HW, Karssen CM (1989) In vivo inhibition of seed development and reserve protein accumulation in recombinants of abscisic acid biosynthesis and responsiveness mutants in *Arabidopsis thaliana*. Plant Physiol 90:463–469

Kreis M, Shewry PR (1992) The control of protein synthesis in developing barley seeds. In: Shewry PR (ed) Barley: genetics, biochemistry, molecular biology and biotechnology. Alden Press, Oxford, pp 319–333

Lanfermeijer FC, van Oene MA, Borstlap AC (1992) Compartmental analysis of amino-acid release from attached and detached pea seed coats. Planta 187:75–82

Li G, Chandrasekharan MB, Wolffe AP, Hall TC (2001) Chromatin structure and phaseolin gene regulation. Plant Mol Biol 46:121–129

Mena M, Vicente-Carbajosa J, Schmidt RJ, Carbonero P (1998) An endosperm-specific DOF protein from barley, highly conserved in wheat, binds to and activates transcription from the prolamine-box of a native B-hordein promoter in barley endosperm. Plant J 16:53–62

Miflin BJ, Lea PJ (1977) Amino acid metabolism. Annu Rev Plant Physiol 28:99–329

Miflin BJ, Field JM, Shewry PR (1983) Cereal storage proteins and their effects on technological properties. In: Daussant J, Mosse J, Vaughan J (eds) Seed proteins. Academic Press, London, pp 255–319

Millet M-O, Montembauld A, Autran JC (1991) Hordein compositional differences in various anatomical regions of the kernel between two different barley types. Sci Aliment 11:155–161

Miranda M, Borisjuk L, Tewes A, Heim U, Sauer N, Weber H (2001) Amino acid permeases in developing seeds of *Vicia faba* L.: expression precedes storage protein synthesis and is regulated by amino acid supply. Plant J 28:61–72

Moreno-Risueno MA, González N, Díaz I, Parcy F, Carbonero P, Vicente-Carbajosa J (2008) FUSCA3 from barley unveils a common transcriptional regulation of seed-specific genes between cereals and *Arabidopsis*. Plant J 53:882–894

Müller M, Knudsen S (1993) The nitrogen response of a barley C-hordein promoter is controlled by positive and negative regulation of the GCN4 and endosperm box. Plant J 4:343–355

Müntz K (1989) Intracellular protein sorting and the formation of protein reserves in storage tissue cells of plant seeds. Biochem Physiol Pflanzen 185:315–335

Müntz K (1998) Deposition of storage proteins. Plant Mol Biol 38:77–99

Müntz K (2007) Protein dynamics and proteolysis in plant vacuoles. J Exp Bot 58:2391–2407

Ninfa AJ, Jiang P (2005) PII signal transduction proteins: sensors of alpha-ketoglutarate that regulate nitrogen metabolism. Curr Opin Microbiol 8:168–173

Okumoto S, Schmidt R, Tegeder M, Fischer WN, Rentsch D, Frommer WB, Koch W (2002) High affinity amino acid transporters specifically expressed in xylem parenchyma and developing seeds of *Arabidopsis*. J Biol Chem 277:45338–45346

Oñate L, Vicente-Carbajosa J, Lara P, Diaz I, Carbonero P (1999) Barley BLZ2: a seed-specific bZIP protein that interacts with BLZ1 in vivo and activates transcription from the GCN4-like motif of B-hordein promoters in barley endosperm. J Biol Chem 274:9175–9182

Phillips J, Artsaenko O, Fiedler U, Horstmann C, Mock HP, Müntz K, Conrad U (1997) Seed-specific immunomodulation of abscisic acidabscisic acid activity induces a developmental switch. EMBO J 16:4489–4496

Radchuk VV, Sreenivasulu N, Radchuk RI, Wobus U, Weschke W (2005) The methylation cycle and its possible functions in barley endosperm development. Plant Mol Biol 59:289–307

Radchuk R, Radchuk V, Götz KP, Weichert H, Richter A, Emery RJN, Weschke W, Weber H (2007) Ectopic expression of PEP carboxylase in *Vicia narbonensis* seeds: effects of improved nutrient status on seed maturation and transcriptional regulatory networks. Plant J 51:819–839

Rochat C, Boutin JP (1991) Metabolism of phloem-borne amino acids in maternal tissues of fruit of nodulated or nitrate-fed pea plants. J Exp Bot 42:207–214

Rolletschek H, Hajirezaei M, Wobus U, Weber H (2002) Antisense-inhibition of ADP-glucose pyrophosphorylase in *Vicia narbonensis* seeds increases soluble sugars, causes higher uptake of water and amino acids which leads to higher protein content. Planta 214:954–964

Rolletschek H, Borisjuk L, Radchuk R, Miranda M, Heim U, Weber H (2004) Seed-specific expression of a bacterial phosphoenolpyruvate carboxylase in *Vicia narbonensis* increases protein content and improves carbon economy. Plant Biotechnol J 2:211–220

Rolletschek H, Hosein F, Miranda M, Heim U, Götz KP, Schlereth A, Borisjuk L, Saalbach I, Wobus U, Weber H (2005) Ectopic expression of an amino acid transporter (VfAAP1) in seeds of *Vicia narbonensis* and *Pisum sativum* increases storage proteins. Plant Physiol 137:1236–1249

Ross HA, Davies HV (1992) Purification and characterization of sucrose synthase from the cotyledons of *Vicia faba* L. Plant Physiol 100:1008–1013

Ruuska SA, Girke T, Benning C, Ohlrogge JB (2002) Contrapuntal networks of gene expression during *Arabidopsis* seed filling. Plant Cell 14:1191–1206

Salon C, Munier-Jolain NG, Duc G, Voisin AS, Grandgirard D (2001) Grain legume seed filling in relation to nitrogen acquisition: a review and prospects with particular reference to pea. Agronomie 21:539–552

Schmidt RJ, Burr FA, Aukerman MJ, Burr B (1990) Maize regulatory gene *opaque-2* encodes a protein with a "leucine-zipper" motif that binds to zein DNA. Proc Natl Acad Sci USA 87:46–50

Shen B, Sinkevicius KW, Selinger DA, Tarczynski MC (2006) The homeobox gene GLABRA2 affects seed oil content in *Arabidopsis*. Plant Mol Biol 60:377–387

Shewry PR, Halford NG (2002) Cereal seed storage proteins: structures, properties and role in grain utilization. J Exp Bot 53:947–958

Shewry PR, Tatham AS (1990) The prolamin storage proteins of cereal seeds: structure and evolution. Biochem J 267:1–12

Shewry PR, Napier JA, Tatham AS (1995) Seed storage proteins: structure and biosynthesis. Plant Cell 7:945–956

Shewry PR, Brennan C, Thatam AS, Warburton T, Fido R, Smith D, Griggs D, Cantrell I, Harris N (1996) The development, structure and composition of the barley grain in relation to its end use properties. In: Cereals 96, Proc 46th Australian Cereal Chemistry Conf, Sydney, pp 158–162

Shutov AD, Bäumlein H, Blattner FR, Müntz K (2003) Storage and mobilization as antagonistic functional constraints on seed storage globulin evolution. J Exp Bot 54:1645–1654

Sinclair TR (1998) Historical changes in harvest index and crop nitrogen accumulation. Crop Sci 38:638–643

Singh NK, Shepherd KW, Langridge P, Green LC (1991) Purification and biochemical characterization of triticin, a legume-like protein in the wheat endosperm. J Cereal Sci 3:207–219

Singletary GW, Doehlert DC, Wilson CM, Muhitch MJ, Below FE (1990) Response of enzymes and storage proteins of maize endosperm to nitrogen supply. Plant Physiol 94:858–864

Smeekens S (2000) Sugar-induced signal transduction in plants. Annu Rev Plant Physiol Plant Mol Biol 51:49–81

Smith AJ, Rinne RW, Seif RD (1989) PEP carboxylase and pyruvate kinase involvement in protein and oil biosynthesis during soybean seed development. Crop Sci 29:349–353

Sørensen MB (1992) Methylation of B-hordein genes in barley endosperm is inversely correlated with gene activity and affected by the regulatory gene *Lys3*. Proc Natl Acad Sci USA 89:4119–4123

Sørensen MB, Muller M, Skerritt J, Simpson D (1996) Hordein promoter methylation and transcriptional activity in wild-type and mutant barley endosperm. Mol Gen Genet 250:750–760

Sreenivasulu N, Altschmied L, Radchuk V, Gubatz S, Wobus U, Weschke W (2004) Transcript profiles and deduced changes of metabolic pathways in maternal and filial tissues of developing barley grains. Plant J 37:539–553

Sreenivasulu N, Radchuk V, Strickert M, Miersch O, Weschke W, Wobus U (2006) Gene expression patterns reveal tissue-specific signalling networks controlling programmed cell death and ABA-regulated maturation in developing barley seeds. Plant J 47:310–327

Sugimoto T, Tanaka K, Monma M, Kawamura Y, Saio K (1989) Phosphoenolpyruvate carboxylase level in soybean seed highly correlates to its contents of protein and lipid. Agric Biol Chem 53:885–887

Takahashi M, Uematsu Y Kashiwaba K, Yagasaki K, Hajika M, Matsunaga R, Komatsu K, Ishimoto M (2003) Accumulation of high levels of free amino acids in soybean seeds through integration of mutations conferring seed protein deficiency. Planta 217:577–586

Tegeder M, Wang XD, Frommer WB, Offler EO, Patrick JW (1999) Sucrose transport into developing seeds of *Pisum sativum*. Plant J 18:151–161

Tegeder M, Offler CE, Frommer WB, Patrick JW (2000) Amino acid transporters are localized to transfer cells of developing pea seeds. Plant Physiol 122:319–326

To A, Valon C, Savino G, Guilleminot J, Devic M, Giraudat J, Parcy F (2006) A network of local and redundant gene regulation governs *Arabidopsis* seed maturation. Plant Cell 18:1642–1651

Triboi E, Triboi-Blondel AM (2002) Productivity and grain or seed composition: a new approach to an old problem. Eur J Agron 16:163–186

Tsukagoshi H, Morikami A, Nakamura K (2007) Two B3 domain transcriptional repressors prevent sugar-inducible expression of seed maturation genes in *Arabidopsis* seedlings. Proc Natl Acad Sci USA 104:2543–2547

Turner SR, Barratt DH, Casey R (1990) The effect of different alleles at the *r* locus on the synthesis of seed storage proteins in *Pisum sativum*. Plant Mol Biol 14:793–803

Uauy C, Distelfeld A, Fahima A, Blechl A, Dubcovsky J (2006) A NAC gene regulating senescence improves grain protein, zinc and iron content in wheat. Science 314:1298–1301

Udvardi MK, Kakar K, Wandrey W, Montanari O, Murray J, Andriankaja A, Zhang JY, Benedito V, Hofer JM, Chueng F, Town CD (2007) Legume transcription factors: global regulators of plant development and response to the environment. Plant Physiol 144:538–549

Vertregt N, de Vries P (1987) A rapid method for determining the efficiency of biosynthesis of plant biomass. J Theor Biol 128:109–119

Vicente-Carbajosa J, Carbonero P (2005) Seed maturation: developing an intrusive phase to accomplish a quiescent state. Int J Dev Biol 49:645–651

Vicente-Carbajosa J, Moose SP, Parsons RL, Schmidt RJ (1997) A maize zinc-finger protein binds the prolamin box in zein gene promoters and interacts with the basic leucine zipper transcriptional activator Opaque2. Proc Natl Acad Sci USA 94:7685–7690

Vicente-Carbajosa J, Oñate L, Lara P, Diaz I, Carbonero P (1998) Barley BLZ1: a bZIP transcriptional activator that interacts with endosperm-specific gene promoters. Plant J 5:629–640

Wang TL, Domoney C, Hedley CL, Casey R, Grusak MA (2003) Can we improve the nutritional quality of legume seeds? Plant Physiol 131:886–891

Weber H, Buchner P, Borisjuk L, Wobus U (1996) Sucrose metabolism during cotyledon development of *Vicia faba* L. is controlled by the concerted action of both sucrose-phosphate synthase and sucrose synthase: expression patterns, metabolic regulation and implications on seed development. Plant J 9:841–850

Weber H, Borisjuk L, Wobus U (1997a) Sugar import and metabolism during seed development. Trends Plant Sci 22:169–174

Weber H, Borisjuk L, Heim U, Sauer N, Wobus U (1997b) A role for sugar transporters during seed development: molecular characterization of a hexose and a sucrose carrier in faba bean seeds. Plant Cell 9:895–908

Weber H, Heim U, Golombek S, Borisjuk L, Wobus U (1998a) Assimilate uptake and the regulation of seed development. Seed Sci Res 8:331–345

Weber H, Heim U, Golombek S, Borisjuk L, Manteuffel R (1998b) Expression of a yeast-derived invertase in developing cotyledons of *Vicia narbonensis* alters the carbohydrate state and affects storage functions. Plant J 16:163–172

Weber H, Golombek S, Heim U, Rolletschek H, Gubatz S, Wobus U (2000) Antisense-inhibition of ADP-glucose pyrophosphorylase in developing seeds of *Vicia narbonensis* moderately decreases starch but increases protein content and affects seed maturation. Plant J 24:33–43

Weber H, Borisjuk L, Wobus U (2005) Molecular physiology of legume seed development. Annu Rev Plant Biol 56:253–279

Weigelt K, Küster H, Radchuk R, Müller M, Weichert H, Fait A, Fernie A, Saalbach I, Weber H (2008) Increasing amino acid supply in pea embryos reveals specific interactions of N and C metabolism and highlights the importance of mitochondrial metabolism. Plant J 55:909–926

Weigelt K, Küster H, Rutten T, Fait A, Fernie AR, Miersch O, Wasternack C, Emery RJN, Desel C, Hosein F, Müller M, Saalbach I, Weber H (2009) ADP-Glucose pyrophosphorylase deficient pea embryos reveal specific transcriptional and metabolic changes of C:N metabolism and stress responses. Plant Physiol 149:395–411

Weschke W, Panitz R, Sauer N, Wang Q, Neubohn B, Weber H, Wobus U (2000) Sucrose transport into barley seeds: molecular characterisation of two transporters and implications for seed development and starch accumulation. Plant J 21:455–467

Wiesner H (2007) Chemistry of gluten proteins. Food Microbiol 24:115–119

Wobus U, Weber H (1999) Seed maturation: genetic programmes and control signals. Curr Opin Plant Biol 2:33–38

Zhang H, Sreenivasulu N, Weschke W, Stein N, Rudd S, Radchuk V, Potokina E, Scholz U, Schweizer P, Zierold U, Langridge P, Varshney RK, Wobus U, Graner A (2004) Large-scale analysis of the barley transcriptome based on expressed sequence tags. Plant J 40:276–290

Chapter 6
Fatty Acid Biosynthesis and Regulation in Plants

R. Rajasekharan and V. Nachiappan

6.1 Introduction

Lipids are essential components of all living cells and obligate components of biological membranes. They serve as energy reserves and second messengers. Lipids include oils and fats that are glycerol esters of fatty acids (triacylglycerols, TAG), and are derived mainly from plant and animal sources. Fatty acids are saturated or unsaturated monobasic carboxylic acids. Occurrence of free fatty acids is minimal in biological systems. However, fatty acids that are biosynthetically linked with glycerol (glycerolipids) and sphingosine (sphingolipids) are known to generate membrane and storage lipids (triacylglycerol). The synthesized membrane and storage lipids are hydrolyzed (lipolysis) to fatty acids that are oxidized to give acetyl-CoA, an important intermediate for energy production (tricarboxylic acid cycle) that serves as a precursor for fatty acid biosynthesis (Fig. 6.1). The structures of these fatty acids vary in the length of acyl chains, number and position of double bonds and the functional groups, such as hydroxy, epoxy and acetylenic groups. Vegetable oils are the major source of edible lipids, accounting for more than 75% of the total fats consumed worldwide (Luehs and Friedt 1993). About 75% of the oil extracted is from the endosperm of oilseeds like soybean, peanut or oilseed rape, whereas the remaining 25% is from the pericarp of oil fruits, mainly oil palm and olive (Salas et al. 2000). The oil content in seeds of

R. Rajasekharan
Department of Biochemistry, Indian Institute of Science, Bangalore 560012, India
School of Science, Monash University Sunway Campus, 46150, Bandar Sunway, Selangor, Malaysia
e-mail: ram.raja@artsci.monash.edu.my

V. Nachiappan
Department of Biochemistry, Bharathidasan University, Tiruchirappalli 620024, India

Fig. 6.1 An overview of fatty acid metabolism. Acetyl-CoA serves as a precursor for de novo fatty acid biosynthesis. Fatty acids are assembled into membrane and storage lipids that can be recycled. They may be rapidly degraded by β-oxidation into acetyl-CoA, providing energy and a precursor for secondary metabolites

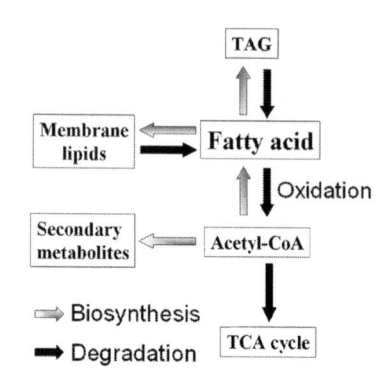

different plant species varies from 4 to 60% of their dry weight. About 95% of the storage lipids in the seeds is present as triacylglycerol. Leaves, roots and other vegetative tissues usually contain less than 10% lipids by dry weight.

6.2   Fatty Acid Biosynthesis

6.2.1   De Novo Fatty Acid Biosynthesis

Fatty acids are synthesized in the stroma of plastids, in both leaves and developing seeds (Weaire and Kekwick 1975; Ohlrogge et al. 1979). The reactions involved in fatty acid biosynthesis in plants are schematically represented in Fig. 6.2. Acetyl-CoA is obtained by the degradation of photosynthate via glycolysis in the cytosol, followed by the action of pyruvate dehydrogenase complex in mitochondria and plastids (Camp and Randall 1985). De novo fatty acid biosynthesis involves two enzymes, acetyl-CoA carboxylase (ACCase; EC 6.4.1.2) and fatty acid synthase (FAS). Acetyl-CoA carboxylase catalyzes the ATP-driven condensation of a molecule of carbon dioxide obtained from bicarbonate with acetyl-CoA to produce malonyl-CoA (Fig. 6.3). This enzyme has been well studied and is an important regulatory enzyme in fatty acid metabolism. Plant acetyl-CoA carboxylases are well characterized and their molecular structures have been studied extensively (Samols et al. 1988; Knowles 1989). Their cytosolic and plastidic isoforms are multifunctional proteins (Ke et al. 2000) of 220 kDa, containing at least three domains, namely biotin carboxyl carrier protein, biotin carboxylase and carboxyl-transferase. In dicotyledons, the cytosolic isoform is a multifunctional enzyme, while the plastid-localized isoform is a multienzyme complex that is similar to those of prokaryotes (Samols et al. 1988) and produces malonyl-CoA for fatty acid elongation and flavonoid production (Sasaki et al. 1993).

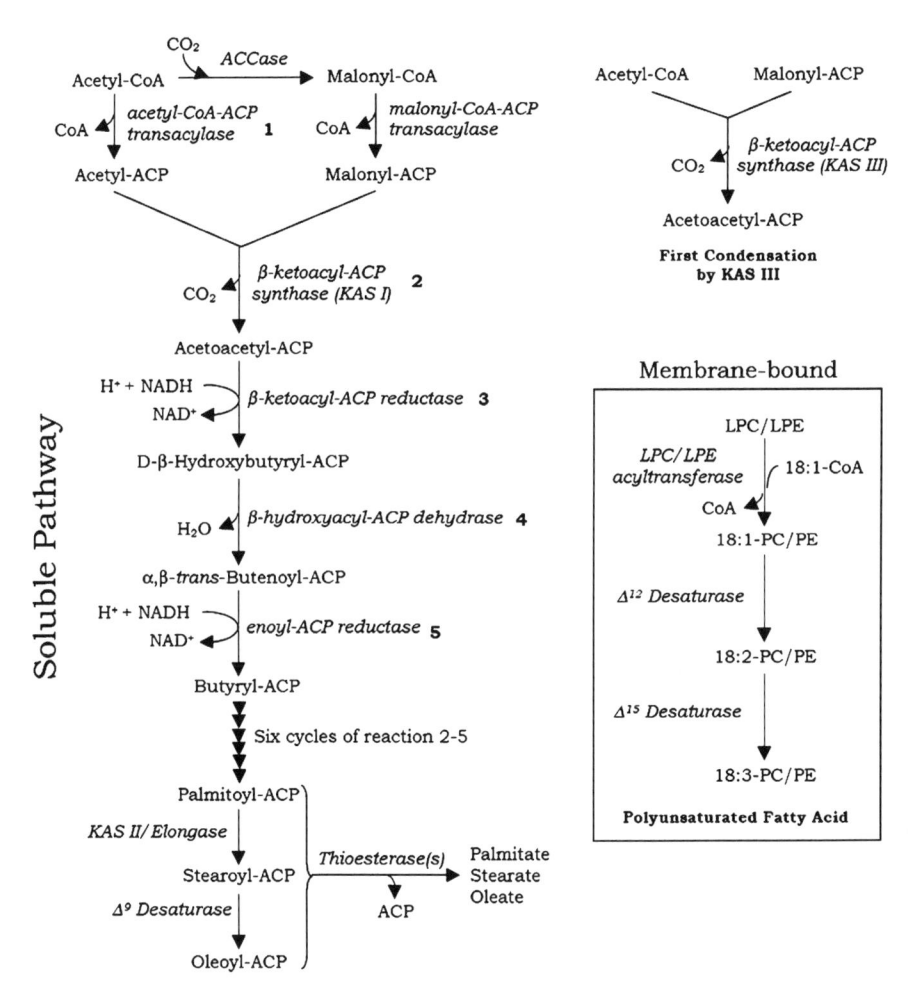

Fig. 6.2 The reactions of fatty acid biosynthesis in plants. Fatty acids are synthesized from acetyl-CoA. In reaction *1*, a malonyl group is derived from acetyl-CoA by ACCase (acetyl-CoA carboxylase) and is transferred to ACP (acyl carrier protein). The first condensation reaction between malonyl-ACP and acetyl-CoA is catalyzed by KAS III (β-ketoacyl-ACP synthase). In reaction *2*, the acetyl group from acetyl-CoA is primed with ACP, and is condensed subsequently with malonyl-ACP by KAS I. This is followed by *3* reduction, *4* dehydration and *5* second reduction. Seven rounds of cyclic reactions (*2–5*) produce palmitoyl-ACP. It is elongated to stearoyl-ACP by KAS II, which is further desaturated to oleoyl-ACP by soluble Δ^9-desaturase. The fatty acids are released from ACP by thioesterases. *Box*: further desaturation of fatty acids using phospholipid backbone is shown at the *right*. Abbreviations: LPC, lysophosphatidylcholine; LPE, lysophosphatidylethanolamine; PC, phosphatidylcholine; PE, phosphatidylethanolamine; 18:1, oleic acid; 18:2, linoleic acid; 18:3, linolenic acid

Regulation of ACCase in dicot plants

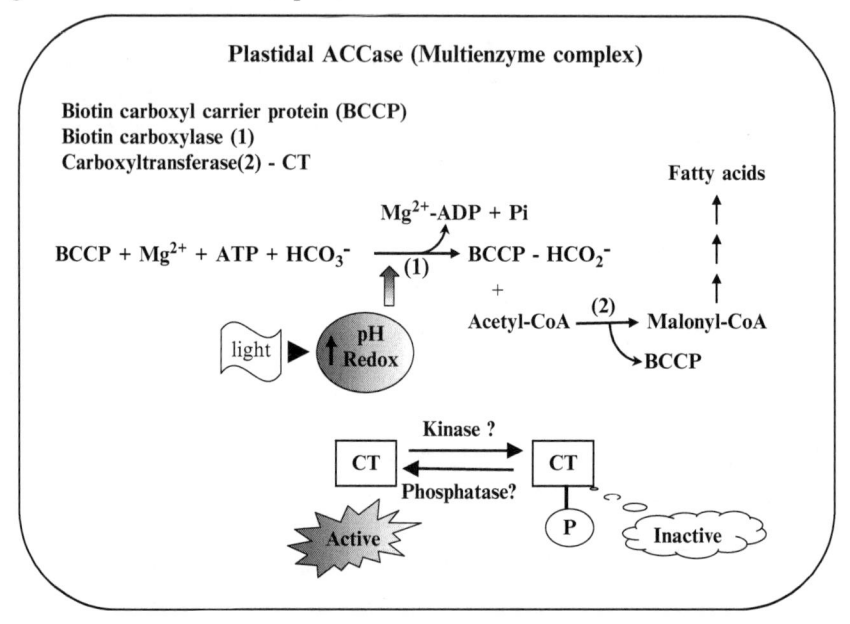

Fig. 6.3 Regulation of plastidial ACCase. Illumination increases pH and redox potential in the plastids, which activates the carboxylase. ACCase is covalently regulated by reversible phosphorylation

6.2.2 Regulation of ACCase

ACCase is covalently and allosterically regulated. Covalent modification is mediated by phosphorylation-dephosphorylation (Savage and Ohlrogge 1999). AMP-dependent protein kinase phosphorylates and switches off the enzyme (Fig. 6.3), whereas an unknown phosphatase activates it. ATP activates while AMP and ADP inhibit the enzyme allosterically. Thus, the carboxylase is inactivated when the energy charge is low (Hunter and Ohlrogge 1998). ACCase is also allosterically stimulated by citrate (Konishi et al. 1996). It can partly reverse the inhibition produced by phosphorylation. The stimulatory effect of citrate on the carboxylase is antagonized by palmitoyl-CoA (Herbert et al. 1996). Illumination increases pH and redox potential in the plastids, thereby activating carboxylase (Fig. 6.3; Kozaki and Sasaki 1999).

The malonyl-CoA produced by acetyl-CoA carboxylase is transferred to a small acidic protein, acyl carrier protein (ACP), by malonyl-CoA:ACP transacylase to form malonyl-ACP and to reach the FAS complex, where palmitic acid is synthesized by a series of cyclic reactions (Fig. 6.2). The fatty acid synthesis starts with the condensation of malonyl-ACP with acetyl-CoA by the action of β-ketoacyl-ACP synthase III (KAS III) to produce acetoacetyl-ACP (Jaworski et al. 1989). It is reduced by β-keto-ACP reductase, an enzyme usually dependent on NADH, to form β-hydroxyacyl-ACP (Harwood 1988). The latter is then dehydrated by β-hydroxyacyl-ACP dehydrase and finally reduced by enoyl-ACP reductase, resulting in a four-carbon butyryl-ACP, which is used for further chain elongation in a cyclic fashion by the serial action of the enzymes mentioned above. Further condensations are catalyzed by KAS I, which uses acyl-ACPs as the primary substrate instead of acetyl-CoA (Shimakata and Stumpf 1983). This enzyme is involved in most of the condensation reactions necessary for the production of acyl-chains. The final product, palmitoyl-ACP, can be elongated to stearoyl-ACP by the condensing enzyme KAS II. This is an important step that determines the ratio of C16/C18 fatty acids in the resulting pool (Stymne and Stobart 1987). This ratio is directly related to the degree of unsaturation. Palmitate (C16:0) is the most commonly found saturated fatty acid, while C18 fatty acids are mostly unsaturated.

The process of fatty acid synthesis ends either in the release of acyl moieties from the ACP derivatives, by the action of plastidic thioesterases, or in the transfer of acyl moieties into lysolipids by acyltransferases. Most of the species have thioesterases showing high substrate specificity for saturated or unsaturated C18 acyl-ACPs (Shimakata and Stumpf 1983; Imai et al. 1992). There are two predominant acyl-ACP thioesterase activities, one specific for palmitoyl-ACP and the other for oleoyl-ACP (Jones et al. 1995). This forms the basis for obtaining oil with altered fatty acid composition from transgenic crops expressing modified acyl-ACP thioesterases (Facciotti et al. 1999; Othman et al. 2000). Palmitate and stearate are the major saturated acyl products of FAS. The release of acyl moieties from ACP derivatives by the action of plastidic acyltransferase is not well characterized. More effort will be required to understand fatty acid termination by acyltransferase(s).

6.3   Fatty Acid Elongation

Malonyl-CoA produced from cytosolic ACCase is used for fatty acid elongation (Baud et al. 2003). Very long chain fatty acids (VLCFAs) are synthesized by the sequential addition of C2 moieties, which are derived from malonyl-CoA, to the pre-existing C18 fatty acids. ACP is used in initial fatty acid synthesis, whereas CoA is involved during fatty acid elongation (FAE). VLCFAs have chain lengths ranging from 20 to 30 carbons. The production of C20:0–C26:0 VLCFAs takes place independent of fatty acid synthase by membrane-bound elongases on the cytosolic face of endoplasmic reticulum. VLCFAs are synthesized by a microsomal FAE system and each cycle requires four enzymatic reactions, namely condensation, reduction, dehydration and further reduction (Kunst et al. 1992).

FAE1 is seed-specific and is involved in the synthesis of eicosenoic (C20:1) and erucic (C22:1) acids. In fatty acid elongation, β-ketoacyl-CoA synthase (KCS) seems to be a rate-limiting step (Ghanevati and Jaworski 2001). Mining of the Arabidopsis genome sequence database revealed 20 genes homologous to seed-specific *FAE1 KCS*. Eight of the 20 putative KCSs were cloned and expressed in yeast. Five of the eight (At1g71160, At1g19440, At1g07720, At5g04530 and At4g34250) had little or no KCS activity with C16 to C20 substrates, while three demonstrated an activity with C16, C18 and C20 saturated acyl-CoA substrates. At1g01120 KCS (KCS1) and At2g26640 KCS showed broad substrate specificities when assayed with saturated and monounsaturated C16 to C24 acyl-CoAs, while At4g34510 KCS was specific for saturated fatty acyl-CoA substrates (James and Dooner 1990; Lemieux et al. 1990). VLCFA are utilized for the production of waxes embedded in suberin and cutin, these being epicuticular waxes that cover the aerial surfaces of plants. They also act as precursors for other aliphatic hydrocarbons found in waxes, including alkanes, primary and secondary alcohols, ketones, aldehydes and acylesters (Post-Beittenmiller 1996).

6.4 Fatty Acid Desaturation

Plants regulate the synthesis of unsaturated fatty acids in response to nutritional status and environmental stresses. This has a profound effect on the fluidity and functionality of biological membranes (Falcone et al. 2004). Most vegetable oils are rich in unsaturated fatty acids such as oleate and linoleate. There is a very active stearoyl-ACP Δ^9-desaturase in the stroma of most plant species that produces oleate as the major product (Haralampidis et al. 1998). This enzyme has been crystallized and its structure and reaction mechanism have been studied extensively (Broun et al. 1998). Further desaturation of oleate to produce polyunsaturated fatty acids (PUFAs) takes place in the plastidial membranes and in the endoplasmic reticulum. The plastid-localized Δ^{12}- and Δ^{15}-desaturases are involved in the production of thylakoid membrane lipids (Ohlrogge et al. 1979), while endoplasmic reticulum-localized desaturases contribute to the production of PUFA-rich oil. The desaturation of oleate (C18:1) in the endoplasmic reticulum involves initial transport of oleate from the plastid into the cytosolic acyl-CoA pool, followed by its incorporation into lysophosphatidylcholine to form phosphatidylcholine (PC). The oleate can be further desaturated to linoleate (C18:2) and then to linolenate (C18:3) by Δ^{12}- and Δ^{15}-desaturases respectively (Fig. 6.2, box).

6.5 Unusual Fatty Acids

Biosynthesis of unusual fatty acids (UFAs) has attracted considerable interest as renewable feed stocks for the chemical, pharmaceutical and nutraceutical industries. The presence of functional groups such as hydroxy, epoxy and acetylenic groups make the fatty acids unusual because their structure is different from the common fatty acids found in membranes. Examples of UFAs include medium-chain fatty acid (lauric acid, 12:0), hydroxylated fatty acid (ricinoleic acid, 12OH-18:1) and fatty acids with the double bond in different positions, such as petroselinic acid, 18:1, Δ^6 and erucic acid, 22:1, ω^9 (Van de Loo et al. 1995). Lauric acid biosynthesis requires a medium-chain specific acyl–acyl carrier protein thioesterase, whereas ricinoleic acid biosynthesis requires a unique 18:1 hydroxylase. Genes encoding these enzymes were isolated from California bay trees (*UcFatB1*; Voelker et al. 1992) and castor bean (*FAH12*; Van de Loo et al. 1995) respectively. Expression of lauroyl-ACP thioesterase under the constitutive cauliflower mosaic virus 35S promoter led to the accumulation of lauric acid in seeds but not in leaves (Eccleston et al. 1996). Castor bean accumulates 90% of 12-hydroxy oleic acid due to the presence of 12-hydroxylase. However, heterologous expression of this gene in other crop plants did not result in high accumulation of UFAs (Bafor et al. 1991). Failure to accumulate UFAs could be due to poor synthesis and inability to incorporate UFAs into TAG, or rapid degradation by β-oxidation (Bafor et al. 1993).

6.6   Assembly of Fatty Acids

The synthesized fatty acids are acylated to the glycerol backbone in the endoplasmic reticulum to produce glycerolipids through a series of acylation and hydrolysis reactions (Kennedy 1961). The first step in this pathway involves acylation of glycerol 3-phosphate to produce lysophosphatidic acid (LPA), catalyzed by glycerol 3-phosphate acyltransferase (G3PAT). The isoforms of G3PAT present in plastids and endoplasmic reticulum show preference towards saturated fatty acids of different chain length (Stymne and Stobart 1987; Frentzen et al. 1987). LPA acyltransferase (LPAAT) converts LPA (acyl-CoA:1-monoacylglycerol 3-phosphate) to phosphatidic acid (PA), a branch-point for anionic phospholipid synthesis. To date, five LPAAT activities have been characterized in the plastids and mitochondria of plant cells (Kim et al. 2005). In general, this enzyme has a strong selectivity for oleoyl-CoA in most plants (Frentzen and Wolter 1998). However, the plastidial LPAAT utilizes palmitoyl-ACP to produce PA with a saturated fatty acid at position *sn-2*, characteristic of lipids synthesized by the prokaryotic pathway. The PA produced by the eukaryotic pathway invariably has C18:1 at position *sn-2* and is either metabolized to phosphatidylglycerol (Sun et al. 1988) or dephosphorylated to diacylglycerol, from which glycodiacylglycerols are synthesized, the latter being characteristic lipids in thylakoid membranes (Douce and Joyard 1990). In contrast to the plastidial LPAAT, the microsomal LPAAT prefers linoleoyl and oleoyl groups compared to saturated acyl groups. Oilseed rape normally does not accumulate trilaurin. However, coexpression of a coconut (*Cocos nucifera*) LPA acyltransferase (preference for 12:0-CoA) with California bay thioesterase resulted in its accumulation (Bourgis et al. 1999; Knutzon et al. 1999). Phosphatidic acid is converted to DAG by PA phosphatase, a probable rate-limiting step in TAG formation. The DAG formed from PA is further converted to TAG or is used in a reversible PC-DAG interconversion to enter membrane lipid biosynthesis. In addition, DAG is also an important branch-point for the production of zwitterionic phosphoglycerides or glycosylglyceride.

In addition to a membrane-bound pathway, TAG can be synthesized by successive esterification of monoacylglycerol (MAG) by cytosolic acyl-CoA:MAG acyltransferase (Tumaney et al. 2001) and DAG acyltransferase (Saha et al. 2006). MAG is formed from LPA by soluble LPA phosphatase (Shekar et al. 2002). The formation of TAG from DAG is catalyzed by acyl-CoA-dependent DAG acyltransferase, which is an integral endoplasmic reticulum protein (Lung and Weselake 2006). In addition to these acyl-CoA-dependent enzymes, microsomal preparations from oilseeds synthesize TAG in the absence of acyl-CoA, using phospholipids as acyl donors and DAG as an acceptor. This reaction is catalyzed by phospholipid: diacylglycerol acyltransferase (PDAT). However, the in vivo function of PDAT in TAG accumulation remains obscure (Dahlqvist et al. 2000).

The acyl composition of lipids is determined by substrate specificity of acyltransferases and the concentration of available acyl donors. Acyltransferases are responsible for the synthesis of both membrane and storage lipids. A wide variety of in vitro studies have indicated that these enzymes show considerable selectivity, both for the activated acyl substrate (acyl-CoA) and the molecule to which they transfer the acyl group (acyl acceptor). The selectivity of the acyltransferases might therefore be a key factor in the channelling of UFAs into storage lipids, and not to the membranes. Furthermore, it has been proposed that membrane lipid and triacylglycerol assembly could be carried out by distinct subsets of enzymes, located in separate domains of the cell (Gangar et al. 2002).

6.7 Conclusions

It is possible to alter the fatty acid composition of any lipid molecule to generate desired lipids for food, feed and oleochemical industries. This becomes increasingly important, especially in the current scenario, where the global demand for plant oils has brought about a compelling need to intensify our efforts to genetically modify oilseeds, mainly to enhance oil yield and to generate novel oils with a specific fatty acid composition. Future research in lipid biochemistry can therefore be expected to align with this global requirement.

References

Bafor M, Smith MA, Jonsson L, Stobart K, Stymne S (1991) Ricinoleic acid biosynthesis and triacylglycerol assembly in microsomal preparations from developing castor-bean endosperms. Biochem J 280:507–514

Bafor M, Smith M, Jonsson L, Stobart K, Stymne S (1993) Biosynthesis of vernoleate (cis-12-epoxyoctadeca-cis-9-enoate) in microsomal preparations from developing endosperm of *Euphorbia lagascae*. Arch Biochem Biophys 303:145–151

Baud S, Guyon V, Kronenberger J, Wuillème S, Miquel M, Caboche M, Lepiniec L, Rochat C (2003) Multifunctional acetyl-CoA carboxylase 1 is essential for very long chain fatty acid elongation and embryo development in Arabidopsis. Plant J 33:75–86

Bourgis F, Kader JC, Barret P, Renard M, Robinson D, Robinson C, Delseny M, Roscoe TJ (1999) A plastidial lysophosphatidic acid acyltransferase from oilseed rape. Plant Physiol 120:913–922

Broun P, Shanklin J, Whittle E, Somerville C (1998) Catalytic plasticity of fatty acid modification enzymes underlying chemical diversity of plant lipids. Science 282:1315–1317

Camp PJ, Randall DD (1985) Purification and characterization of the pea chloroplast pyruvate dehydrogenase complex. A source of acetyl-CoA and NADH for fatty acid biosynthesis. Plant Physiol 77:571–577

Dahlqvist A, Stahl U, Lenman M, Banas A, Lee M, Sandager L, Ronne H, Stymne S (2000) Phospholipid-diacylglycerol acyltransferase: an enzyme that catalyzes the acyl-CoA-independent formation of triacylglycerol in yeast and plants. Proc Natl Acad Sci USA 97:6487–6492

Douce R, Joyard J (1990) Biochemistry and function of the plastid envelope. Annu Rev Cell Biol 6:173–216

Eccleston VS, Cranmer AM, Voelker TA, Ohlrogge JB (1996) Medium-chain fatty acid biosynthesis and utilization in *Brassica napus* plants expressing lauroyl-acyl carrier protein thioesterase. Planta 198:46–53

Facciotti MT, Bertain PB, Yuan L (1999) Improved stearate phenotype in transgenic canola expressing a modified acyl-acyl carrier protein thioesterase. Nature Biotechnol 17:593–597

Falcone DL, Ogas JP, Somerville CR (2004) Regulation of membrane fatty acid composition by temperature in mutants of Arabidopsis with alterations in membrane lipid composition. BMC Plant Biol 17:4–17

Frentzen M, Wolter FP (1998) Molecular Biology of acyltransferases involved in glycerolipids synthesis. In: Harwood JL (ed) Plant lipid biosynthesis. Fundamentals and agricultural applications. Cambridge University Press, Cambridge, pp 247–272

Frentzen M, Nishida I, Murata N (1987) Properties of the plastidal acyl-(acyl-carrier-protein): glycerol-3-phopshate acyltransferase from the chilling-sensitive plant, squash (*Cucurbita moschata*). Plant Cell Physiol 28:1195–1201

Gangar A, Raychaudhuri S, Rajasekharan R (2002) Alteration in the cytosolic triacylglycerol biosynthetic machinery leads to decreased cell growth and triacylglycerol synthesis in oleaginous yeast. Biochem J 365:577–589

Ghanevati M, Jaworski JG (2001) Active-site residues of a plant membrane-bound fatty acid elongase beta-ketoacyl-CoA synthase, FAE1 KCS. Biochim Biophys Acta 1530:77–85

Haralampidis K, Milioni D, Sanchez J, Baltrusch M, Heinz E, Hatzopoulos P (1998) Temporal and transient expression of stearoyl-ACP carrier protein desaturase gene during olive fruit development. J Exp Bot 49:1661–1669

Harwood JL (1988) Fatty acid metabolism. Annu Rev Plant Physiol Plant Mol Biol 39:101–138

Herbert D, Price LJ, Alban C, Dehaye L, Job D, Cole DJ, Pallett KE, Harwood JL (1996) Kinetic studies on two isoforms of acetyl-CoA carboxylase from maize leaves. Biochem J 318:997–1006

Hunter SC, Ohlrogge JB (1998) Regulation of spinach chloroplast acetyl-CoA carboxylase. Arch Biochem Biophys 359:170–178

Imai H, Nishida I, Murata N (1992) Acyl-(acyl-carrier protein) hydrolase from squash cotyledons specific to long-chain fatty acids: purification and characterization. Plant Mol Biol 20:199–206

James DW Jr, Dooner HK (1990) Isolation of EMS-induced mutants in Arabidopsis altered in seed fatty acid composition. Theor Appl Genet 80:241–245

Jaworski JG, Clough RC, Barnum SR (1989) A cerulenin insensitive short chain 3-ketoacyl acyl carrier protein synthase in *Spinacia oleracea*. Plant Physiol 90:41–44

Jones A, Davies HM, Voelker TA (1995) Palmitoyl-acyl carrier protein (ACP) thioesterase and the evolutionary origin of plant acyl-ACP thioesterases. Plant Cell 7:359–371

Ke J, Wen T-N, Nikolau BJ, Wurtele ES (2000) Coordinate regulation of the nuclear and plastidic genes coding for the subunits of the heteromeric acetyl-coenzyme A carboxylase. Plant Physiol 122:1057–1071

Kennedy EP (1961) Biosynthesis of complex lipids. Fed Proc Am Soc Exp Biol 20:934–940

Kim HU, Li Y, Huang AHC (2005) Ubiquitous and endoplasmic reticulum-located lysophosphatidyl acyltransferases, LPAT2, is essential for female but not male gametophyte development in Arabidopsis. Plant Cell 17:1073–1089

Knowles JR (1989) The mechanism of biotin-dependent enzymes. Annu Rev Biochem 58: 195–221

Knutzon DS, Hayes TR, Wyrick A, Xiong H, Maelor-Davies H, Voelker TA (1999) Lysophosphatidic acid acyltransferase from coconut endosperm mediates the insertion of laurate at the sn-2 position of triacylglycerols in lauric rapeseed oil and can increase total laurate levels. Plant Physiol 120:739–746

Konishi T, Shinohara K, Yamada K, Sasaki Y (1996) Acetyl-CoA carboxylase in higher plants: most plants other than Gramineae have both the prokaryotic and the eukaryotic forms of this enzyme. Plant Cell Physiol 37:117–122

Kozaki A, Sasaki Y (1999) Light-dependent changes in redox status of the plastidic acetyl-CoA carboxylase and its regulatory component. Biochem J 339:541–546

Kunst L, Taylor DC, Underhill EW (1992) Fatty acid elongation in developing seeds of *Arabidopsis thaliana*. Plant Physiol Biochem 30:425–434

Lemieux B, Miquel M, Somerville C, Browse J (1990) Mutants of Arabidopsis with alterations in seed lipid fatty acid composition. Theor Appl Genet 80:234–240

Luehs W, Friedt W (1993) Non-food uses of vegetable oils and fatty acids. In: Murphy DJ (ed) Designer oil crops, breeding, processing and biotechnology. VCH Verlagsgesellschaft, Weinheim, pp 73–130

Lung SC, Weselake RJ (2006) Diacylglycerol acyltransferase: a key mediator of plant triacylglycerol synthesis. Lipids 41:1073–1088

Ohlrogge JB, Kuhn DN, Stumpf PK (1979) Subcellular localization of acyl carrier protein in leaf protoplasts of *Spinacia oleracea*. Proc Natl Acad Sci USA 76:1194–1198

Othman A, Lazarus C, Fraser T, Stobart K (2000) Cloning of a palmitoyl-acyl carrier protein thioesterase from oil palm. Biochem Soc Trans 28:619–622

Post-Beittenmiller D (1996) Biochemistry and molecular biology of wax production in plants. Annu Rev Plant Physiol Plant Mol Biol 47:405–430

Saha S, Enugutti B, Rajakumari S, Rajasekharan R (2006) Cytosolic triacylglycerol biosynthetic pathway in oilseeds: Molecular cloning and expression of peanut cytosolic diacylglycerol acyltransferase. Plant Physiol 141:1533–1543

Salas JJ, Sánchez J, Umi S, Ramli US, Manaf AM, Williams M, Harwood JL (2000) Biochemistry of lipid metabolism in olive and other oil fruits. Prog Lipid Res 39:151–180

Samols D, Thornton CG, Murtif VL, Kumar GK, Haase FC, Wood HG (1988) Evolutionary conservation among biotin enzymes. J Biol Chem 263:6461–6464

Sasaki Y, Hakamada K, Suama Y, Nagano Y, Furusawa I, Matsuno R (1993) Chloroplast-encoded protein as a subunit of acetyl-CoA carboxylase in pea plant. J Biol Chem 268:25118–25123

Savage LJ, Ohlrogge JB (1999) Phosphorylation of pea chloroplast acetyl-CoA carboxylase. Plant J 18:521–527

Shekar S, Tumaney AW, Rao TJ, Rajasekharan R (2002) Isolation of lysophosphatidic acid phosphatase from developing peanut cotyledons. Plant Physiol 128:988–996

Shimakata T, Stumpf PK (1983) Purification and characterization of β-ketoacyl-ACP synthetase I from *Spinacia oleracea* leaves. Arch Biochem Biophys 220:39–45

Stymne S, Stobart AK (1987) Triacylglycerol biosynthesis. In: Stumpf PK, Conn EE (eds) The Biochemistry of Plants, vol 9. Academic Press, New York, pp 175–214

Sun C, Cao YZ, Huang AHC (1988) Acyl coenzyme A preference of the glycerol phosphate pathway in the microsomes from the maturing seeds of palm, maize and B. napus. Plant Physiol 88:56–60

Tumaney AW, Shekar S, Rajasekharan R (2001) Identification, purification, and characterization of monoacylglycerol acyltransferase from developing peanut cotyledons. J Biol Chem 276:10847–10852

Van de Loo FJ, Broun P, Turner S, Somerville C (1995) An oleate 12-hydroxylase from *Ricinus communis* L. is a fatty acyl desaturase homolog. Proc Natl Acad Sci USA 92:6743–6747

Voelker TA, Worrell AC, Anderson L, Bleibaum J, Fan C, Hawkins DJ, Radke SE, Davies HM (1992) Fatty acid biosynthesis redirected to medium chains in transgenic oilseed plants. Science 257:72–74

Weaire PJ, Kekwick RGO (1975) The synthesis of fatty acids in avocado mesocarp and cauliflower bud tissue. Biochem J 146:425–437

Chapter 7
Biosynthesis and Regulation of Carotenoids in Plants—Micronutrients, Vitamins and Health Benefits

C.I. Cazzonelli, N. Nisar, D. Hussain, M.E. Carmody, and B.J. Pogson

7.1 Introduction: Carotenoid Biosynthesis in Higher Plants

Carotenoid enzymes are labile in vitro, confounding many classical biochemical approaches to understand their biosynthesis. Elucidation of the biosynthetic pathway has relied largely upon molecular genetics and it is only since the 1990s that carotenoid biosynthesis was described at the molecular level. It has since been reviewed very thoroughly and, therefore, will be described only briefly here (Armstrong 1997; Britton 1998; Hirschberg 1998, 2001; Cunningham and Gantt 1998; Cuttriss and Pogson 2004; Fraser and Bramley 2004; Tian and DellaPenna 2004; DellaPenna and Pogson 2006; Sandmann et al. 2006; Lu and Li 2008). There are a few reports describing regulatory processes which control carotenoid biosynthesis, transcript abundance as well as plastid organelle storage, and this topic has been reviewed (Bramley 2002; DellaPenna and Pogson 2006; Lu and Li 2008). Examples of carotenoid regulatory processes, where appropriate, have been cited throughout this chapter.

Carotenoids are terpenoids, a large class of plant isoprenoid pigments which participate in essential processes such as respiration (ubiquinone), photosynthesis (carotenoids, chlorophylls, plastoquinone) and regulation of growth and development (strigolactones, cytokinins, brassinosteroids, gibberellins and abscisic acid; Fig. 7.1). Secondary isoprenoid metabolites contain commercial value as flavours, pigments, polymers or drugs. However, only a limited supply of these compounds is usually available from natural plant sources. Two distinct pathways have evolved for the synthesis of the universal precursors of all isoprenoid products, namely a 5-carbon isoprene unit, isopentenyl diphosphate (IPP), and its allylic isomer dimethylallyl diphosphate (DMAPP). The IPP and DMAPP used for carotenoid

C.I. Cazzonelli, N. Nisar, D. Hussain, M.E. Carmody, and B.J. Pogson
Research School of Biology, Biochemistry and Molecular Biology, The Australian National University, Building 41, Linnaeus Way, Canberra ACT0200, Australia
e-mail: barry.pogson@anu.edu.au

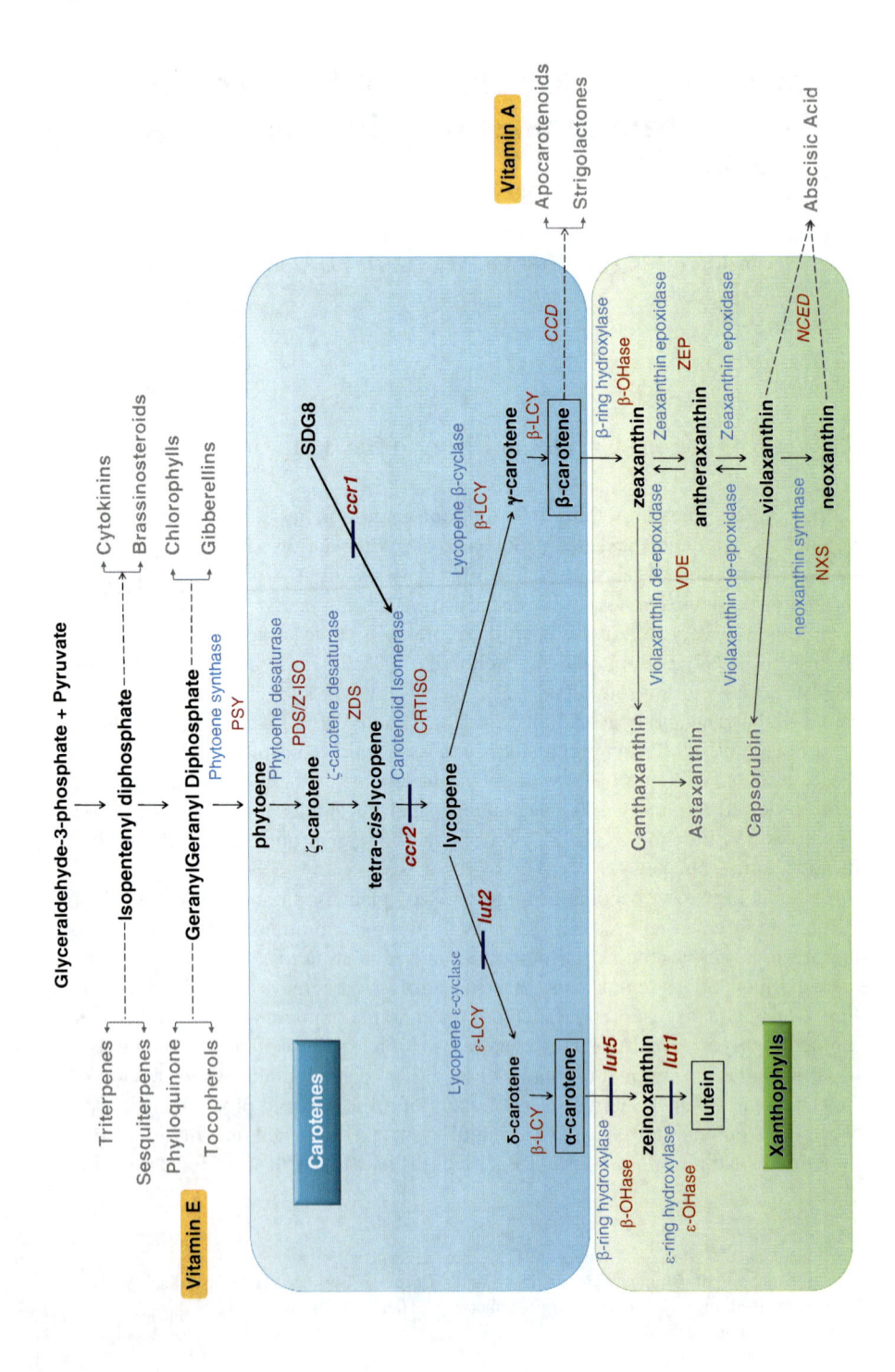

biosynthesis are derived from a plastid-localised methylerythritol phosphate pathway (MEP; Phillips et al. 2008). The MEP pathway uses glyceraldehyde-3-phosphate and pyruvate as initial substrates (Fig. 7.1).

The first committed step in carotenoid biosynthesis is the condensation of two geranylgeranyl diphosphate (GGPP) molecules to form phytoene. Phytoene is a colourless carotenoid produced as a 15-*cis* isomer by the enzyme phytoene synthase (PSY) and usually does not normally accumulate in plant tissues (Beyer 1989; Beyer and Kleinig 1991). Carotenoid pathway flux is often controlled by multiple *PSYs*. In *Arabidopsis*, there is only one *PSY* gene, two in tomato, where there are tissue-specific isoforms, and three in rice (Bartley and Scolnik 1993; Fraser et al. 1999; Welsch et al. 2000, 2008; Giorio et al. 2008). The *PSY3* isoform identified recently in rice was postulated to be involved in feedback regulation and plays a specialised role in abiotic stress-induced abscisic acid (ABA) formation (Welsch et al. 2008).

Phytoene yields tetra-*cis*-lycopene by two similar desaturases—e.g. phytoene desaturase synthase (PDS) and ζ-carotene desaturase (ZDS)—which introduces a series of four double bonds in a *cis*-configuration by dehydrogenation reactions. All bacterial and plant desaturases contain a flavin-binding site near the N-terminus. Desaturation is linked to the respiratory redox chain (Nievelstein et al. 1995). Evidence for a quinone requirement has been demonstrated in daffodil and *Arabidopsis* (Beyer 1989; Mayer et al. 1992; Norris et al. 1995). The cloning of the *IMMUTANS* gene has eluded to the molecular basis of the desaturation reaction catalysed by PDS (Carol et al. 1999; Wu et al. 1999). IMMUTANS contains lesions in a plastid targeted alternative oxidase (*PTOX*) gene, which results in the white 'ghost' tomato phenotype (Josse et al. 2000) and variegated sectors lacking functional chloroplasts in *Arabidopsis* (Carol et al. 1999). *PTOX* gene mutants accumulate phytoene in leaf and fruit tissues and links carotenoid desaturation to chloroplast respiratory activity in a non-essential way (Aluru et al. 2001; Yu et al. 2007).

In contrast, bacteria require only a single desaturase (crtI) to introduce four double bonds and produce a different lycopene isomer, referred to as all-*trans*-lycopene (Bartley et al. 1999). Bacterial cyclases appear unable to accept tetra-*cis*-lycopene as a substrate, instead requiring all-*trans*-lycopene to proceed (Schnurr et al. 1996), although this is contradicted in isolated membranes of daffodil (Beyer et al. 1994). This isomer anomaly suggested a higher plant requirement for a specific isomerase enzyme (Beyer and Kleinig 1991), which was postulated since the discovery of the tetra-*cis*-lycopene accumulating *tangerine* tomato mutant

---

Fig. 7.1 Carotenoid biosynthetic pathway showing key enzymatic steps towards plant hormone and vitamin A biosynthesis in higher plants. The pathway shows the primary steps found in nearly all plant species. The desaturases introduce a series of four double bonds in a *cis*-configuration, which are isomerized to the all-*trans*-configuration by the carotenoid isomerase. *Arabidopsis* lutein deficient mutations (*ccr2*, *ccr1*, *lut1*, *lut2* and *lut5*) are shown in an *italic red font*. GGPP, geranylgeranyl pyrophosphate; PSY, phytoene synthase; PDS, phytoene desaturase; ZDS, ζ-carotene desaturase; CRTISO, carotenoid isomerase; βLCY, β-cyclase; εLCY, ε-cyclase; βOHase, β-hydroxylase; εOHase, ε-hydroxylase; ZEP, zeaxanthin epoxidase; VDE, violaxanthin de-epoxidase; NXS, neoxanthin synthase; SDG8, histone methyltransferase; NCED, 9-*cis*-epoxycarotenoid dioxygenase; CCD, carotenoid cleavage dioxygenase

(Tomes et al. 1953; Isaacson et al. 2002). A carotenoid isomerase (*CRTISO*) gene capable of isomerising *cis*-bonds to the all-*trans*-lycopene was identified in tomato and *Arabidopsis* (Isaacson et al. 2002; Park et al. 2002). CRTISO shows 20–30% identity to the bacterial carotenoid desaturases, crtN and crtI, particularly at conserved motifs such as the di-nucleotide flavin-binding domain (characteristic of desaturases and cyclases). Despite this high level of identity, CRTISO demonstrated no desaturase or cyclase activity in an *E. coli* expression system (Park et al. 2002). A similar level of identity in the tomato gene led Isaacson and colleagues (2002) to suggest that, although ZDS could use *cis*-isomers as substrates, it could not alter the isomer state and therefore CRTISO may operate in vivo as part of a complex with ZDS. Characterization of the maize (*Zea mays*) *pale yellow9* (*y9*) locus has identified a new isomerase (Z-ISO) required in plant carotenoid biosynthesis. Z-ISO activity occurs upstream of *CRTISO* and catalyses the *cis* to *trans* conversion of the 15-*cis*-bond in 9,15,9¢-tri-*cis*-zeta-carotene, the product of phytoene desaturase, to form 9,9-di-*cis*-zeta-carotene, the substrate of ZDS (Fig. 7.1; Li et al. 2007). The geometrical isomer substrate requirement of ZDS in evolutionarily distant plants may suggest that Z-ISO activity is not unique to maize and may be present in all higher plants.

A very intriguing and recent insight into the regulation of isomerase activity was alluded to by the discovery of the *ccr1* regulatory mutant which contains reduced lutein as a result of down-regulation of *CRTISO* in *Arabidopsis* (Cazzonelli et al. 2009a; Fig. 7.1). *ccr1* was identified to contain a mutation in a SET2 domain, histone methyltransferase chromatin modifying enzyme known as SDG8 (Cazzonelli et al. 2009b). *SDG8* was shown to be required for maintaining active marks of histone methylation surrounding the *CRTISO* promoter; this is not only the first report of a carotenoid biosynthetic regulatory mutant, but also the first implication that epigenetic regulatory mechanisms may control carotenoid composition and flux during plant development (Cazzonelli et al. 2009a, c).

After the formation of lycopene, the carotenoid biosynthetic pathway splits into two branches, distinguished by different cyclic end groups, namely beta or epsilon (Fig. 7.1). Two beta rings form the β,β branch (β-carotene and its derivatives) with one beta and one epsilon forming the β,ε branch (α-carotene and its derivatives). Lycopene β-cyclase (βLCY) introduces a β-ionone ring to either end of lycopene to produce β-carotene. Homozygous *βLCY* mutants in *Arabidopsis* are lethal. The β-cyclase and ε-cyclase enzymes each introduce a ring to form α-carotene. The *Arabidopsis* lycopene ε-cyclase (*εLCY; lut2*) gene can catalyse only cyclisation of one end group, and the *lut2* mutant is semi-dominant, does not accumulate zeinoxanthin or any other ε-ring carotenoid, and β-cyclisation is not interrupted (Pogson et al. 1996). The *εLCY* gene has been postulated to be rate-limiting in the production of lutein and may be regulated by metabolic feedback (Cuttriss et al. 2007). Lettuce (*Lactuca sativa*) appears to be unique among higher plants in that its εLCY can catalyse formation of a bicyclic ε,ε-carotene and its hydroxylated derivative lactucaxanthin (Phillip and Young 1995; Cunningham and Gantt 2001).

α-carotene and β-carotene are further hydroxylated to produce the oxygenated α-carotene derivatives lutein and zeaxanthin respectively. These reactions are catalysed by the ε-ring hydroxylase (ε-OHase) and the β-ring hydroxylase (β-Ohase; Fig. 7.1). Lutein is the most abundant carotenoid found in leaf tissues of plants and is the product of the β,ε branch. There are some mutations reported while investigating the role of lutein biosynthesis, like *lut1* (Tian et al. 2004), *lut2* (Cunningham et al. 1996; Pogson et al. 1996), *ccr2* (Isaacson et al. 2002; Park et al. 2002) and *lut5* (Kim and DellaPenna 2006). Interestingly, manipulation of lycopene in higher plants can alter lutein biosynthesis (Misawa et al. 1994) and epigenetic regulation of *CRTISO* transcript abundance has been reported to be a rate-limiting step in lutein production (Cazzonelli et al. 2009a). Overexpression of the εLCY gene in *Arabidopsis* has been shown to enhance lutein levels and causes a marginal, but significant, increase in the rate of induction of non-photochemical quenching, despite a reduction in the xanthophyll cycle pool size (Pogson and Rissler 2000). Zeaxanthin can be epoxidised to produce violaxanthin by zeaxanthin epoxidase (ZEP). Under high-light stress, this reaction is reversed by violaxanthin deepoxidase (VDE; Pfundel and Bilger 1994). Violaxanthin is converted to neoxanthin by neoxanthin synthase (NXS). Neoxanthin is the last carotenoid of the β,β branch of the classical biosynthetic pathway. Both 9-*cis*-neoxanthin and 9-*cis*-violaxanthin can be cleaved by 9-*cis*-epoxycarotenoid dioxygenase (NCED). These cleavage products are further modified to ABA (Seo and Koshiba 2002; Chinnusamy et al. 2008).

7.2 Carotenoids and Plant Development

The function and roles of carotenoids during plant development have been reviewed extensively (Niyogi 1999; Cuttriss and Pogson 2004; Cheng 2006; DellaPenna and Pogson 2006; Howitt and Pogson 2006; Pogson et al. 2006; Sandmann et al. 2006; Giuliano and Diretto 2007; Lu and Li 2008; Giuliano et al. 2008; Fraser et al. 2009). The following section highlights some of the more recent advances by which carotenoid research has provided novel insights and considerations for plant biology and metabolic engineering.

Carotenoid pigments and their corresponding enzymes are located in the plastid, although all known carotenoid biosynthetic genes are encoded by the nuclear genome. Plant organelle structures, such as the plastids, are required in order to synthesise, store and retain carotenoid metabolites (Lopez-Juez and Pyke 2005). All plastid types are derived from the pluripotent progenitor proplastid, which is found in undifferentiated meristem tissue. Proplastids are colourless but have the potential to form differentiated plastids capable of storing chlorophylls and carotenoids (Kirk and Tilney-Bassett 1978). Carotenoids are synthesised de novo in nearly all types of differentiated plastids of fruits, flowers, roots and seeds, accumulating largely in chloroplasts (green photosynthetic plastids) and chromoplasts (coloured plastids), but also in etioplasts (dark-grown precursors of the chloroplast), amyloplasts (starch-storing plastids), leucoplasts (colourless plastids) and elaioplasts (lipid-storing plastids). Several recent studies have provided links between changes in

carotenoid composition and plastid biogenesis, morphology and protein translocation (Lu et al. 2006; Cuttriss et al. 2007; Tzvetkova-Chevolleau et al. 2007).
Understanding how carotenoids are sequestered within plastid types will ultimately
improve metabolic engineering attempts to enhance a plant composition of essential dietary micronutrients, such as the carotenes.

Within the chloroplast, carotenoids harvest light energy and are essential components of the photosynthesis apparatus (Niyogi 2000; Lu and Li 2008). They
absorb light in the blue region of the spectrum (400 to 600 nm), and the energy
absorbed can be transferred to chlorophylls. Thus, carotenoids serve as light harvesting accessory pigments. In addition, carotenoids play an essential and important
role to protect the photosynthetic machinery from photoxidation via energy dissipation, free radical detoxification and limiting damage to membranes (Pogson et al.
2008). In the absence of pigments within the chloroplasts, plants will ultimately
suffer severe photooxidative damage and reactive oxygen species will not be
quenched, which results in death to the plant (Pogson et al. 1998). Improving
light harvesting and photoprotection against excess light energy has the potential
to improve both growth and yield of important crop species.

Carotenoid pigments serve as precursors in the production of plant phytohormones and signalling compounds. Recent advances in understanding apical dominance has eluded to a novel root transmissible signal hormone which inhibits shoot
branching in *Arabidopsis*, pea, petunia and rice (Beveridge et al. 1996, 2000;
Morris et al. 2001; Stirnberg et al. 2002; Sorefan et al. 2003; Booker et al. 2004;
Schwartz et al. 2004; Snowden et al. 2005; Gomez-Roldan et al. 2008; Umehara
et al. 2008). Shoot branching genes encoding CAROTENOID CLEAVAGE-DIOX-
YGENASES, *CCD7* and *CCD8* (Sorefan et al. 2003; Booker et al. 2004; Snowden
et al. 2005; Zou et al. 2006; Arite et al. 2007), appear to be essential for synthesis of
a branching inhibitor hormone, which was recently eluded to being a member of the
strigolactone class of metabolites (Gomez-Roldan et al. 2008; Umehara et al. 2008).
Previously, strigolactones were associated with functions in the rhizosphere, where
they stimulate germination of parasitic plant seeds, such as those of *Striga*, and also
influence hyphal branching in mycorrhizae (Cook et al. 1972; Akiyama et al. 2005;
Pichersky 2008). β-Carotene (Fig. 7.1) has been proposed as an initial substrate for
strigolactone biosynthesis (Matusova et al. 2005; Rani et al. 2008), but the complete
biochemistry of strigolactones is yet to been described.

Another recent discovery has implicated yet another novel mobile carotenoid-
derived signalling compound required for normal root and shoot development. The
Arabidopsis BYPASS1 (*BPS1*) gene encodes a protein with no homology to any
functionally characterized proteins. The roots of *BYPASS1* (*bps1*) mutants produce
a graft transmissible signal which is capable of arresting shoot development (Van
Norman et al. 2004). The carotenoid biosynthesis inhibitor, fluridone, can partially
rescue both leaf and root defects in *bps1*. Studies indicate that the mobile signal is
neither ABA- nor strigalactone-related, and that production of the signal does not
require the activity of any single carotenoid cleavage dioxygenase. In addition,
2-(4-chlorophenylthio)-triethylamine hydrochloride (CPTA), a lycopene cyclase
inhibitor, shows that signal production requires synthesis of β-carotene (Fig. 7.1)

and its derivatives. However, its identity remains unknown (Van Norman et al. 2004; Van Norman and Sieburth 2007).

Finally, ABA, one of the most heavily researched phytohormones, is derived from the carotenoid biosynthetic pathway. ABA is involved in multiple functions and regulates many developmental processes, such as transpiration through stomatal guard cells, seed germination and dormancy, and plays important roles in response to abiotic stresses, such as drought (Rock and Zeevaart 1991; Seo and Koshiba 2002; Nambara and Marion-Poll 2005; Wilson et al. 2009).

A major function of carotenoids in flower and fruit tissues is to provide bright colours, scent and flavour in order to attract insects and animals for pollination and seed dispersal. Such pollinators are ultimately essential for plant reproduction. Generally, it has been recognized that the type of pigment, shape of cells accumulating the pigment, and thickness of petal tissues are the important factors that influence flower colours. The petals of chrysanthemum (Kishimoto et al. 2004) and marigold (Moehs et al. 2001) contain predominantly lutein (yellow in colour), while only a few plant species are known to produce ketocarotenoid pigments such as astaxanthin (3,3′-dihydroxy-4,4′-diketo-β,β-carotene) which furnishes an attractive orange–red colour, as in the petals of *Adonis aestivalis* (Cunningham and Gantt 2005). The red style branches of *Crocus sativus* accumulate the unique apocarotenoids, crocetin glycosides, picrocrocin and safranal, from which the spice saffron is derived. They are responsible for the colour, taste and aroma of saffron and are produced by the cleavage of zeaxanthin (Bouvier et al. 2003; Schwab et al. 2008).

Other carotenoid-derived volatiles, such as α- and β-ionone and their related compounds and derivatives, are also important components of aroma and taste produced during flower and fruit development (Goff and Klee 2006). Colour modification in different fruits, vegetables and flowers has been accomplished by natural selection—for example, the orange carrot and yellow bell pepper. Genetic manipulation of carotenoid pigmentation in flower petals of ornamental crops was also achieved through the overexpression of the *CrtW* gene (β-carotene ketolase from marine bacteria) in *Lotus japonicas*, which changed the flower from pink to red due to ketocarotenoid production (Suzuki et al. 2007).

The elucidation of novel functions for carotenoids and their derivatives in shoot and root development, plastid biogenesis, chloroplast-to-nucleus signalling, abiotic and biotic stress tolerance, as well as in floral pigmentation and fruit development is an ongoing process. Undoubtedly, new insights will reveal new molecular screening tools, and genetic engineering for improved nutrition will become more predictable.

7.3 Health Benefits of Carotenoid-Derived Vitamins and Nutrients

Plants and microorganisms provide a primary natural source of essential vitamins for human diets. These micronutrients are necessary for human health and development. There has been a quest over the last decade to discover genes and biochemical processes which control an individual plant's production of these health-promoting micronutrients. They are generally classified into fat and water soluble vitamins, depending on their biochemical properties. The group of fat soluble vitamins consists of provitamin A (typically α- or β-carotene), vitamins D (calciol), E (tocopherols and tocotrienols) and K1 (phylloquinone). The water soluble class of vitamins include vitamin B1 (thiamine), B2 (riboflavin, nicotinamide, folate and pantothenate), B6 (pyridoxal), B12 (cobalamine), C (ascorbate) and H (biotin). A dietary sufficiency of these essential micronutrients is not typically a major hurdle for residents of wealthy nations. They can be obtained from a balanced diet or via vitamin and mineral supplements. However, people from developing nations rely upon only a few staple crops, such as rice, maize, wheat and cassava, all of which are poor sources of essential nutrients. Even if people consume large quantities of these foods, they are still at high risk of malnutrition.

Biofortification is a sustainable means to enrich the nutrient content of crops. The goal is to provide a balanced group of micronutrients in staple crops, as opposed to adding them by fortification during processing in products, such as bread flour. Target crops include wheat and rice enriched with iron, vitamin A (prevents blindness), vitamin E (antioxidant activities) and folate (a B-vitamin which prevents congenital disorders such as spina bifida and facial deformities). There are a number of wheat, rice and maize varieties available to plant breeders and, when equipped with the appropriate molecular tools, they can select for varieties containing optimal levels of some micronutrients for bread wheat, pasta and noodles. For particular micronutrients in specific species, genetic modification is essential—for example, the production of golden rice which enhances provitamin A content (Paine et al. 2005). Fruits and vegetables also have the potential to be fortified with bone-building calcium and the nutrient content can be enhanced though genetic modification. The reason for all this scientific effort being targeted at fruits, vegetables and grains is that they can provide a rich source of a number of vitamins and minerals essential to human health (reviewed in Davies 2007).

The health benefits associated with vitamins, carotenoid metabolites and other micronutrients, as well as their modification in plants have been reviewed extensively (Herbers 2003; Fraser and Bramley 2004; Krinsky and Johnson 2005; DellaPenna and Pogson 2006; Davies 2007; Tanaka and Ohmiya 2008; Giuliano et al. 2008). The nutraceutical industry has targeted carotenoids mainly on the basis of their antioxidant properties, some having provitamin A activity, and their potential role in limiting age-related macular degeneration of the eye leading to blindness. Five carotenoids are manufactured synthetically on an industrial scale, namely lycopene, β-carotene, canthaxanthin, zeaxanthin and astaxanthin. These nutraceuticals are used in a range of food products and cosmetics, such as vitamin supplements, and health products, and as feed additives for poultry, livestock, fish and crustaceans, which are essential for the animals' growth, health and reproduction (reviewed in Del Campo et al. 2007; Jackson et al. 2008). The following section reviews only the medical benefits and molecular approaches for biosynthesis associated with (1) lutein, zeaxanthin and prevention of macular degeneration of the eye, (2) β-carotene and the biosynthesis of provitamin A, and (3) antioxidant properties of other xanthophylls and xanthophyll derivatives.

7.3.1 Zeaxanthin, Lutein and Prevention of Macular Degeneration

Both lutein and zeaxanthin are necessary for efficient photoprotection in plants and have been implicated in protecting against age-related blindness due to macular degeneration (AMD) in humans (reviewed in Coleman and Chew 2007). Macular degeneration is the leading cause of blindness in the developed world and results in the loss of central and detailed vision especially in the elderly over 60 years of age. Within the central macula, zeaxanthin is the dominant carotenoid, whereas in the peripheral retina, lutein predominates. These pigments are able to absorb blue light which damages photoreceptors and pigmentary epithelium. Due to their antioxidative properties, they can reduce changes in membrane permeability via quenching reactive oxygen species and free radicals. A higher dietary intake of lutein and zeaxanthin was independently associated with decreased likelihood of having neovascular age-related macular degeneration, geographic atrophy, and large or extensive intermediate drusen, while other nutrients such as vitamin A, α-tocopherol and vitamin C did not show a similar association (SanGiovanni et al. 2007). Zeaxanthin is the pigment which gives corn its characteristic colour. Some yellow fruits and vegetables and almost any green vegetable are considered a good source of lutein and may have some zeaxanthin. These may include kale, spinach, turnip greens, collard greens, romaine lettuce, broccoli, zucchini, garden peas and brussels sprouts.

A number of biotechnological approaches are underway to establish algal production systems to replace industrial synthesis of lutein and zeaxanthin (Del Campo et al. 2007). The unicellular green alga *Dunaliella salina* is currently being cultivated as a source of β-carotene by the natural products industry in Australia and Israel. A novel mutant (*zea1*) of the halotolerant unicellular green alga *D. salina* is impaired in the zeaxanthin epoxidation reaction, thereby lacking a number of the beta-branch xanthophylls. The *zea1* mutant lacks neoxanthin, violaxanthin and antheraxanthin, but constitutively accumulates zeaxanthin with no adverse effects upon photosynthesis or growth of the *zea1* strain (Jin et al. 2003).

The staple crop potato, which accumulates high concentrations of lutein and violaxanthin, was altered genetically by suppressing the potato *zeaxanthin epoxidase* gene, which blocked the conversion from zeaxanthin to violaxanthin and increased tuber concentrations of zeaxanthin (four- to 130-fold) and slightly enhanced antheraxanthin without affecting the level of lutein. Surprisingly, manipulation of this enzyme also resulted in elevated transcript levels of the first step in the carotenoid pathway and a concomitant two to threefold increase in α-tocopherol (vitamin E; Römer et al. 2002). In addition to the potential health benefits displayed by zeaxanthin, it may limit lipid peroxidation and enhance the maintenance of membrane fluidity and thermostability (Tardy and Havaux 1997; Gruszecki et al. 1999). Therefore, manipulation of zeaxanthin and lutein may not only improve the nutritional value of the foods, but may also improve plant vigour.

Lutein is the most abundant carotenoid found in plants and its levels are remarkably stable, with no known natural mutants being identified to date. Mutant screening has previously identified lutein deficient mutants in *Arabidopsis* (*ccr1*, *ccr2*, *lut2* and *lut1*) which exhibit altered non-photochemical quenching and light harvesting complexes (Pogson et al. 1996, 1998; Park et al. 2002; Cazzonelli et al. 2009a). There are examples of plant species which contain higher concentrations of α-carotene (Fig. 7.1; e.g. *Coffea canephora*), the precursor to lutein (Simkin et al. 2008), and it is likely that variations in εLCY amino acid composition may be responsible for modifying the εLCY and/or βLCY enzyme activity, thereby causing an increased flux of carotenoids through the lutein branch of the pathway (Cunningham and Gantt 2001; Howitt et al. 2009). As a result, natural variation in the amount of lutein amongst different wheat varieties and novel molecular markers (linked with phytoene synthase and epsilon cyclase biosynthetic genes) have been developed, which may now provide the potential to breed new varieties to produce grains rich in lutein (Howitt et al. 2009). Alternatively, further insight into chromatin modifying enzymes (e.g. SDG8; Fig. 7.1) which alter the accessibility of transcription factors to the *CRTISO* promoter, and affect not only plant development but also lutein composition, may provide the switch to increase antioxidant levels in staple food crops (Cazzonelli et al. 2009a).

7.3.2 *β-Carotene and the Biosynthesis of Vitamin A*

Vitamin A deficiency (VAD) is the leading worldwide cause of preventable blindness, affecting 127 million children and 7 million pregnant women worldwide and increasing the risk of disease and death from severe infections (West 2003). Vitamin A (retinol) is generated from non-hydroxylated β-ring containing provitamin A carotenoids in the diet. Three carotenoids have provitamin A activity, namely β-carotene, α-carotene and β-cryptoxanthin. β-carotene is cleaved to form two retinals (vitamin A aldehyde) by human carotenoid cleavage enzymes such as β-carotene 15,15'-dioxygenase (von Lintig and Vogt 2000; Wyss et al. 2000). The deduced amino acid sequence of β-carotene 15,15'-dioxygenase shows homology to the plant carotenoid cleavage enzymes, in particular, the viviparous 14 (Vp14) carotenoid dioxygenases from maize, which is involved in the synthesis of the stress responsive hormone ABA (Schwartz et al. 2003). Considerable health benefits could be obtained through the production of β-carotene in agronomic crop species. Moreover, in contrast to excess dietary vitamin A (which can be toxic at high levels), excess dietary β-carotene has no harmful effects, making transgenic plants with increased β-carotene a safe and effective means of vitamin A delivery.

Several approaches have been employed to increase the levels of provitamin A in crop plants. For example, a new source of β-carotene was developed in orange-fleshed sweet potato (Hagenimana et al. 1999). Interspecific hybridizations have served to enhance carotene content twofold by crossing cassava with its wild-type relative, *Manihot dichotoma* (Nassar 2004). In addition to these species, similar opportunities exist for crop breeders to produce and disseminate high β-carotene varieties of beans, maize and bananas which are nutritionally superior, well-adapted to local growing conditions, and more profitable for the release to farmers in developing countries (Bouis 2007).

There are many examples of carotenoid enhancement in plants through genetic manipulation (Table 7.1). 'Golden Rice' is a well-known example of micronutrient improvement through genetic engineering aimed at addressing the problem of vitamin A deficiency in developing countries. *PSY* from daffodil (*Narcissus pseudonarcissus*) was overexpressed in the endosperm of rice seed, which does not usually accumulate carotenoids, and transgenic seeds accumulated small amounts of phytoene (Burkhardt et al. 1997). Subsequently, coexpression of PSY, a bacterial phytoene desaturase (*CrtI*, from *Erwinia uredovora*) and β-cyclase (from daffodil) were expressed in the endosperm of rice and, depending on the transgenic event, transgenic seed accumulated β-carotene, lutein and zeaxanthin at variable levels (Ye et al. 2000). Further studies revealed that the presence of β-cyclase was not necessary; only overexpression of *PSY* and *CrtI* was required for accumulation of β-carotene, lutein and zeaxanthin in rice endosperm (Beyer et al. 2002). From these experiments, a high β-carotene accumulating line was chosen to develop 'Golden Rice' (Beyer et al. 2002). Subsequent studies have shown that the source of the *PSY* gene has a significant impact on carotenoid accumulation. 'Golden Rice 2', which

Table 7.1 Manipulation of carotenoid content in crop species and *Arabidopsis*

Pigment and plant species	Tissue type	Gene manipulated	Enhanced carotenoid	References
Lutein				
Arabidopsis	Leaf	εLCY	Lutein	(Pogson and Rissler 2000)
Zeaxanthin				
Potato	Stem	*ZEP*	Zeaxanthin	(Römer et al. 2002)
β-Carotene				
Cabbage	Flower	*Or*	β-Carotene	(Lu et al. 2006; Giuliano et al. 2008; Zhou et al. 2008)
Potato	stem	*DXS*	Phytoene	(Morris et al. 2006)
Golden rice	Seed	*PSY* & *CrtI*	β-Carotene	(Ye et al. 2000)
Canola	Seed	*CrtI*	β-Carotene	(Ravanello et al. 2003)
Arabidopsis	Seed	*CrtI*	β-Carotene	(Lindgren et al. 2003)
Tomato	Fruit	*CrtI*	β-Carotene	(Römer et al. 2000)
Tobacco	Leaves	*CrtI*	β-Carotene	(Misawa et al. 1993)
Lotus	Flowers	*CrtW*	Ketocarotenoids	(Suzuki et al. 2007)
Astaxanthin				
Carrot	Root	*CrtO*	Astaxanthin	(Jayaraj et al. 2008)
Tobacco	Leaves and flowers	*CrtO*	Astaxanthin	(Mann et al. 2000)

overexpresses *PSY* from maize and *CrtI*, accumulated 23-fold more carotenoids (37 μg/g) than was the case for 'Golden Rice' which overexpressed *PSY* from daffodil (Paine et al. 2005). Golden rice lines in elite Indica and Japonica variety backgrounds have since been produced and, in order to avoid the complication of antibiotic resistance selection, phosphomannose isomerase was used as a selectable marker. The elite lines are now ready for risk assessment, nutritional evaluation and breeding for local adaptation (Hoa et al. 2003). The varieties will be freely provided without any commercial licensing agreements and the farmers will be free to collect their own seed and breed it further into their own varieties.

Another means of enhancing the β-carotene in plants is to manipulate the storage of pigments (Zhou et al. 2008). The identification and characterization of a novel, naturally occurring mutation in a gene, *Or* (orange-curd), results in high β-carotene levels in the typically white cauliflower curd (now orange in the *Or* mutants). The study of this gene reveals that creating a metabolite sink to accumulate β-carotene in chromoplasts is an important mechanism by which to control carotenoid accumulation in plants (Lu and Li 2008). The *Or* gene encodes a DnaJ cysteine-rich domain-containing protein, which appears to mediate the differentiation of pro-plastids in apical shoot and inflorescence meristematic tissues of curd into chromoplasts for associated carotenoid accumulation (Li and Du 2001; Lu et al. 2006). Chromoplasts accumulate carotenoids in lipoprotein structures (Bartley and Scolnik 1995; Vishnevetsky et al. 1999) and sequester them as crystals, which may serve to allow additional carotenoid biosynthesis.

Possibly the most dramatic example of carotenoid manipulation was performed in canola (*Brassica napus*). Canola was transformed with a triple construct containing genes for *PSY*, *CrtI* (from a bacterial source) and lycopene cyclase (either plant or bacterial origin), and carotenoid content was enhanced (Ravanello et al. 2003). Seed-specific overexpression of the bacterial *PSY* gene results in a remarkable 50-fold increase in carotenoid accumulation (Shewmaker et al. 1999). In addition, the enhanced expression of lycopene β-cyclase diverted xanthophyll formation away from α-carotene and lutein (Fig. 7.1), shifting incorporation towards β-carotene formation. In *Arabidopsis*, a similar approach resulted in a 43-fold increase of β-carotene and other carotenoids and chlorophyll in seed tissues; however, there was a delay in germination due to greater concentrations of ABA (Lindgren et al. 2003). There was also alterations in fatty acid composition and a reduction of tocopherol and chlorophyll, showing that manipulation of carotenoid accumulation often has unexpected effects on other cellular processes (Shewmaker et al. 1999). In order for carotenoids to accumulate to such high concentrations, there must be an adequate sink. In this case, the hydrophobic molecules can be sequestered in cellular vesicles or lipid globules within the plastid, which is perhaps why such high levels were accumulated in seeds of an oil crop. The introduction of a bacterial gene, as opposed to overexpression of the endogenous equivalent, may also have bypassed the usual regulatory controls which prevent such high levels of carotenoid accumulation.

Manipulation of the carotenoid biosynthetic pathway in several transgenic plants has resulted in unexpected up-regulation of other carotenogenic genes not targeted by the transgene. Constitutive expression of the bacterial phytoene desaturase from *Erwinia uredovora*, *CrtI*, in tobacco (Misawa et al. 1994) produced mutant lines which were resistant to herbicides which induce bleaching, such as norflurazon. Plants accumulated high concentrations of β-carotene and its derivatives at the cost of reduced lutein which originates from α-carotene (Misawa et al. 1993). In tomato fruit, the enhancement of β-carotene content was also achieved after transformation with the *CrtI* gene. Although cyclisation of lycopene to β-carotene was not altered, β-carotene concentrations were unexpectedly up-regulated, while there was an overall decrease in carotenoid content of fruit tissues (Römer et al. 2000; Fraser et al. 2001). In contrast, overexpression of *βLCY* increased β-carotene and total carotenoid content. It was postulated that the unexpected carotenoid compositions might be due to feedback regulation by β-carotene, or one of its metabolites (Römer et al. 2000).

In general, the results of carotenoid engineering are predictable to a point. Beyond that point, it is clear that further understanding of the regulatory mechanisms which finely tune carotenoid flux and control feedback regulation is necessary in order to not perturb other carotenoids essential for photoprotection, phytohormone biosynthesis, growth and development in plants.

7.3.3 Antioxidant Properties of Other Xanthophylls and Xanthophyll Derivatives

In addition to lutein and zeaxanthin, the ketocarotenoids (e.g. astaxanthin, capsanthin, capsorubin and canthaxanthin) display antioxidant properties and are a beneficial human dietary component (Jyonouchi et al. 1995; Zhu et al. 2008). A major value of astaxanthin is to provide it as an additive for aquaculture feed, where it gives salmon their characteristic pink colour (Lorenz and Cysewski 2000). Astaxanthin is produced by a limited number of organisms, including certain marine bacteria, yeast and some green algae (Johnson and Schroeder 1995). However, for commercial purposes, it is usually synthesised chemically at a high cost. This prompted researchers to engineer tobacco flowers producing astaxanthin by the introduction of the *Haematococcus pluvialis* β-carotene ketolase (*CrtO*) gene (Mann et al. 2000). Trangenic tobacco plants resulted in an increased accumulation of ketocarotenoids, including astaxanthin, in the chromoplasts of the nectary tissue in their flowers, changing the flower colour from yellow to red (Mann et al. 2000). Additional biotechnological approaches using marine alga and other crops are underway in a number of institutions. The overexpression of *CrtO* in carrot roots also results in the overproduction of astaxanthin (Jayaraj et al. 2008).

A transgenic line suppressing potato zeaxanthin epoxidase was used to overexpress the *crtO* β-carotene ketolase gene from the cyanobacterium *Synechocystis*, and tubers were shown to accumulate ketocarotenoids (∼10% of total carotenoid pool) such as astaxanthin, echinenone and ketozeaxanthin (Gerjets and Sandmann

2006). Ketocarotenoids have also been enhanced in nectarines or leaves of tobacco and tomato as a consequence of the expression of both the *crtW* and *CrtZ* (3,3 β-hydroxylase) genes (Ralley et al. 2004). The capsanthin-capsorubin synthase (*CCS*) gene from pepper is an enzyme which converts antheraxanthin and violax-anthin to the red xanthophylls (ketocarotenoids), such as capsanthin and capsor-ubin. It is likely that such key genes will offer opportunities to enhance ketocarotenoid production in crop species (Hornero-Mendez et al. 2000).

Zeaxanthin can also be cleaved to form picrocrocin and safranal, which are responsible for the taste and aroma of saffron. Saffron is derived from the dried red stigmas of *Crocus sativus* and is used as flavouring and colouring agents. There is a growing interest in the impact of saffron carotenoids on human health due to their high antioxidant capacity (Verma and Upadhyaya 1998).

7.4 Conclusions and Future Prospects

The carotenoid biosynthetic pathway is now well understood. The major challenge is to provide insight into the regulatory aspects of carotenoid biosynthesis during nuclear transcription, protein translocation, plastid biogenesis and plant develop-ment. Understanding needs to be improved of how phytohormones, abiotic stress and metabolic feedback affect carotenoid composition and regulation. As know-ledge of the regulatory processes increases, researchers will be able to make informed decisions about the effects of manipulating the pathway. This will allow the development of the next generation of crops which enhance and build upon the benefits of 'Golden Rice'. The result could entail significant health benefits for society by providing essential antioxidant micronutrients and vitamin A deriva-tives. As the health benefits from these crops can be obtained by changing the composition of foods already eaten without altering the eating habits of the con-sumer, they will have a much greater chance of being accepted. Biotechnological approaches and traditional breeding practices both play a prominent role for increased production of natural carotenoids of interest in plants and, hence, improvements to human health in developing countries.

References

Akiyama K, Matsuzaki K, Hayashi H (2005) Plant sesquiterpenes induce hyphal branching in arbuscular mycorrhizal fungi. Nature 435:824–827

Aluru MR, Bae H, Wu D, Rodermel SR (2001) The Arabidopsis *immutans* mutation affects plastid differentiation and the morphogenesis of white and green sectors in variegated plants. Plant Physiol 127:67–77

Arite T, Iwata H, Ohshima K, Maekawa M, Nakajima M, Kojima M, Sakakibara H, Kyozuka J (2007) DWARF10, an RMS1/MAX4/DAD1 ortholog, controls lateral bud outgrowth in rice. Plant J 51:1019–1029

Armstrong GA (1997) Genetics of eubacterial carotenoid biosynthesis: a colorful tale. Annu Rev Microbiol 51:629–659

Bartley GE, Scolnik PA (1993) cDNA cloning, expression during development, and genome mapping of *PSY2*, a second tomato gene encoding phytoene synthase. J Biol Chem 268:25718–25721

Bartley G, Scolnik P (1995) Plant carotenoids: pigments for photoprotection, visual attraction, and human health. Plant Cell 7:1027–1038

Bartley GE, Scolnik PA, Beyer P (1999) Two *Arabidopsis thaliana* carotene desaturases, phytoene desaturase and ζ-carotene desaturase, expressed in *Escherichia coli*, catalyze a poly-*cis* pathway to yield pro-lycopene. Eur J Biochem 259:396–402

Beveridge CA, Ross JJ, Murfet IC (1996) Branching in pea (action of genes *Rms3* and *Rms4*). Plant Physiol 110:859–865

Beveridge CA, Symons GM, Turnbull CG (2000) Auxin inhibition of decapitation-induced branching is dependent on graft-transmissible signals regulated by genes *Rms1* and *Rms2*. Plant Physiol 123:689–698

Beyer P (1989) Carotene biosynthesis in daffodil chromoplasts: on the membrane-integral desaturation and cyclization reactions. In: Boyer CD, Shannon JC, Hardison RC (eds) Physiology, biochemistry and genetics of nongreen plastids. Am Soc Plant Physiol, Rockville, MD, pp 157–170

Beyer P, Kleinig H (1991) Carotenoid biosynthesis in higher plants - membrane-bound desaturation and cyclization reactions in chromoplast membranes from *Narcissus Peudonarcissus*. Biol Chem Hoppe-Seyler 372:527–527

Beyer P, Nievelstein V, Albabili S, Bonk M, Kleinig H (1994) Biochemical aspects of carotene desaturation and cyclization in chromoplast membranes from *Narcissus pseudonarcissus*. Pure Appl Chem 66:1047–1056

Beyer P, Al-Babili S, Ye XD, Lucca P, Schaub P, Welsch R, Potrykus I (2002) Golden rice: introducing the β-carotene biosynthesis pathway into rice endosperm by genetic engineering to defeat vitamin A deficiency. J Nutr 132:506S–510S

Booker J, Auldridge M, Wills S, McCarty D, Klee H, Leyser C (2004) MAX3/CCD7 is a carotenoid cleavage dioxygenase required for the synthesis of a novel plant signaling molecule. Curr Biol 14:1232–1238

Bouis HE (2007) Micronutrient fortification of plants through plant breeding: can it improve nutrition in man at low cost? Proc Nutr Soc 62:403–411

Bouvier F, Dogbo O, Camara B (2003) Biosynthesis of the food and cosmetic plant pigment bixin (annatto). Science 300:2089–2091

Bramley PM (2002) Regulation of carotenoid formation during tomato fruit ripening and development. J Exp Bot 53:2107–2113

Britton G (1998) Overview of carotenoid biosynthesis, vol 3. Birkhäuser, Basel

Burkhardt PK, Beyer P, Wunn J, Kloti A, Armstrong GA, Schledz M, von Lintig J, Potrykus I (1997) Transgenic rice (*Oryza sativa*) endosperm expressing daffodil (*Narcissus pseudonarcissus*) phytoene synthase accumulates phytoene, a key intermediate of provitamin A biosynthesis. Plant J 11:1071–1078

Carol P, Stevenson D, Bisanz C, Breitenbach J, Sandmann G, Mache R, Coupland G, Kuntz M (1999) Mutations in the *Arabidopsis* gene *immutans* cause a variegated phenotype by inactivating a chloroplast terminal oxidase associated with phytoene desaturation. Plant Cell 11:57–68

Cazzonelli C, Cuttriss A, Cossetto S, Pye W, Crisp P, Whelan J, Finnegan E, Turnbull C, Pogson B (2009a) Regulation of carotenoid composition and shoot branching in *Arabidopsis* by a chromatin modifying histone methyltransferase, SDG8. Plant Cell 21(1):39–53

Cazzonelli CI, Millar T, Finnegan J, Pogson BJ (2009b) Promoting gene expression in plants by permissive histone lysine methylation. Plant Signaling Behavior 4(6):484–488

Cazzonelli CI, Yin K, Pogson BJ (2009c) Potential implications for epigenetic regulation of carotenoid biosynthesis during root and shoot development. Plant Signaling Behavior 4(4):339–341

Cheng Q (2006) Structural diversity and functional novelty of new carotenoid biosynthesis genes. J Ind Microbiol Biotechnol 33:552–559

Chinnusamy V, Gong Z, Zhu JK (2008) Abscisic acid-mediated epigenetic processes in plant development and stress responses. J Integr Plant Biol 50:1187–1195

Coleman H, Chew E (2007) Nutritional supplementation in age-related macular degeneration. Curr Opin Ophthalmol 18:220–223

Cook C, Whichard L, Wall M, Egley G, Coggon P (1972) Germination stimulants. II. The structure of strigol - a potent seed germination stimulant for witchweed (*Striga lutea* Lour). J Am Chem Soc 94:6198–6199

Cunningham FJ, Gantt E (1998) Genes and enzymes of carotenoid biosynthesis in plants. Annu Rev Plant Physiol Plant Mol Biol 49:557–583

Cunningham FX Jr, Gantt E (2001) One ring or two? Determination of ring number in carotenoids by lycopene epsilon-cyclases. Proc Natl Acad Sci USA 98:2905–2910

Cunningham FX Jr, Gantt E (2005) A study in scarlet: enzymes of ketocarotenoid biosynthesis in the flowers of *Adonis aestivalis*. Plant J 41:478–492

Cunningham FX, Pogson B, Sun ZR, McDonald KA, DellaPenna D, Gantt E (1996) Functional analysis of the beta and epsilon lycopene cyclase enzymes of *Arabidopsis* reveals a mechanism for control of cyclic carotenoid formation. Plant Cell 8:1613–1626

Cuttriss AJ, Pogson BJ (2004) Carotenoids. In: Davies KM (ed) Plant pigments and their manipulation, vol 14. CRC Press, Boca Raton, FL, pp 57–91

Cuttriss AJ, Chubb A, Alawady A, Grimm B, Pogson B (2007) Regulation of lutein biosynthesis and prolamellar body formation in *Arabidopsis*. Funct Plant Biol 34:663–672

Davies KM (2007) Genetic modification of plant metabolism for human health benefits. Mutat Res 622:122–137

Del Campo JA, Garcia-Gonzalez M, Guerrero MG (2007) Outdoor cultivation of microalgae for carotenoid production: current state and perspectives. Appl Microbiol Biotechnol 74:1163–1174

DellaPenna D, Pogson BJ (2006) Vitamin synthesis in plants: tocopherols and carotenoids. Annu Rev Plant Biol 57:711–738

Fraser PD, Bramley PM (2004) The biosynthesis and nutritional uses of carotenoids. Prog Lipid Res 43:228–265

Fraser PD, Kiano JW, Truesdale MR, Schuch W, Bramley PM (1999) Phytoene synthase-2 enzyme activity in tomato does not contribute to carotenoid synthesis in ripening fruit. Plant Mol Biol 40:687–698

Fraser PD, Bramley P, Seymour GB (2001) Effect of the *Cnr* mutation on carotenoid formation during tomato fruit ripening. Phytochemistry 58:75–79

Fraser PD, Enfissi EM, Bramley PM (2009) Genetic engineering of carotenoid formation in tomato fruit and the potential application of systems and synthetic biology approaches. Arch Biochem Biophys 483(2):196–204

Gerjets T, Sandmann G (2006) Ketocarotenoid formation in transgenic potato. J Exp Bot 57:3639–3645

Giorio G, Stigliani AL, D'Ambrosio C (2008) Phytoene synthase genes in tomato (*Solanum lycopersicum* L.) - new data on the structures, the deduced amino acid sequences and the expression patterns. FEBS J 275:527–535

Giuliano G, Diretto G (2007) Of chromoplasts and chaperones. Trends Plant Sci 12:529–531

Giuliano G, Tavazza R, Diretto G, Beyer P, Taylor MA (2008) Metabolic engineering of carotenoid biosynthesis in plants. Trends Biotechnol 26:139–145

Goff SA, Klee HJ (2006) Plant volatile compounds: sensory cues for health and nutritional value? Science 311:815–819

Gomez-Roldan V, Fermas S, Brewer PB, Puech-Pages V, Dun EA, Pillot JP, Letisse F, Matusova R, Danoun S, Portais JC, Bouwmeester H, Becard G, Beveridge CA, Rameau C, Rochange SF (2008) Strigolactone inhibition of shoot branching. Nature 455:189–194

Gruszecki WI, Grudzinski W, Banaszek-Glos A, Matula M, Kernen P, Krupa Z, Sielewiesiuk J (1999) Xanthophyll pigments in light-harvesting complex II in monomolecular layers: localisation, energy transfer and orientation. Biochim Biophys Acta-Bioenerg 1412:173–183

Hagenimana V, Carey EE, Gichuki ST, Oyunga MA, Imungi JK (1999) Carotenoid contents in fresh, dried and processed sweetpotato products. Ecol Food Nutr 37:455–473

Herbers K (2003) Vitamin production in transgenic plants. J Plant Physiol 160:821–829

Hirschberg J (1998) Molecular biology of carotenoid biosynthesis. In: Britton G, Liaaen-Jensen S, Pfander H (eds) Biosynthesis and metabolism, vol 3. Birkhäuser, Basel, pp 149–194

Hirschberg J (2001) Carotenoid biosynthesis in flowering plants. Curr Opin Plant Biol 4:210–218

Hoa TT, Al-Babili S, Schaub P, Potrykus I, Beyer P (2003) Golden Indica and Japonica rice lines amenable to deregulation. Plant Physiol 133:161–169

Hornero-Mendez D, de Guevara RGL, Minguez-Mosquera MI (2000) Carotenoid biosynthesis changes in five red pepper (*Capsicum annuum* L.) cultivars during ripening. Cultivar selection for breeding. J Agric Food Chem 48:3857–3864

Howitt CA, Pogson BJ (2006) Carotenoid accumulation and function in seeds and non-green tissues. Plant Cell Environ 29:435–445

Howitt C, Cavanagh C, Bowerman A, Cazzonelli C, Rampling L, Mimica J, Pogson B (2009) Regulation of wheat endosperm lutein content by sequence alterations in *eLCY*, and activation of a cryptic exon in phytoene synthase. Funct Integr Genomics 9:363–373

Isaacson T, Ronen G, Zamir D, Hirschberg J (2002) Cloning of *tangerine* from tomato reveals a carotenoid isomerase essential for the production of beta-carotene and xanthophylls in plants. Plant Cell 14:333–342

Jackson H, Braun CL, Ernst H (2008) The chemistry of novel xanthophyll carotenoids. Am J Cardiol 101:50D–57D

Jayaraj J, Devlin R, Punja Z (2008) Metabolic engineering of novel ketocarotenoid production in carrot plants. Transgenic Res 17:489–501

Jin E, Feth B, Melis A (2003) A mutant of the green alga *Dunaliella salina* constitutively accumulates zeaxanthin under all growth conditions. Biotechnol Bioeng 81:115–124

Johnson E, Schroeder W (1995) Microbial carotenoids. Biochem Eng Biotechnol 53:119–178

Josse EM, Simkin AJ, Gaffe J, Laboure AM, Kuntz M, Carol P (2000) A plastid terminal oxidase associated with carotenoid desaturation during chromoplast differentiation. Plant Physiol 123:1427–1436

Jyonouchi H, Sun S, Tomita Y, Gross MD (1995) Astaxanthin, a carotenoid without vitamin A activity, augments antibody responses in cultures including T-helper cell clones and suboptimal doses of antigen. J Nutr 125:2483–2492

Kim J, DellaPenna D (2006) Defining the primary route for lutein synthesis in plants: the role of *Arabidopsis* carotenoid β-ring hydroxylase CYP97A3. Proc Natl Acad Sci USA 103: 3474–3479

Kirk J, Tilney-Bassett R (1978) Proplastids, etioplasts, amyloplasts, chromoplasts and other plastids. In: Kirk JTO, Tilney-Bassett RAE (eds) The plastids. Their chemistry, structure, growth and inheritance, Elsevier/North Holland Biomedical Press, Amsterdam, pp 217–239

Kishimoto S, Maoka T, Nakayama M, Ohmiya A (2004) Carotenoid composition in petals of chrysanthemum (*Dendranthema grandiflorum* (Ramat.) Kitamura). Phytochemistry 65: 2781–2787

Krinsky NI, Johnson EJ (2005) Carotenoid actions and their relation to health and disease. Mol Aspects Med 26:459–516

Li J, Du LF (2001) A new approach to detect plant thykaloid phosphoprotein in vivo. Prog Biochem Biophys 28:740–743

Li F, Murillo C, Wurtzel ET (2007) Maize *Y9* encodes a product essential for 15-*cis*-zeta-carotene isomerization. Plant Physiol 144:1181–1189

Lindgren LO, Stalberg KG, Hoglund A-S (2003) Seed-specific overexpression of an endogenous *Arabidopsis* phytoene synthase gene results in delayed germination and increased levels of carotenoids, chlorophyll, and abscisic acid. Plant Physiol 132:779–785

Lopez-Juez E, Pyke KA (2005) Plastids unleashed: their development and their integration in plant development. Int J Dev Biol 49:557–577

Lorenz RT, Cysewski GR (2000) Commercial potential for *Haematococcus* microalgae as a natural source of astaxanthin. Trends Biotechnol 18:160–167

Lu S, Li L (2008) Carotenoid metabolism: biosynthesis, regulation, and beyond. J Integr Plant Biol 50:778–785

Lu S, Van Eck J, Zhou X, Lopez A, O'Halloran D, Cosman K, Conlin B, Paolillo D, Garvin D, Vrebalov J, Kochian L, Kupper H, Earle E, Cao J, Li L (2006) The cauliflower *Or* gene encodes a DnaJ cysteine-rich domain-containing protein that mediates high-levels of β-carotene accumulation. Plant Cell 18:3594–3605

Mann V, Harker M, Pecker I, Hirschberg J (2000) Metabolic engineering of astaxanthin production in tobacco flowers. Nature Biotechnol 18:888–892

Matusova R, Rani K, Verstappen FW, Franssen MC, Beale MH, Bouwmeester HJ (2005) The strigolactone germination stimulants of the plant-parasitic *Striga* and *Orobanche* spp. are derived from the carotenoid pathway. Plant Physiol 139:920–934

Mayer MP, Nievelstein V, Beyer P (1992) Purification and characterization of a NADPH dependent oxidoreductase from chromoplasts of *Narcissus pseudonarcissus*: a redox-mediator possibly involved in carotene desaturation. Plant Physiol Biochem 30:389–398

Misawa N, Yamano S, Linden H, de Felipe MR, Lucas M, Ikenaga H, Sandmann G (1993) Functional expression of the *Erwinia uredovora* carotenoid biosynthesis gene *CrtI* in transgenic plants showing an increase of β-carotene biosynthesis activity and resistance to the bleaching herbicide norflurazon. Plant J 4:833–840

Misawa N, Masamoto K, Hori T, Ohtani T, Boger P, Sandmann G (1994) Expression of an *Erwinia* phytoene desaturase gene not only confers multiple resistance to herbicides interfering with carotenoid biosynthesis but also alters xanthophyll metabolism in transgenic plants. Plant J 6:481–489

Moehs C, Tian L, Osteryoung K, DellaPenna D (2001) Analysis of carotenoid biosynthetic gene expression during marigold petal development. Plant Mol Biol 45:281–293

Morris SE, Turnbull CG, Murfet IC, Beveridge CA (2001) Mutational analysis of branching in pea. Evidence that Rms1 and Rms5 regulate the same novel signal. Plant Physiol 126:1205–1213

Morris WL, Ducreux LJM, Hedden P, Millam S, Taylor MA (2006) Overexpression of a bacterial 1-deoxy-D-xylulose 5-phosphate synthase gene in potato tubers perturbs the isoprenoid metabolic network: implications for the control of the tuber life cycle. J Exp Bot 57:3007–3018

Nambara E, Marion-Poll A (2005) Abscisic acid biosynthesis and catabolism. Annu Rev Plant Biol 56:165–185

Nassar NMA (2004) Cassava: Some ecological and physiological aspects related to plant breeding. Geneconserve 13:229–245

Nievelstein V, Vandekerckhove J, Tadros MH, Lintig JV, Nitschke W, Beyer P (1995) Carotene desaturation is linked to a respiratory redox pathway in *Narcissus-Pseudonarcissus* chromoplast membranes - involvement of a 23-KDa oxygen-evolving-complex-like protein. Eur J Biochem 233:864–872

Niyogi KK (1999) Photoprotection revisited: Genetic and molecular approaches. Annu Rev Plant Physiol Plant Mol Biol 50:333–359

Niyogi KK (2000) Safety valves for photosynthesis. Curr Opin Plant Biol 3:455–460

Norris SR, Barrette TR, DellaPenna D (1995) Genetic dissection of carotenoid synthesis in *Arabidopsis* defines plastoquinone as an essential component of phytoene desaturation. Plant Cell 7:2139–2149

Paine JA, Shipton CA, Chaggar S (2005) A new version of Golden Rice with increased pro-vitamin A content. Nature Biotechnol 23:482–487

Park H, Kreunen SS, Cuttriss AJ, DellaPenna D, Pogson BJ (2002) Identification of the carotenoid isomerase provides insight into carotenoid biosynthesis, prolamellar body formation, and photomorphogenesis. Plant Cell 14:321–332

Pfundel E, Bilger W (1994) Regulation and possible function of the violaxanthin cycle. Photosynth Res 42:89–109

Phillip D, Young AJ (1995) Occurrence of the carotenoid lactucaxanthin in higher plant LHC II. Photosynth Res 43:273–282

Phillips MA, Leon P, Boronat A, Rodriguez-Concepcion M (2008) The plastidial MEP pathway: unified nomenclature and resources. Trends Plant Sci 13:619–623

Pichersky E (2008) Raging hormones in plants. Nature Chem Biol 4:584–586

Pogson BJ, Rissler HM (2000) Genetic manipulation of carotenoid biosynthesis and photoprotection. Philos Trans R Soc Lond B Biol Sci 355:1395–1403

Pogson B, McDonald KA, Truong M, Britton G, DellaPenna D (1996) *Arabidopsis* carotenoid mutants demonstrate that lutein is not essential for photosynthesis in higher plants. Plant Cell 8:1627–1639

Pogson BJ, Niyogi KK, Bjorkman O, DellaPenna D (1998) Altered xanthophyll compositions adversely affect chlorophyll accumulation and nonphotochemical quenching in *Arabidopsis* mutants. Proc Natl Acad Sci USA 95:13324–13329

Pogson BJ, Rissler HM, Frank HA (2006) The roles of carotenoids in photosystem II of higher plants. In: Wydrzynski T, Satoh K (eds) Photosystem II: The water/plastoquinone oxidoreductase in photosynthesis, vol 22. Springer, Dordrecht, pp 516–536

Pogson BJ, Woo NS, Forster B, Small ID (2008) Plastid signalling to the nucleus and beyond. Trends Plant Sci 13:602–609

Ralley L, Enfissi EMA, Misawa N, Schuch W, Bramley PM, Fraser PD (2004) Metabolic engineering of ketocarotenoid formation in higher in higher plants. Plant J 39:477–486

Rani K, Zwanenburg B, Sugimoto Y, Yoneyama K, Bouwmeester HJ (2008) Biosynthetic considerations could assist the structure elucidation of host plant produced rhizosphere signalling compounds (strigolactones) for arbuscular mycorrhizal fungi and parasitic plants. Plant Physiol Biochem 46:617–626

Ravanello MP, Ke D, Alvarez J, Huang B, Shewmaker CK (2003) Coordinate expression of multiple bacterial carotenoid genes in canola leading to altered carotenoid production. Metab Eng 5:255–263

Rock CD, Zeevaart JA (1991) The *aba* mutant of *Arabidopsis thaliana* is impaired in epoxy-carotenoid biosynthesis. Proc Natl Acad Sci USA 88:7496–7499

Römer S, Fraser PD, Kiano JW, Shipton CA, Misawa N, Schuch W, Bramley PM (2000) Elevation of the provitamin A content of transgenic tomato plants. Nature Biotechnol 18:666–669

Römer S, Lubeck J, Kauder F, Steiger S, Adomat C, Sandmann G (2002) Genetic engineering of a zeaxanthin-rich potato by antisense inactivation and co-suppression of carotenoid epoxidation. Metab Eng 4:263–272

Sandmann G, Römer S, Fraser PD (2006) Understanding carotenoid metabolism as a necessity for genetic engineering of crop plants. Metab Eng 8:291–302

SanGiovanni JP, Chew EY, Clemons TE, Ferris III FL, Gensler G, Lindblad AS, Milton RC, Seddon JM, Sperduto RD (2007) The relationship of dietary carotenoid and vitamin A, E and C intake with age-related macular degeneration in a case-control study: AREDS Report No 22. Arch Ophthalmol 125:1225–1232

Schnurr G, Misawa N, Sandmann G (1996) Expression, purification and properties of lycopene cyclase from *Erwinia uredovora*. Biochem J 315:869–874

Schwab W, Davidovich-Rikanati R, Lewinsohn E (2008) Biosynthesis of plant-derived flavor compounds. Plant J 54:712–732

Schwartz SH, Tan BC, McCarty DR, Welch W, Zeevaart JAD (2003) Substrate specificity and kinetics for VP14, a carotenoid cleavage dioxygenase in the ABA biosynthetic pathway. Biochim Biophys Acta 1619:9–14

Schwartz SH, Qin X, Loewen MC (2004) The biochemical characterization of two carotenoid cleavage enzymes from *Arabidopsis* indicates that a carotenoid-derived compound inhibits lateral branching. J Biol Chem 279:46940–46945

Seo M, Koshiba T (2002) Complex regulation of ABA biosynthesis in plants. Trends Plant Sci 7:41–48

Shewmaker C, Sheehy JA, Daley M, Colburn S, DY K (1999) Seed-specific overexpression of phyoene synthase: increase in carotenoids and other metabolic effects. Plant J 20:401–412

Simkin AJ, Moreau H, Kuntz M, Pagny G, Lin C, Tanksley S, McCarthy J (2008) An investigation of carotenoid biosynthesis in *Coffea canephora* and *Coffea arabica*. J Plant Physiol 165: 1087–1106

Snowden KC, Simkin AJ, Janssen BJ, Templeton KR, Loucas HM, Simons JL, Karunairetnam S, Gleave AP, Clark DG, Klee HJ (2005) The *Decreased apical dominance 1/Petunia hybrida CAROTENOID CLEAVAGE DIOXYGENASE8* gene affects branch production and plays a role in leaf senescence, root growth, and flower development. Plant Cell 17:746–759

Sorefan K, Booker J, Haurogne K, Goussot M, Bainbridge K, Foo E, Chatfield S, Ward S, Beveridge C, Rameau C, Leyser O (2003) *MAX4* and *RMS1* are orthologous dioxygenase-like genes that regulate shoot branching in *Arabidopsis* and pea. Genes Dev 17:1469–1474

Stirnberg P, van De Sande K, Leyser HM (2002) *MAX1* and *MAX2* control shoot lateral branching in *Arabidopsis*. Development 129:1131–1141

Suzuki S, Nishihara M, Nakatsuka T, Misawa N, Ogiwara I, Yamamura S (2007) Flower color alteration in *Lotus japonicus* by modification of the carotenoid biosynthetic pathway. Plant Cell Rep 26:951–959

Tanaka Y, Ohmiya A (2008) Seeing is believing: engineering anthocyanin and carotenoid biosynthetic pathways. Curr Opin Biotechnol 19(2):190–197

Tardy F, Havaux M (1997) Thylakoid membrane fluidity and thermostability during the operation of the xanthophyll cycle in higher-plant chloroplasts. Biochim Biophys Acta 1330:179–193

Tian L, DellaPenna D (2004) Progress in understanding the origin and functions of carotenoid hydroxylases in plants. Arch Biochem Biophys 430:22–29

Tian L, Musetti V, Kim J, Magallanes-Lundback M, DellaPenna D (2004) The *Arabidopsis LUT1* locus encodes a member of the cytochrome P450 family that is required for carotenoid epsilon-ring hydroxylation activity. Proc Natl Acad Sci USA 101:402–407

Tomes ML, Quackenbush FL, Nelsom OE, North B (1953) The inheritance of carotenoid pigment systems in the tomato. Genetics 38:117–127

Tzvetkova-Chevolleau T, Hutin C, Noel LD, Goforth R, Carde JP, Caffarri S, Sinning I, Groves M, Teulon JM, Hoffman NE, Henry R, Havaux M, Nussaume L (2007) Canonical signal recognition particle components can be bypassed for posttranslational protein targeting in chloroplasts. Plant Cell 19:1635–1648

Umehara M, Hanada A, Yoshida S, Akiyama K, Arite T, Takeda-Kamiya N, Magome H, Kamiya Y, Shirasu K, Yoneyama K, Kyozuka J, Yamaguchi S (2008) Inhibition of shoot branching by new terpenoid plant hormones. Nature 455:195–200

Van Norman JM, Sieburth LE (2007) Dissecting the biosynthetic pathway for the *bypass1* root-derived signal. Plant J 49:619–628

Van Norman JM, Frederick RL, Sieburth LE (2004) *BYPASS1* negatively regulates a root-derived signal that controls plant architecture. Curr Biol 14:1739–1746

Verma PK, Upadhyaya KC (1998) A multiplex RT-PCR assay for analysis of relative transcript levels of different members of multigene families: application to *Arabidopsis* calmodulin gene family. Biochem Mol Biol Int 46:699–706

Vishnevetsky M, Ovadis M, Vainstein A (1999) Carotenoid sequestration in plants: the role of carotenoid-associated proteins. Trends Plant Sci 4:232–235

von Lintig J, Vogt K (2000) Molecular identification of an enzyme cleaving β-carotene to retinal. J Biol Chem 275:11915–11920

Welsch R, Beyer P, Hugueney P, Kleinig H, von Lintig J (2000) Regulation and activation of phytoene synthase, a key enzyme in carotenoid biosynthesis, during photomorphogenesis. Planta 211:846–854

Welsch R, Wust F, Bar C, Al-Babili S, Beyer P (2008) A third phytoene synthase is devoted to abiotic stress-induced abscisic acid formation in rice and defines functional diversification of phytoene synthase genes. Plant Physiol 147:367–380

West KP (2003) Vitamin A deficiency disorders in children and women. Food Nutr Bull 24:78–90

Wilson P, Estavillo G, Field K, Pornsiriwong P, Carroll A, Howell K, Woo N, Lake J, Smith S, Millar A, von Caemmerer S, Pogson B (2009) The nucleotidase/phosphatase, SAL1, is a negative regulator of drought tolerance in *Arabidopsis*. Plant J 58(2):299–317

Wu D, Wright DA, Wetzel C, Voytas DF, Rodermel S (1999) The *IMMUTANS* variegation locus of *Arabidopsis* defines a mitochondrial alternative oxidase homolog that functions during early chloroplast biogenesis. Plant Cell 11:43–56

Wyss A, Wirtz G, Woggon W, Brugger R, Wyss M, Friedlein A, Bachmann H, Hunziker W (2000) Cloning and expression of β-carotene, 15,15′-dioxygenase. Biochem Biophys Res Comm 271:334–336

Ye X, Al-Babili S, Klöti A, Zhang J, Lucca P, Beyer P, Potrykus I (2000) Engineering the provitamin A (-carotene) biosynthetic pathway into (carotenoid-free) rice endosperm. Science 287:303–305

Yu F, Fu A, Aluru M, Park S, Xu Y, Liu H, Liu X, Foudree A, Nambogga M, Rodermel S (2007) Variegation mutants and mechanisms of chloroplast biogenesis. Plant Cell Environ 30: 350–365

Zhou X, Van Eck J, Li L, El-Gewely MR (2008) Use of the cauliflower *Or* gene for improving crop nutritional quality. Biotechnol Annu Rev 14:171–190

Zhu C, Naqvi S, Capell T, Christou P (2008) Metabolic engineering of ketocarotenoid biosynthesis in higher plants. Arch Biochem Biophys 483:182–190

Zou J, Zhang S, Zhang W, Li G, Chen Z, Zhai W, Zhao X, Pan X, Xie Q, Zhu L (2006) The rice *HIGH-TILLERING DWARF1* encoding an ortholog of *Arabidopsis* MAX3 is required for negative regulation of the outgrowth of axillary buds. Plant J 48:687–698

Chapter 8
Biosynthesis and Regulation of Alkaloids

G. Guirimand, V. Courdavault, B. St-Pierre, and V. Burlat

8.1 Introduction

Plant secondary metabolites, also commonly named plant natural products, nowadays encompassed about 100,000 chemically identified, low molecular weight compounds. These molecules are commonly synthesised in a plant-, organ- and even cell-specific manner. Alkaloids constitute a very chemically diverse group of secondary metabolites with an estimated 12,000 different molecules sharing as a unique common feature the presence of a nitrogen atom within a heterocyclic ring. Many alkaloids are toxic, in agreement with some function in plant defence. This toxicity has long been understood by humans, and some alkaloids are very well known for their human health benefit, leading to the development of many recent research projects towards understanding the architecture and the regulation of their long and complex biosynthetic pathways. The purpose of this chapter is to give an overview of some of the major advances obtained recently with the most prominent model species, including the architecture and the spatial organisation of biosynthetic pathways, the crystallisation and modelling of alkaloid biosynthetic enzymes and the transcription factor regulatory networks of alkaloid biosynthesis. The ultimate goal of such research is the application of the discoveries to develop metabolic engineering strategies to overcome the usually very low yield of production for these biomolecules. In the meantime, some peculiarities of plant physiology have been elucidated in these highly

G. Guirimand, V. Courdavault, and B. St-Pierre
Université François Rabelais de Tours, EA 2106 "Biomolécules et Biotechnologies Végétales", IFR 135, "Imagerie Fonctionnelle", 37200, Tours, France
e-mail: vincent.courdavault@univ-tours.fr

V. Burlat
Surfaces Cellulaires et Signalisation chez les Végétaux, UMR 5546 CNRS - UPS - Université de Toulouse, Pôle de Biotechnologie Végétale, 24 chemin de Borde-Rouge, BP 42617 Auzeville, 31326, Castanet-Tolosan, France
Université François Rabelais de Tours, EA 2106 "Biomolécules et Biotechnologies Végétales", IFR 135, "Imagerie Fonctionnelle", 37200, Tours, France
e-mail: burlat@scsv.ups-tlse.fr

E-C. Pua and M.R. Davey (eds.),
Plant Developmental Biology – Biotechnological Perspectives: Volume 2,
DOI 10.1007/978-3-642-04670-4_8, © Springer-Verlag Berlin Heidelberg 2010

specialised plant species, illustrating the fundamental interest of studying such secondary metabolic pathways beyond their primary interest for industrial application.

8.2 Chemical Diversity and Biosynthesis

Alkaloids are classified in several families that present totally different biosynthetic pathways. Four major families, for which the biosynthesis and the regulation are more particularly studied, are discussed in this chapter, these being monoterpene indole alkaloids (MIA), benzylisoquinoline alkaloids (BIA), tropane and nicotine alkaloids (TNA) and purine alkaloids (PA; Fig. 8.1, Table 8.1). Despite their chemical diversity, alkaloids share the fact that they originate commonly from primary metabolites such as amino acids or bases (Fig. 8.1). Except for *Nicotiana tabacum*, no genome sequencing project exists for the major alkaloid-producing plants. Therefore, most of the enzymatic steps have been identified using classical biochemical and molecular biology studies. The ongoing elucidation of some biosynthetic pathways illustrates the recruitment of enzymes belonging to recurrent multigene families such as cytochrome P450 monooxygenases, acetyl transferases or methyltransferases. Recent expressed sequence tags (EST) transcriptomic projects focusing on MIA-producing species (Murata et al. 2006, 2008; Rischer et al. 2006; Shukla et al. 2006), BIA-producing species (Morishige et al. 2002; Carlson et al. 2006; Ziegler et al. 2006; Kato et al. 2007; Zulak et al. 2007) and TNA-producing species (Li et al. 2006) have been helpful for the identification of some missing enzymatic steps, and should accelerate this process in the near future (Table 8.2). Metabolic profiling studies of the major alkaloid-producing species have also been performed recently (Dräger 2002; Yamazaki et al. 2003; Schmidt et al. 2007; Hagel et al. 2008), sometimes in association with transcriptomic studies (Rischer et al. 2006; Ziegler et al. 2006; Zulak et al. 2007; Hagel and Facchini 2008). Together, these approaches now allow tremendous progress in understanding these complex alkaloid biosynthetic pathways.

The purpose of this section is more to give an overlook and a link to references, including major recent reviews on the different biosynthetic pathways, than to give an exhaustive detailed description of all these complicated biochemical processes.

8.2.1 Biosynthesis of Monoterpene Indole Alkaloids (MIA)

The approximately 2,000 MIA chemical structures described so far are widespread in a large number of plant species (Ziegler and Facchini 2008). Some of these molecules are of interest to human health, such as the anticancer drugs vinblastine and vincristine and the antihypertensive drug ajmalicine specifically produced in *Catharanthus roseus*, the anti-arrythmic ajmaline produced in *Rauvolfia serpentina*, or the anticancer compound camptothecin produced mostly in *Camptotheca acuminata* (Fig. 8.1, Table 8.1). These molecules are part of the large array of MIA

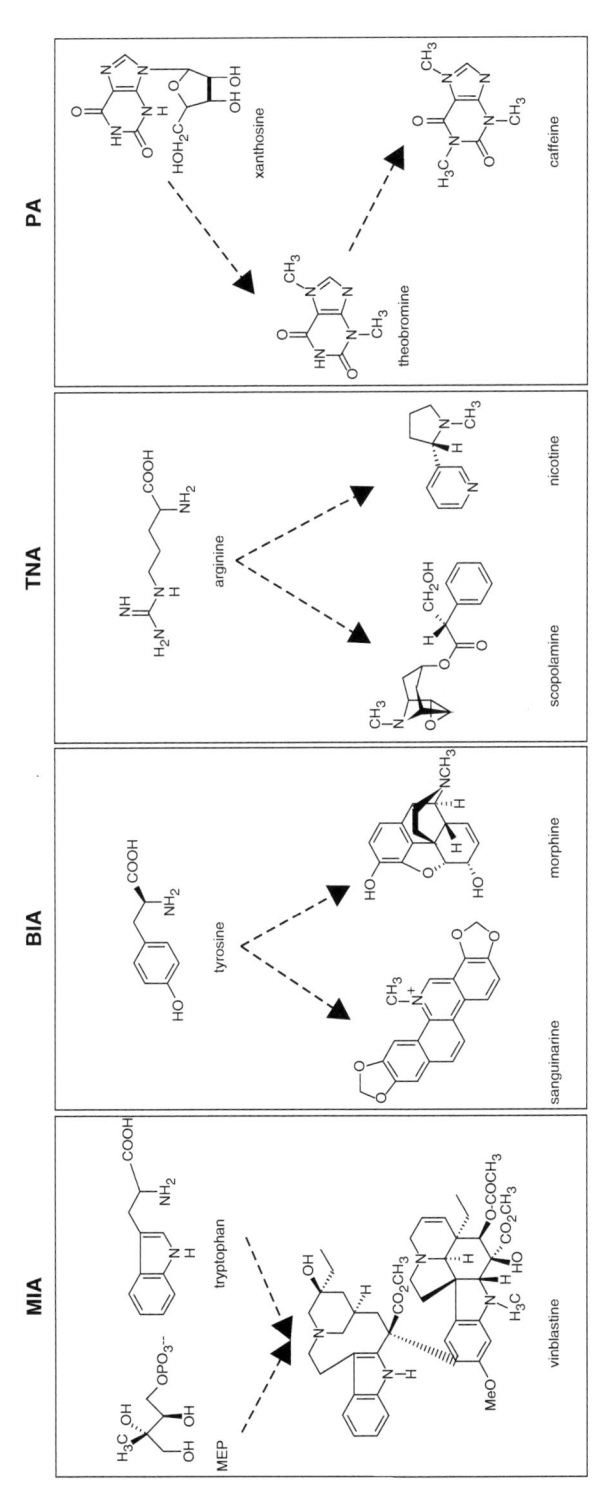

Fig. 8.1 Chemical structures of precursors and examples of alkaloids from the four families where the major advances in the elucidation and regulation of the biosynthetic pathways have been obtained. MIA, monoterpene indole alkaloids; BIA, benzylisoquinoline alkaloids; TNA, tropane and nicotine alkaloids; PA, purine alkaloids; MEP, 2-C-methyl-D-erythritol 4-phosphate

Table 8.1 Sources, chemical diversity and biological activity of four major families of alkaloids

Alkaloid family	Family (plant species)	Active compounds	Pharmacological activity	Target
Monoterpene indole alkaloids (MIA)	Apocynaceae			
	Catharanthus roseus	Vinblastine, vincristine	Anticancer	Tubulin
		Ajmalicine	Antihypertensive	α Adrenergic receptor
	Rauvolfia serpentina	Ajmaline	Anti-arrythmic	Na$^+$ channels
	Nyssaceae			
	Camptotheca acuminata	Camptothecin	Anticancer	DNA topoisomerase I
Benzylisoquinoline alkaloids (BIA)	Papaveraceae			
	Papaver somniferum	Codeine	Antitussive, analgesic	Nicotinic acetylcholine (nACh) receptor
		Morphine	Analgesic, narcotic	μ3 Opioid receptor
		Papaverine	Spasmolytic, vasodilators	c-AMP phosphodiesterase
	Eschscholzia californica	Sanguinarine	Antibacterial, proapoptotic	FtsZ (bacterial cytokinesis), mitoch. pathway
	Ranunculaceae			
	Thalictrum flavum	Berberine	Antibacterial, antimicrobial	DNA
	Coptis japonica	Berberine, sanguinarine		
Tropane and nicotine alkaloids (TNA)	Solanaceae			
	Hyoscyamus niger	Hyoscyamine	Anticholinergic, narcotic, myorelaxant	Muscarinic receptor
	Datura stramonium	Scopolamine	Anticholinergic, narcotic, myorelaxant	Muscarinic receptor
	Atropa belladonna	Hyoscyamine	Anticholinergic, narcotic, myorelaxant	Muscarinic receptor
	Nicotiana tabacum	Nicotine	Neurostimulant, insecticide	nACh receptor
Purine alkaloids (PA)	Coffeeae			
	Coffea arabica	Caffeine	Central nervous system stimulant	Adenosine A$_1$ & A$_{2A}$ receptors; phosphodiesterase
	Theaceae			

Camellia sinensis	Caffeine, theophylline	Central nervous system stimulant	Adenosine A_1 & A_{2A} receptors; phosphodiesterase
Byttnerioideae			
Theobroma cacao	Theobromine, caffeine	Central nervous system stimulant	Adenosine A_1 & A_{2A} receptors; phosphodiesterase

Table 8.2 Molecular resources for alkaloid biosynthesis

Family (plant species)	NCBI hints (nt seq)	NCBI hints (EST)	Alkaloid biosynthetic enzymes (GenBank access numbers, full enzyme name and abbreviation)
Alkaloid family: monoterpene indole alkaloids (MIA)			
Apocynaceae			
Catharanthus roseus	20,820	19,899	25 [CAA09804(1-deoxyxylulose 5-phosphate synthase: dxs); ABI35993(dxs2); AAF65154(1-deoxy-D-xylulose-5-phosphate reductoisomerase: dxr); ACI16377 (4-diphosphocytidyl-methylerythritol 2-phosphate synthase: cms); ABI35992(4-diphosphocytidyl-2-C-methyl-D-erythritol kinase: cmk); AAF65155(2C-methyl-D-erythritol 2,4-cyclodiphosphate synthase: mecs); AAO24774(4-hydroxy-3-methyl-2-butenyl diphosphate synthase: hds); ABI30631(1-hydroxy-2-methyl-butenyl 4-diphosphate reductase: hdr); ABW98669(isopentenyl pyrophosphate:dimethylallyl pyrophosphate isomerase: idi); ACC77966(geranyl pyrophosphate synthase: gpps); CAC80883(CYP76B6=geraniol 10-hydroxylase: g10h); Q05001(NADPH cytochrome P450 reductase: cpr); AAQ55962(10-hydroxygeraniol oxidoreductase: 10hgo); ABW38009(loganic acid methyltransferase: lamt); AAA33106 (CYP72A1=secologanin synthase: sls); CAC29060(anthranilate synthase α subunit: asa); AAA33109(tryptophan decarboxylase: tdc); CAA43936(strictosidine synthase: str); AAF28800(strictosidine β-D-glucosidase: sgd); AAO13736(minovincinine 19-hydroxy-O-acetyltransferase: mat); CAB56503(CYP71D12=tabersonine 16-hydroxylase: t16h); ABR20103(16-hydroxytabersonine O-methyltransferase: 16omt); O04847(desacetoxyvindoline-4-hydroxylase: d4h); AAC99311 (deacetylvindoline 4-O-acetyltransferase: dat); CAJ84723 (peroxidase 1: prx1)]
Rauvolfia serpentina	17	-	7 [CAA44208(strictosidine synthase: str1); CAC83098(strictosidine β-D-glucosidase: sgd1); AAF22288(polyneuridine aldehyde esterase: pnae); Q70PR7(vinorine synthase: vs); AAF03675(raucaffricine β-D-glucosidase: rg); AAW88320 (acetylajmalan acetylesterase: aae); AAX11684(perakine reductase: pr)]
Nyssaceae			
Camptotheca acuminata	31	-	10 [ABC86579(1-deoxy-D-xylulose-5-phosphate reductoisomerase: dxr); O48964 (isopentenyl pyrophosphate:dimethylallyl pyrophosphate isomerase: idi1); O48965 (idi2); AAV64030(5-enolpyruvylshikimate 3-phosphate synthase: epsps); AAU84988(anthranilate synthase α subunit: asa1); AAU84989(asa2); AAB97526 (tryptophan synthase β subunit: tsb); AAB39708(tryptophan decarboxylase: tdc1); AAB39709(tdc2); AAQ20892(10-hydroxygeraniol oxidoreductase: 10hgo)]

Alkaloid family: benzylisoquinoline alkaloids (BIA)

Papaveraceae			
Papaver somniferum	20,458	20,340	16 [P54768(tyrosine/DOPA decarboxylase: tydc1); P54770(tydc3); AAX56303(S-norcoclaurine synthase: ncs1); AAX56304.(ncs2); AAQ01669(norcoclaurine 6-O-methyltransferase: 6omt); AAP45316(coclaurine N-methyltransferase: cnmt); AAF61400(CYP80B1 renamed CYP80B3=N-methylcoclaurine 3'-hydroxylase); AAP45313(3'-hydroxy-N-methylcoclaurine 4'-O-methyltransferase: 4'omt1); AAP45314(4'omt2); AAQ01668(reticuline 7-O-methyltransferase: 7omt); P93479 (berberine bridge enzyme: bbe); AAY79177(tetrahydroprotoberberine-cis-N-methyltransferase: tnmt); ABR14720(salutaridine synthase: ss); ABC47654 (salutaridine reductase: salr); Q94FT4(salutaridinol 7-O-acetyltransferase: salat); AAF13738(codeinone reductase: cor1)]
Eschscholzia californica	9,161	9,083	5 [P30986(berberine bridge enzyme: bbe1); BAE79723(reticuline-7-O-methyltransferase: 7omt); O64899(CYP80B1 renamed CYP80B3=N-methylcoclaurine 3'-hydroxylase); AAC39454(CYP82B1=N-methylcoclaurine 3'-hydroxylase); O64900(CYP80B2=N-methylcoclaurine 3'-hydroxylase)]
Ranunculaceae			
Thalictrum flavum	10	–	9 [AAG60665(tyrosine/dopa decarboxylase: tydc1); AAR22502(norcoclaurine synthase: ncs); AAU20765(norcoclaurine 6-O-methyltransferase: 6omt); AAU20766 (coclaurine N-methyltransferase: cnmt); AAU20767(CYP80B3=N-methylcoclaurine 3'-hydroxylase); AAU20768(3'-hydroxy-N-methylcoclaurine 4'-O-methyltransferase: 4'omt); AAU20769(berberine bridge enzyme: bbe); AAU20770(scoulerine 9-O-methyltransferase: somt); AAU20771 (CYP719A1=canadine synthase)]
Coptis japonica	135	37	9 [BAF45337(norcoclaurine synthase: ncs); Q9LEL6(norcoclaurine 6-O-methyltransferase: 6omt); Q948P7(coclaurine N-methyltransferase: cnmt); Q9LEL5 (3'-hydroxy-N-methylcoclaurine 4'-O-methyltransferase: 4'omt); Q39522(scoulerine 9-O-methyltransferase: somt); BAB68769(CYP719A=methylenedioxy bridge-forming enzyme); Q8H9A8(columbamine O-methyltransferase: coomt); BAB12433 (CYP80B2=N-methylcoclaurine-3'-hydroxylase); BAF80448 (CYP80G2=corytuberine synthase)]

(continued)

Table 8.2 (continued)

Family (plant species)	NCBI hints (nt seq)	NCBI hints (EST)	Alkaloid biosynthetic enzymes (GenBank access numbers, full enzyme name and abbreviation)
Alkaloid family: tropane and nicotine alkaloids (TNA)			
Solanaceae			
Hyoscyamus niger	23	-	5 [BAA82263(putrescine N-methyltransferase: pmt); BAA85844(tropinone reductase: tr1); P50164(tr2); ABD39696(CYP80F1=littorine mutase/monooxygenase); P24397 (hyoscyamine 6-β-hydroxylase: h6h)]
Atropa belladonna	58	-	3 [BAA82264(putrescine N-methyltransferase: pmt1); BAA82262(pmt2); BAA78340 (hyoscyamine 6-β-hydroxylase: h6h)]
Datura stramonium	31	-	6 [P50134(ornithine decarboxylase: odc); CAB64599(arginine decarboxylase: adc1); CAE47481(putrescine N-methyltransferase: pmt); P50162(tropinone reductase: tr1); P50163(tr2); P50165(trh)]
Nicotiana tabacum	1,673,039	240,440	4 [AAQ14852(ornithine decarboxylase: odc); AAF14881(putrescine N-methyltransferase: pmt); ABI93948(methylputrescine oxidase: mpo1); DQ131886 (CYP82E4v1=nicotine N-demethylase)]
Coffeeae			
Coffea arabica	6,079	1,577	10 [BAB39215(xanthosine methyltransferase: xmt1); BAC75665(xmt2); BAB39216(7-methylxanthine N-methyltransferase: mxmt1); BAC75664(mxmt2); BAC75663(3,7-dimethylxanthine N-methyltransferase: dxmt1); BAC43756(theobromine synthase: cts1); BAC43757(cts2); BAC43755(7-methylxanthosine synthase: xrs1); BAC43760 (caffeine synthase: ccs1); BAC43761(ctcs7)]
Camellia sinensis	3,732	3,288	1 [ABP98983(caffeine synthase: tcs)]
Byttnerioideae			
Theobroma cacao	7,469	6,790	1 [BAE79730(caffeine synthase: bcs1)]

that a single plant species is able to produce. For example, there are more than 130 MIA in *C. roseus* (van der Heijden et al. 2004). The elucidation of MIA biosynthetic pathways in the three mentioned species has undergone major recent progress with the identification of 42 clones corresponding to 31 enzymatic steps (Table 8.2). In *C. roseus* alone, 27 enzymatic steps have been studied (25 cDNA clones and two additional enzymatic activities with no assigned clone). In these species, the pathways share a common origin with strictosidine synthase (STR), catalysing the condensation of the indole precursor tryptamine with the terpenoid precursor secologanin to form the first MIA, strictosidine. The upstream biosynthesis of the indole precursor derived from the shikimate pathway via tryptophan, and of the terpenoid precursor originating from the methyl erythritol phosphate (MEP) pathway, is also shared within these plant species. Strictosidine β-glucosidase (SGD), catalysing the deglucosylation of strictosidine, is the last common enzyme for the biosynthesis of 2,000 MIA, since the resulting aglycon is the starting point for many different species-specific lateral MIA pathways, with the observed possibility for a given species to harbour more than one of these pathways (e.g. *C. roseus*, reviewed in van der Heijden et al. 2004). Many enzymatic steps are yet to be discovered. More details on these pathways and the identified enzymatic steps are available in recent reviews (Lorence and Nessler 2004; van der Heijden et al. 2004; Stöckigt and Panjikar 2007; Mahroug et al. 2007; Ziegler and Facchini 2008).

8.2.2 Biosynthesis of Benzylisoquinoline Alkaloids (BIA)

BIA constitute a diverse class of more than 2,500 compounds with, for some of them, potent pharmacological properties and socio-economic importance. In opium poppy (*Papaver somniferum*) alone, more than 80 alkaloids have been identified. BIA have also been widely studied in other members of the Papaveraceae, such as *Eschscholzia californica*, or members of the Ranunculaceae such as *Thalictrum flavum* and *Coptis japonica* (Table 8.1). The biosynthesis of BIA starts with the condensation of two tyrosine derivatives to produce (S)-norcoclaurine. Four characterised enzymatic steps are necessary to produce (S)-reticuline, the central precursor of the five major BIA subpathways, leading to palmatine, berberine, sanguinarine, laudanine and codeine/morphine respectively. Overall, regardless of the plant model species, a total of 39 clones and 19 enzymatic steps have been characterised (Table 8.2), making the BIA biosynthetic pathway one of the best characterised plant natural product complex pathways (reviewed in Ziegler and Facchini 2008).

8.2.3 Biosynthesis of Tropane and Nicotine Alkaloids (TNA)

TNA are widely used in medicine as nonselective muscarinic antagonists affecting peripheral and central nervous systems (Table 8.1). This alkaloid family is found mainly in Solanaceae species, such as *Hyoscyamus niger*, *Datura stramonium*, *Atropa belladonna* or *Nicotiana tabacum*, and accounts for more than 200 different compounds (Dräger 2002, 2006; Oksman-Caldentey 2007). TNA are amongst the most studied alkaloids and their pharmacological effects are well documented. However, their biosynthetic pathways are still only partially understood. TNA are derived from the amino acids ornithine and arginine (Fig. 8.1), and the early biosynthetic steps leading to *N*-methylputrescine formation have been elucidated in several species (Table 8.2). The oxidative deamination of *N*-methylputrescine leads to *N*-methylpyrrolium cation, which constitutes a branching point towards tropane alkaloids and nicotine alkaloids respectively. The final steps of both pathways are partially characterised, but molecular information concerning the central steps is still missing. Overall, seven enzymatic steps of the tropane alkaloid biosynthetic pathway and four enzymatic steps of the nicotinic alkaloid biosynthetic pathway have been characterised. More details on these pathways and on the identified enzymatic steps are available in recent reviews (Dräger 2006; Oksman-Caldentey 2007).

8.2.4 Biosynthesis of Purine Alkaloids (PA)

PA are natural products derived from purine nucleotides (Fig. 8.1; Ashihara et al. 2008). The main PA are caffeine and theobromine that affect the central nervous system as neurostimulants and are synthesised by several plants, including *Coffea arabica*, *Camellia sinensis* or *Theobroma cacao* (Table 8.1). The initial precursor of PA is xanthosine, which is supplied by at least four different pathways. The main caffeine biosynthetic pathway has four enzymatic steps, comprising three characterised *S*-adenosylmethionine-dependant *N*-methylation reactions and one uncharacterised nucleosidase reaction (Table 8.2). However, according to structural studies of *N*-methyltransferases involved in caffeine biosynthesis, it has been suggested that the ribose hydrolysis could be performed by xanthosine 7 *N*-methyltransferase (McCarthy and McCarthy 2007). The detail of this pathway has been reviewed recently (Ashihara et al. 2008).

8.3 Spatial Organisation of Alkaloid Biosynthesis

The spatial organisation of alkaloid biosynthesis has been recently investigated extensively using in situ hybridisation and immunocytochemistry methods (reviewed in De Luca and St-Pierre 2000; Kutchan 2005; Mahroug et al. 2007; Ziegler and Facchini 2008). An astonishing complexity has been uncovered showing multicellular organisations as a recurrent common feature. These types of organisation implicate the necessity of intercellular translocation processes. In *C. roseus*, a series of publications showed the sequential involvement of internal phloem associated parenchyma (IPAP), epidermis and laticifers-idioblasts during

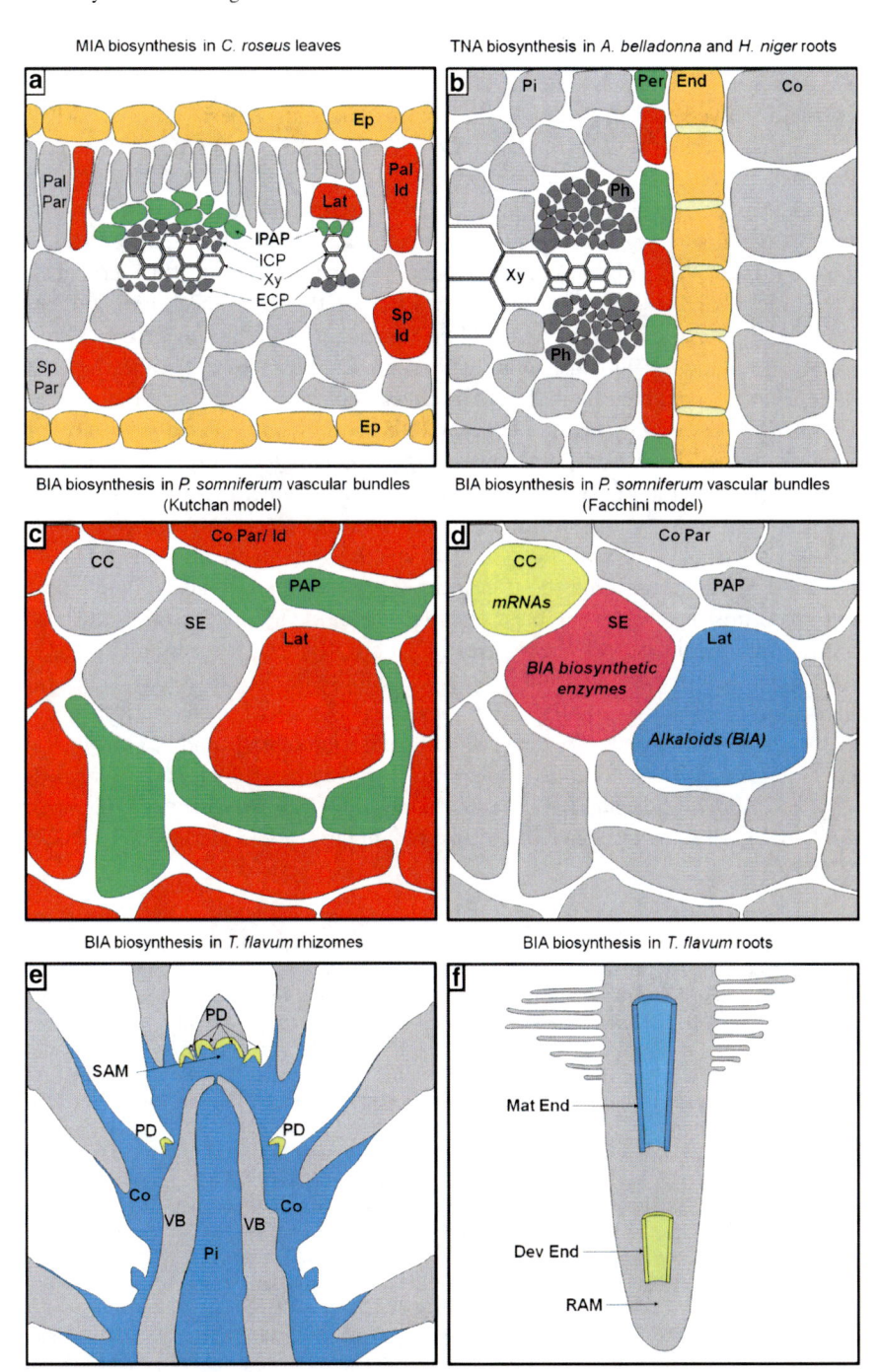

Fig. 8.2 Examples of multicellular organisation of alkaloid biosynthesis implicating the necessity of intercellular translocation events. Six models of spatial organisation of alkaloid biosynthesis in higher plants are presented in **a–f**. In **a–c**, the *green–orange–red* traffic light signal colour code

MIA biosynthesis in aerial organs (Fig. 8.2a; St-Pierre et al. 1999; Irmler et al. 2000; Burlat et al. 2004; Oudin et al. 2007). The IPAP cells harbour the expression of genes involved in early steps of monoterpenoid biosynthesis, i.e. four MEP pathway genes and *geraniol 10-hydroxylase* (G10H, CYP76B6) encoding the first committed enzyme in monoterpenoid biosynthesis (Burlat et al. 2004; Oudin et al. 2007; Guirimand et al. 2009). The intermediate steps leading to the synthesis of the two MIA precursors, tryptamine and secologanin, and to their subsequent condensation to form the first MIA strictosidine, occur within the epidermis (St-Pierre et al. 1999; Irmler et al. 2000). Finally, the last two steps in the biosynthesis of vindoline, one of the monomeric MIA precursors of the dimeric MIA vinblastine, are localised to specialised laticifer-idioblast cells (St-Pierre et al. 1999). Recently, these results were elegantly completed by an RT-PCR analysis of laser capture microdissected *C. roseus* leaf cells (Murata and De Luca 2005) and by an EST analysis study of epidermis-enriched fractions obtained using an original carborundum abrasion technique (Levac et al. 2008; Murata et al. 2008). Altogether these results suggest that, to ensure a continuity in the metabolic flux along the MIA pathway, it is necessary to consider the translocation of an unknown monoterpenoid intermediate from IPAP to epidermis, and the translocation of an unknown MIA intermediate from epidermis to laticifer-idioblast cells (Fig. 8.2a). The identification of these shuttling intermediates will necessitate the localisation of two subsequent enzymatic steps within two different cell types. Similarly, in the roots of several solanaceous species, the shuttle of intermediate TNA between the pericycle and the endodermis is suggested by the specific localisation of early and late enzymatic steps (putrescine *N*-methyltransferase and hyoscyamine 6-hydroxylase respectively) to the pericycle, and the specific expression of an intermediate enzymatic step (tropinone reductase 1) within the neighbouring endodermis (Fig. 8.2b;

Fig. 8.2 (continued) illustrates (respectively) early, intermediate and late biosynthetic steps sequentially occurring in different cell types along a given pathway. These three models implicate the intercellular translocation of metabolic intermediates. In **d–f**, the yellow, magenta and blue colour code shows the common localisation of all the mRNAs (*yellow*), enzymes (*magenta*) and alkaloids (*blue*) from a given pathway in different cell types. The Facchini model (**d**) implicates the translocation of enzymes from companion cells to sieve elements and the translocation of alkaloids from sieve elements to laticifers. The models in **d–f** illustrate that a same pathway can be localised to different cell types according to the plant species and/or to the organ considered. IPAP, internal phloem associated parenchyma; Ep, epidermis; Lat, laticifers; Pal Id, palisade idioblasts; Sp Id, spongy parenchyma idioblasts; Pal Par, palisadic parenchyma; Sp par, spongy parenchyma; ICP, internal conducting phloem; ECP, external conducting phloem; Xy, xylem; Pi, pith; Per, pericycle; End, endodermis; Co, cortex; CC, companion cells; SE, sieve elements; Co Par/Id, cortex parenchyma/Idioblasts; PAP, phloem associated parenchyma; PD, protoderm; SAM, shoot apical meristem; VB, vascular bundles; RAM, root apical meristem; Dev End, developing endodermis; Mat End, mature endodermis. These models were drawn from results taken from the reports by **a** St-Pierre et al. (1999), Irmler et al. (2000), Burlat et al. (2004) and Oudin et al. (2007); **b** Hashimoto et al. (1991), Nakajima and Hashimoto (1999) and Suzuki et al. (1999a, b); **c** Bock et al. (2002) and Weid et al. (2004); **d** Bird et al. (2003) and Samanani et al. (2005); **e, f** Samanani et al. (2005)

Hashimoto et al. 1991; Nakajima and Hashimoto 1999; Suzuki et al. 1999a, b). In this case, given the proximity of conducting tissues, it is hypothesised that the precursor of TNA (arginine) could come from the phloem, and that final TNA products, such as scopolamine, could flow to the aerial parts of the plant through the xylem (Fig. 8.2b; Nakajima and Hashimoto 1999). Four models of the spatial organisation of BIA synthesis in opium poppy and in a species (*T. flavum*) of Ranunculaceae also illustrate the complexity of these compartmentations. In opium poppy, two different models have been proposed (Fig. 8.2c, Kutchan 2005; Fig. 8.2d, Facchini and St-Pierre 2005). In one model, a rationale similar to that described for MIA and TNA biosynthesis is proposed, since a cell-specific separation occurs between early and late steps of different branches of the BIA pathway. According to immunolocalisation studies performed on five biosynthetic enzymes, this model proposed that BIA synthesis commonly starts in the phloem parenchyma with the synthesis of (S)-reticuline. Different situations are then observed for three BIA subpathways leading to laudanine, codeine/morphine and scoulerine respectively. The synthesis of laudanine appears to occur within the same phloem parenchyma cells, whereas the biosynthesis of scoulerine is localised to leaf idioblasts and root cortex cells. Finally, an intermediate step in the codeine pathway is also localised in phloem parenchyma, whereas the final step occurs in laticifers (Fig. 8.2c; Bock et al. 2002; Weid et al. 2004; reviewed in Kutchan 2005). However, Facchini and St-Pierre (2005) proposed a totally different type of compartmentation in the same species with a so-called tale of three cell types, implicating the transcription of seven BIA biosynthesis genes within companion cells, the immunolocalisation of the corresponding enzymes to the sieve elements, and the accumulation of BIA in neighbouring laticifers (Fig. 8.2d; Bird et al. 2003; Samanani et al. 2006; reviewed in Ziegler and Facchini 2008). The discrepancy between both models has been attributed tentatively to differences in cultivars and/ or in developmental stages. The last example of BIA synthesis in *T. flavum* illustrates that several steps of a pathway may be compartmentalised in a different manner within different plant species (compare *P. somniferum*, Fig. 8.2c, d and *T. flavum*, Fig. 8.2e, f) and even within different organs (rhizomes versus roots) of the same plant species (Fig. 8.2e, f; Samanani et al. 2005).

Alkaloid biosynthesis also involves subcellular compartmentation of biosynthetic enzymes, as is the case for most natural product biosynthetic pathways. Beside cytosol, organelles such as ER (either lumen or membranes), plastids (either stroma or thylakoids), mitochondria and vacuoles have been implicated in various alkaloid biosynthetic pathways (reviewed in Facchini and St-Pierre 2005; Mahroug et al. 2007; Ziegler and Facchini 2008). These results are based on in silico analysis of biosynthetic enzyme sequences and also on more formal experimental evidence. Part of these experiments corresponded to density gradient analysis, and a few examples of direct localisation of enzymes by immunogold (McKnight et al. 1991; Bock et al. 2002; Alcantara et al. 2005; Samanani et al. 2006; Oudin et al. 2007; Guirimand et al. 2009) and by GFP-fusion image analyses (Bird and Facchini 2001; Costa et al. 2008; Guirimand et al. 2009) have been published. This emphasizes that a systematic (re)evaluation of the subcellular localisation of all the enzymes available in a given alkaloid pathway constitutes a future challenge that should enable the drawing of more complete spatial compartmentation models that integrate both cellular and subcellular levels.

Together, these results suggest the recurrent necessity of transmembrane and intercellular translocation processes during alkaloid biosynthesis. Members of the ATP binding cassette (ABC) transporter superfamily have been demonstrated to be able to recruit various alkaloids such as vinblastine, hyoscyamine, scopolamine or berberine (Kolaczkowski et al. 1996; Sakai et al. 2002; Goossens et al. 2003; Shitan et al. 2003). The possibility that alkaloid intermediates and/or enzymes flow through the symplasm should also be considered, even though no experimental proofs are available. Finally, the organisation of clusters of biosynthetic enzymes in metabolic channels has also been purported in BIA synthesis (Samanani et al. 2006).

8.3.1 Crystallisation and Three-Dimensional Structure of Alkaloid Biosynthetic Enzymes

The molecular architecture of some alkaloid biosynthetic enzymes has recently been investigated, providing details on their catalytic mechanism and opening new perspectives on the production of alkaloid derivatives with improved properties using enzyme engineering. The TNA biosynthetic pathway provides the first two examples of elucidation of three-dimensional architecture of plant alkaloid biosynthetic enzymes by determination of the crystal structures of two tropinone reductases from *D. stramonium* (Nakajima et al. 1998). Similarly, elucidation of the structure of two *N*-methyltransferases from the caffeine (PA) biosynthetic pathway of *Coffea canephora* led to the identification of critical residues for substrate selectivity and catalysis (McCarthy and McCarthy 2007). The MIA biosynthetic pathway shows the broadest characterisation of alkaloid biosynthetic enzyme structure. In *R. serpentina*, in addition to the preliminary X-ray analysis performed on crystals of raucaffricine glucosidase and perakine reductase (Rosenthal et al. 2006; Ruppert et al. 2006), X-ray crystallography allowed the elucidation of STR and SGD structures, as well as vinorine synthase located in the ajmaline branch of MIA biosynthesis (Ma et al. 2005, 2006; Barleben et al. 2007). STR represents a novel six-bladed β-propeller fold protein that catalyses a Pictet-Spengler condensation between secologanin and tryptamine in a highly substrate-specific manner. The elucidation of the architecture of the substrate binding pockets and of catalytic residues led to the development of structure-based engineering of STR aiming at a redesign of the substrate binding pockets (Chen et al. 2006; Bernhardt et al. 2007; Loris et al. 2007). Several STR variants generated by site-directed mutagenesis displayed altered substrate specificity and accommodated analogs of both tryptamine and secologanin. The three-dimensional structure of SGD appears as a typical $(\beta/\alpha)_8$ barrel fold as encountered in the glucosidase family 1. Using site-directed mutagenesis and structural analysis, catalytic residues have been identified that are involved in the deglucosylation of strictosidine and the conformation of the catalytic pocket (Barleben et al. 2007). Furthermore, steady-state kinetics of SGD with various strictosidine analogs enabled the identification of the substrate preference of SGD at two positions of strictosidine, opening new opportunities for SGD redesign (Yerkes et al. 2008). Finally, crystal structure analysis of vinorine synthase provides the first example of the three-dimensional organisation of an enzyme of the BAHD superfamily (Ma et al. 2005). This is of particular interest, since the vinorine synthase structure could be used for homology-based modelling of other BAHD enzymes occurring in MIA and BIA biosynthetic pathways. Indeed, homology-based modelling has already been applied successfully to the identification of catalytic residue of polyneuridine aldehyde esterase (PNAE), an αβ-hydrolase superfamily member from the ajmaline biosynthetic pathway of *R. serpentina*, using the X-ray crystallographic structure of hydroxynitrile lyase, another αβ-hydrolase superfamily member (Mattern-Dogru et al. 2002). In *Papaver bracteatum*, the structure of salutaridine reductase implicated in BIA biosynthesis was also analysed by homology modelling using the X-ray structure of human carbonyl reductase 1 (Geissler et al. 2007).

The elucidation of the structure of alkaloid biosynthetic enzymes using X-ray crystallography and homology modelling, as well as the identification of key residues for reaction catalysis and substrate specificity, open new enzyme redesign perspectives for the production of large libraries of alkaloid analogs of potential interest.

8.3.2 *Transcription Factor Regulatory Networks of Alkaloid Biosynthesis*

Transcription factors (TF) are thought to play a key role in regulating fluxes through alkaloid pathways by controlling the levels of pathway gene expression, transporters, and the differentiation of specialised cellular structures where the alkaloids are synthesised and accumulate. TF usually exert their control by modulating the expression level of multiple pathway genes. In several species, alkaloid accumulation is preceded by the coordinated induction of several pathway genes, a result of their regulation by specific TF. Methyl jasmonate (MeJA), one of the major internal signal molecule in plant defence response, is known to induce a number of secondary metabolisms, including nicotine (Shoji et al. 2008), MIA (van der Fits and Memelink 2000) and BIA (Färber et al. 2003). For instance, the majority of MIA pathway genes tested are induced by MeJA in *C. roseus* cell cultures (van der Fits and Memelink 2000; Oudin et al. 2007). Detailed analysis of the STR gene promoter led to the identification of several TF involved in elicitor and jasmonate responses. The jasmonate- and elicitor-responsive element (JERE) was found to interact with two TF, octadecanoid-responsive *Catharanthus* AP2-domain proteins

(ORCAs; Menke et al. 1999; van der Fits and Memelink 2000). The expression of the ORCA2 and ORCA3 genes themselves is rapidly induced by MeJA (Menke et al. 1999; van der Fits and Memelink 2001). In nicotine biosynthesis, MeJA response is controlled by homologs of the *Arabidopsis* MeJA-signal transduction pathway (Chini et al. 2007; Thines et al. 2007; Katsir et al. 2008), namely the NtCOI1 subunit of E3-ubiquitin ligase complex, the NtJAZ repressor family (Shoji et al. 2008), as well as by the transcription factor NtORC1, an homolog of ORCA3, and NtJAP1 (De Sutter et al. 2005). NtORC1 and NtJAP1 were shown to up-regulate the promoter of the PMT gene encoding the first committed step in nicotine biosynthesis (De Sutter et al. 2005). In BIA biosynthesis, a WRKY protein was isolated recently as a transcriptional regulator of the berberine pathway in *C. japonica* (Kato et al. 2007). Silencing of CjWRKY1 down-regulated the expression of several berberine pathway genes, whereas over-expression of CjWRKY1 up-regulated the berberine pathway but did not regulate primary metabolism genes, including the tyrosine biosynthetic genes. Interestingly, over-expression of an Arabidopsis WRKY TF was found to enhance BIA concentration up to 30-fold in California poppy cell cultures (Apuya et al. 2008). This TF was isolated in a study that involved screening of regulatory factors isolated from *Arabidopsis* to identify those that modulate the expression of genes encoding enzymes involved in the biosynthesis of BIA in *P. somniferum* and *E. californica*. In opium poppy, the over-expression of selected regulatory factors was found to increase the concentrations of codeinone reductase, 3′-hydroxy-*N*-methylcoclaurine 4′-*O*-methyltransferase and norcoclaurine 6-*O*-methyltransferase transcripts 10- to 100-fold, and to enhance BIA production up to tenfold (Apuya et al. 2008).

8.3.3 Metabolic Engineering of Alkaloid Biosynthesis

Progress in the knowledge of alkaloid pathways at the gene level allows several attempts to improve alkaloid factories. Various objectives have been considered to improve the economic value of medicinal plants or cell cultures, including yield improvement of active or total alkaloids by enhancing flux in the pathway from primary precursor, by overcoming the rate-limiting step, and by reducing flux to competing pathways. Metabolic capacity to produce alkaloids has also been introduced into new species by gene transfer or, in contrast, blocked to eliminate toxic or undesired alkaloids.

Significant yield improvement has been achieved in opium poppy recently by deregulating either an early or a late metabolic step. In an elite commercial line of *P. somniferum*, over-expression of the CYP80B3 gene encoding the early-step *N*-methylcoclaurine 3′-hydroxylase resulted in an increase in morphine reaching 450% under greenhouse conditions (Frick et al. 2007). Over-expression of the penultimate step, codeinone reductase, also resulted in an increase in total alkaloids and morphine of up to 20% under field conditions (Larkin et al. 2007). In MIA biosynthesis, attempts to deregulate the indole branch have been disappointing, probably owing to the dual origin of this class of alkaloids and the complexity of the regulatory networks. Over-expression of tryptophan decarboxylase (TDC) in *C. roseus* cell culture increased the concentration of only tryptamine (Whitmer et al. 2002), while over-expression of both TDC and STR resulted in increased tryptamine and MIA (Canel et al. 1998; Geerlings et al. 1999). Similarly, in *C. roseus* hairy roots, inducible expression of both a feedback insensitive anthranilate synthase from Arabidopsis and TDC led to a marked increase in tryptophan and tryptamine pools, but only modest change in alkaloid pools (Hong et al. 2006). The terpenoid branch has been shown to be limiting in these transgenic lines by precursor feeding experiments (Whitmer et al. 2002; Peebles et al. 2006), suggesting that further improvement would need to target this part of the pathway.

Genetic manipulation of the tropane alkaloid pathway is a good example to improve the metabolic profile of a medicinal plant. Compared to the more valuable scopolamine, hyoscyamine usually accumulates to greater concentrations in various solanaceous plants (e.g. *Atropa, Datura, Hyoscyamus*). Improved conversion of the precursor hyoscyamine to scopolamine has been achieved by over-expression of hyoscyamine 6-hydroxylase (Yun et al. 1992; Hashimoto et al. 1993). Recently, biotransformation of hyoscyamine into scopolamine has been reported in transgenic tobacco cell cultures by over-expressing the H6H gene from *Hyoscyamus muticus* (Moyano et al. 2007). In some cases, incomplete knowledge of metabolic pathways may be overcome by using a more conventional approach. For instance, chemical mutagenesis of seeds has resulted in the isolation of a *P. somniferum* mutant, *top1*, which accumulates thebaine, the starting material for several analgesic semi-synthetic derivatives (Milgate et al. 2004).

Blocking an undesired production of alkaloid may be achieved by silencing an essential enzymatic step early in the pathway. For instance, engineering of decaffeinated coffee plants is in progress in *Coffea canophora*, by silencing 7-methyl-xanthine methyltransferase (Ashihara et al. 2006). Due to the high conservation of the amino acid sequence, the two other *N*-methyltransferases were also down-regulated. Conversely, new metabolic capacity for caffeine production was introduced into tobacco by expressing the three *N*-methyltransferase genes. The transgenic plants accumulated caffeine, which apparently deters insect pests and enhances resistance to bacterial and viral infections (Uefuji et al. 2005; Kim and Sano 2008). In tobacco, RNAi-induced suppression of nicotine demethylase activity was shown to reduce the levels of *N'*-nitrosonornicotine, a carcinogen in cured tobacco leaves (Lewis et al. 2008).

8.4 Conclusions

Progress in metabolomic and transcriptomic analyses as well as system biology approaches are likely to yield detailed analysis of bottlenecks in metabolic pathways and more rational strategies to improve alkaloid biosynthesis. The importance of cellular and subcellular compartmentation, including metabolic channels, has been so far overlooked in improving alkaloid yield. The targeting of enzymes to proper compartments is likely to be an important factor. Engineering at the enzyme structural level by site-directed mutagenesis will also likely be an important avenue to create a new level of metabolic diversity by generating new semi-synthetic alkaloid structures from synthetic precursors and modified enzymes with broad substrate specificity.

References

Alcantara J, Bird DA, Franceschi VR, Facchini PJ (2005) Sanguinarine biosynthesis is associated with the endoplasmic reticulum in cultured opium poppy cells after elicitor treatment. Plant Physiol 138:173–183

Apuya NR, Park J-H, Zhang L, Ahyow M, Davidow P, Van Fleet J, Rarang JC, Hippley M, Johnson TW, Yoo H-D, Trieu A, Krueger S, Wu C-y, Lu Y-p, Flavell RB, Bobzin SC (2008) Enhancement of alkaloid production in opium and California poppy by transactivation using heterologous regulatory factors. Plant Biotechnol J 6:160–175

Ashihara H, Zheng XQ, Katahira R, Morimoto M, Ogita S, Sano H (2006) Caffeine biosynthesis and adenine metabolism in transgenic *Coffea canephora* plants with reduced expression of *N*-methyltransferase genes. Phytochemistry 67:882–886

Ashihara H, Sano H, Crozier A (2008) Caffeine and related purine alkaloids: biosynthesis, catabolism, function and genetic engineering. Phytochemistry 69:841–856

Barleben L, Panjikar S, Ruppert M, Koepke J, Stöckigt J (2007) Molecular architecture of strictosidine glucosidase: the gateway to the biosynthesis of the monoterpenoid indole alkaloid family. Plant Cell 19:2886–2897

Bernhardt P, McCoy E, O'Connor SE (2007) Rapid identification of enzyme variants for reengineered alkaloid biosynthesis in periwinkle. Chem Biol 14:888–897

Bird DA, Facchini PJ (2001) Berberine bridge enzyme, a key branch-point enzyme in benzylisoquinoline alkaloid biosynthesis, contains a vacuolar sorting determinant. Planta 213:888–897

Bird DA, Franceschi VR, Facchini P (2003) A tale of three cell types: alkaloid biosynthesis is localized to sieve elements in opium poppy. Plant Cell 15:2626–2635

Bock A, Wanner G, Zenk MH (2002) Immunocytological localization of two enzymes involved in berberine biosynthesis. Planta 216:57–63

Burlat V, Oudin A, Courtois M, Rideau M, St-Pierre B (2004) Co-expression of three MEP pathway genes and geraniol 10-hydroxylase in internal phloem parenchyma of *Catharanthus roseus* implicates multicellular translocation of intermediates during the biosynthesis of monoterpene indole alkaloids and isoprenoid-derived primary metabolites. Plant J 38:131–141

Canel C, Lopes-Cardoso I, Whitmer S, van der Fits L, Pasquali G, van der Heijden R, Hoge JHC, Verpoorte R (1998) Effects of over-expression of strictosidine synthase and tryptophan decarboxylase on alkaloid production by cell cultures of *Catharanthus roseus*. Planta 205:414–419

Carlson JE, Leebens-Mack JH, Wall PK, Zahn LM, Mueller LA, Landherr LL, Hu Y, Ilut DC, Arrington JM, Choirean S, Becker A, Field D, Tanksley SD, Ma H, dePamphilis CW (2006) EST database for early flower development in California poppy (*Eschscholzia californica* Cham., Papaveraceae) tags over 6,000 genes from a basal eudicot. Plant Mol Biol 62:351–369

Chen S, Galan MC, Coltharp C, O'Connor SE (2006) Redesign of a central enzyme in alkaloid biosynthesis. Chem Biol 13:1137–1141

Chini A, Fonseca S, Fernández G, Adie B, Chico JM, Lorenzo O, García-Casado G, López-Vidriero I, Lozano FM, Ponce MR, Micol JL, Solano R (2007) The JAZ family of repressors is the missing link in jasmonate signalling. Nature 448:666–671

Costa MM, Hilliou F, Duarte P, Pereira LG, Almeida I, Leech M, Memelink J, Barceló AR, Sottomayor M (2008) Molecular cloning and characterization of a vacuolar class III peroxidase involved in the metabolism of anticancer alkaloids in *Catharanthus roseus*. Plant Physiol 146:403–417

De Luca V, St-Pierre B (2000) The cell and developmental biology of alkaloid biosynthesis. Trends Plant Sci 5:168–173

De Sutter V, Vanderhaeghen R, Tilleman S, Lammertyn F, Vanhoutte I, Karimi M, Inzé D, Goossens A, Hilson P (2005) Exploration of jasmonate signalling via automated and standardized transient expression assays in tobacco cells. Plant J 44:1065–1076

Dräger B (2002) Analysis of tropane and related alkaloids. J Chromatogr A 978:1–35

Dräger B (2006) Tropinone reductases, enzymes at the branch point of tropane alkaloid metabolism. Phytochemistry 67:327–337

Facchini PJ, St-Pierre B (2005) Synthesis and trafficking of alkaloid biosynthetic enzymes. Curr Opin Plant Biol 8:657–666

Färber K, Schumann B, Miersch O, Roos W (2003) Selective desensitization of jasmonate and pH-dependent signaling in the induction of benzophenanthridine biosynthesis in cells of *Eschscholzia californica*. Phytochemistry 62:491–500

Frick S, Kramell R, Kutchan TM (2007) Metabolic engineering with a morphine biosynthetic P450 in opium poppy surpasses breeding. Metab Eng 9:169–176

Geerlings A, Hallard D, Caballero AM, Cardoso IL, van der Heijden R, Verpoorte R (1999) Alkaloid production by a *Cinchona officinalis* "Ledgeriana" hairy root culture containing constitutive expression constructs of tryptophan decarboxylase and strictosidine synthase cDNAs from *Catharanthus roseus*. Plant Cell Rep 19:191–196

Geissler R, Brandt W, Ziegler J (2007) Molecular modeling and site-directed mutagenesis reveal the benzylisoquinoline binding site of the short-chain dehydrogenase/reductase salutaridine reductase. Plant Physiol 143:1493–1503

Goossens A, Häkkinen ST, Laakso I, Oksman-Caldentey KM, Inzé D (2003) Secretion of secondary metabolites by ATP-binding cassette transporters in plant cell suspension cultures. Plant Physiol 131:1161–1164

Guirimand G, Burlat V, Oudin A, Lanoue A, St-Pierre B, Courdavault V (2009) Optimization of the transient transformation of *Catharanthus roseus* cells by particle bombardment and its application to the subcellular localization of hydroxymethylbutenyl 4-diphosphate synthase and geraniol 10-hydroxylase. Plant Cell Rep 28:1215–1234

Hagel JM, Facchini PJ (2008) Plant metabolomics: analytical platforms and integration with functional genomics. Phytochem Rev 7:479–497

Hagel JM, Weljie AM, Vogel HJ, Facchini PJ (2008) Quantitative 1H nuclear magnetic resonance metabolite profiling as a functional genomics platform to investigate alkaloid biosynthesis in opium poppy. Plant Physiol 147:1805–1821

Hashimoto T, Hayashi A, Amano Y, Kohno J, Iwanari H, Usuda S, Yamada Y (1991) Hyoscyamine 6β-hydroxylase, an enzyme involved in tropane alkaloid biosynthesis, is localized at the pericycle of the root. J Biol Chem 266:4648–4653

Hashimoto T, Yun DJ, Yamada Y (1993) Production of tropane alkaloids in genetically engineered root cultures. Phytochemistry 32:713–718

Hong SB, Peebles CAM, Shanks JV, San KY, Gibson SI (2006) Expression of the Arabidopsis feedback-insensitive anthranilate synthase holoenzyme and tryptophan decarboxylase genes in *Catharanthus roseus* hairy roots. J Biotechnol 122:28–38

Irmler S, Schröder G, St-Pierre B, Crouch NP, Hotze M, Schmidt J, Strack D, Matern U, Schröder J (2000) Indole alkaloid biosynthesis in *Catharanthus roseus*: new enzyme activities and identification of cytochrome P450 CYP72A1 as secologanin synthase. Plant J 24:797–804

Kato N, Dubouzet E, Kokabu Y, Yoshida S, Taniguchi Y, Dubouzet JG, Yazaki K, Sato F (2007) Identification of a WRKY protein as a transcriptional regulator of benzylisoquinoline alkaloid biosynthesis in *Coptis japonica*. Plant Cell Physiol 48:8–18

Katsir L, Schilmiller AL, Staswick PE, He SY, Howe GA (2008) COI1 is a critical component of a receptor for jasmonate and the bacterial virulence factor coronatine. Proc Natl Acad Sci USA 105:7100–7105

Kim Y-S, Sano H (2008) Pathogen resistance in transgenic tobacco plants producing caffeine. Phytochemistry 69:882–888

Kolaczkowski M, van der Rest M, Cybularz-Kolaczkowska A, Soumillion J-P, Konings WN, Goffeau A (1996) Anticancer drugs, ionophoric peptides, and steroids as substrates of the yeast multidrug transporter Pdr5p. J Biol Chem 271:31543–31548

Kutchan TM (2005) A role for intra- and intercellular translocation in natural product biosynthesis. Curr Opin Plant Biol 8:292–300

Larkin PJ, Miller JAC, Allen RS, Chitty JA, Gerlach WL, Frick S, Kutchan TM, Fist AJ (2007) Increasing morphinan alkaloid production by over-expressing codeinone reductase in transgenic *Papaver somniferum*. Plant Biotechnol J 5:26–37

Levac D, Murata J, Kim WS, De Luca V (2008) Application of carborundum abrasion for investigating the leaf epidermis: molecular cloning of *Catharanthus roseus* 16-hydroxytabersonine-16-*O*-methyltransferase. Plant J 53:225–236

Lewis RS, Jack AM, Morris JW, Robert VJM, Gavilano LB, Siminszky B, Bush LP, Hayes AJ, Dewey RE (2008) RNA interference (RNAi)-induced suppression of nicotine demethylase activity reduces levels of a key carcinogen in cured tobacco leaves. Plant Biotechnol J 6:346–354

Li R, Reed DW, Liu E, Nowak J, Pelcher LE, Page JE, Covello PS (2006) Functional genomic analysis of alkaloid biosynthesis in *Hyoscyamus niger* reveals a cytochrome P450 involved in littorine rearrangement. Chem Biol 13:513–520

Lorence A, Nessler CL (2004) Camptothecin, over four decades of surprising findings. Phytochemistry 65:2735–2749

Loris EA, Panjikar S, Ruppert M, Barleben L, Unger M, Schübel H, Stöckigt J (2007) Structure-based engineering of strictosidine synthase: auxiliary for alkaloid libraries. Chem Biol 14:979–985

Ma X, Koepke J, Panjikar S, Fritzsch G, Stöckigt J (2005) Crystal structure of vinorine synthase, the first representative of the BAHD superfamily. J Biol Chem 280:13576–13583

Ma X, Panjikar S, Koepke J, Loris E, Stöckigt J (2006) The structure of *Rauvolfia serpentina* strictosidine synthase is a novel six-bladed beta-propeller fold in plant proteins. Plant Cell 18:907–920

Mahroug S, Burlat V, St-Pierre B (2007) Cellular and sub-cellular organization of the monoterpenoid indole alkaloid pathway in *Catharanthus roseus*. Phytochem Rev 6:363–381

Mattern-Dogru E, Ma X, Hartmann J, Decker H, Stöckigt J (2002) Potential active-site residues in polyneuridine aldehyde esterase, a central enzyme of indole alkaloid biosynthesis, by modelling and site-directed mutagenesis. Eur J Biochem 269:2889–2896

McCarthy AA, McCarthy JG (2007) The structure of two *N*-methyltransferases from the caffeine biosynthetic pathway. Plant Physiol 144:879–889

McKnight TD, Bergey DR, Burnett RJ, Nessler CL (1991) Expression of enzymatically active and correctly targeted strictosidine synthase in transgenic tobacco plants. Planta 185:148–152

Menke FLH, Champion A, Kijne JW, Memelink J (1999) A novel jasmonate- and elicitor-responsive element in the periwinkle secondary metabolite biosynthetic gene *Str* interacts with a jasmonate- and elicitor-inducible AP2-domain transcription factor, ORCA2. EMBO J 18:4455–4463

Milgate AG, Pogson BJ, Wilson IW, Kutchan TM, Zenk MH, Gerlach WL, Fist AJ, Larkin PJ (2004) Morphine-pathway block in top1 poppies. Nature 431:413–414

Morishige T, Dubouzet E, Choi KB, Yazaki K, Sato F (2002) Molecular cloning of columbamine *O*-methyltransferase from cultured *Coptis japonica* cells. Eur J Biochem 269:5659–5667

Moyano E, Palazón J, Bonfill M, Osuna L, Cusidó RM, Oksman-Caldentey K-M, Piñol MT (2007) Biotransformation of hyoscyamine into scopolamine in transgenic tobacco cell cultures. J Plant Physiol 164:521–524

Murata J, De Luca V (2005) Localization of tabersonine 16-hydroxylase and 16-OH tabersonine-16-*O*-methyltransferase to leaf epidermal cells defines them as a major site of precursor biosynthesis in the vindoline pathway in *Catharanthus roseus*. Plant J 44:581–594

Murata J, Bienzle D, Brandle JE, Sensen CW, De Luca V (2006) Expressed sequence tags from Madagascar periwinkle (*Catharanthus roseus*). FEBS Lett 580:4501–4507

Murata J, Roepke J, Gordon H, De Luca V (2008) The leaf epidermome of *Catharanthus roseus* reveals its biochemical specialization. Plant Cell 20:524–542

Nakajima K, Hashimoto T (1999) Two tropinone reductases, that catalyze opposite stereospecific reductions in tropane alkaloid biosynthesis, are localized in plant root with different cell-specific patterns. Plant Cell Physiol 40:1099–1107

Nakajima K, Yamashita A, Akama H, Nakatsu T, Kato H, Hashimoto T, Oda J, Yamada Y (1998) Crystal structures of two tropinone reductases: different reaction stereospecificities in the same protein fold. Proc Natl Acad Sci USA 95:4876–4881

Oksman-Caldentey KM (2007) Tropane and nicotine alkaloid biosynthesis-novel approaches towards biotechnological production of plant-derived pharmaceuticals. Curr Pharm Biotechnol 8:203–210

Oudin A, Mahroug S, Courdavault V, Hervouet N, Zelwer C, Rodríguez-Concepción M, St-Pierre B, Burlat V (2007) Spatial distribution and hormonal regulation of gene products from methyl erythritol phosphate and monoterpene-secoiridoid pathways in *Catharanthus roseus*. Plant Mol Biol 65:13–30

Peebles CAM, Hong SB, Gibson SI, Shanks IV, San KY (2006) Effects of terpenoid precursor feeding on *Catharanthus roseus* hairy roots over-expressing the alpha or the alpha and beta subunits of anthranilate synthase. Biotechnol Bioeng 93:534–540

Rischer H, Oresic M, Seppänen-Laakso T, Katajamaa M, Lammertyn F, Ardiles-Diaz W, Van Montagu MC, Inzé D, Oksman-Caldentey KM, Goossens A (2006) Gene-to-metabolite networks for terpenoid indole alkaloid biosynthesis in *Catharanthus roseus* cells. Proc Natl Acad Sci USA 103:5614–5619

Rosenthal C, Mueller U, Panjikar S, Sun L, Ruppert M, Zhao Y, Stöckigt J (2006) Expression, purification, crystallization and preliminary X-ray analysis of perakine reductase, a new member of the aldo-keto reductase enzyme superfamily from higher plants. Acta Crystallogr Sect F Struct Biol Cryst Commun 62:1286–1289

Ruppert M, Panjikar S, Barleben L, Stöckigt J (2006) Heterologous expression, purification, crystallization and preliminary X-ray analysis of raucaffricine glucosidase, a plant enzyme specifically involved in *Rauvolfia* alkaloid biosynthesis. Acta Crystallogr Sect F Struct Biol Cryst Commun 62:257–260

Sakai K, Shitan N, Sato F, Ueda K, Yazaki K (2002) Characterization of berberine transport into *Coptis japonica* cells and the involvement of ABC protein. J Exp Bot 53:1879–1886

Samanani N, Park SU, Facchini PJ (2005) Cell type-specific localization of transcripts encoding nine consecutive enzymes involved in protoberberine alkaloid biosynthesis. Plant Cell 17:915–926

Samanani N, Alcantara J, Bourgault R, Zulak KG, Facchini PJ (2006) The role of phloem sieve elements and laticifers in the biosynthesis and accumulation of alkaloids in opium poppy. Plant J 47:547–563

Schmidt J, Boettcher C, Kuhnt C, Kutchan TM, Zenk MH (2007) Poppy alkaloid profiling by electrospray tandem mass spectrometry and electrospray FT-ICR mass spectrometry after [*ring*-^{13}C$_6$]-tyramine feeding. Phytochemistry 68:189–202

Shitan N, Bazin I, Dan K, Obata K, Kigawa K, Ueda K, Sato F, Forestier C, Yazaki K (2003) Involvement of CjMDR1, a plant multidrug-resistance-type ATP-binding cassette protein, in alkaloid transport in *Coptis japonica*. Proc Natl Acad Sci USA 100:751–756

Shoji T, Ogawa T, Hashimoto T (2008) Jasmonate-induced nicotine formation in tobacco is mediated by tobacco *COI1* and *JAZ* genes. Plant Cell Physiol 49:1003–1012

Shukla AK, Shasany AK, Gupta MM, Khanuja SP (2006) Transcriptome analysis in *Catharanthus roseus* leaves and roots for comparative terpenoid indole alkaloid profiles. J Exp Bot 57:3921–3932

Stöckigt J, Panjikar S (2007) Structural biology in plant natural product biosynthesis-architecture of enzymes from monoterpenoid indole and tropane alkaloid biosynthesis. Nat Prod Rep 24:1382–1400

St-Pierre B, Vazquez-Flota FA, De Luca V (1999) Multicellular compartmentation of *Catharanthus roseus* alkaloid biosynthesis predicts intercellular translocation of a pathway intermediate. Plant Cell 11:887–900

Suzuki K, Yamada Y, Hashimoto T (1999a) Expression of *Atropa belladonna* putrescine *N*-methyltransferase gene in root pericycle. Plant Cell Physiol 40:289–297

Suzuki K, Yun DJ, Chen XY, Yamada Y, Hashimoto T (1999b) An *Atropa belladonna* hyoscyamine 6β-hydroxylase gene is differentially expressed in the root pericycle and anthers. Plant Mol Biol 40:141–152

Thines B, Katsir L, Melotto M, Niu Y, Mandaokar A, Liu G, Nomura K, He SY, Howe GA, Browse J (2007) JAZ repressor proteins are targets of the SCF^{COI1} complex during jasmonate signalling. Nature 448:661–665

Uefuji H, Tatsumi Y, Morimoto M, Kaothien-Nakayama P, Ogita S, Sano H (2005) Caffeine production in tobacco plants by simultaneous expression of three coffee *N*-methyltransferases and its potential as a pest repellent. Plant Mol Biol 59:221–227

van der Fits L, Memelink J (2000) ORCA3, a jasmonate-responsive transcriptional regulator of plant primary and secondary metabolism. Science 289:295–297

van der Fits L, Memelink J (2001) The jasmonate-inducible AP2/ERF-domain transcription factor ORCA3 activates gene expression via interaction with a jasmonate-responsive promoter element. Plant J 25:43–53

van der Heijden R, Jacobs DI, Snoeijer W, Hallard D, Verpoorte R (2004) The *Catharanthus* alkaloids: pharmacognosy and biotechnology. Curr Med Chem 11:607–628

Weid M, Ziegler J, Kutchan TM (2004) The roles of latex and the vascular bundle in morphine biosynthesis in the opium poppy, *Papaver somniferum*. Proc Natl Acad Sci USA 101: 13957–13962

Whitmer S, van der Heijden R, Verpoorte R (2002) Effect of precursor feeding on alkaloid accumulation by a tryptophan decarboxylase over-expressing transgenic cell line T22 of *Catharanthus roseus*. J Biotechnol 96:193–203

Yamazaki Y, Urano A, Sudo H, Kitajima M, Takayama H, Yamazaki M, Aimi N, Saito K (2003) Metabolite profiling of alkaloids and strictosidine synthase activity in camptothecin producing plants. Phytochemistry 62:461–470

Yerkes N, Wu JX, McCoy E, Galan MC, Chen S, O'Connor SE (2008) Substrate specificity and diastereoselectivity of strictosidine glucosidase, a key enzyme in monoterpene indole alkaloid biosynthesis. Bioorg Med Chem Lett 18:3095–3098

Yun D-J, Hashimoto T, Yamada Y (1992) Metabolic engineering of medicinal plants: transgenic *Atropa belladonna* with an improved alkaloid composition. Proc Natl Acad Sci USA 89: 11799–11803

Ziegler J, Facchini PJ (2008) Alkaloid biosynthesis: metabolism and trafficking. Annu Rev Plant Biol 59:735–769

Ziegler J, Voigtländer S, Schmidt J, Kramell R, Miersch O, Ammer C, Gesell A, Kutchan TM (2006) Comparative transcript and alkaloid profiling in Papaver species identifies a short chain dehydrogenase/reductase involved in morphine biosynthesis. Plant J 48:177–192

Zulak KG, Cornish A, Daskalchuk TE, Deyholos MK, Goodenowe DB, Gordon PM, Klassen D, Pelcher LE, Sensen CW, Facchini PJ (2007) Gene transcript and metabolite profiling of elicitor-induced opium poppy cell cultures reveals the coordinate regulation of primary and secondary metabolism. Planta 225:1085–1106

Chapter 9
Molecular Biology and Biotechnology of Flower Pigments

K.M. Davies and K.E. Schwinn

9.1 Introduction

Three groups of pigments account for the great majority of flower colours, namely flavonoids, carotenoids and betalains. Chlorophylls can also provide the main flower pigments, but only in a very few species. The flavonoids are a large subgroup of phenylpropanoids, with those involved in flower colour being water-soluble and generally located in the vacuole. The predominant flavonoid pigments are the anthocyanins, which are the basis for nearly all orange, pink, red, purple, blue and blue-black flower colours. Carotenoids are lipid-soluble, plastid-located terpenoids, which for pigmentation of flowers (and fruit) accumulate to high concentrations in specialized plastids called chromoplasts. They are the basis of pigmentation in most yellow flowers, and can also generate orange, red, bronze and brown colours. The carotenoids also participate in the harvesting and dissipation of light energy in chloroplasts, protecting the photosynthetic machinery from photo-oxidation. The betalains are vacuolar-located nitrogenous compounds that produce yellow, orange, red and purple flower colours. Their occurrence in plants is restricted to the Caryophyllales, in most families of which the betalains completely replace anthocyanins as the key floral pigments. The molecular basis for the mutual exclusion of betalains and anthocyanins in the Caryophyllales remains to be determined. Many flower colours and floral pigmentation patterns are the result of the co-occurrence of both anthocyanins and carotenoids in petals.

The biosynthetic pathways for flavonoids and carotenoids have been characterized extensively, and this review contains only a brief overview. In contrast, there are still steps in betalain biosynthesis for which molecular characterization of the corresponding genes is lacking. For flavonoids, there is also now considerable

K.M. Davies and K.E. Schwinn
New Zealand Institute for Plant & Food Research Limited, Private Bag 11600, Palmerston North, 4442, New Zealand
e-mail: kevin.davies@plantandfood.co.nz

knowledge of the transcription factors (TFs) that regulate the activity of the biosynthetic pathway, which are reviewed in Section 9.3. There are surprisingly few data on TFs regulating carotenoid biosynthesis in flowers, and none for betalain biosynthesis.

Throughout the review, the state of biotechnologies for pigment pathways is also considered. The focus is on genetic modification (GM) approaches. However, it is important to note that the pigment-related gene sequences, especially those encoding TFs, also offer much promise for non-GM plant breeding, such as by marker assisted selection. Indeed, marker genes have already been identified for control of anthocyanin production in apple (Allan et al. 2008) and grape (Walker et al. 2007), and for carotenoid-based flesh colour in watermelon (Bang et al. 2007).

9.2 Pigment Biosynthetic Pathways and Their Genetic Modification

9.2.1 Flavonoids

Flavonoid biosynthesis is a many-branched sub-section of the phenylpropanoid pathway, producing coloured and colourless compounds with diverse biological functions. The core pathway, which is common to most plants, is well defined (Fig. 9.1). Gene sequences are available for all the biosynthetic steps to the principal coloured compounds and also for many of the enzymes that perform secondary modifications of those structures. Extensive recent reviews of the molecular biology of the general flavonoid pathway are available (Grotewold 2006; Davies and Schwinn 2006).

Anthocyanins are the most significant flavonoid pigments, with aurones, chalcones and some flavonols playing a limited role as primary pigments in flower colour. Flavonoids are based on a core 15-carbon (C_{15}) structure of two aromatic rings (the A and B rings) joined by a third ring of three carbons and one oxygen (the C-ring). The degree of oxidation of the C-ring defines the various flavonoid types. The anthocyanin is formed by glycosylation of a core anthocyanidin structure; in the great majority of cases, the initial O-glycosylation is at the C-3 position by uridine diphosphate-glucose (UDPG):flavonoid 3-O-glucosyltransferase (F3GT). Subsequent further modification by hydroxylation, acylation and methylation generates the large number of known anthocyanin variant structures (>530; Andersen and Jordheim 2006). Although over 31 different core anthocyanidins have been identified, the six common ones (Fig. 9.1) are the basis of about 90% of the known anthocyanins (Andersen and Jordheim 2006). Some of the less common anthocyanidins contribute important flower pigments in a small number of species, with the 3-deoxyanthocyanins and 6- and 8-hydroxyanthocyanins being of particular interest due to the bright orange and red colours they generate. However, although there

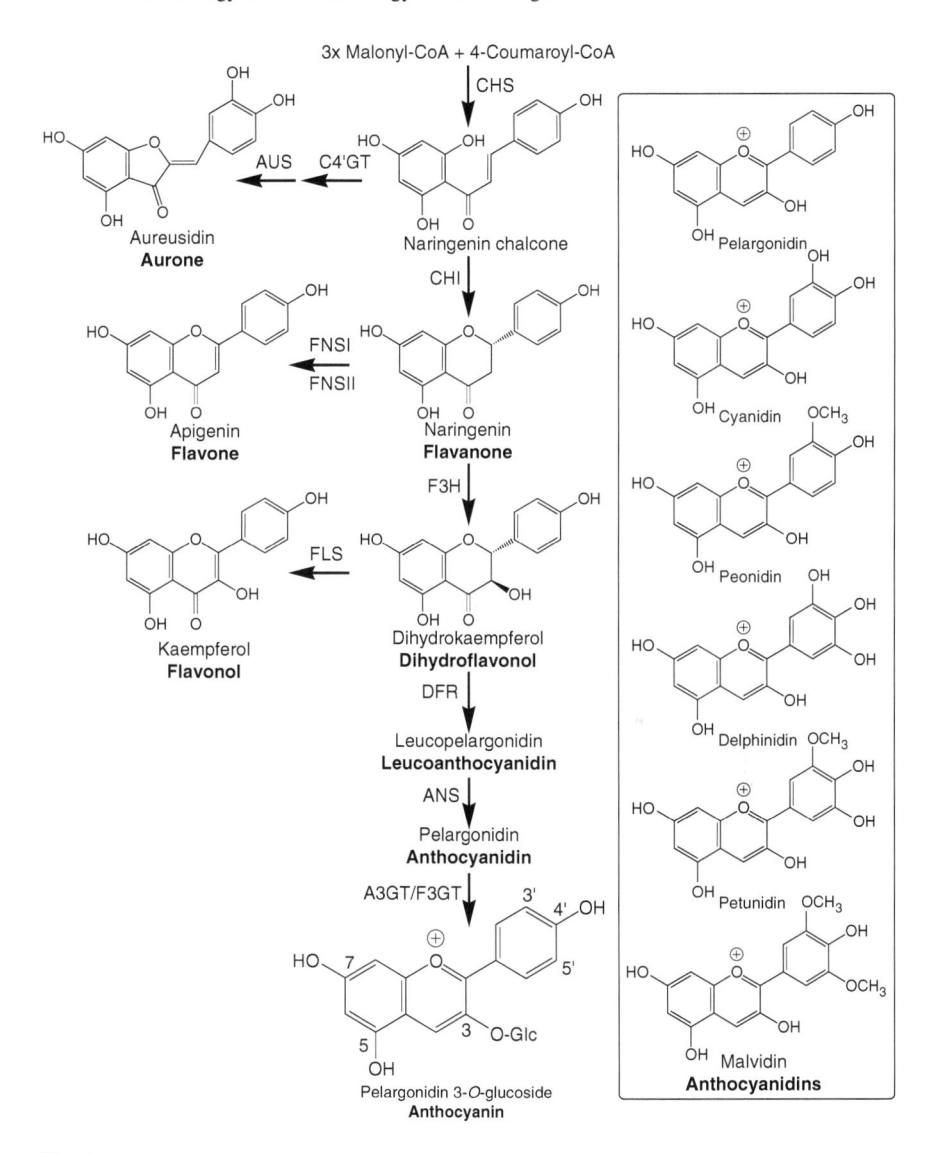

Fig. 9.1 A section of the flavonoid biosynthetic pathway leading to the anthocyanins and other flavonoids found in flowers. For ease of presentation, only the route for flavonoids with 4'-hydroxylation of the B-ring is shown. DNA sequences are available for all of the enzymes shown. CHS, chalcone synthase; CHI, chalcone isomerase; F3H, flavanone 3-hydroxylase; DFR, dihydroflavonol 4-reductase; ANS, anthocyanidin synthase; F3GT, uridine diphosphate-glucose (UDPG):flavonoid 3-*O*-glucosyltransferase; A3GT, UDPG:anthocyanidin 3-*O*-glucosyltransferase; C4'GT, UDPG: chalcone 4'-*O*-glucosyltransferase; AUS, aureusidin synthase; FLS, flavonol synthase; FNSI and FNSII, flavone synthase. The six common anthocyanidins are shown on the *right*. The numbering of carbon positions is shown on the anthocyanin structure

is information on the biosynthetic pathway of some of these, there is no reported success to date with regard to GM of their production.

Flavonoids are likely synthesized on the cytoplasmic face of the endoplasmic reticulum (ER), but the pathway end-products accumulate mostly in the vacuole. For anthocyanins, vacuolar location is essential for their function as pigments, and the pH and other environmental conditions within the vacuole have a major effect on the resulting colour (Andersen and Jordheim 2006). There are various models for how flavonoids traffic from the surface of the ER to the tonoplast, based around either protein chaperones or transport within vesicles budded off the ER (Grotewold and Davies 2008).

In the following sections, we consider in brief some of the flavonoid biosynthetic steps particularly relevant to flower colour and its GM. For each, we consider what stage towards predictive metabolic engineering the GM technologies have reached.

9.2.1.1   Core Biosynthetic Steps of the Flavonoid Pathway

Flavonoid production has been reduced in flowers of several species by the inhibition of one of the core biosynthetic steps, chalcone synthase (CHS), chalcone isomerase (CHI), flavanone 3-hydroxylase (F3H), dihydroflavonol 4-reductase (DFR) or anthocyanidin synthase (ANS). Species successfully modified include carnation, chrysanthemum, cyclamen, gentian, gerbera, lisianthus, petunia, rose, tobacco and torenia (reviewed in Davies and Schwinn 2006; To and Wang 2006). Sense-suppression, antisense RNA and RNAi have all been used successfully, with RNAi of core biosynthetic genes emerging as a powerful and reliable technology for reducing flavonoid production. When directly compared for targeting ANS in torenia, RNAi gave a higher efficiency of down-regulation than sense-suppression or antisense, and the GM phenotype was shown to be stable for at least 3 years (Nakamura et al. 2006). Also, RNAi has been the only approach effective for suppression of CHI (Nishihara et al. 2005). Furthermore, allele-specific inhibition of CHS has been demonstrated using RNAi constructs for the 3'-untranslated region (Fukusaki et al. 2004). Conversely, Gonzalez et al. (2008) were able to inhibit the activity of all four members of a multigene family in arabidopsis using an RNAi to a conserved sequence (for the TF genes *Pap1*, *AtMyb90*, *AtMyb113*, *AtMyb114*). The use of tissue-specific gene promoters for RNAi allows for reduction of anthocyanins in specific regions of the plant, and may generate pigment patterning in the flower (Nakatsuka et al. 2007a). Another application of RNAi for ornamentals is as a biolistics-based transient inhibition system to assay gene function or test colour modification strategies (Shang et al. 2007).

In almost all experiments with RNAi, sense-suppression or antisense, lines with the predicted pale or white flower colour phenotypes have been obtained. However, there have also been unexpected phenotypes. In some species, inhibiting flavonoid production results in female or male sterility (Taylor and Grotewold 2005), and in tomato the production of parthenocarpic fruit (Schijlen et al. 2007). Down-regulation of the biosynthetic genes has resulted in both ordered and erratic petal pigmentation patterns but, to date, only for species that naturally have patterned varieties. It is of note that sequence-specific degradation of CHS also accounts for some 'natural' flower colour patterns (see Sect. 9.3). In some cases, accumulated substrate of down-regulated enzymes has been redirected into other products, including other biosynthetic pathways. For example, the down-regulation of F3H in carnation resulted in an increase in methylbenzoate levels and more fragrant flowers (Zuker et al. 2002). Likewise, reducing competing biosynthetic routes can lead to an increase in anthocyanin levels. Inhibition of the production of flavonol synthase (FLS) in lisianthus, petunia or tobacco resulted in an increase in anthocyanin content in the flowers (Holton et al. 1993; Nielsen et al. 2002; Davies et al. 2003). The opposite approach of over-production of enzymes that compete for substrate with core pathway enzymes can reduce anthocyanin concentrations, and has resulted in lines with pale flower colours but not yet white flowers. Examples include the introduction of stilbene synthase (Fischer et al. 1997) or chalcone reductase (CHR, polyketide reductase) to compete with CHS (Davies et al. 1998), flavone synthase (FNS) to compete with F3H (Nakatsuka et al. 2006), and anthocyanidin reductase to compete with F3GT (Xie et al. 2003). As expected, stilbenes, $6'$-deoxychalcones and flavan-3-ols (or their derivatives) accumulated respectively, and in the case of $6'$-deoxychalcones imparted pale yellow pigmentation to the petals.

When GM for flower colour was first attempted, it was thought that particular biosynthetic steps, such as CHS, may be rate-determining, and that over-production of these could enable higher anthocyanin concentrations in plants. This has not proved the case for flowers that are already strongly pigmented by anthocyanins. However, there has been success in species or cultivars that initially had low anthocyanin concentrations or were anthocyanin-free. Good examples of increasing flavonoid concentrations by over-production of biosynthetic enzymes are the development of anthocyanin-pigmented cultivars of forsythia by over-production of both DFR and ANS (Rosati et al. 2003), and a 78-fold increase in flavonols in tomato fruit by over-production of CHI (Muir et al. 2001).

9.2.1.2   Anthocyanin B-Ring Hydroxylation

The degree of hydroxylation of the anthocyanin B-ring is a major determinant of the resulting colour type. Pelargonidin-, cyanidin- or delphinidin-derived anthocyanins tend to give pink to red, red to magenta, and purple to blue colours respectively. The $4'$-hydroxyl group is incorporated during formation of the pathway precursor 4-coumarate, while the addition of the $3'$ and $5'$ hydroxyl groups is usually catalyzed by flavonoid $3'$-hydroxylase (F3'H) and flavonoid $3',5'$-hydroxylase (F3'5'H), acting on anthocyanin precursors. In general, the mix of different anthocyanidin types in a flower is determined by the activity of these two enzymes. A clear approach for flavonoid biotechnology is to introduce or down-regulate the corresponding genes.

Of particular interest is the prospect of introducing delphinidin production using a F3'5'H transgene to form blue flower colours in species such as carnation, chrysanthemum, gerbera and rose. However, our knowledge of the chemical basis of blue flower colours suggests introduction of delphinidin-based anthocyanins may not be sufficient in itself to confer true blue colours. Nor is the introduction of a F3'5'H transgene alone enough to confer high levels of delphinidin production. F3'5'H transgenes in cyanidin-accumulating carnation, rose or tobacco lines caused only relatively low levels of delphinidin accumulation and minor changes in flower colour (Shimada et al. 1999; Katsumoto et al. 2007). This was due to competition for substrate with the F3'H, and preferences of the other biosynthetic enzymes for 3'- or 3'4'-hydroxylated substrates. Extensive, iterative experiments have shown that a combination of transgenes is required to achieve substantial delphinidin-derived anthocyanin concentrations (if possible) in a suitable mutant background. The preferred option, and one that was possible in carnation, is to use a plant line that accumulates dihydrokaempferol (DHK) (a DFR/F3'H double mutant) and introduce a DFR that prefers the 3'4'5'-hydroxylated precursor dihydromyricetin (DHM), along with the F3'5'H transgene. When such mutants cannot be found, they may be generated using RNAi. The delphinidin-producing roses developed for commercial release are F3'H mutants that contain an RNAi transgene for the endogenous DFR, as well as sense viola F3'5'H and iris DFR transgenes (Katsumoto et al. 2007). Interestingly, the viola F3'5'H cDNA gave good delphinidin production in roses, but the F3'5'H cDNAs from petunia, gentian and butterfly pea did not express successfully (Katsumoto et al. 2007). Possible high-efficiency F3'5'H proteins, such as that from *Campanula medium*, have also been reported (Okinaka et al. 2003).

The delphinidin-producing lines of rose and carnation are mauve/violet, rather than blue, illustrating that introducing delphinidin-derived anthocyanins is only one step towards obtaining the often desired true blue colours. Indeed, species such as cyclamen, impatiens, lisianthus, pelargonium and tulip naturally produce delphinidin-derived anthocyanins but lack blue flowers. For blue flower colours to form, either a very high vacuolar pH is required or a combination of pH >5.5 and intra- or inter-molecular (copigmentation) interactions within the vacuole (Andersen and Jordheim 2006). Some of the most attractive blue flower colours are based on complex anthocyanin structures. This may include multiple monosaccharide additions, acyl substitutions, the covalent attachment of copigments, the formation of anthocyanin–metal ion complexes, or the development of large anthocyanin–copigment complexes. Although the requirements for generating blue flower colours are now known, GM of either vacuolar pH or the specific anthocyanin structures required is a significant challenge. Multiple transgenes would be needed, and DNA sequences are not available for several of the required enzyme activities. Furthermore, some of the modification enzymes characterized to date showed high substrate specificity.

Pelargonidin-based anthocyanins are the basis of bright orange to red flower colours in a range of species, with the orange colours probably occurring when vacuolar pH is low and pelargonidin-derived anthocyanins predominate. GM of the activity of either DFR or the B-ring hydroxylases has been applied successfully for modifying accumulation of pelargonidin-based anthocyanins. Knocking out F3′H and/or F3′5′H activities may promote DHK and, hence, pelargonidin accumulation. The introduction of a transgene for a DFR with a strong activity with DHK can overcome a block caused by the substrate preference of the endogenous DFR of some species. Combinations of these approaches have been used in African daisy (Seitz et al. 2007), petunia (Tsuda et al. 2004), torenia (Suzuki et al. 2000) and tobacco (Nakatsuka et al. 2007b).

9.2.1.3 Aurones and Chalcones

The chalcones are the first coloured flavonoids produced in the pathway (Fig. 9.1). On rare occasions, they are the basis of yellow flower colours, most notably in carnation, but they are usually converted to colourless flavanones by CHI. Either inhibition of CHI activity (Nishihara et al. 2005), or using CHR to generate 6′-deoxychalcones (Davies et al. 1998) can confer chalcone accumulation. However, only pale yellow flower colours have resulted. In a few species, chalcones are converted to bright yellow aurone pigments by the vacuolar-located aureusidin synthase, via a glycosylated intermediate generated by UDPG: chalcone 4′-O-glucosyltransferase (Ono et al. 2006). The identification of DNA sequences for these two aurone biosynthetic enzymes should enable the introduction of yellow flower colours into the wide range of ornamentals species that currently lack these pigments. In the only published example of their use, their co-introduction into torenia, in combination with RNAi-mediated down-regulation of anthocyanin biosynthesis, did indeed result in yellow flowers (Ono et al. 2006).

9.2.2 Carotenoids

Most carotenoids have a C_{40} backbone and up to 15 conjugated double bonds. More than 700 naturally occurring carotenoids have been identified, and they are divided into two principal groups, the hydrocarbon carotenes and their oxygenated derivatives, the xanthophylls (Fraser and Bramley 2004). In addition to providing pigments to chloroplasts and chromoplasts, they are substrates for the biosynthesis of abscisic acid (ABA). They share precursors with a range of compounds, including gibberellins (GAs), chlorophylls, tocopherols, triterpenes and sesquiterpenes. The core carotenoid pathway is common to all plants, as well as some bacteria and algae. Many species-specific or rare plant carotenoids have also been characterized, some of which are important in providing colour. Notable, due to the associated genes having been isolated, are capsanthin and capsorubin of capsicums, and astaxanthin and other ketocarotenoids—for example, from *Adonis aestivalis*.

The biosynthesis of carotenoids is now well defined, and has been the subject of recent reviews (Fraser and Bramley 2004; DellaPenna and Pogson 2006). The first specific step is formation of the C_{40} carotene, phytoene, from the C_{20} compound geranylgeranyl pyrophosphate (GGPP), which itself is derived from the C_5 compound isopentenyl diphosphate (IPP) produced by the 2-C-methyl-D-erythritol 4-phosphate (MEP) pathway. Starting from phytoene, there are a series of desaturation, isomerization, cyclization and oxygenation reactions to form different carotenoids (Fig. 9.2). The desaturation reactions increase the number of conjugated double bonds in the carotene background and form the chromophore, with the first coloured compound being ζ-carotene. The desaturases form part of a respiratory chain in the plasma membrane that requires plastaquinone and plastid terminal oxidase. The pathway branches after lycopene according to the lycopene cyclase activity that is present. Multigene families for the carotenoid biosynthetic enzymes have been found in most of the plant species examined to date, and it has been suggested that these may provide separate pathways in chloroplasts and chromoplasts (Galpaz et al. 2006).

A strong drive for GM of the carotenoid pathway is the desire to improve human-health properties of crop plants, especially increasing levels of provitamin A (β-carotene) or other bioactive carotenoids. There is also interest in developing plant-based production systems for extracted carotenoids for the food and aquaculture industries as alternatives to synthetic or bacterially sourced pigments. Potential targets for GM of flower colour include redirecting the pathway into desired compounds in species that already have carotenoid-pigmented flowers, introducing novel carotenoid types from other species, and increasing carotenoid concentrations overall.

DNA sequences for the biosynthetic enzymes acting from IPP through to lutein and neoxanthin have been characterized, as well as for additional activities from specific plant species or bacteria. In theory, predictive GM of the pathway should be possible. Using some of the multifunctional bacterial desaturases, introduction of only a few transgenes should be sufficient to generate de novo accumulation of coloured carotenoids. Furthermore, as some of the intermediate compounds such as lycopene are brightly coloured, simple interruption of an active pathway to xanthophylls should allow for generation of novel colours. However, GM of carotenoid biosynthesis poses several problems, many of which do not apply to the flavonoid pathway. Potential issues include location of the biosynthetic pathway in plastids, the requirement for chromoplasts for high carotenoid accumulation, availability of precursors shared with other pathways, degradation/turnover rates, and the essential role of general carotenoid biosynthesis for plant function. Thus, despite the availability of carotenoid biosynthetic gene sequences from plants and bacteria for many years, until recently there were far fewer examples of effective GM of carotenoid biosynthesis than for the flavonoid pathway, and almost no reports concerning flower colour. This has changed over the last few years, with many examples published for food crops and the first examples for engineering flower colour (reviewed in Giuliano et al. 2008). Table 9.1 presents recent examples of successful GM for carotenoid production in food and ornamental crops. These examples fall into four groups of approaches, namely increasing activity of biosynthetic enzymes, inhibiting activity of specific enzymes to redirect substrate in the pathway,

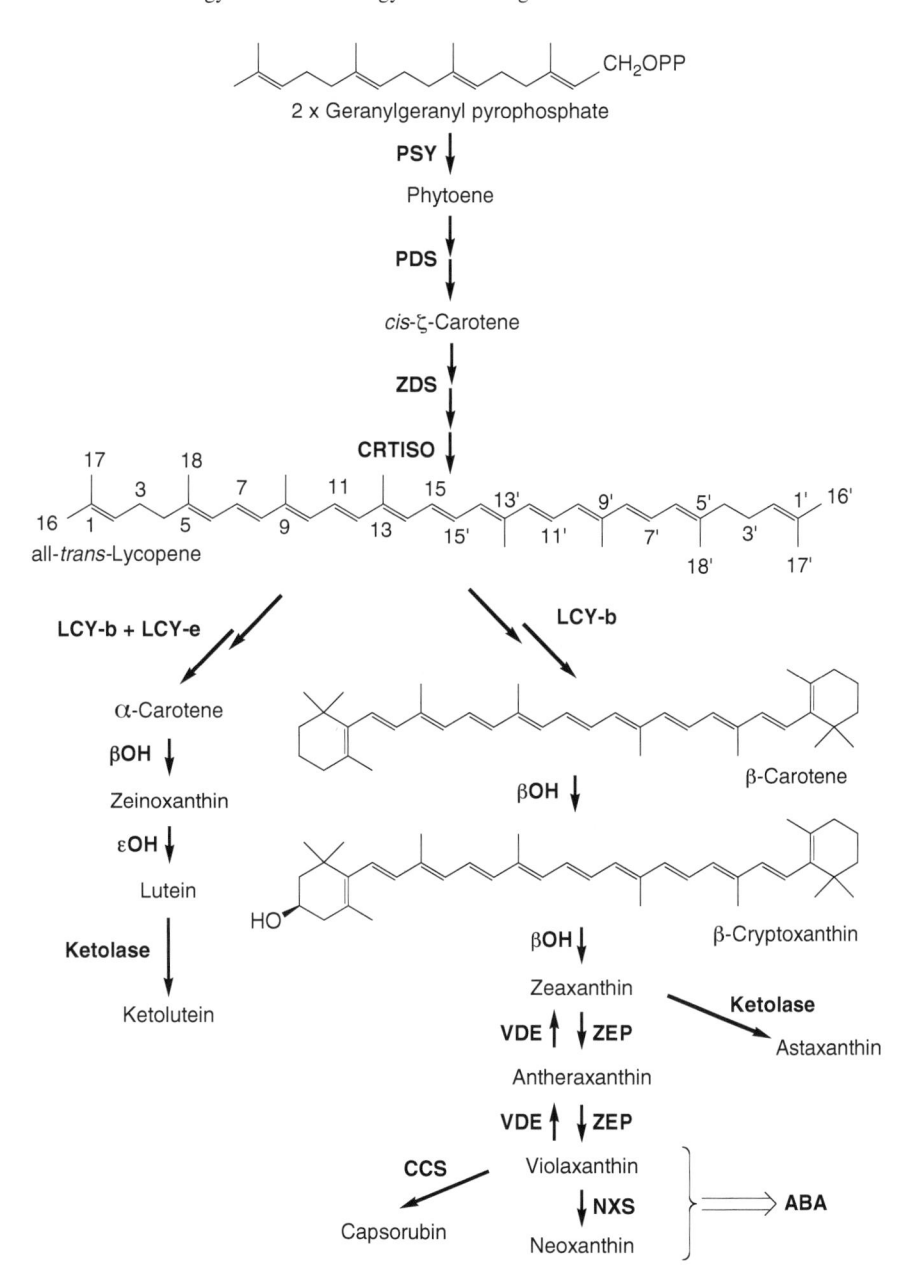

Fig. 9.2 Carotenoid biosynthesis pathway showing routes to compounds common in plants, as well as some ketocarotenoids. PSY, phytoene synthase; PDS, phytoene desaturase; ZDS, ζ-carotene desaturase; CRTISO, carotenoid isomerase; LCY-b, lycopene β-cyclase; LCY-e, lycopene ε-cyclase; εOH, ε-hydroxylase; βOH, β-hydroxylase; ZEP, zeaxanthin epoxidase; VDE, violaxanthin de-epoxidase; NXS, neoxanthin synthase. The numbering of key carbons is shown on the lycopene structure. Only some of the proposed biosynthetic routes to ketocarotenoids are shown

Table 9.1 Selected examples of in planta GM of carotenoid biosynthesis

Approach	Species	Phenotypic change	Reference
Seed-specific over-expression of bacterial geranylgeranyl diphosphate synthase, PSY and PDS and plant or bacterial LCY-b	Canola	50-fold increase in seed carotenoid levels (principally carotenes), an orange colour imparted and altered α- to β-carotene ratio	(Ravanello et al. 2003)
Endosperm-specific over-expression of daffodil or maize PSY and bacterial CDS	Rice	Development of Golden Rice Indica and Japonica cultivars	(Hoa et al. 2003; Paine et al. 2005)
Fruit-specific over-expression of bacterial PSY or PDS	Tomato	Increase in fruit carotenoid (PSY) or β-carotene (PDS) levels	(Römer et al. 2000; Fraser et al. 2002)
Over-expression of bacterial PSY	Potato	Tuber carotenoid content increased, particularly β-carotene and lutein	(Ducreux et al. 2005)
Fruit-specific over-expression of LCY-b or LCY-b and β-OH	Tomato	Fruit β-carotene, β-cryptoxanthin and zeaxanthin levels increased, lycopene levels decreased	(Rosati et al. 2000; Dharmapuri et al. 2002; D'Ambrosio et al. 2004)
Inhibition of gene activity for β-OH	Potato	Increase in β-carotene levels	(van Eck et al. 2007)
Inhibition of gene activity for LCY-e	Potato	Total tuber carotenoid levels increased up to 2.5-fold and β-carotene levels increased up to 14-fold	(Diretto et al. 2006)
Inhibition of gene activity for LCY-b	Tomato	Loss of β-carotene, small increase in fruit lycopene content	(Rosati et al. 2000)
Inhibition of gene activity for ZEP	Potato	Tuber zeaxanthin levels increased up to 130-fold	(Römer et al. 2002)
Over-expression of plant or bacterial β-carotene ketolase, including in a potato transgenic with ZEP gene activity inhibited	Carrot, potato, tobacco, *Lotus japonicus*	Formation of ketocarotenoids in roots, tubers, leaves and/or flowers	(Mann et al. 2000; Gerjets and Sandmann 2006; Morris et al. 2006; Gerjets et al. 2007; Suzuki et al. 2007; Zhu et al. 2007; Jayaraj et al. 2008)
Over-expression of bacterial β-carotene ketolase and carotenoid-3,3'-β-hydroxylase	Tobacco, tomato	Formation of ketocarotenoids in leaves of tobacco, and at low levels in leaves and fruit of tomato	(Ralley et al. 2004)

(continued)

Table 9.1 (continued)

Approach	Species	Phenotypic change	Reference
Over-expression of the *Orange* gene	Cauliflower, potato	Increased chromoplast numbers and carotenoid levels	(Lu et al. 2006; Lopez et al. 2008)
Inhibition of CmCCD4a gene activity	Chrysanthemum	Increase in carotenoid levels, conversion of white flowers to yellow	(Ohmiya et al. 2006)
Fruit-specific over-expression of tomato geraniol synthase	Tomato	50% reduction in lycopene, accumulation of flavour-related monoterpenes	(Davidovich-Rikanati et al. 2007)

introducing biosynthetic activities novel to the species and preventing carotenoid degradation.

9.2.2.1   Increasing Activity of Carotenoid Biosynthetic Enzymes

Use of transgenes for plant or bacterial phytoene synthase (PSY) and phytoene desaturase (PDS) has been effective in greatly increasing concentrations of β-carotene in food staples. This was initially for the generation of 'Golden rice' cultivars, which have now achieved >30 μg g^{-1} dry seed for β-carotene, and has since been applied to other staple food crops, with some GM potato cultivars reaching levels of >10 μg g^{-1} dry weight in tubers (Ducreux et al. 2005). One finding of importance to predictive GM of carotenoid biosynthesis is that the source of the biosynthetic gene, in terms of bacterial versus plant and specific plant source, can influence the phenotypes arising. This is clear from an extensive comparison made of different PSY transgenes for increasing carotenoid biosynthesis in rice endosperm (Paine et al. 2005). Other findings of importance are that the background activities of the endogenous biosynthetic enzymes strongly influence the transgenic phenotypes. Activity of biosynthetic transgenes can induce changes in the regulation of the endogenous carotenoid biosynthetic genes, while tissue-specific expression of transgenes is desirable to avoid pleotropic effects, such as dwarfing due to reduced GA production from disrupted plastid function or precursor allocation.

9.2.2.2   Inhibiting Activity of Carotenoid Biosynthetic Enzymes

Inhibiting the activity of lycopene β-cyclase (LCY-b) may promote lycopene accumulation at the levels that would provide strong red colouration in fruit, tubers or flowers, but no GM examples of this have been published. In tomato, LCY-b inhibition raised the (already high) lycopene concentrations only slightly (Rosati et al. 2000). Inhibition of lycopene ε-cyclase (LCY-e) in potato raised β-carotene concentrations significantly but, surprisingly, without any reduction in lutein concentrations (Diretto et al. 2006). Tuber-specific changes in transcript abundance for several carotenoid biosynthetic genes occurred, and total carotenoid levels increased 2.5-fold, suggesting feedback regulation within the pathway.

9.2.2.3 Introducing Novel Carotenoid Biosynthetic Activities

Some bacteria and algae, and a few plants produce carotenoids with modified ring structures, called ketocarotenoids. One of the most significant of these is the red astaxanthin, which is used in large quantities as an aquaculture feed. Introduction of the key enzyme β-carotene ketolase (4,4′-β-oxygenase) has conferred ketocarotenoid production into a range of species. In addition to astaxanthin, ketocarotenoids such as 4-ketolutein and 4-ketozeaxanthin have been found in the transgenics. Of particular note for this review are the introduction of orange flower colour into yellow-flowered *Lotus japonicus* using a bacterial ketolase (*crtW*; Suzuki et al. 2007), and the engineering of ketocarotenoid production in the nectary of tobacco (*Nicotiana tabacum*) and petals of *Nicotiana glauca* using bacterial- (*crtO*, *crtW*), algal- or plant-sourced ketolases (Mann et al. 2000; Gerjets et al. 2007; Zhu et al. 2007). Over-expressing an algal ketolase (*bkt1*) in carrot caused a dramatic conversion of up to 70% of total carotenoids into ketocarotenoids, with ketocarotenoids accumulating up to 270 μg g^{-1} fresh weight (Jayaraj et al. 2008). Transcript levels of the endogenous β-carotene hydroxylases increased in response to the transgene in carrot, suggesting feedback regulation within the pathway. The ketocarotenoids in the transgenics of some species are esterified, including examples of modification of the 'unnatural' substrates by endogenous biosynthetic enzymes of species that normally do not produce ketocarotenoids.

9.2.2.4 Carotenoid Degradation

Carotenoids may be subject to rapid turnover, a process that has not been shown be a factor in anthocyanin-based flower colours. Ohmiya et al. (2006) demonstrated that in all chrysanthemum cultivars they examined the white flower colour was maintained through carotenoid degradation by a carotenoid cleavage dioxygenase (CmCCD4a), rather than being due to lack of synthesis. Inhibition of CmCCD4a gene activity using RNAi converted the petal colour of a white-flowered chrysanthemum cultivar to yellow. As yet, it is not known how generally important an imbalance between carotenoid synthesis and degradation is in establishing white or pale flower colours.

  The degradative enzymes are also important with regard to formation of apocarotenoids by cleavage of carotenoids. Water-soluble apocarotenoids are pigments in roots, fruits and other tissues in a range of species. The carotenoid cleavage enzymes, and associated cDNAs, that form red apocarotenoids from lycopene and zeaxanthin have been characterized and, in theory, could be used to confer apocarotenoid biosynthesis to flowers (Bouvier et al. 2003a, b).

9.2.2.5 Sinks for Carotenoid Sequestration

Where the aim is to introduce novel carotenoid-based colours into species that lack significant carotenoid production in the flowers, the lack of floral plastids of sufficient number and type may be a problem. Not only may this limit the total amount of carotenoids that may be accumulated, but it is possible that a lack of plastid sink sites could limit the biosynthetic activity of the pathway through feedback regulation. However, likewise, it is possible that increasing carotenoid biosynthetic activity could promote plastid division and differentiation, or that increasing available plastid sink capacity could promote carotenoid biosynthesis. The linkage between plastid biogenesis and carotenoid biosynthesis is not clear, but there has been some recent experimental success in this regard. The identification and characterization of the gene behind the high-carotenoid *Orange* phenotype of cauliflower, and its successful application in GM of potato, suggests that creating a metabolic sink to sequester carotenoids provides an important mechanism to promote carotenoid accumulation (Lu et al. 2006). The *Or* gene encodes a DnaJ cysteine-rich domain-containing protein that may mediate the differentiation of proplastids and/or non-coloured plastids into chromoplasts, with an associated increase in carotenoid accumulation. Similarly, ABA deficiency may cause increased plastid division and associated increases in carotenoid content—for example, *high-pigment3* mutant of tomato (Galpaz et al. 2008). These results suggest that carotenoid precursor supply may be less of a problem than overall plastid volume and plastid differentiation. However, there has also been some success in the metabolic engineering of carotenoid precursor supply (Enfissi et al. 2005).

One advantage carotenoid biosynthesis offers for GM is that the plastid localisation of the pathway facilitates the limitation of transgene flow through introduction of the transgene to the plastid genome. Chloroplast transformation has recently been used to engineer plants for increased β-carotene production through LCY-b over-expression (Wurbs et al. 2007).

9.2.3 Betalains

The base chromophore of betalains is betalamic acid. Conjugation with amino acids/amines yields the yellow betaxanthins. Condensation with *cyclo*-dihydroxyphenylalanine (*cyclo*-DOPA) yields the first of the red-violet betacyanins, betanidin and isobetanidin. Like anthocyanins, the betacyanins are commonly glycosylated. The glycosylation may occur on the betanidin and isobetanidin products or, in some species, on *cyclo*-DOPA prior to condensation (Sasaki et al. 2004). The initial betacyanins are often modified by further glycosylation and acylation.

The betalain biosynthetic pathway is relatively simple, with only a few enzyme-catalyzed steps (Fig. 9.3). In the commonly proposed model, the two key enzymes are a bifunctional tyrosinase of the polyphenol oxidase (PPO) enzyme group, and the DOPA-4,5-extradiol dioxygenase (DOD; Strack et al. 2003). The tyrosinase is required both for the conversion of the aromatic amino acid tyrosine to L-DOPA,

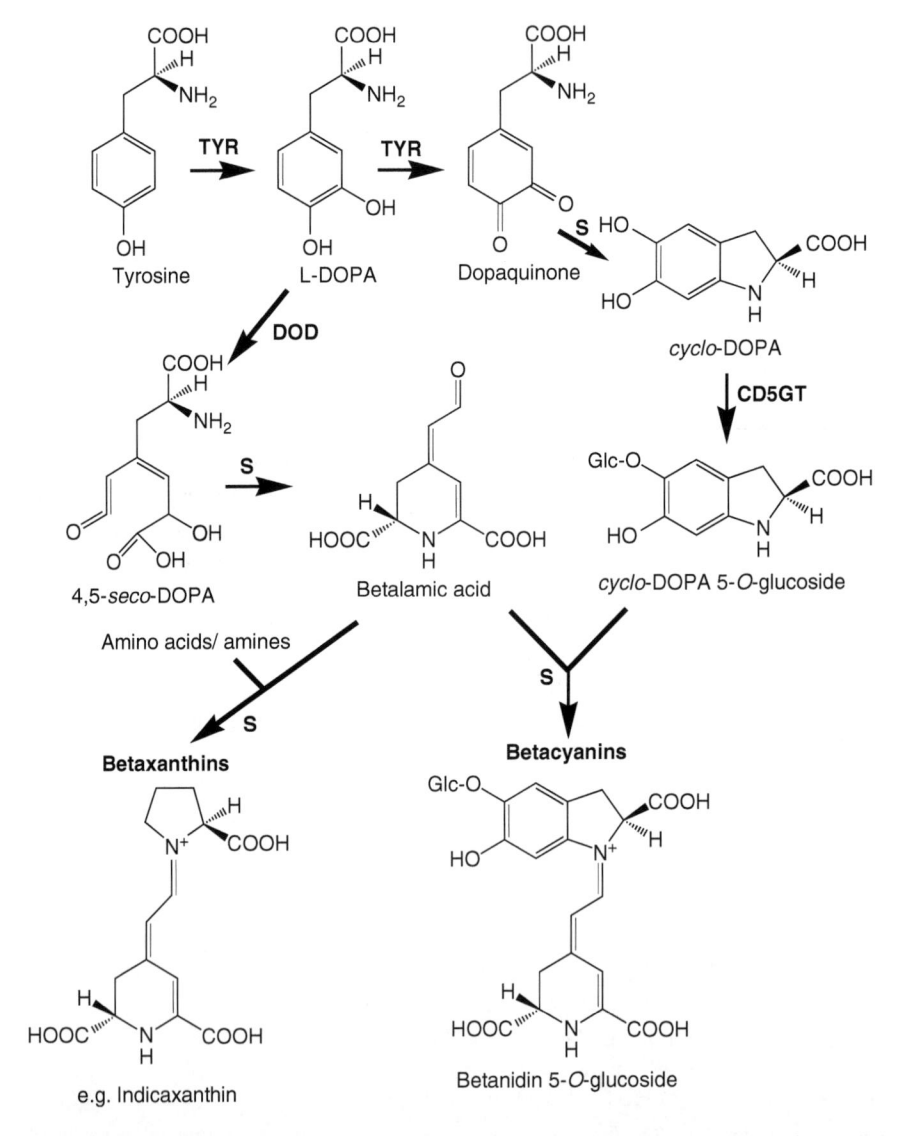

Fig. 9.3 Biosynthesis of betalains, showing the route to betacyanins with O-glucosylation at the C-5 position. TYR, tyrosinase; CD5GT, *cyclo*-DOPA glucosyltransferase; DOD, DOPA-4,5-extradiol dioxygenase; S, spontaneous conversion. The glucosylation is shown on *cyclo*-DOPA, but in some species may occur after condensation with betalamic acid

the substrate of DOD, and the conversion of L-DOPA to *cyclo*-DOPA. In theory, the activities of the tyrosinase and DOD should be sufficient to permit formation of betaxanthins, if sufficient tyrosine is available. Formation of betacyanins requires the additional activity of *O*-glycosyltransferases (GTs), and may also involve acyltransferases. A cDNA for DOD has been isolated and characterized from plants (Christinet et al. 2004). However, DNA sequences have not been published from plants for the tyrosinase and there are no reports of the GM of the pathway.

9.3 Regulation of Floral Pigmentation

There is rich diversity in floral pigmentation and its patterning (Fig. 9.4). Flower colour patterns may be due to viral infection or genetic instability and heterogeneity resulting from, for example, transposons, paramutation and chimeral dispositions (Itoh et al. 2002; Koseki et al. 2005; Olbricht et al. 2006). The genetic basis of

Fig. 9.4 Examples of patterning of floral pigmentation in flowers of **a** *Gazania*, **b** painted tongue (*Salpiglossis sinuata*), **c** *Miltoniopsis* and **d** *Brassidium* (a hybrid between the genera *Brassia* and *Oncidium*). Bar = ~5 cm

several of these unstable patterns is well understood. Of particular note is the recent demonstration that short-interfering RNAs targeting CHS underlie the 'Red Star' pattern of petunia (Koseki et al. 2005), and may be responsible for some picotee patterns in the same species (Saito et al. 2006). However, most flower colour patterns are established through genetic programmes orchestrated by developmental cues, having evolved through selection driven by mutualistic and parasitic interactions with other organisms. Pigmentation patterns (in both visible and UV light) with other floral traits are used by pollinators (and other organisms) to distinguish one species from another, and can serve to signal to pollinators where the reward is within a single flower, which flowers within an inflorescence have a reward, or even the quality or quantity of the reward (Weiss 1995; Schaefer et al. 2004). They can also be used to deceive—for example, by reward-less flowers to lure pollinators. Impressive examples of deceptive signalling are found in the Orchidaceae, the largest family in the angiosperms. In the genus *Ophrys*, there are species that have evolved in floral colour patterning, form and scent to resemble female insects, for pollination through attempted copulation by the males of the species (Schlüter and Schiestl 2008).

The developmental control of floral pigmentation may involve not only the timing and the level of production (coordinated with other floral traits like scent), but also restricting production to particular organs and specific regions or cells within those organs. It may include the non-senescence-related degradation of pigments associated with pollination. In many plants, there is the need for coordinate regulation of more than one pigment pathway (commonly flavonoid and carotenoid) to lay down the pigmentation pattern or overall flower colour. In the following sections, the transcriptional regulation of pigment biosynthetic genes is discussed, the developmental cues that are involved and biotechnological perspectives. The focus is on flavonoid biosynthesis because knowledge is most advanced for this pathway.

9.3.1 Transcriptional Control

The first plant transcription factor (TF) identified and isolated was C1, an R2R3-MYB TF from maize that controls anthocyanin pigmentation in the kernel (Cone et al. 1986; Paz-Ares et al. 1986). Since then, studies with antirrhinum, petunia and, in more recent years, arabidopsis and morning glory have led to transcriptional regulation of anthocyanin production being one of the best-characterized examples of TF combinatorial control in plants. For anthocyanin biosynthesis, the MYB protein does not activate transcription on its own but rather needs to interact with a TF of the basic helix-loop-helix (bHLH) type. The interaction occurs between the DNA-binding domain of the MYB protein and the N-terminal region of the bHLH protein. The transcriptional complex formed with these proteins is proposed also to contain a protein of the WD40 or WD-repeat (WDR) type, and may also require the activity of additional interacting proteins (Grotewold 2006; Hernandez et al. 2007). The bHLH component of the complex is thought to be in dimer form (Feller et al. 2006).

Although the role of the different components of the transcriptional complex is still being determined, a model is emerging. The MYB protein has a key role in promoter recognition and binding. A major role of the bHLH protein role may be to assist the MYB, either by stabilizing the protein complex or by promoting the activation process (Hernandez et al. 2004; Grotewold 2006). In particular, the bHLH domain, as shown for the maize protein R, recruits an EMSY-related factor to the MYB/bHLH complex that is proposed to be active in chromatin functions (Hernandez et al. 2007). A role in chromatin remodelling has also been proposed for the WDR protein (Broun 2005). Alternatively (or additionally), it may stabilize the complex, interacting with the bHLH protein (Baudry et al. 2004).

The regulation of anthocyanin pigmentation also involves proteins that have a repressive function, inhibiting pigmentation. They are single-repeat MYB proteins, and have been identified in petunia and arabidopsis, where they inhibit floral and vegetative/seed pigmentation respectively (Koes et al. 2005; Dubos et al. 2008; Matsui et al. 2008). They may bind the bHLH factors, functioning as a passive repressor by preventing the activation complexes from forming (Koes et al. 2005; Dubos et al. 2008). The arabidopsis protein has an identified repression domain. Alternatively, it has been postulated that the protein integrates into the complexes, turning them into direct transcriptional repressors of target gene expression (Matsui et al. 2008).

The characterization of the anthocyanin-related TFs is now starting to reveal the regulatory mechanism of some developmentally programmed floral pigmentation patterns, in model species such as antirrhinum and petunia. In antirrhinum, the activities of different alleles of the MYB activators, *Rosea1* and *Rosea2*, determine the strength of pigmentation and spatial distribution between major domains of the petal lobe, and a third MYB gene, *Venosa*, determines the venation pattern (Schwinn et al. 2006).

TF genes have been characterized not only for regulation of anthocyanin production, but also for regulation of the biosynthesis of other flavonoids, in particular proanthocyanidins, flavone *C*-glycosides and flavonols (Grotewold 2006). In contrast, no TFs have yet been reported for the betalain pathway and there is limited information on TFs regulating carotenoid production. Transcriptional control is likely the main regulatory step for betalain production in flowers, as it is for anthocyanins. Certainly, transcript levels of the known betalain biosynthetic genes increase in parallel with accumulation of pigment (Christinet et al. 2004; Sasaki et al. 2005). The regulation of carotenoid production may be complex, with interactions with plastid biogenesis and differentiation, and post-transcriptional regulation prevalent even in highly carotenogenic tissues such as tomato fruit (Kahlau and Bock 2008). However, for the production of large amounts of carotenoids in chromoplasts for petal pigmentation, transcriptional regulation is probably the principal control point. As with anthocyanins and betalains, transcript levels of the carotenoid biosynthetic genes increase greatly during flower development of species studied. TF genes have not been reported that regulate carotenoid biosynthesis in flowers, but there has been progress recently in identifying TFs that bind to promoter elements of carotenoid biosynthetic genes in general. Specifically, a member of the APETALA2 TF family, AtRAP2.2, has been shown to bind to a *cis*-element of the PSY and PSD genes of arabidopsis, and may act to maintain basal levels of carotenoid biosynthetic gene activity in leaves (Welsch et al. 2007).

9.3.2 Developmental Signalling

The direct activators of anthocyanin biosynthetic genes, Myb and bHLH, must themselves be regulated by signals further up a signalling pathway. What these signals may be and whether they are common among most plants is not clear, but there are preliminary data for the role of some of the endogenous and environmental signals.

In floral development of at least some species, the hormone gibberellic acid (GA), produced by anthers, plays a role both in petal cell expansion and pigmentation (both carotenoid- and anthocyanin-based; Weiss 2000). In petunia, the action of GA includes the induction of expression of two genes encoding anthocyanin-related TFs. Light is commonly required for anthocyanin pigmentation. However, there are few data for control of the anthocyanin-related TFs in flowers, although much is known of light regulation of vegetative pigmentation, particularly relating to the central role of the COP signalosome in arabidopsis (Wei and Deng 2003).

Of great interest is how the 'ABCE' TFs specify floral organ identity and control both early and late processes in floral organogenesis. Regarding pigmentation, loss of activity of the petunia E-class gene *floral binding protein2* results in small, aberrant green petals that fail to produce anthocyanins or express CHS (Angenent et al. 1994; Ferrario et al. 2003). In antirrhinum, activity of the B-class TF DEFICIENS is required until late stages of development to maintain petal (and stamen) differentiation, and loss of DEF function causes reduced pigmentation and lower transcript abundance for anthocyanin biosynthetic genes (Bey et al. 2004). It is not known whether these TFs activate the flavonoid biosynthetic genes or initiate a regulatory cascade. It has been postulated that the ABCE TFs exert their control over many downstream genes involved in organogenesis in a direct way rather than through intermediary TFs, given the dearth of TFs identified in transcript profiling approaches to determine the target genes of these TFs (Zik and Irish 2003; Bey et al. 2004). In arabidopsis, it has been shown recently that the C-class gene *Agamous* controls stamen maturation through direct activation of the gene encoding the enzyme catalyzing the first step in the pathway that produces the hormone jasmonic acid (Ito et al. 2007).

For regulation of carotenoid production in flowers, there is little molecular information. However, there is clearly a link between plastid biogenesis and chromoplast pigmentation in fruit that is likely to apply to flowers (see Sect. 9.2.2.5).

9.3.3 Biotechnology Applications of TFs

There is tremendous interest across the gamut of all crops in using TF-based GM strategies to modify traits. The appeal centres on the prospect of finding single controlling switches for traits that are based on the expression of multiple genes. That TFs may provide such gene technology is supported by data suggesting that variation in TF gene activity underlies much of the diversity in plant traits (Cronk 2001), including anthocyanin pigmentation (Schwinn et al. 2006).

It was first shown in maize that ectopic expression of one Myb and one bHLH from the anthocyanin-related gene families was necessary and sufficient to induce anthocyanin production in a wide range of tissues (Ludwig et al. 1990). Since then, there have been many experiments using flavonoid-related TFs in transgenic plants. However, there are surprisingly few published examples for modifying anthocyanin biosynthesis in ornamentals. The majority of the experiments have been conducted to define TF function in model species, or for modification of traits such as PA biosynthesis in crop plants (reviewed in Davies and Schwinn 2006). The first published use of a TF to modify pigmentation in an ornamental crop species was in petunia with the maize bHLH factor LC. A *35S:Lc* transgene conferred deep purple pigmentation to the vegetative tissue (Bradley et al. 1998). *Lc* was also effective in inducing anthocyanin pigmentation in the ornamental foliage plant *Caladium bicolour* using the *35S* promoter (Li et al. 2005) and in a white-flowered petunia cultivar when expressed under a petal-specific promoter (Chi-Chun et al. 2007), but gave no phenotypic change in lisianthus or pelargonium (Bradley et al. 1999). An interesting variation of this approach is the use of the *Tag1* transposon from arabidopsis with a maize gene closely related to *Lc, R. 35S:Tag1-R* expressed in non-ornamental tobacco-generated plants with a variety of floral pigmentation patterns (that varied both within and between lines) due to the occasional excision of *Tag1* restoring the functional *R* gene (Liu et al. 2001). The only published examples for the Myb genes in ornamental crops are using *Rosea1* from antirrhinum and *PAP1* (*AtMYB75*) from arabidopsis. *35S:Rosea1* in lisianthus induced anthocyanin production at an earlier stage of floral development and increased anthocyanin concentrations in the sepals, but had no effect on pigmentation of the open flower or the vegetative tissue (Schwinn et al. 2001). *35S:PAP1* greatly increased production of various phenylpropanoids in arabidopsis or tobacco (Borevitz et al. 2000). Moyal Ben Zvi et al. (2008) reasoned that PAP1 may allow for co-regulation of colour and scent in ornamentals. When expressed in petunia, *35S:PAP1* did indeed increase production of scent compounds (up to tenfold), resulting in more deeply pigmented flowers that were noticeably more fragrant.

It is clear from the experiments to date that TFs can provide powerful gene technology for controlling both the amount of anthocyanin produced in plants and its spatial distribution. However, there is still an element of 'trial and error' in the use of TFs in GM. It has been found that in a given species one TF transgene may have no phenotypic effect, while a closely related TF transgene induces a visual change. A transgene for a bHLH may prove more effective than one for a MYB, or vice versa. The reasons for these variations are not certain, but may reflect differences in activation strength and gene targets of the individual TF proteins and/or hierarchies of regulation among the TFs. Multigene families are common for the anthocyanin-related TFs, and different members of the gene families have indeed been shown to vary in activation characteristics (Schwinn et al. 2006).

TFs with an inhibitory action may also be used to modify anthocyanin production. These may be plant genes with a repressor function, such as FaMYB1 of strawberry (Aharoni et al. 2001). Alternatively, repressor motifs may be added to TFs that are normally activators of gene expression. Chimeric repressor silencing technology (CRES-T) is based on adding the EAR-motif repression domain from the arabidopsis TF SUPERMAN. Proof of concept was shown using EAR-PAP1, which inhibits anthocyanin production in arabidopsis seedlings (Hiratsu et al. 2003). The authors of the initial study, and colleagues are currently applying CRES-T with various arabidopsis TFs to a range of ornamental plants (e.g. torenia and lisianthus), and the transgenic lines generated to date include examples with altered floral pigmentation patterning and morphology (http://www.cres-t.org/fiore/public_db/index.shtml). While repressor TFs allow for a dominant trait transgene, there is also the potential for unintended negative consequences for the production of related compounds, due to the repression of genes that supply precursors to multiple branches of the phenylpropanoid pathway. Indeed, the arabidopsis plants expressing the repressor version of PAP1 were unable to produce proanthocyanidins in the seed (Matsui et al. 2004), despite PAP1 normally being involved only in anthocyanin biosynthesis (Gonzalez et al. 2008).

Almost all the experiments to date for modification of the regulation of pigment biosynthesis concern the flavonoid pathway. Successful reports for GM of carotenoid biosynthesis by altering the regulatory pathways are limited to some of the genes several stages up in the regulatory cascade, and which consequently affect multiple pathways. For example, fruit-specific RNAi of the photomorphogenesis regulatory gene DET1 in tomato produces plants having fruit with increased concentrations of both carotenoids and flavonoids (Davuluri et al. 2005). One of the few examples of GM using a transgene for a TF that directly regulates carotenoid biosynthetic genes is the over-expression of AtRAP2.2 in arabidopsis. However, no change in either AtRAP2.2 transcript or protein level occurred in the transgenics, probably reflecting a complex regulatory network that may include proteasome-mediated regulation of AtRAP2.2 protein levels (Welsch et al. 2007).

9.4 Concluding Comments

Knowledge of the biosynthetic pathways for flavonoids and carotenoids is well advanced, with corresponding DNA sequences available for all of the main biosynthetic steps. There are also extensive data on the molecular basis of the regulation of flavonoid biosynthesis, an area for which details are only now emerging concerning the carotenoid pathway. For both pathways, several of the outstanding questions are now focused on the cell biology of the biosynthetic pathway. For flavonoids, these include what are the steps from the site of synthesis on the ER to stable accumulation in the vacuole? What are the structures of the biosynthetic complexes and how important is metabolic channelling? For carotenoids, there are questions around the linkage to chromoplast development, metabolic flux regulation within the pathway, substrate supply, and relationship to other pathways that share precursors. There are also many secondary modification enzymes yet to be characterized that contribute to the great diversity of known carotenoid and anthocyanin structures.

Gene technology for flavonoids is close to providing reliable tools for the predictive metabolic engineering of the pathway in many species. There are proven approaches to reducing or increasing anthocyanin amounts, as well as changing the type of anthocyanin produced. Nevertheless, it is generally not yet possible to generate new flower varieties of particular colours, such as sky blue or orange, as there is no proven approach to engineer major changes in vacuolar pH, or to predictably trigger the formation of anthocyanin complexes or anthocyanin–metal interactions. The use of TF sequences in molecular breeding of ornamentals is in the early days, as it is for all crops, but the research to date with the flavonoid pathway has shown the great potential of TF technology.

Despite the apparent simplicity of the betalain biosynthetic pathway, it is still only partly characterized, and DNA sequences for the key step of L-DOPA formation have not been isolated. There are few molecular data on the regulation of the pathway, and no significant reports on the GM of betalain production. In contrast, GM of carotenoid production has moved ahead rapidly in the last few years, with strategies based around over-expression of multiple biosynthetic genes for increasing levels of human health-related carotenoids in food crops such as rice and potato. While this research has provided some standard approaches to increasing carotenoid levels in specific tissues (e.g. the rice grain or potato tuber), its general applicability to flowers is not yet known. Also, there is a lack of understanding of how endogenous carotenoid biosynthesis is regulated and how this regulatory network may influence the performance of transgenes in different species. However, the recent success with the *Or* gene illustrates the potential for new gene technology with wide applicability for genetic modification of carotenoid production.

References

Aharoni A, de Vos CHR, Wein M, Sun Z, Greco R, Kroon A, Mol JNM, O'Connell AP (2001) The strawberry FaMYB1 transcription factor suppresses anthocyanin and flavonol accumulation in transgenic tobacco. Plant J 28:319–332

Allan AC, Hellens RP, Laing WA (2008) MYB transcription factors that colour our fruit. Trends Plant Sci 13:99–102

Andersen ØM, Jordheim M (2006) The anthocyanins. In: Andersen ØM, Markham KR (eds) The flavonoids. Chemistry, biochemistry, and applications. CRC Press, Boca Raton, FL, pp 471–552

Angenent G, Franken J, Busscher M, Weiss D, van Tunen A (1994) Co-suppression of the petunia homeotic gene *fbp2* affects the identity of the generative meristem. Plant J 5:33–44

Bang H, Kim S, Leskovar D, King S (2007) Development of a codominant CAPS marker for allelic selection between canary yellow and red watermelon based on SNP in lycopene β-cyclase (*LCYB*) gene. Mol Breed 20:63–72

Baudry A, Heim MA, Dubreucq B, Caboche M, Weisshaar B, Lepiniec L (2004) TT2, TT8, and TTG1 synergistically specify the expression of *BANYULS* and proanthocyanidin biosynthesis in *Arabidopsis thaliana*. Plant J 39:366–380

Bey M, Stuber K, Fellenberg K, Schwarz-Sommer Z, Sommer H, Saedler H, Zachgo S (2004) Characterization of *Antirrhinum* petal development and identification of target genes of the class B MADS box gene *DEFICIENS*. Plant Cell 16:3197–3215

Borevitz JO, Xia Y, Blount J, Dixon RA, Lamb C (2000) Activation tagging identifies a conserved MYB regulator of phenylpropanoid biosynthesis. Plant Cell 12:2383–2394

Bouvier F, Dogbo O, Camara B (2003a) Biosynthesis of the food and cosmetic plant pigment bixin (annatto). Science 300:2089–2091

Bouvier F, Suire C, Mutterer J, Camara B (2003b) Oxidative remodeling of chromoplast carotenoids: identification of the carotenoid dioxygenase *CsCCD* and *CsZCD* genes involved in crocus secondary metabolite biogenesis. Plant Cell 15:47–62

Bradley JM, Davies KM, Deroles SC, Bloor SJ, Lewis DH (1998) The maize *Lc* regulatory gene up-regulates the flavonoid biosynthetic pathway of *Petunia*. Plant J 13:381–392

Bradley JM, Deroles SC, Boase MR, Bloor S, Swinny E, Davies KM (1999) Variation in the ability of the maize *Lc* regulatory gene to upregulate flavonoid biosynthesis in heterologous systems. Plant Sci 140:31–39

Broun P (2005) Transcriptional control of flavonoid biosynthesis: a complex network of conserved regulators involved in multiple aspects of differentiation in *Arabidopsis*. Curr Opin Plant Biol 8:272–279

Chi-Chun Y, Jing-Juan Y, Qian Z, Deng-Yun Z, Guang-Ming A (2007) Influence of maize *Lc* regulatory gene on flower colour of transgenic tobacco and petunia. Chinese J Agric Biotechnol 4:145–150

Christinet L, Burdet FX, Zaiko M, Hinz U, Zrÿd J-P (2004) Characterization and functional identification of a novel plant 4,5-extradiol dioxygenase involved in betalain pigment biosynthesis in *Portulaca grandiflora*. Plant Physiol 134:265–274

Cone KC, Burr FA, Burr B (1986) Molecular analysis of the maize anthocyanin regulatory locus *C1*. Proc Natl Acad Sci USA 83:9631–9635

Cronk QC (2001) Plant evolution and development in a post-genomic context. Nature Rev Genet 2:607–619

D'Ambrosio C, Giorio G, Marino I, Merendino A, Petrozza A, Salfi L, Stigliani AL, Cellini F (2004) Virtually complete conversion of lycopene into β-carotene in fruits of tomato plants transformed with the tomato lycopene β-cyclase (*tlcy*-b) cDNA. Plant Sci 166:207–214

Davidovich-Rikanati R, Sitrit Y, Tadmor Y, Iijima Y, Bilenko N, Bar E, Carmona B, Fallik E, Dudai N, Simon JE, Pichersky E, Lewinsohn E (2007) Enrichment of tomato flavor by diversion of the early plastidial terpenoid pathway. Nature Biotechnol 25:899–901

Davies KM, Schwinn KE (2006) Molecular biology and biotechnology of flavonoid biosynthesis. In: Andersen ØM, Markham KR (eds) The flavonoids. Chemistry, biochemistry, and applications. CRC Press, Boca Raton, FL, pp 143–218

Davies KM, Bloor SJ, Spiller GB, Deroles SC (1998) Production of yellow colour in flowers: redirection of flavonoid biosynthesis in *Petunia*. Plant J 13:259–266

Davies KM, Schwinn KE, Deroles SC, Manson DG, Lewis DH, Bloor SJ, Bradley JM (2003) Enhancing anthocyanin production by altering competition for substrate between flavonol synthase and dihydroflavonol 4-reductase. Euphytica 131:259–268

Davuluri GR, van Tuinen A, Fraser PD, Manfredonia A, Newman R, Burgess D, Brummell DA, King SR, Palys J, Uhlig J, Bramley PM, Pennings HMJ, Bowler C (2005) Fruit-specific RNAi-mediated suppression of *DET1* enhances carotenoid and flavonoid content in tomatoes. Nature Biotechnol 23:890–895

DellaPenna D, Pogson BJ (2006) Vitamin synthesis in plants: tocopherols and carotenoids. Annu Rev Plant Biol 57:711–738

Dharmapuri S, Rosati C, Pallara P, Aquilani R, Bouvier F, Camara B, Giuliano G (2002) Metabolic engineering of xanthophyll content in tomato fruits. FEBS Lett 519:30–34

Diretto G, Tavazza R, Welsch R, Pizzichini D, Mourgues F, Papacchioli V, Beyer P, Giuliano G (2006) Metabolic engineering of potato tuber carotenoids through tuber-specific silencing of lycopene epsilon cyclase. BMC Plant Biol 6:13

Dubos C, Le Gourrierec J, Baudry A, Huep G, Lanet E, Debeaujon I, Routaboul J-M, Alboresi A, Weisshaar B, Lepiniec L (2008) MYBL2 is a new regulator of flavonoid biosynthesis in *Arabidopsis thaliana*. Plant J 55:940–953

Ducreux LJM, Morris WL, Hedley PE, Shepherd T, Davies HV, Millam S, Taylor MA (2005) Metabolic engineering of high carotenoid potato tubers containing enhanced levels of β-carotene and lutein. J Exp Bot 56:81–89

Enfissi EMA, Fraser PD, Lois LM, Boronat A, Schuch W, Bramley PM (2005) Metabolic engineering of the mevalonate and non-mevalonate isopentenyl diphosphate-forming pathways for the production of health-promoting isoprenoids in tomato. Plant Biotechnol J 3:17–27

Feller A, Hernandez JM, Grotewold E (2006) An ACT-like domain participates in the dimerization of several plant basic-helix-loop-helix transcription factors. J Biol Chem 281: 28964–28974

Ferrario S, Immink R, Shchennikova A, Busscher-Lange J, Angenent G (2003) The MADS box gene *FBP2* is required for SEPALLATA function in petunia. Plant Cell 15:914–925

Fischer R, Budde I, Hain R (1997) Stilbene synthase gene expression causes changes in flower colour and male sterility in tobacco. Plant J 11:489–498

Fraser PD, Bramley PM (2004) The biosynthesis and nutritional uses of carotenoids. Prog Lipid Res 43:228–265

Fraser PD, Romer S, Shipton CA, Mills PB, Kiano JW, Misawa N, Drake RG, Schuch W, Bramley PM (2002) Evaluation of transgenic tomato plants expressing an additional phytoene synthase in a fruit-specific manner. Proc Natl Acad Sci USA 99:1092–1097

Fukusaki E, Kawasaki K, Kajiyama S, An C, Suzuki K, Tanaka Y, Kobayashi A (2004) Flower color modulations of *Torenia hybrida* by downregulation of chalcone synthase genes with RNA interference. J Biotechnol 111:229–240

Galpaz N, Ronen G, Khalfa Z, Zamir D, Hirschberg J (2006) A chromoplast-specific carotenoid biosynthesis pathway is revealed by cloning of the tomato *white-flower* locus. Plant Cell 18:1947–1960

Galpaz N, Wang Q, Menda N, Zamir D, Hirschberg J (2008) Abscisic acid deficiency in the tomato mutant *high-pigment 3* leading to increased plastid number and higher fruit lycopene content. Plant J 53:717–730

Gerjets T, Sandmann G (2006) Ketocarotenoid formation in transgenic potato. J Exp Bot 57:3639–3645

Gerjets T, Sandmann M, Zhu C, Sandmann G (2007) Metabolic engineering of ketocarotenoid biosynthesis in leaves and flowers of tobacco species. Biotechnol J 2:1263–1269

Giuliano G, Tavazza R, Diretto G, Beyer P, Taylor MA (2008) Metabolic engineering of carotenoid biosynthesis in plants. Trends Biotechnol 26:139–145

Gonzalez A, Zhao M, Leavitt JM, Lloyd AM (2008) Regulation of the anthocyanin biosynthetic pathway by the TTG1/bHLH/Myb transcriptional complex in Arabidopsis seedlings. Plant J 53:814–827

Grotewold E (2006) The genetics and biochemistry of floral pigments. Annu Rev Plant Biol 57:761–780

Grotewold E, Davies K (2008) Trafficking and sequestration of anthocyanins. Nat Prod Comm 3:1251–1258

Hernandez JM, Heine GF, Irani NG, Feller A, Kim M-G, Matulnik T, Chandler VL, Grotewold E (2004) Different mechanisms participate in the R-dependent activity of the R2R3 MYB transcription factor C1. J Biol Chem 279:48205–48213

Hernandez JM, Feller A, Morohashi K, Frame K, Grotewold E (2007) The basic helix-loop-helix domain of maize R links transcriptional regulation and histone modifications by recruitment of an EMSY-related factor. Proc Natl Acad Sci USA 104:17222–17227

Hiratsu K, Matsui K, Koyama T, Ohme-Takagi M (2003) Dominant repression of target genes by chimeric repressors that include the EAR motif, a repression domain, in *Arabidopsis*. Plant J 34:733–739

Hoa TTC, Al-Babili S, Schaub P, Potrykus I, Beyer P (2003) Golden indica and Japonica rice lines amenable to deregulation. Plant Physiol 133:161–169

Holton TA, Brugliera F, Tanaka Y (1993) Cloning and expression of flavonol synthase from *Petunia hybrida*. Plant J 4:1003–1010

Ito T, Ng K-H, Lim T-S, Yu H, Meyerowitz EM (2007) The homeotic protein AGAMOUS controls late stamen development by regulating a jasmonate biosynthetic gene in Arabidopsis. Plant Cell 19:3516–3529

Itoh Y, Higeta D, Suzuki A, Yoshida H, Ozeki Y (2002) Excision of transposable elements from the chalcone isomerase and dihydroflavonol 4-reductase genes may contribute to the variegation of the yellow-flowered carnation (*Dianthus caryophyllus*). Plant Cell Physiol 43:578–585

Jayaraj J, Devlin R, Punja Z (2008) Metabolic engineering of novel ketocarotenoid production in carrot plants. Transgenic Res 17:489–501

Kahlau S, Bock R (2008) Plastid transcriptomics and translatomics of tomato fruit development and chloroplast-to-chromoplast differentiation: chromoplast gene expression largely serves the production of a single protein. Plant Cell 20:856–874

Katsumoto Y, Fukuchi-Mizutani M, Fukui Y, Brugliera F, Holton TA, Karan M, Nakamura N, Yonekura-Sakakibara K, Togami J, Pigeaire A, Tao G-Q, Nehra NS, Lu C-Y, Dyson BK, Tsuda S, Ashikari T, Kusumi T, Mason JG, Tanaka Y (2007) Engineering of the rose flavonoid biosynthetic pathway successfully generated blue-hued flowers accumulating delphinidin. Plant Cell Physiol 48:1589–1560

Koes R, Verweij W, Quattrocchio F (2005) Flavonoids: a colorful model for the regulation and evolution of biochemical pathways. Trends Plant Sci 10:236–242

Koseki M, Goto K, Masuta C, Kanazawa A (2005) The star-type color pattern in *Petunia hybrida* 'Red Star' flowers is induced by sequence-specific degradation of chalcone synthase RNA. Plant Cell Physiol 46:1879–1883

Li SJ, Deng XM, Mao HZ, Hong Y (2005) Enhanced anthocyanin synthesis in foliage plant *Caladium bicolor*. Plant Cell Rep 23:716–720

Liu D, Galli M, Crawford NM (2001) Engineering variegated floral patterns in tobacco plants using the arabidopsis transposable element *Tag1*. Plant Cell Rep 42:419–423

Lopez AB, van Eck J, Conlin BJ, Paolillo DJ, O'Neill J, Li L (2008) Effect of the cauliflower *Or* transgene on carotenoid accumulation and chromoplast formation in transgenic potato tubers. J Exp Bot 59:213–223

Lu S, van Eck J, Zhou X, Lopez AB, O'Halloran DM, Cosman KM, Conlin BJ, Paolillo DJ, Garvin DF, Vrebalov J, Kochian LV, Kupper H, Earle ED, Cao J, Li L (2006) The Cauliflower *Or* gene encodes a DnaJ cysteine-rich domain-containing protein that mediates high levels of β-carotene accumulation. Plant Cell 18:3594–3605

Ludwig SR, Bowen B, Beach L, Wessler SR (1990) A regulatory gene as a novel visible marker for maize transformation. Science 247:449–450

Mann V, Harker M, Pecker I, Hirschberg J (2000) Metabolic engineering of astaxanthin production in tobacco flowers. Nature Biotechnol 18:888–892

Matsui K, Tanaka H, Ohme-Takagi M (2004) Suppression of the biosynthesis of proanthocyanidin in *Arabidopsis* by a chimeric PAP1 repressor. Plant Biotechnol J 2:487–493

Matsui K, Umemura Y, Ohme-Takagi M (2008) AtMYBL2, a protein with a single MYB domain, acts as a negative regulator of anthocyanin biosynthesis in Arabidopsis. Plant J 55:954–967

Morris WL, Ducreux LJM, Fraser PD, Millam S, Taylor MA (2006) Engineering ketocarotenoid biosynthesis in potato tubers. Metab Eng 8:253–263

Moyal Ben Zvi M, Negre-Zakharov F, Masci T, Ovadis M, Shklarman E, Ben-Meir H, Tzfira T, Dudareva N, Vainstein A (2008) Interlinking showy traits: co-engineering of scent and colour biosynthesis in flowers. Plant Biotechnol J 6:403–415

Muir SR, Collins GJ, Robinson S, Hughes S, Bovy A, De Vos CHR, van Tunen AJ, Verhoeyen ME (2001) Overexpression of petunia chalcone isomerase in tomato results in fruit containing increased levels of flavonols. Nature Biotechnol 19:470–474

Nakamura N, Fukuchi-Mizutani M, Miyazaki K, Suzuki K, Tanaka Y (2006) RNAi suppression of the anthocyanidin synthase gene in *Torenia hybrida* yields white flowers with higher frequency and better stability than antisense and sense suppression. Plant Biotechnol 23:13–17

Nakatsuka T, Nishihara M, Mishiba K, Yamamura S (2006) Heterologous expression of two gentian cytochrome P450 genes can modulate the intensity of flower pigmentation in transgenic tobacco plants. Mol Breed 17:91–99

Nakatsuka T, Pitaksutheepong C, Yamamura S, Nishihara M (2007a) Induction of differential flower pigmentation patterns by RNAi using promoters with distinct tissue-specific activity. Plant Biotechnol Rep 1:251–257

Nakatsuka T, Abe Y, Kakizaki Y, Yamamura S, Nishihara M (2007b) Production of red-flowered plants by genetic engineering of multiple flavonoid biosynthetic genes. Plant Cell Rep 26:1951–1959

Nielsen K, Deroles SC, Markham KR, Bradley MJ, Podivinsky E, Manson D (2002) Antisense flavonol synthase alters copigmentation and flower color in lisianthus. Mol Breed 9:217–229

Nishihara M, Nakatsuka T, Yamamura S (2005) Flavonoid components and flower color change in transgenic tobacco plants by suppression of chalcone isomerase gene. FEBS Lett 579:6074–6078

Ohmiya A, Kishimoto S, Aida R, Yoshioka S, Sumitomo K (2006) Carotenoid cleavage dioxygenase (CmCCD4a) contributes to white color formation in chrysanthemum petals. Plant Physiol 142:1193–1201

Okinaka Y, Shimada Y, Nakano-Shimada R, Ohbayashi M, Kiyokawa S, Kikuchi Y (2003) Selective accumulation of delphinidin derivatives in tobacco using a putative flavonoid 3′,5′-hydroxylase cDNA from *Campanula medium*. Biosci Biotechnol Biochem 67:161–165

Olbricht K, Plaschil S, Pohlheim F (2006) Causes of flower colour patterns with a focus on chimeral patterns. In: da Silva JAT (ed) Floriculture, ornamental and plant biotechnology. Advances and topical issues. Global Science Books, London, pp 311–319

Ono E, Fukuchi-Mizutani M, Nakamura N, Fukui Y, Yonekura-Sakakibara K, Yamaguchi M, Nakayama T, Tanaka T, Kusumi T, Tanaka Y (2006) Yellow flowers generated by expression of the aurone biosynthetic pathway. Proc Natl Acad Sci USA 103:11075–11080

Paine JA, Shipton CA, Chaggar S, Howells RM, Kennedy MJ, Vernon G, Wright SY, Hinchliffe E, Adams JL, Silverstone AL, Drake R (2005) Improving the nutritional value of Golden Rice through increased pro-vitamin A content. Nature Biotechnol 23:482–487

Paz-Ares J, Wienand U, Peterson PA, Saedler H (1986) Molecular cloning of the *c* locus of *Zea mays*: a locus regulating the anthocyanin pathway. EMBO J 5:829–833

Ralley L, Enfissi EMA, Misawa N, Schuch W, Bramley PM, Fraser PD (2004) Metabolic engineering of ketocarotenoid formation in higher plants. Plant J 39:477–486

Ravanello MP, Ke DY, Alvarez J, Huang BH, Shewmaker CK (2003) Coordinate expression of multiple bacterial carotenoid genes in canola leading to altered carotenoid production. Metab Eng 5:255–263

Römer S, Fraser PD, Kiano JW, Shipton CA, Misawa N, Schuch W, Bramley PM (2000) Elevation of the provitamin A content of transgenic tomato plants. Nature Biotechnol 18:666–669

Römer S, Lubeck J, Kauder F, Steiger S, Adomat C, Sandmann G (2002) Genetic engineering of a zeaxanthin-rich potato by antisense inactivation and co-suppression of carotenoid epoxidation. Metab Eng 4:263–272

Rosati C, Aquilani R, Dharmapuri S, Pallara P, Marusic C, Tavazza R, Bouvier F, Camara B, Giuliano G (2000) Metabolic engineering of β-carotene and lycopene content in tomato fruit. Plant J 24:413–419

Rosati C, Simoneau P, Treutter D, Poupard P, Cadot Y, Cadic A, Duron M (2003) Engineering of flower color in forsythia by expression of two independently-transformed dihydro-flavonol 4-reductase and anthocyanidin synthase genes of flavonoid pathway. Mol Breed 12:197–208

Saito R, Fukuta N, Ohmiya A, Itoh Y, Ozeki Y, Kuchitsu K, Nakayama M (2006) Regulation of anthocyanin biosynthesis involved in the formation of marginal picotee petals in *Petunia*. Plant Sci 170:828–834

Sasaki N, Adachi T, Koda T, Ozeki Y (2004) Detection of UDP-glucose:cyclo-DOPA 5-*O*-glucosyltransferase activity in four o'clocks (*Mirabilis jalapa* L.). FEBS Lett 568:159–162

Sasaki N, Wada K, Koda T, Kasahara K, Adachi T, Ozeki Y (2005) Isolation and characterization of cDNAs encoding an enzyme with glucosyltransferase activity for *cyclo*-DOPA from four o'clocks and feather cockscombs. Plant Cell Physiol 46:666–670

Schaefer HM, Schaefer V, Levey DJ (2004) How plant-animal interactions signal new insights in communication. Trends Ecol Evol 19:577–584

Schijlen EGWM, de Vos CHR, Martens S, Jonker HH, Rosin FM, Molthoff JW, Tikunov YM, Angenent GC, van Tunen AJ, Bovy AG (2007) RNA interference silencing of chalcone synthase, the first step in the flavonoid biosynthesis pathway, leads to parthenocarpic tomato fruits. Plant Physiol 144:1520–1530

Schlüter PM, Schiestl FP (2008) Molecular mechanisms of floral mimicry in orchids. Trends Plant Sci 13:228–235

Schwinn K, Davies K, Alm V, Mackay S, Martin C (2001) Regulation of anthocyanin biosynthesis in antirrhinum. Acta Hort 560:201–206

Schwinn K, Venail J, Shang YJ, Mackay S, Alm V, Butelli E, Oyama R, Bailey P, Davies K, Martin C (2006) A small family of MYB-regulatory genes controls floral pigmentation intensity and patterning in the genus *Antirrhinum*. Plant Cell 18:831–851

Seitz C, Vitten M, Steinbach P, Hartl S, Hirsche J, Rathje W, Treutter D, Forkmann G (2007) Redirection of anthocyanin synthesis in *Osteospermum hybrida* by a two-enzyme manipulation strategy. Phytochemistry 68:824–833

Shang Y, Schwinn K, Hunter DA, Waugh TL, Bennett MJ, Pathirana NN, Brummell DA, Jameson PE, Davies KM (2007) Methods for transient assay of gene function in floral tissues. BMC Plant Methods 3:1

Shimada Y, Nakano-Shimada R, Ohbayashi M, Okinaka Y, Kiyokawa S, Kikuchi Y (1999) Expression of chimeric P450 genes encoding flavonoid-3′,5′-hydroxylase in transgenic tobacco and petunia plants. FEBS Lett 461:241–245

Strack D, Vogt T, Schliemann W (2003) Recent advances in betalain research. Phytochemistry 62:247–269

Suzuki K, Xue HM, Tanaka Y, Fukui Y, Fukuchi-Mizutani M, Murakami Y, Katsumoto Y, Tsuda S, Kusumi T (2000) Flower color modifications of *Torenia hybrida* by cosuppression of anthocyanin biosynthesis genes. Mol Breed 6:239–246

Suzuki S, Nishihara M, Nakatsuka T, Misawa N, Ogiwara I, Yamamura S (2007) Flower color alteration in *Lotus japonicus* by modification of the carotenoid biosynthetic pathway. Plant Cell Rep 26:951–959

Taylor LP, Grotewold E (2005) Flavonoids as developmental regulators. Curr Opin Plant Biol 8:317–323

To K-Y, Wang C-K (2006) Molecular breeding of flower color. In: Silva JAT (ed) Floriculture, ornamental and plant biotechnology, vol I. Global Science Books, London, pp 300–310

Tsuda S, Fukui Y, Nakamura N, Katsumoto Y, Yonekura-Sakakibara K, Fukuchi-Mizutani M, Ohira K, Ueyama Y, Ohkawa H, Holton TA, Kusumi T, Tanaka Y (2004) Flower color modification of *Petunia hybrida* commercial varieties by metabolic engineering. Plant Biotechnol 21:377–386 van Eck J, Conlin B, Garvin DF, Mason H, Navarre DA, Brown CR (2007) Enhancing β-carotene content in potato by RNAi-mediated silencing of the β-carotene hydroxylase gene. Am J Potato Res 84:331–342

Walker AR, Lee E, Bogs J, McDavid DAJ, Thomas MR, Robinson SP (2007) White grapes arose through the mutation of two similar and adjacent regulatory genes. Plant J 49:772–785

Wei N, Deng XW (2003) The COP9 signalosome. Annu Rev Cell Dev Biol 19:261–286

Weiss MR (1995) Floral color change: a widespread functional convergence. Am J Bot 82: 167–185

Weiss D (2000) Regulation of flower pigmentation and growth: multiple signaling pathways control anthocyanin synthesis in expanding petals. Physiol Plant 110:152–157

Welsch R, Maass D, Voegel T, DellaPenna D, Beyer P (2007) Transcription factor RAP2.2 and its interacting partner SINAT2: stable elements in the carotenogenesis of arabidopsis leaves. Plant Physiol 145:1073–1085

Wurbs D, Ruf S, Bock R (2007) Contained metabolic engineering in tomatoes by expression of carotenoid biosynthesis genes from the plastid genome. Plant J 49:276–288

Xie D-Y, Sharma SB, Paiva NL, Ferreira D, Dixon RA (2003) Role of anthocyanidin reductase, encoded by BANYULS in plant flavonoid biosynthesis. Science 299:396–399

Zhu C, Gerjets T, Sandmann G (2007) Nicotiana glauca engineered for the production of ketocarotenoids in flowers and leaves by expressing the cyanobacterial crtO ketolase gene. Transgenic Res 16:813–821

Zik M, Irish VF (2003) Global identification of target genes regulated by APETALA3 and PISTILLATA floral homeotic gene action. Plant Cell 15:207–222

Zuker A, Tzfira T, Ben-Meir H, Ovadis M, Shklarman E, Itzhaki H, Forkmann G, Martens S, Neta-Sharir I, Weiss D, Vainstein A (2002) Modification of flower color and fragrance by antisense suppression of the flavanone 3-hydroxylase gene. Mol Breed 9:33–41

Chapter 10
Biosynthesis and Regulation of Flower Scent

B. Piechulla and U. Effmert

10.1 Introduction

Scent emission and perception allow inter- and intra-organismic communication over a long distance. The biologically active chemical compounds of such interactions are of small molecular weight (usually less than 300 Dalton), and have a high vapor pressure. The aliphatic and often lipophilic characters of the molecules support the emission from tissues. The volatiles can act through airflow in the atmosphere, as well as through diffusion in aqueous habitats. These properties allow living organisms to rely on these volatile molecules for communication. In seed plants, a complex strategy to ensure reproduction and preservation of species has evolved, which includes flower–animal interactions determined by defined floral traits like color, size, shape, texture, and volatile emission. Floral volatiles represent a crucial element of pollination syndromes, facilitating the attraction of specific pollinators over a wide distance. Floral volatiles underlie natural variations in the number and relative abundance between populations, within populations, within a plant, within a flower, and within different organs and tissues of the flower, which may reflect additional important functions, like defense against enemies and pathogens. An interesting phenomenon is based on mimicked odors to defend and attract plant-interacting organisms (pseudocopulation). In addition, many examples demonstrate that insects use floral scents to communicate with members of their community (e.g., beehives and ant colonies).

B. Piechulla and U. Effmert
University of Rostock, Institute of Biological Sciences, Albert-Einstein-Str. 3, 18059 Rostock, Germany
e-mail: birgit.piechulla@uni-rostock.de

10.2   Functions of Floral Scents

10.2.1   Floral Scents for Pollination

Flowers can present animals/pollinators with a virtually unlimited range of species-specific odors. Some compounds are nearly ubiquitous, while others are found only in certain species. This universality, variation, and diversity is contrasted by species-specific and compound-specific flower–animal pollination systems (pollination syndromes). Depending on the plant, species-specific pollinators respond to floral odors. A multivariate analysis of various floral volatile traits was used to characterize distinct groups of pollination syndromes. The survey showed trends of chemical profiles of floral scents that can be attributed to particular animal groups visiting the flowers, but there are no clear-cut boundaries (Dobson 2006).

10.2.2   Floral Scents with Diverse Functions

A large proportion of the volatiles within a scent mixture do not correlate with pollinator attraction. There are two conflicting evolutionary pressures facing the plant, namely, volatiles that are needed to advertise an attractive reward to pollinators and those to protect the flower from overexploitation by non-pollinating insects or destructive pollen-feeding animals, and visits by ovipositing animals, pathogens, herbivores, and other enemies. Although not many studies have addressed the latter possibilities, it is conceivable that floral scent compounds are involved in defense reactions by functioning as, for example, insect repellents and/or antimicrobial compounds. Detailed analysis of volatiles and their spatial allocation in different organs and tissues of the flower is important to understand the complex species-specific host-seeking and host-avoidance strategies. *Mirabilis jalapa* emits dominantly trans-β-ocimene, and in minor concentrations myrcene from the petaloid lobes for pollination, while the defense compound (E)-β-farnesene is localized in the abaxial trichomes of the petals (Effmert et al. 2005a, 2006). Sesquiterpene lactones secreted by anther glands in *Helianthus maximiliani* and terpenoid aldehydes in *Gossypium hirsutum* act detrimentally to the larvae of flower-feeding insects (Dobson and Bergström 2000). In the sunflower moth, *Homoseosoma electellum*, pollen volatiles physiologically affect virgin females by triggering them to initiate calling behavior earlier, resulting in a higher rate of egg maturation. In addition, pollen odor contains a volatile oviposition stimulant that enhances the female's localization of newly opened sunflower heads. Some volatiles might also be involved in the initiation of calling behavior and oviposition by the European sunflower moth *Homoseosoma nebulellum*. Deterrent compounds of pollen odor may also influence pollen selection. Defensive chemicals, such as the lactone protoanemonin in *Ranunculus acris*, or 2-undecanone, 2-tridecanone and α-methyl ketones in *Rosa rugosa*, are preferentially found in pollen odor (Bergström et al.

1995). Other defense compounds are sesquiterpene lactones, which possess activity against fungi and bacteria (Picman 1986). Compounds often have dual functions, and it is a matter of concentration or dose that initiates a biological reaction. For example, eugenol attracts a variety of insects to *R. acris*, but it also possesses antimicrobial activity.

10.3 Patterns of Floral Emission

Most flowers do not employ their entire surface, but use only certain flower parts, floral organs, or even confined areas of a floral organ with distinct morphological characteristics, for volatile production and emission (Bergstöm et al. 1995; Flamini et al. 2003; Dötterl and Jürgens 2005). Although petals often represent the main volatile source, sepals, stamina (anthers, pollen), or pistils (styli, stigmata) can also contribute to, or even dominate, the floral bouquet (Custódio et al. 2006; Effmert et al. 2006). The most apparent morphological feature of emitting tissues is a rugose epidermis often with cells exhibiting a conical or bullate appearance (Effmert et al. 2006; Bergougnoux et al. 2007). The most sophisticated emitting floral tissue is represented by osmophores. These glandular-like floral tissues have been found to be part of the perianth, bracts, appendices of peduncles, or anthers (Effmert et al. 2006). Floral trichomes can also emit volatiles, as shown for *Antirrhinum majus*. However, in many flowers, volatiles released from trichomes do not significantly add to the floral bouquet (Sexton et al. 2005; Effmert et al. 2006).

Besides these spatial differences, floral volatile emission follows temporal variations (Fig. 10.1). Although constant emitters like flowers of *Lathyrus odoratus* (Sexton et al. 2005), *Clarkia breweri* (Pichersky et al. 1994), or *Nicotiana otophora* (Loughrin et al. 1990) are well known, many flowering plants exhibit diurnal (Helsper et al. 1998; Kolosova et al. 2001; Hendel-Rahmanim et al. 2007), crepuscular (Effmert et al. 2005a; Kaiser 2006), or nocturnal (Matile and Altenburger 1988; Loughrin et al. 1990, 1991; Kolosova et al. 2001; Effmert et al. 2008) emission patterns that reflect the time of the plants' main pollinator activities (Levin et al. 2001; Theis and Raguso 2005). Approximately 8% of all flowering plants exhibit a scent emanation that reaches a maximum at night. Many plant species have adapted efficiently so as to be exclusively night-scented. Therefore, it is possible that plants that have been described as scentless may, in fact, be found to be scented at another time of the day. *Aerangis confusa* has been considered scentless during the day, but emits a typical 'white-floral' scent after sunset (Kaiser 2006). Other plant species, such as *Masdevallia laucheana* and *Constantia cipoensis*, have been shown to emanate fragrance only for 1 h during twilight, while *Cattleya luteola* is fragrant only between 4 and 6 a.m., which correlates with the short period of pollination (5.30 to 5.45 a.m.; Kaiser 2006). The precise timing of floral volatile emission is a special phenomenon often controlled by an endogenous clock, as demonstrated for *Cestrum nocturnum* (Overland 1960), *Stephanotis floribunda* (Altenburger and Matile 1990; Pott et al. 2002), *Hoya carnosa* (Altenburger

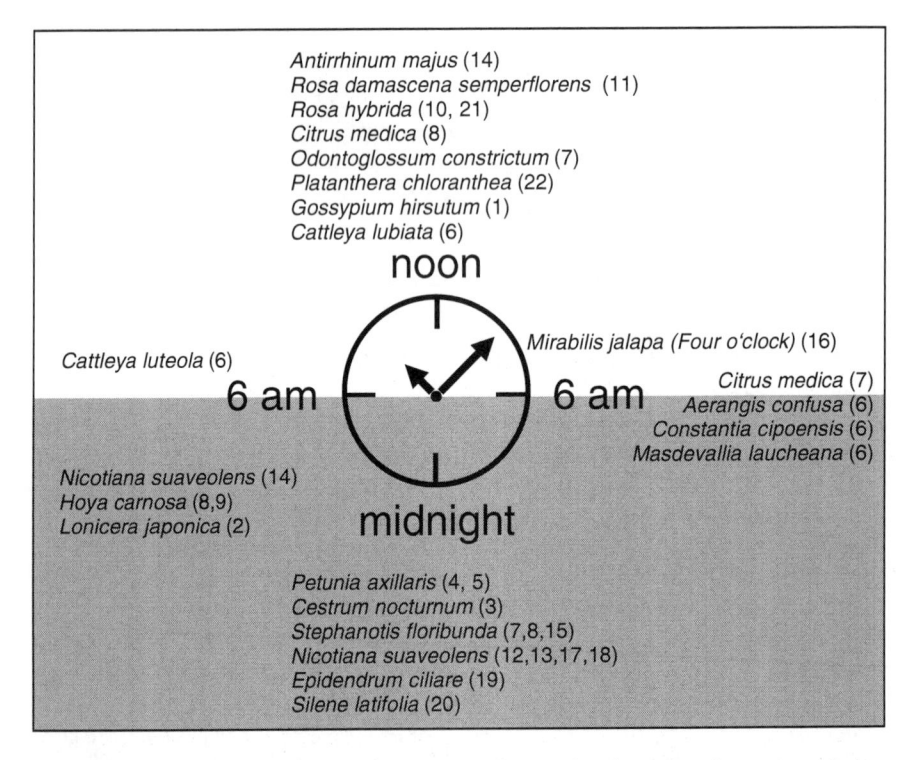

Fig. 10.1 Scent clock: plants are arranged according to the time of major scent emission. *1* Loughrin et al. (1994), *2* Miyake et al. (1998), *3* Overland (1960)*, *4* Hoballah et al. (2005), *5* Oyama-Okuba et al. (2005), *6* Kaiser (2006), *7* Altenburger and Matile (1990)*, *8* Matile and Altenburger (1988), *9* Altenburger and Matile (1988)*, *10* Helsper et al. (1998)*, *11* Picone et al. (2004), *12* Loughrin et al. (1993), *13* Loughrin et al. (1991), *14* Dudareva et al. (2000), *15* Pott et al. (2002), *16* Effmert et al. (2005a), *17* Effmert et al. (2008), *18* Roeder et al. (2007)*, *19* Kaiser (1993), *20* Jürgens et al. (2002), *21* Hendel-Rahmanim et al. (2007), *22* Nilsson (1978; *, regulated by the circadian clock)

and Matile 1988), *Nicotiana sylvestris* and *N. suaveolens* (Loughrin et al. 1991), and *Rosa hybrida* (Helsper et al. 1998). This so-called circadian clock is characterized by (1) a recurring rhythm depending on an external signaling cycle (Zeitgeber) within 24 h, (2) a resynchronized rhythm if the Zeitgeber is shifted, (3) a persistent rhythm with a 'free-running period' of ca. 24 h (circadian) under constant conditions like continuous light, when the Zeitgeber is missing, and (4) a temperature compensation of the 'free-running period'. Rhythmicity does not necessarily capture all components of floral mixtures to the same extent. While the majority of volatiles may conform to a nocturnal or diurnal rhythm, some volatiles keep an inverse pattern, or might even be emitted constantly (Loughrin et al. 1991; Nielsen et al. 1995). These temporal variations within a volatile mixture are sometimes linked to spatial variations (Dötterl and Jürgens 2005).

The onset of volatile emission usually corresponds with flower anthesis (Pichersky et al. 1994; Effmert et al. 2008). In flowers with several days of lifespan, the rhythm of volatile release recurs until the amount of volatiles is reduced during senescence, or until successful pollination. A distinct emission can already emerge in young flowers, as documented for *N. suaveolens* (Effmert et al. 2008), but it may also peak only in mature flowers as shown for *A. majus*, where the release of methyl benzoate, as well as terpenoids, reaches a maximum at day 5 to 7 after anthesis (flower lifespan ca. 12 days; Dudareva et al. 2003; Nagegowda et al. 2008). Not all components of a floral volatile mixture appear at the same time during floral development. In *A. majus* flowers, the myrcene and linalool emission is more persistent compared to the emission of methyl benzoate, (E)-β-ocimene, or neroli-dol (Dudareva et al. 2003; Nagegowda et al. 2008). Flowers of *N. alata* released considerable amounts of the sesquiterpene nerolidol at day 5 post-anthesis, while other major constituents like the monoterpenes β-linalool and 1,8-cineol were released at day 2 post-anthesis (Ganz and Piechulla, unpublished data).

Flowering plants are exposed to a constantly changing environment. Hence, abiotic and biotic factors affect the floral metabolism. Increasing temperatures resulted in significantly greater volatile emission (Hanstedt et al. 1994). Elevated temperatures enhanced terpenoid but not benzenoid emission, indicating that temperature has an impact on the biosynthetic pathway, and not only on volatile emanation (Nielsen et al. 1995). Under field conditions, this effect is often superimposed by elevated light intensities, which have a positive effect on volatile emission (Pecetti and Tava 2000). The length of the photoperiod has little influence, as shown for *Mahonia japonica*, where the emission of most floral volatiles remained unchanged (Picone et al. 2002). One of the most influential biotic factors governing floral volatile emission is pollination. In flowers of *Clarkia breweri*, *Cirsium arvense*, *Cirsium repandum*, and *Antirrhimum majus*, volatile emission declined rapidly shortly after pollination (Theis and Raguso 2005). In flowers of *Ophrys sphegodes*, a sexually deceptive orchid, the amount of volatiles decreased only slightly, but a significant increase in all-trans-farnesyl hexanoate has been detected after pollination by a solitary male bee (Schiestl and Ayasse 2001). Floral herbivory on immature *N. attenuata* flowers resulted in a significant decline of benzyl acetone emission, although this has been attributed to a significant reduction in the corolla mass (Euler and Baldwin 1996). However, leaf herbivory did not have a significant impact on floral scent emission in *N. suaveolens* (Effmert et al. 2008).

10.4 Biosynthetic Pathways and Key Enzymes

More than 2,000 volatile compounds have been known to be emitted from flowers of 991 plant species (compilation of compounds in 'SCENTbase' and 'Super Scent'; Knudsen et al. 2006; Dunkel et al. 2009). Although the overall diversity of floral volatiles is greater than that detected in vegetative tissue, the biosynthetic pathways involved in both tissues are found to be terpenoid biosynthesis, phenylpropanoid biosynthesis, and fatty acid ester synthesis. This suggests that derivatization and modification reactions are well established in plants.

10.4.1 Terpenoids

Terpenes are formed from C5 building blocks. C5 compound biosynthesis occurs in the cytosol via the acetate-mevalonate pathway, and in plastids from pyruvate via methyl erythritol phosphate (MEP). The C5 compounds, isopentenyl pyrophosphate (IPP) and the isomer dimethyl allyl pyrophosphate (DMAP), are combined to geranylpyrophosphate (GPP) by a GPP synthase. Floral GPP synthases, which are short-chain prenyltransferases, have been isolated from *Antirrhinum majus* and *C. breweri*, and are heterodimeric enzymes (Tholl et al. 2004). Another group of floral enzymes, farnesyl pyrophosphate (FPP) synthases that add an additional C5 unit to GPP, are relevant for volatile sesquiterpene synthesis.

The most common single compounds in floral scent are monoterpenes, such as limonene, (E)-β-ocimene, myrcene, linalool, and α- and β-pinene. GPP is the substrate for monoterpene synthases. The linalool synthase (LIS) was initially isolated from *C. breweri* flowers (Pichersky et al. 1995). This enzyme catalyzes the reaction from GPP to the acyclic monoterpene linalool, without major side products. Therefore, LIS has been considered as a monoproduct enzyme. A multiproduct monoterpene synthase is the cineol synthase (CIN). CIN from *N. suaveolens* synthesizes cineol as a major product, together with seven cylic and acyclic side products (α- and β-pinene, sabinene, myrcene, (E)-β-ocimene, α-terpineole; Roeder et al. 2007). The development of multiproduct enzymes during evolution provides the advantage of simultaneous product synthesis and emission of several volatiles. Beside LIS and CIN, other floral monoterpene synthases have been isolated. These include ocimene, myrcene and nerolidol synthases (Dudareva et al. 2003; Nagegowda et al. 2008). The monoterpenes initially synthesized can be modified further (e.g., acetylation) to form other floral volatiles (Shalit 2003). Irregular terpenoids, such as ionones, are cleavage products of carotenoids.

10.4.2 Benzenoids and Phenylpropanoids

The synthesis of benzenoids and phenylpropanoids starts with the deamination of the amino acid phenylalanine. Benzenoids (C6–C1) are widespread in floral scents. Their synthesis requires the elimination of a C2 unit. It is not clear as to whether this loss occurs from a phenylpropanoid precursor (C6–C3), or prior to phenylalanine in the shikimate pathway. The benzenoid and phenylpropanoid pathway is presently being elucidated, and a few genes/enzymes have been identified. These genes include BPBT (benzoyl-CoA:benzyl alcohol/2-phenylethanol benzoyltransferase), IGS (isoeugenol synthase), PAAS (phenylacetaldehyde synthase), BA2H (benzoic acid 2-hydroxylase), BZL (benzoate:Co A ligase), C4H (cinnamic acid-4-hydroxylase), and SA GTase (UDP-glucose:salicylic acid glucosyltransferase; Boatright et al. 2004).

10.4.3 Aliphatic Compounds

This group of compounds (C1 to C25) includes the abundant C6 and C9 aldehydes and alcohols. They are synthesized predominantly from derivatives of fatty acids. Fatty acids are synthesized in plastids. The starting unit is acetyl-CoA, which is the acceptor of a C2 unit from malonyl-CoA. A multienzyme complex eliminates in two reduction reactions, and by elimination reactions, oxygen and double bonds from the new product molecule. Several rounds of C2 unit additions and reduction reactions result in the formation of middle- and long-chain fatty acids.

Many primary products can be modified to increase volatility, as well as to increase the number of diverse compounds with varied olfactory properties. Such modifications or derivatizations are catalyzed by specific enzymes or group of enzymes. In the past decade, an increasing number of floral scent-synthesizing enzymes or genes have been isolated. These include terpene synthases, carboxyl methyltransferases, acyltransferases, and acetyltransferases (Table 10.1).

Typical are methylation reactions. Methylation of hydroxyl groups and hydroxyl groups of carboxyl groups can be distinguished. Many plant compounds contain hydroxyl groups that can be methylated by O′methyltransferases (type I MTs) to reveal the methoxy groups. In general, O′MTs utilize S-adenosylmethionine (SAM) as methyl donor. Members of this MT family catalyze, for example, the formation of methyl eugenol in C. breweri flowers, or methyl orcinol in Rosa chinensis (Dudareva et al. 2004). The methylation of the hydroxyl group within the carboxyl group results in the formation of esters. Dominant scent compounds of this class are methyl benzoate and methyl salicylate. The enzymes that catalyze this reaction are type III methyltransferases (SABATH methyltransferases). Some accept solely benzoic acid (BAMT), or prefer salicylic acid (2-hydroxy benzoic acid) compared to benzoic acid (SAMTs and BSMTs; Effmert et al. 2005b). Interestingly, active pocket amino acids are highly conserved for the SAMT-type enzymes, while several mutations that result in amino acid changes in the substrate-binding site of the BSMT enzymes allow the binding and catalysis with a wider spectrum of benzoic acid derivatives.

Oxidation reactions result in the introduction of hydroxyl groups. The reactions are catalyzed by cytochrome P450 enzymes, and many of these enzymes have been characterized in plants. The skeletons of monoterpenes and sesquiterpenes are often modified by hydroxylation (e.g., menthol and carvone synthesis). However, a cyt P450 enzyme has not been isolated that catalyzes the derivatization of floral monoterpenes or sesquiterpenes. Similarly, such enzymes have not been reported that are involved in floral phenylpropanoid and fatty acid modifications.

Acylation reactions (including acetylation, butanoylation, and benzoyl acylation) are also common to make compounds more volatile. The basic reaction is the transfer of the acyl group from an acyl-CoA intermediate to the hydroxyl group of

Table 10.1 Floral scent-synthesizing genes/enzymes

Enzyme	Plant species	References
Carboxyl methyltransferases (SAMT, BSMT, BAMT)	*Clarkia breweri, Antirrhinum majus, Stephanotis floribunda, Petunia hybrida, Nicotiana suaveolens, Hoya carnosa, Arabidopsis thaliana, Arabidopsis lyrata*	Summarized in Effmert et al. (2005b)
Linalool synthase (LIS)	*C. breweri*	Pichersky et al. (1995)
Cineol synthase (CIN)	*N. suaveolens*	Roeder et al. (2007)
	Citrus unshiu	Shimada et al. (2005)
Ocimene/myrcene synthase (OCS/MYR)	*C. unshiu*	Shimada et al. (2005)
	A. majus	Dudareva et al. (2003)
Nerolidol synthase (NER)	*A. majus*	Nagegowda et al. (2008)
GPP synthase	*A. majus, C breweri*	Tholl et al. (2004)
IPP isomerase	*A. thaliana*	Phillips et al. (2008)
Acetyl-CoA:benzylalcohol acetyltransferase (BEAT)	*C. breweri*	Dudareva et al. (1998)
Benzyl-CoA: benzylalcoholbenzoyl transferase (BEBT)	*C. breweri*	D'Auria et al. (2002)
Benzoyl-CoA:benzyl alcohol/ 2-phenylethanol benzoyltransferase (BPBT)	*P. hybrida*	Boatright et al. (2004)
Benzoate:CoA ligase (BZL)	*P. hybrida*	Boatright et al. (2004)
Cinnamic acid-4-hydroxylase (C4H)	*P. hybrida*	Boatright et al. (2004)
UDP-glucose:salicylic acid glucosyltransferase (SA Gtase)	*P. hybrida*	Boatright et al. (2004)
Isoeugenol synthase (IGS)	*P. hybrida*	Boatright et al. (2004)
Phenylacetaldehyde synthase (PAAS)	*P. hybrida*	Boatright et al. (2004)
Benzoic acid 2-hydroxylase (BA2H)	*P. hybrida*	Boatright et al. (2004)

an alcohol. A recently discovered plant enzyme family that catalyzes such reactions is BAHD acyltransferase. These BAHD enzymes have been isolated from *C. breweri*, and have been shown to produce benzyl acetate or benzyl benzoate, or acetylate citronellol and geraniol in *R. hybrida* (Dudareva et al. 1998; Shalit 2003).

10.5 Regulation of Floral Volatile Biosynthesis

10.5.1 Regulation at the Molecular Level

Floral volatile emission seems to rely on de novo-synthesized products, and the site of biosynthesis also represents the site of emission. In the past decade, evidence from several lines of study have revealed that volatile emission directly corresponded with spatial, temporal, and developmental expression patterns of related floral genes. Furthermore, biosynthesis of floral volatiles was regulated at the transcriptional and/or post-translational level, as well as by the availability of the substrates. It has been reported that genes responsible for floral volatile synthesis are expressed exclusively in floral tissue, more specifically, in the emitting flower part (Wang et al. 1997; Pott et al. 2002, 2004). Cellular immunolocalization demonstrated that corresponding volatile-producing enzymes are detected mostly in epidermal cells, or are membrane-bound (Rohrbeck et al. 2006; Scalliet et al. 2006). At the subcellular level, GFP-fusion to two terpene synthases located the *Am*NES/LIS-1 in the cytosol producing nerolidol, and the *Am*NES/LIS-2 in plastids producing linalool (Nagegowda et al. 2008).

In nocturnally emitting flowers of *S. floribunda* and *N. suaveolens*, and in diurnally emitting flowers of *A. majus*, the circadian-controlled rhythm of the methyl benzoate release correlates with a circadian-controlled oscillation of the steady-state mRNA levels of the floral methyltransferases *Sf*SAMT, *Ns*BSMT, and *Am*BAMT, respectively (Kolosova et al. 2001; Pott et al. 2002; Effmert et al. 2005b). The relative transcript level of *Ns*BSMT reached its maximum at the day of anthesis (Effmert et al. 2005b), whereas *Am*BAMT showed the highest transcript level at day 4 post-anthesis (Dudareva et al. 2000). Protein levels of these methyltransferases did not show similar pronounced daily oscillation (Effmert et al. 2008), but methylation activities in turn oscillated (Kolosova et al. 2001; Pott et al. 2004).

These results indicate the importance of post-translational modifications and/or the availability of substrates. The determination of substrate concentrations revealed that rhythms in enzyme activities depended on the substrate availability. For example, SAMT in *S. floribunda* flowers in planta methylates benzoic acid, although the in vitro catalytic efficiency for salicylic acid is much greater, because of the substrate salicylic acid that is by far underrepresented in the floral tissue (Pott et al. 2004; Effmert et al. 2005b). Similarly, the mRNA level of an alcohol acetyl transferase (*Rh*AAT) expressed in petals of *R. hybrida* followed a diurnal rhythm, which appeared to be controlled by the circadian clock. However, as a result of substrate shortage, the emission of geranyl acetate, as well as germacrene D, ceased under continuous light, indicating that geraniol and germacrene D synthesis is regulated differentially (Hendel-Rahmanim et al. 2007). In contrast, 1,8-cineole synthase *Ns*CIN isolated from *N. suaveolens* flowers displayed a pronounced oscillation at the transcript level, and this rhythmicity is controlled by the circadian clock (Roeder et al. 2007).

10.5.2   Mechanisms of Regulation

Little is known about the mechanisms of regulation and signaling at the molecular level of floral volatile synthesis. Microarray analysis allowed the identification of a cDNA encoding a floral transcription factor (*odo1*), which has been shown to be upregulated in a fragrant cultivar of wild-type petunia compared to a non-fragrant cultivar (Verdonk et al. 2005). *Odo1* belongs to the R2R3-type of MYB transcription factors, which are involved in anthocyanin and phenylpropanoid biosynthesis. Because of the amino acid variation in the R2R3 domain, *Odo1* clusters together with two MYBs of *Arabidopsis thaliana* and one MYB of *Pimpinella brachicarpa* in a new subgroup of MYB proteins. Verdonk et al. (2005) showed that *odo1* suppression modulated the expression of genes belonging to the shikimate pathway, but did not influence the expression of genes responsible for anthocyanin biosynthesis. *Odo1*-suppressed RNAi petunia lines showed partly a dramatic decrease in the emission of floral volatiles such as benzyl benzoate, benzyl acetate, vanillin, and isoeugenol, all of which originated from intermediates of the shikimate pathway, but the purple color of the flower tube remained unchanged. In contrast, Ben Zvi et al. (2008) showed a close link between the scent and anthocyanin biosynthesis. Constitutive overexpression of the anthocyanin pigment1 (*pap1*) MYB transcription factor resulted in an enhanced purple pigmentation in transgenic petunia flowers, and a dramatic increase in the production of nocturnally emitted volatiles. Additional supply of the shikimate pathway intermediate phenylalanine, which is crucial for benzenoid synthesis, abolished nocturnal rhythms of those volatiles in *pap1*-transgenic flowers. These results suggest that phenylalanine is the limiting factor for benzenoid production at daytime when phenylalanine concentrations are down-regulated (Ben Zvi et al. 2008). The constitutive overexpression of *pap1-myb* was superimposed on all subsequent regulatory components (e.g., *odo1*). A linkage between color and scent was also supported by Zuker et al. (2002), who demonstrated that suppression of a flavanone-3-hydrolase, a key enzyme in anthocyanin biosynthesis in *Dianthus caryophyllus*, resulted in a complete loss of petal color, and a marked increase in methyl benzoate emission.

Another factor involved in the regulation of scent emission is the phytohormone ethylene, an important regulator during plant tissue senescence and fruit ripening. Treatment of petunia flowers with exogenous ethylene reduced the emission of seven major volatiles, including methyl benzoate. It also caused a rapid decline in mRNA levels of *Ph*BSMT1 and *Ph*BSMT2 in different flower parts like stigmata and styles, ovaries, petal tubes, and petal limbs after 10 h of treatment (Underwood et al. 2005). The decline of mRNA levels in wild-type flowers in response to exogenous ethylene could also be observed for *Ph*CFAT, an acyltransferase involved in the biosynthesis of the floral volatile isoeugenol (Dexter et al. 2007). Considering that pollination induces ethylene production in different flower parts starting with stamen and style tissue, followed by ovary and the corolla tissue (Jones and Woodson 1999), it can be concluded that post-pollination processes regulated by ethylene signaling include the down-regulation of volatile biosynthesis and emission (Underwood et al. 2005).

10.6 Biotechnological Aspects

Unraveling biosynthetic pathways of floral volatiles and their respective genes and enzymes provides the opportunity to genetically engineer floral scent production. Application of this technique might have practical potential when, for example, suboptimal pollination rates or even lack of natural pollination could be counter-balanced by reintroducing or improving scent synthesis and emission (Pichersky and Dudareva 2007). Furthermore, ornamental industries, especially the cut-flower industries, have an increasing interest in a genetic approach toward floral scent modulation. During decades of conventional breeding, floral scent has been sacrificed for showy colors or shapes of flowers, long vase life, disease resistance, and endurance of shipment around the world. To date, many commercial flowers lack floral fragrance, although humans still associate flowers with sensual pleasures, and a pleasant fragrance with wellbeing (Vainstein et al. 2001; Pichersky and Dudareva 2007). Driven by this rediscovered desire of consumers, and also driven by the commercial interest of the producer, ornamental industries have to face the challenge to reintroduce floral scents. This implies not only the recovery of native floral volatile traits, but also the modulation of the composition of floral bouquets and timing of volatile production and emission.

In principle, all metabolic pathways of floral volatile biosynthesis are amenable for bioengineering. Advanced functional genomic strategies like high-throughput DNA sequencing (Guterman et al. 2002), targeted transcriptome analyses (Verdonk et al. 2005), and proteomic technologies (Dafny-Yelin et al. 2005) allow the isolation and characterization of a rapidly increasing number of floral genes involved in volatile biosynthesis and related regulatory processes (see Sect. 10.4). Consequently, the increasing number of target genes and a better understanding of pathways have promoted floral scent engineering. However, it is still only just out of infancy, compared to the field of genetic engineering in food crops.

Metabolic engineering of floral scent has been performed using different approaches (Table 10.2). The introduction of a single or multiple transgenes encoding enzymes that are not expressed, or even absent, in the target species has yielded the emission of desired novel volatiles (Lücker et al. 2001, 2004a, b; El Tamer et al. 2003). The wild-type tobacco (*N. tabacum*) was transformed with transgenes encoding three monoterpene synthases native to *Citrus lemon* driven by the cauliflower mosaic virus (CaMV) 35S constitutive promoter. Transgenic plants were shown to emit the main products β-pinene, (+)-limonene and γ-terpinene in a non-tissue-specific manner. However, these transgenic plants emitted (+)-(*E*)-isopiperitenol after they were transformed with an additional transgene encoding limonene-3-hydroxylase that catalyzes hydroxylation of (+)-limonene (Lücker et al. 2004b). In transgenic *Dianthus caryophyllus* (carnation), a transgene comprising the β-linalool-synthase cDNA driven by the CaMV 35S promoter was shown to be expressed in flowers (Lavy et al. 2002). Aranovich et al. (2007) reported transgenic *Eustoma grandiflorum* expressing *C. breweri* benzyl alcohol acetyltransferase (BEAT), but benzyl acetate emission was not be observed. Acetate products could be detected in floral and green tissues after feeding the substrates, such as benzyl alcohol, hexanol, or cinnamyl alcohol. Another approach of metabolic engineering involves the down- and up-regulation of native genes associated with floral volatile production. Down-regulation of genes has been achieved by RNA interference (RNAi) techniques (Verdonk et al. 2005; Orlova et al. 2006; Dexter et al. 2007), antisense inhibition of the target gene (Zuker et al. 2002), or virus-based

Table 10.2 Overview on genetic engineering of floral scent (based on Pichersky and Dudareva 2007; Dudareva and Pichersky 2008)

Gene[a]	Target species	Technique used	Effect	References
CbLIS	Petunia x hybrida	Single-gene introduction	No emission	Lücker et al. (2001)
CbLIS	Dianthus caryophyllus	Single-gene introduction	Emission	Lavy et al. (2002)
DcF3H	D. caryophyllus	Antisense suppression	Up-regulation	Zuker et al. (2002)
FaSAAT	P. hybrida	Single-gene introduction	No emission	Beekwilder et al. (2004)
ClLIM, ClPIN, ClTER	Nicotiana tabacum	Triple-gene introduction (single-gene introduction and crossings)	Emission	Lücker et al. (2004a)
MspLIM3H	Transgenic ClLIM/PIN/ TER:N. tabacum	Final quadruple-gene introduction	Emission	Lücker et al. (2004b)
PhODO1	P. hybrida	RNAi	Down-regulation	Verdonk et al. (2005)
PhBSMT	P. hybrida	RNAi	Down-regulation	Underwood et al. (2005)
RhAAT	P. hybrida	Single-gene introduction	Emission	Guterman et al. (2006)
PhBPBT	P. hybrida	RNAi	Down-regulation	Orlova et al. (2006)
PhCFAT	P. hybrida	RNAi	Down-regulation	Dexter et al. (2007)
CbBEAT	Eustoma grandiflora	Single-gene introduction	No emission	Aranovich et al. (2007)
PhCHS	P. hybrida	Virus-induced gene silencing	Up-regulation	Spitzer et al. (2007)
AthPAP1	P. hybrida	Overexpression	Up-regulation	Ben Zvi et al. (2008)

[a]*CbLIS*, linalool synthase (*Clarkia breweri*); *DcF3H*, flavanone 3-hydroxylase (*D. caryophyllus*); *FaSAAT*, strawberry alcohol acyltransferase (*Fragaria x ananassa*); *ClLIM*, limonene synthase (*Citrus lemon*); *ClPIN*, pinene synthase (*C. lemon*); *ClTER*, γ-terpinene synthase (*C. lemon*); *MspLIM3H*, limonene-3-hydroxylase (*Mentha spicata* 'Crispa'); *PhODO1*, transcription factor ODORANT1 (*P. hybrida*); *PhBSMT*, benzoic acid/salicylic acid carboxyl methyltransferase (*P. hybrida*); *RhAAT*, alcohol acetyltransferase (*Rosa x hybrida*); *PhBPBT*, benzylalcohol/phenylethanol benzoyltransferase (*P. hybrida*); *PhCFAT*, coniferyl alcohol acyltransferase (*P. hybrida*); *CbBEAT*, benzyl alcohol acetyltransferase (*C. breweri*); *PhCHS*, chalcone synthase (*P. hybrida*); *AthPAP1*, *Anthocyanin* Pigment1 *MYB* transcription factor (*Arabidopsis thaliana*)

gene silencing methods (Spitzer et al. 2007). In a recent study, Ben Zvi et al. (2008) reported a new approach of metabolic engineering by the introduction of foreign regulators, which were superimposed on the native downstream regulators of volatile biosynthesis. Results showed that modulation of regulatory components changed the rhythms of volatile production and emission.

10.7 Conclusions

Successful floral scent engineering allows (1) the introduction of novel scent components, (2) the enhancement of underrepresented components in a floral bouquet, (3) a decrease in the amount of unpleasant or removal of unwanted components, and (4) the modulation of floral scent traits in seed plants, including changes in temporal emission patterns of volatiles. Although progress in floral scent engineering has been considerable, floral scent production remains relatively unpredictable. Floral volatile biosynthesis is a complex network of overlapping and competing pathways, in which the regulatory mechanisms are poorly understood (Dudareva and Pichersky 2008). Up-regulating and overexpression of genes, as well as the introduction of genes of an underrepresented pathway, might result in substrate shortage (Beekwilder et al. 2004; Aranovich et al. 2007; Ben Zvi et al. 2008), or disposal of toxic gene products by glucosylation (Lücker et al. 2001). Down-regulation of native genes may lead to unexpected results due to the re-channeling of metabolites (Zuker et al. 2002). Nevertheless, transgenic flowers releasing appropriate amounts of engineered metabolites may enhance human pleasure. As the flower is the target organ for floral scent production, modulation of transgene expression may be achieved more efficiently by the use of floral promoters instead of the CaMV 35S promoter. Floral scent engineering not only has a good potential in floral biotechnology, but it can also serve as an important tool for the elucidation of floral volatile metabolism.

References

Altenburger R, Matile P (1988) Circadian rhythmicity of fragrance emission in flowers of *Hoya carnosa* R. Br. Planta 174:248–252

Altenburger R, Matile P (1990) Further observations on rhythmic emission of fragrance in flowers. Planta 180:194–197

Aranovich D, Lewinsohn E, Zaccai M (2007) Post-harvest enhancement of aroma in transgenic lisanthus (*Eustoma grandiflorum*) using the *Clarkia breweri* benzyl alcohol acetyltransferase (BEAT) gene. Postharv Biol Technol 473:255–260

Beekwilder J, Alvarez-Huerta M, Neef E, Verstappen FWA, Bouwmeester HJ, Aharoni A (2004) Functional characterization of enzymes forming volatile esters from strawberry and banana. Plant Physiol 134:1865–1875

Ben Zvi MM, Negre-Zakharov F, Masci T, Ovadis M, Shklarman E, Ben-Meir H, Tzfira T, Dudareva N, Vainstein A (2008) Interlinking showy traits: co-engineering of scent and colour biosynthesis in flowers. Plant Biotechnol J 6:403–415

Bergougnoux V, Caissard JC, Jullien F, Magnard JL, Scalliet G, Cock JM, Hugueney P, Baudino S (2007) Both the adaxial and abaxial epidermal layers of the rose petal emit volatile scent compounds. Planta 226:853–866

Bergström G, Dobson HEM, Groth I (1995) Spatial fragrance patterns within the flowers of *Ranunculus acris* (Ranunculaceae). Plant Syst Evol 195:221–242

Boatright J, Negre F, Chen X, Kish CM, Wood B, Peel G, Orlova I, Gang D, Rhodes D, Dudareva N (2004) Understanding in vivo benzenoid metabolism in petunia petal tissue. Plant Physiol 135:1993–2011

Custódio L, Serra H, Ngueira JMF, Gonçalves S, Romano A (2006) Analysis of the volatiles emitted by whole flowers and isolated flower organs of the carob tree using HS-SPME-GC/MS. J Chem Ecol 32:929–942

Dafny-Yelin M, Guterman I, Ovadis NMM, Shalit M, Pichersky E, Zamir D, Lewinsohn E, Weiss ZAD, Vainstein A (2005) Flower proteome: changes in protein spectrum during advanced changes in rose petal development. Planta 222:37–46

D'Auria JC, Chen F, Pichersky E (2002) Characterization of an acyltransferase capable of synthesizing benzylbenzoate and other volatile esters in flowers and damaged leaves of *Clarkia breweri*. Plant Physiol 130:466–476

Dexter R, Qualley A, Kish DM, Ma CJ, Koeduka T, Nagegowda DA, Dudareva N, Pichersky E, Clark D (2007) Characterization of a petunia acetyltransferase involved in the biosynthesis of the floral volatile isoeugenol. Plant J 49:265–275

Dobson HEM (2006) Relationship between floral fragrance composition and type of pollinator. In: Dudareva N, Pichersky E (eds) Biology of floral scent. Taylor & Francis Group, Boca Raton, FL, pp 147–198

Dobson HEM, Bergström G (2000) The ecology and evolution of pollen odors. Plant Syst Evol 222:63–87

Dötterl S, Jürgens A (2005) Spatial fragrance patterns in flowers of *Silene latifolia*: lilac compounds as olfactory nectar guides? Plant Syst Evol 255:99–109

Dudareva N, Pichersky E (2008) Metabolic engineering of plant volatiles. Curr Opin Biotechnol 19:1–9

Dudareva N, D'Auria JC, Nam KH, Raguso RA, Pichersky E (1998) Acetyl-CoA:benzylalcohol acetyltransferase – an enzyme involved in floral scent production in *Clarkia breweri*. Plant J 14:297–304

Dudareva N, Murfitt LM, Mann CJ, Gorenstein N, Kolosova N, Kish CM, Bonham C, Wood K (2000) Developmental regulation of methyl benzoate biosynthesis and emission in snapdragon flowers. Plant Cell 12:949–961

Dudareva N, Martin D, Kish CM, Kolosova N, Gorenstein N, Fäldt J, Miller B, Bohlmann J (2003) (*E*)-β-ocimene and myrcene synthase genes of floral scent biosynthesis in snapdragon: function and expression of three terpene synthase genes of a new terpene synthase subfamily. Plant Cell 15:1227–1241

Dudareva N, Pichersky E, Gershenzon J (2004) Biochemistry of plant volatiles. Plant Physiol 135:1893–1902

Dunkel M, Schmidt U, Struck S, Berger L, Gruening B, Hossbach J, Jäger I, Effmert U, Piechulla B, Erikson R, Knudsen J, Preissner R (2009) Super Scent – a database of flavours and scents. Nucleic Acid Res 37:D291–D294

Effmert U, Große J, Röse URS, Ehrig F, Kägi R, Piechulla B (2005a) Volatile composition, emission pattern and localization of floral scent emission in *Mirabilis jalapa* (Nyctaginaceae). Am J Bot 92:2–12

Effmert U, Saschenbrecker S, Ross J, Negre F, Fraser CM, Noel JP, Dudareva N, Piechulla B (2005b) Floral benzenoid carboxyl methyltransferases: from *in vitro* to *in planta* function. Phytochemistry 66:1211–1230

Effmert U, Buss D, Rohrbeck D, Piechulla B (2006) Localization of the synthesis and emission of scent compounds within the flower. In: Dudareva N, Pichersky E (eds) Biology of floral scent. Taylor & Francis Group, Boca Raton, FL, pp 105–124

Effmert U, Dinse C, Piechulla B (2008) Influence of green leaf herbivory by *Manduca sexta* on floral volatiles emission by *Nicotiana suaveolens*. Plant Physiol 146:1996–2007

El Tamer MK, Smeets M, Holthuysen N, Lücker J, Tang A, Roozen J, Bouwmeester HJ, Voragen AGJ (2003) The influence of monoterpene synthase transformation on the odour of tobacco. J Biotechnol 106:15–21

Euler M, Baldwin IT (1996) The chemistry of defense and apparency in the corollas of *Nicotiana attenuata*. Oecologia 107:102–112

Flamini G, Cioni PL, Morelli I (2003) Differences in the fragrances of pollen, leaves and floral parts of garland (*Chrysanthemum coronarium*) and composition of the essential oils from flowerheads and leaves. J Agric Food Chem 51:2267–2271

Guterman I, Shalit M, Menda N, Piestun D, Dafny-Yelin M, Shalev G, Bar E, Davydov O, Ovadis M, Emanuel M, Wang J, Adam Z, Pichersky E, Lewinsohn E, Zamir D, Vainstein A, Weiss D (2002) Rose scent: genomics approach to discovering novel floral fragrance-related genes. Plant Cell 14:2325–2338

Guterman I, Masci T, Chen X, Negre F, Pichersky E, Dudareva N, Weiss D, Vainstein A (2006) Generation of phenylpropanoid pathway-derived volatiles in transgenic plants: rose alcohol acetyltransferase produces phenylethyl acetate and benzyl acetate in petunia flowers. Plant Mol Biol 60:555–563

Hanstedt L, Jacobsen HB, Olsen CE (1994) Influence of temperature on the rhythmic emission of volatiles from *Ribes nigrum* flowers *in situ*. Plant Cell Environ 17:1069–1072

Helsper JPFG, Davies JA, Bourmeester HJ, Krol AF, van Kampen MH (1998) Circadian rhythmicity in emission of volatile compounds by flowers of *Rosa hybrida* L. cv. Honesty. Planta 207:88–95

Hendel-Rahmanim K, Masci T, Vainstein A, Weiss D (2007) Diurnal regulation of scent emission in rose flowers. Planta 226:1491–1499

Hoballah ME, Suurman J, Turlings TCJ, Guerin PM, Connetable S, Kuhlemeier C (2005) The composition and timing of flower odour emission by wild *Petunia axillaris* coincides with the antennal perception and nocturnal activity of the pollinator *Manduca sexta*. Planta 222:141–150

Jones ML, Woodson WR (1999) Interorgan signaling following pollination in carnations. J Am Soc Hort Sci 124:598–604

Jürgens A, Witt T, Gottsberger G (2002) Flower scent composition in night-flowering *Silene* species (Caryophyllaceae). Biochem System Ecol 30:383–397

Kaiser R (1993) The scent of orchids. Elsevier, Amsterdam

Kaiser R (2006) Meaningful scents around the world. Wiley-VCH, Zürich

Knudsen JT, Eriksson R, Gershenzon J, Stahl B (2006) Diversity and distribution of floral scent. Bot Rev 72:1–120

Kolosova N, Gorenstein N, Kish CM, Dudareva N (2001) Regulation of circadian methyl benzoate emission in diurnally and nocturnally emitting plants. Plant Cell 13:2333–2347

Lavy M, Zuker A, Lewinsohn E, Larkov O, Ravid U, Vainstein A, Weiss D (2002) Linalool and linalool oxide production in transgenic carnation flowers expressing the *Clarkia breweri* linalool synthase gene. Mol Breed 9:103–111

Levin RA, Raguso RA, McDade LA (2001) Fragrance chemistry and pollinator affinities in Nyctaginaceae. Phytochemistry 58:429–440

Loughrin JH, Hamilton-Kemp TR, Andersen RA, Hildebrand DF (1990) Volatiles from flowers of *Nicotiana sylvestris*, *N. otophora* and *Malus* x *domestica*: headspace components and day/night changes in their relative concentrations. Phytochemistry 29:2473–2477

Loughrin JH, Hamilton-Kemp TR, Andersen RA, Hildebrand DF (1991) Circadian rhythm of volatile emission from flowers of *Nicotiana sylvestris* and *N. suaveolens*. Physiol Plant 83:492–496

Loughrin JH, Hamilton-Kemp TR, Burton HR, Andersen RA (1993) Effect of diurnal sampling on the headspace composition of detached *Nicotiana suaveolens* flowers. Phytochemistry 32:1417–1419

Loughrin JH, Manukian A, Heath RR, Turlings TCJ, Tumlinson JH (1994) Diurnal cycle of emission of induced volatile terpenoids by herbivore-injured cotton plants. Proc Natl Acad Sci USA 91:11836–11840

Lücker J, Bouwmeester HJ, Schwab W, Blaas J, van der Plas LHW, Verhoeven HA (2001) Expression of Clarkia S-linalool synthase in transgenic petunia plants results in the accumulation of S-linalyl-β-D-glucopyranoside. Plant J 27:315–324

Lücker J, Schwab W, van Hautum B, Blaas J, van der Plas LHW, Bouwmeester HJ, Verhoeven HA (2004a) Increased and altered fragrance of tobacco plants after metabolic engineering using three monoterpene synthases from lemon. Plant Physiol 134:510–519

Lücker J, Schwab W, Franssen MCR, van der Plas LHW, Bouwmeester HJ, Verhoeven HA (2004b) Metabolic engineering of monoterpene biosynthesis: two-step production of (+)-*trans*-isopiperitenol by tobacco. Plant J 39:135–145

Matile P, Altenburger R (1988) Rhythms of fragrance emission in flowers. Planta 174:242–247

Miyake T, Yamaoka R, Yahara T (1998) Floral scents of hawkmoth-pollinated flowers in Japan. J Plant Res 111:199–205

Nagegowda DA, Gutensohn M, Wilkerson CG, Dudareva N (2008) Two nearly identical terpene synthases catalyze the formation of nerolidol and linalool in snapdragon flowers. Plant J 55:224–239

Nielsen JK, Jacobsen HB, Friis P, Hansen K, Møller J, Olsen CE (1995) Asynchronous rhythms in the emission of volatiles from *Hesperis matronalis* flowers. Phytochemistry 38:847–851

Nilsson LA (1978) Pollination ecology and adaptation in *Platanthera chlorantha* (Orchidaceae). Bot Notiser 131:35–51

Orlova I, Marshall-Colón A, Schnepp J, Wood B, Varbanova M, Fridman E, Blakeslee JJ, Peer WA, Murphy AS, Rhodes D, Pichersky E, Dudareva N (2006) Reduction of benzenoid synthesis in petunia flowers reveals multiple pathways to benzoic acid and enhancement in auxin transport. Plant Cell 18:3458–3475

Overland L (1960) Endogenous rhythm in opening and odor of flowers of *Cestrum nocturnum*. Am J Bot 47:378–382

Oyama-Okubo N, Ando T, Watanabe N, Marchesi E, Uchida K, Nkayama M (2005) Emission mechanism of floral scent in *Petunia axillaris*. Biosci Biotechnol Biochem 69:773–777

Pecetti L, Tava A (2000) Effect of flower color and sampling time on volatile emanation in alfalfa flowers. Crop Sci 40:126–130

Phillips MA, D'Auria JC, Gershenzon J, Pichersky E (2008) The Arabidopsis thaliana type I isopentenyl diphosphate isomerases are targeted to multiple subcellular compartments and have overlapping functions in isoprenoid biosynthesis. Plant Cell 20:677–696

Pichersky E, Dudareva N (2007) Scent engineering: toward the goal of controlling how flowers smell. Trends Biotechnol 25:105–110

Pichersky E, Raguso RA, Lewinsohn E, Croteau R (1994) Floral scent production in *Clarkia* (Onagraceae). Localization and developmental modulation of monoterpene emission and linalool synthase activity. Plant Physiol 106:1533–1540

Pichersky E, Lewinsohn E, Croteau R (1995) Purification and characterization of S-linalool synthase, an enzyme involved in the production of floral scent in *Clarkia breweri*. Arch Biochem Biophys 316:803–807

Picman A (1986) Biological activities of sequiterpene lactones. Biochem Syst Ecol 14:255–281

Picone JM, MacTavish HS, Clery RA (2002) Emission of floral volatiles from *Mahonia japonica* (Berberidaceae). Phytochemistry 60:611–617

Picone JM, Clery RA, Watanabe N, MacTavish HS, Turnbull CGN (2004) Rhythmic emission of floral volatiles from *Rosa damascena semperflorens* cv. 'Quatre Saisons'. Planta 219:468–478

Pott MB, Pichersky E, Piechulla B (2002) Evening specific oscillations of scent emission, SAMT enzyme activity, and SAMT mRNA in flowers of *Stephanotis floribunda*. J Plant Physiol 159:925–934

Pott MB, Hippauf F, Saschenbrecker S, Chen F, Ross J, Kiefer I, Slusarenko A, Noel JP, Pichersky E, Effmert U, Piechulla B (2004) Biochemical and structural characterization of benzenoid carboxyl methyltransferases involved in floral scent production in *Stephanotis floribunda* and *Nicotiana suaveolens*. Plant Physiol 135:1946–1955

Roeder S, Hartmann AM, Effmert U, Piechulla B (2007) Regulation of simultaneous synthesis of floral scent terpenoids by the 1,8-cineole synthase of *Nicotiana suaveolens*. Plant Mol Biol 65:107–124

Rohrbeck D, Buss D, Effmert U, Piechulla B (2006) Localization of methyl benzoate synthesis and emission in *Stephanotis floribunda* and *Nicotiana suaveolens* flowers. Plant Biol 8:615–626

Scalliet G, Lionnet C, Le Bechec M, Dutron L, Magnard JL, Baudino S, Bergougnoux V, Jullien F, Chambrier P, Vergne P, Dumas C, Cock JM, Hugueney P (2006) Role of petal-specific orcinol *O*-methyltransferases in the evolution of rose scent. Plant Physiol 140:18–29

Schiestl JP, Ayasse M (2001) Post-pollination emission of a repellent compound in a sexually deceptive orchid: a new mechanism for maximising reproductive success? Oecologia 126: 531–534

Sexton R, Stopford AP, Porter M, Porter AEA (2005) Aroma production from cut sweet pea flowers (*Lathyrus odoratus*): the role of ethylene. Physiol Plant 124:381–389

Shalit M (2003) Volatile ester formation in roses. Identification of an acetyl-coenzyme A geraniol/citronellol acetyltransferase in developing rose petals. Plant Physiol 131:1868–1876

Shimada T, Endo T, Fujii H, Hara M, Omura M (2005) Isolation and characterization of (E)-β-ocimene and 1,8 cineole synthases in *Citrus unshiu* Marc. Plant Sci 168:987–995

Spitzer B, Ben Zvi MM, Ovadis M, Marhevka E, Barkai O, Edelbaum O, Marton I, Masci T, Alon M, Morin S, Rogachev I, Aharoni A, Vainstein A (2007) Reverse genetics of floral scent: application of tobacco rattle virus-based gene silencing in petunia. Plant Physiol 145: 1241–1250

Theis N, Raguso RA (2005) The effect of pollination on floral fragrance in thistles. J Chem Ecol 31:2581–2600

Tholl D, Kish CM, Orlova I, Sherman D, Gershenzon J, Pichersky E, Dudareva N (2004) Formation of monoterpenes in *Antirrhinum majus* and *Clarkia breweri* flowers involves heterodimeric geranyl diphosphate synthases. Plant Cell 16:977–992

Underwood BA, Tieman DM, Shibuya K, Dexter RJ, Loucas HM, Simkin AJ, Sims CA, Schmelz EA, Klee HJ, Clark DG (2005) Ethylene-regulated floral volatile synthesis in petunia corollas. Plant Physiol 138:255–266

Vainstein A, Lewinsohn E, Pichersky E, Weiss D (2001) Floral fragrance. New inroads into an old commodity. Plant Physiol 127:1383–1389

Verdonk JC, Haring MA, van Tunen AJ, Schuurink RC (2005) *ODORANT1* regulates fragrance biosynthesis in petunia flowers. Plant Cell 17:1612–1624

Wang J, Dudareva N, Bhakta S, Raguso RA, Pichersky E (1997) Floral scent production in *Clarkia breweri* (Onagraceae): II. Localization and developmental modulation of the enzyme *S*-adenosyl-L-methionine:(iso)eugenol O-methyltransferase and phenylpropanoid emission. Plant Physiol 114:213–221

Zuker A, Tzfira T, Ben-Meir H, Ovadis M, Shklarman E, Itzhaki H, Forkmann G, Martens S, Neta-Sharir I, Weiss D, Vainstein A (2002) Modification of flower color and fragrance by antisense suppression of the flavone 3-hydroxylase gene. Mol Breed 9:33–41

Part III
Hormonal and Environmental Signalling

Chapter 11
Amino Compound-Containing Lipids: a Novel Class of Signals Regulating Plant Development

R. Ortiz-Castro, A. Méndez-Bravo, and J. López-Bucio

11.1 Introduction

Plants produce compounds of different chemical identity that mediate a range of cellular functions, including volatiles such as ethylene and jasmonate, small organic hormones such as auxins, cytokinins, gibberellins and abscisic acid, brassinosteroids and lipids (Weyers and Paterson 2001). Lipids have long been recognized as signalling molecules that have the capacity of triggering profound physiological responses. In animals, ceramides and sphingosines are lipids that have pro-apoptotic and anti-proliferative actions (Wymann and Schneiter 2008). In plants, ceramides, sphingosines and phosphatidic acid are involved in mediating growth, development and responses to biotic and abiotic stimuli, and their production is regulated by a key set of lipid-modifying enzymes such as phospholipases, lipid kinases and/or phosphatases (Worrall et al. 2003; Wang 2004). Studies on the downstream targets and modes of action of signalling lipids are still in their early stages.

Alkamides and *N*-acyl ethanolamines (NAEs) comprise a group of bioactive acylamides with varied acyl chain length and saturation grade (Chapman 2004; López-Bucio et al. 2006; Morquecho-Contreras and López-Bucio 2007). In mammals, the metabolism of NAEs is part of the endocannabinoid signalling pathway wherein anandamide (NAE 20:4) acts as an endogenous agonist of G protein-coupled cannabinoid receptors that, in turn, regulate a wide array of physiological and behavioural processes, including modulation of neurotransmission in the central nervous system (Wilson and Nicoll 2002), synchronization of embryo development (Paria and Dey 2000) and vasodilation (Kunos et al. 2000). Recently, additional cellular targets of NAEs have been discovered, including ion channels

R. Ortiz-Castro, A. Méndez-Bravo, and J. López-Bucio

Instituto de Investigaciones Químico-Biológicas, Universidad Michoacana de San Nicolás de Hidalgo, CP 58030 Morelia, Michoacán, México
e-mail: jbucio@zeus.umich.mx

(Movahed et al. 2005; Oz et al. 2005) and transcription factors (LoVerme et al. 2006).

Alkamides comprise more than 200 related compounds widely distributed in plants, ranging from lichens to angiosperms. An important role for these compounds in plant morphogenesis has been inferred from pharmacological application and mutant analysis in the model plant *Arabidopsis thaliana* (López-Bucio et al. 2006). In the past 5 years, accumulated evidence points to a role of NAEs and alkamides in diverse physiological processes, including seed germination and regulation of plant architecture. In this chapter, the recent information on NAEs and alkamides in plants is reviewed in the context of their occurrence, metabolism, interactions with other plant growth regulators and functions in plant development. Structurally related compounds from bacteria are the *N*-acyl homoserine lactones (AHLs). These compounds participate in cell-to-cell signalling between bacteria, usually referred to as quorum-sensing. It is proposed that small lipid signalling based on NAEs, alkamides and AHLs might be part of an ancestral inter-kingdom communication system between plants and their associated bacteria to regulate symbiotic and/or pathogenic behaviours.

11.2   Biosynthesis and Metabolism of Acylamides in Plants

Most organisms are known to contain in their inner and outer membranes amphipatic lipids based on one or two amino acids linked to a fatty acid through an amide bond. Three classes of compounds are beginning to be considered as plant signals based on their potent activities and strong morphogenic effects—NAEs, alkamides and AHLs (Table 11.1). NAEs represent compounds with aminoalcohol linked as an amide to the fatty acid. They are likely produced from the hydrolysis of *N*-acyl phosphatidylethanolamines (NAPEs), a minor constituent of cell membranes, by phospholipase D. NAPEs occur naturally in diverse biological systems. For example, NAPEs accumulate during cell injury or stress in animal tissues and dehydration in plant seeds (Hansen et al. 2002; Chapman 2004). The conformation of NAPEs in a lipid bilayer and their interaction with other lipids have been analyzed by spectroscopy studies (Lafrance et al. 1990; Swamy et al. 2000). For long-chain NAPEs, the *N*-acyl moiety is bent and buried in the hydrophobic phase of the phospholipid bilayer. By contrast, for short-chain NAPEs, the *N*-acyl chain remains at the level of the glycerol backbone, exposed to the aqueous milieu. NAPEs are the source of signalling NAEs, including anandamide (*N*-arachidonoyl ethanolamide; Table 11.1), which binds to cannabinoid receptors CB1 and CB2 and regulates important physiological processes in animals. The molecular determinants of the activity of the different types of phospholipases have been the subject of several structural studies (Caramelo et al. 2003; Okamoto et al. 2004). A novel NAPE-hydrolyzing phospholipase D (PLD) was cloned from mammals, which exhibited specificity towards NAPE substrates but not to other common membrane phospholipids, indicating that this is one of the enzymes responsible for converting NAPE to

Table 11.1 Characteristics of NAEs, alkamides and AHLs

Producing organism	Compound	Function
Animals	N-arachidonoyl ethanolamide (Anandamide)	Modulation of neurotransmission in central nervous system Synchronization of embryo development Brain development Cell proliferation Cardiovascular and immune regulation
Arabidopsis thaliana	N-acyl ethanolamine (NAE 12:0)	Regulation of seed germination Regulation of cell division and expansion Regulation of root architecture Control of cytoskeletal structure
Acmella radicans	N-isobutil decanamide	Regulation of cell division Regulation of root architecture Activation of cytokinin-signalling
Pseudomonas aeruginosa	N-3-oxo-dodecanoyl-HL	Virulence and biofilm formation
Pseudomonas fluorescens	N-decanoyl-HL	Population growth

NAE in vivo (Okamoto et al. 2004). Recent reports have shown that NAEs may be generated from NAPEs via phospholipase A (Simon and Cravatt 2006) or phospholipase C (Liu et al. 2006). In plants, PLD-β and PLD-γ catalyze the formation of NAEs from NAPEs in vitro (Pappan et al. 1998). PLDs from plants, animals and fungi share similarities in structure and catalytic mechanisms. However, the plant PLD family is much more complex than those of other organisms; for example, 12 PLD genes are present in *Arabidopsis*, whereas only two PLD genes are in mammals and one in yeast (*Saccharomyces cerevisiae*; Wang 2004). This complexity might account for the critical role of the PLD products phosphatidic acid and NAEs in plant signalling.

The enzymatic machinery for the degradation of NAEs is conserved between animals and plants. For example, an enzyme that rapidly hydrolyzes NAEs into ethanolamine and their corresponding fatty acids has been cloned from mammals. This enzyme, called fatty acid amide hydrolase (FAAH), belongs to a group of proteins containing a conserved amidase sequence (Shresta et al. 2003, 2006). The amidase region of these proteins consists of about 125 amino acids. There is 18.5% identity between the *Arabidopsis* FAAH (*AtFAAH*) and rat FAAH when compared over the entire length of the proteins, whereas there is 37% identity within the amidase region. Functional homologues of *AtFAAH* were also identified in rice (*Oriza sativa*) and *Medicago truncatula*, supporting a common mechanism for the regulation of NAE hydrolysis in diverse plant species (Shresta et al. 2006). The possibility has been considered that a breakdown product of ceramide or other sphingolipids may result in metabolites similar to NAEs or alkamides (Ramírez-Chávez et al. 2004). Interestingly, a fatty acid amidase was identified recently as an alternate enzyme for NAE hydrolysis. This enzyme belongs to the choloylglycine hydrolase family with structural and functional similarity to acid ceramidase (Tsuboi et al. 2005). The combined action of fatty acid hydrolases *AtFAAH* and ceramidases may account for the efficient metabolism of NAEs in different plant tissues.

11.3   Distribution of Acylamides

NAEs have been quantified in seeds of some higher plants, including cotton (*Medicago truncatula*), corn (*Zea mays*), *Arabidopsis*, soybean (*Glycine max*), tomato (*Lycopersicon esculentum*) and pea (*Pisum sativum*). The total NAE content varied among plant species from 500 to 1,600 ng/g fresh weight, with acyl chain length ranging from 12 to 20 carbon atoms (Chapman 2004). NAEs containing 16C and 18C were the most abundant compounds in dry seeds. Total NAE concentrations fall drastically following seed imbibition and during germination. In desiccated *Arabidopsis* seeds, total NAE content was of the order of 2,000 ng/g, and this declined to 500 ng/g within 96 h after sowing (Wang et al. 2006). The elevated concentrations of NAEs in seeds point to the possibility that these lipids may function in processes relevant to seed or seedling development.

Compared to seeds, the vegetative tissues of plants have lower NAE content and their NAE profile also differs, in that NAE 12:0 and NAE 14:0 appear to predominate (Chapman 2004). There is evidence that, in stem and leaves, medium-chain NAEs are synthesized on demand rather than being stored. For instance, the concentration of NAE 14:0 in tobacco (*Nicotiana tabacum*) leaves is approximately 5 ng/g on a fresh weight basis. Interestingly, a 10-min exposure to nanomolar concentrations of two protein elicitors, xylanase and cryptogein, caused a 10- and 50-fold increase in NAE 12:0 and NAE 14:0 respectively (Tripathy et al. 1999). Exogenous application of synthetic NAE 14:0 at a concentration of 1 µM induced a fourfold increase in phenylalanine ammonia lyase (PAL) expression in a manner similar to that elicited by xylanase and cryptogein in both cell suspensions and leaves of tobacco (Tripathy et al. 1999). These results suggest medium-chain NAEs may participate in the signal transduction events leading to plant defence responses.

Alkamides comprise over 200 related compounds that have been found in as many as ten plant families, namely the Aristolochiaceae, Asteraceae, Brassicaceae,

Convolvulaceae, Euphorbiaceae, Menispermaceae, Piperaceae, Poaceae, Rutaceae and Solanaceae. Species containing high concentrations of alkamides are found in the Asteraceae, Piperaceae and Rutaceae (Christensen and Lam 1991; Laurerio-Rosario et al. 1996; Kashiwada et al. 1997). Certain alkamides, such as pellitorin, have been described as allelochemicals. Pellitorin is naturally present in *Stauranthus perforatus*, a rutaceous tree, and exerts a strong allelochemical effect on the growth of weeds (Anaya et al. 2005). At a concentration of 100 mg/ml supplied to the culture medium in vitro, pellitorin caused a 45 and 80% inhibition in root growth of *Amaranthus hypochondriacus* and *Echinocloa crusgalli* respectively. In a glasshouse experiment, the decomposition of leaves and roots of *S. perforatus* incorporated as green manures (2%) to the soil had a significant inhibitory effect on the growth of weeds. The allelopathic action of decomposition of plant tissues was comparable to that of DPCA (dimethyl tetrachloroterephthalate), a commercial herbicide (Anaya et al. 2005). Based on the results discussed above, it is tempting to speculate that pellitorin production by *S. perforatus* might represent a strategy for competition.

Alkamide-accumulating plants can occur in different plant families. Certain medicinal plants, such as *Echinacea angustifolia*, *Echinacea purpurea* and *Heliopsis longipes*, have been used in the past and present by different civilizations. These plants accumulate alkamides in plant tissues (Bauer and Reminger 1989; Molina-Torres et al. 1996). In *H. longipes*, a traditional herb endemic to central México, the alkamide affinin accumulates especially in roots, where it is present in as high as 1% (w/w) on a fresh weight basis (Molina-Torres et al. 1996). In *E. purpurea*, alkamides accumulate preferentially in flower heads and in roots. Their levels are low at the beginning of vegetative growth and increase at the flowering stage (Letchamo et al. 1999; Qu et al. 2005). For total alkamides, concentrations varied from 5 to 27.6 mg/g in roots and from 0.22 to 5.3 mg/g in vegetative tissues (Qu et al. 2005). The reason why certain plants accumulate alkamides is not clear. However, it could provide an advantage for competition, acting as allelochemicals as in the case of pellitorin, or might represent a mechanism to adjust plant growth and development, as their importance in cellular processes is increasingly being appreciated.

11.4 Role of NAEs and Alkamides in Plant Development

The plant kingdom is a vast storehouse of chemical substances manufactured and used by plants as defences against viruses, bacteria, fungi and insects. Physiological and ecological constraints play key roles in plant growth patterns. Plant activity at the cellular level can be classified in general terms as growth (cell division and enlargement) or differentiation (chemical and morphological changes leading to cell specialization). Some plants provision seeds with high concentrations of secondary metabolites, possibly to protect the seed and the rapid growing seedling before it has developed the capacity to synthesize significant quantities on its own

(Herms and Mattson 1992). Growth processes demand particularly high concentrations of limited plant resources. They are highly dependent on protein synthesis for the manufacture of photosynthetic, biosynthetic and regulatory enzymes, as well as for structural protein. The production of secondary metabolites competes directly with protein synthesis and, consequently, with growth. Thus, it is not surprising that plants exposed to pathogens or to sub-lethal abiotic stress conditions exhibit a broad range of morphogenic responses (Potters et al. 2007). Despite the diversity of phenotypes, a generic stress-induced growth response can be recognized that appears to be carefully orchestrated and comprises three components, namely inhibition of cell elongation, localized stimulation of cell division and alteration in cell differentiation (Herms and Mattson 1992; Potters et al. 2007). Although the stress-induced developmental responses seem to be part of a general acclimation strategy, whereby plant growth is redirected to diminish stress exposure, little is known about the molecular mechanism underlying this response. Altered phytohormone synthesis, transport and/or metabolism could be part of the physiological component of plant growth modulation. Five major hormones have been identified, namely auxins, abscicic acid (ABA), cytokinins, ethylene and gibberellin (GA). Other signals are currently being added to this list, including brassinosteroids and certain lipids such as sphingolipids and phosphatidic acid (Weyers and Paterson 2001; Wang 2004). In our studies, we have identified a new class of regulatory molecules in plants, the alkamides, which are structurally similar to NAEs. The alkamides were initially identified in terms of their similarity to ceramides, a group of key signals in yeast and animals (Ramírez-Chávez et al. 2004; López-Bucio et al. 2006). Until recently, NAEs and alkamides had been considered as a class of secondary metabolites that accumulated in particular plant species and tissues. Research using *Arabidopsis thaliana* suggested that they might have a pivotal role in integrating environmental signals into developmental transitions. Alkamides and NAEs appear to deliver a message regulating a particular plant function. The functions include germination and alteration of growth and differentiation in the course of development.

11.4.1   Seed Germination

Seed dormancy and germination are under the control of phytohormones and their signalling pathways. Genetic analysis of seed germination in *Arabidopsis* has revealed that GA and ABA are crucial regulators (Finch-Savage and Leubner-Metzger 2006). GA is regarded widely as a growth-promoting compound that positively regulates germination. By contrast, ABA has historically been considered to function as a germination inhibitor (Razem et al. 2006).

The notion that NAEs plays a role in seed germination is supported by their accumulation in desiccated seeds of a wide range of plant species but, during imbibition and germination, NAE concentrations decrease significantly and remain low during subsequent seedling growth (Venables et al. 2005). These observations suggest that the rapid metabolism of NAEs is a prerequisite for germination. In fact, exogenous application of NAE 10:0, and the alkamides affinin and *N*-isobutyl decanamide regulated in a dose–response manner the germination of seeds from different plant species, with low concentrations (nm range) promoting germination and high concentrations (mM range) drastically inhibiting germination (Morquecho-Contreras and López-Bucio, unpublished data).

Important information about the in vivo role of NAEs in germination came from the manipulated expression of *AtFAAH*, an enzyme that hydrolyzes NAEs into ethanolamine and free fatty acids, in *Arabidopsis thaliana*. In this way, Wang et al. (2006) reported that *AtFAAH* expression and FAAH catalytic activity increased during seed germination and seedling growth, consistent with the timing of NAE depletion. Moreover, the authors identified T-DNA mutants of *A. thaliana* and generated transgenic plants overexpressing *AtFAAH*. It was found that seeds of *AtFAAH* mutants possessed elevated concentrations of endogenous NAEs, and seedling growth was hypersensitive to exogenously applied NAE 12:0. By contrast, seeds and seedlings of *AtFAAH*-overexpressing plants had lower endogenous NAE content and seedlings were less sensitive to exogenous NAE (Wang et al. 2006). Intriguingly, no phenotype on germination was reported for *AtFAAH* mutants or overexpressors, indicating that additional enzymes are likely involved in NAE metabolism with potential redundant functions.

The interaction between NAEs and abscisic acid in regulating seed germination was recently shown by Teaster and co-workers (2007), who reported that NAE and ABA concentrations were depleted during seed germination. Combined application of low concentrations of ABA and NAEs produced a more dramatic reduction in germination than either compound alone. Transcript profiling and gene expression studies in NAE-treated seedlings revealed elevated transcripts for a number of ABA-responsive genes and genes typically enriched in desiccated seeds (Teaster et al. 2007). These data suggest that NAEs act in concert with ABA to regulate seed germination. Whether alkamides interact with ABA or GA to affect germination remains to be determined.

11.4.2 Shoot Development

The shoot represents the aboveground part of higher plants. It is composed of the stem with its branches, and axillary meristems. The stem raises foliage and flowers for optimal light exposure and seed dispersal. The ultimate source of all above-ground organs is a small population of stem cells in the central zone of the shoot apical meristem (Laux and Mayer 1998). Leaf initiation at the shoot apical meristem involves a balance between cell proliferation and commitment to make primordia. *Arabidopsis* has a typical simple leaf, which consists of a petiole and a blade (Fig. 11.1a). The presence of a petiole is presumed to be important in the effective capture of light by ensuring that the leaf blades do not overlap. To produce this leaf shape, the cells on the proximal side of the leaf differentiate into petioles

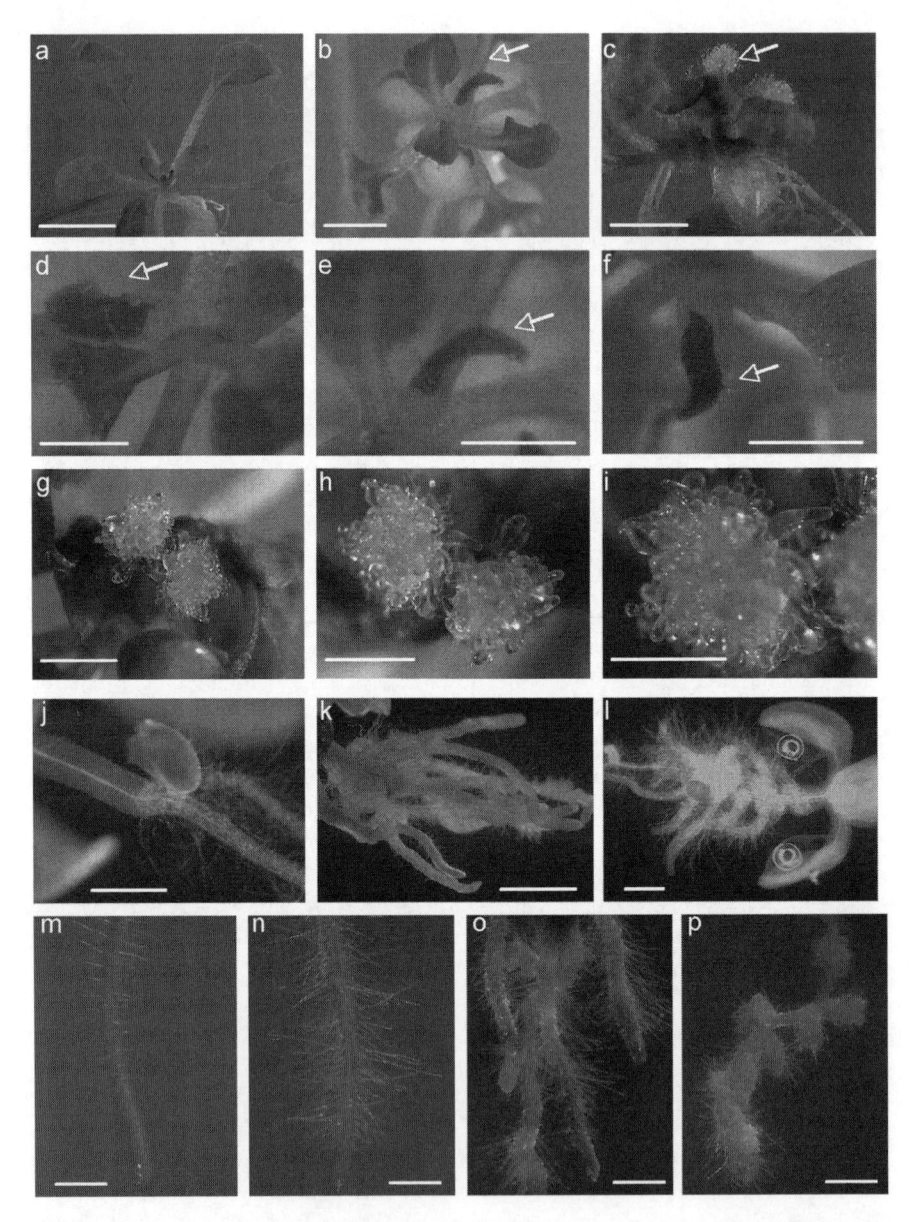

Fig. 11.1 Effects of *N*-isobutyl decanamide on *Arabidopsis* shoot and root development. (**a–c**) Shoot development of 21-day-old *Arabidopsis thaliana* grown on 0, 28 and 56 μM *N*-isobutyl decanamide respectively. Note the formation of blades on petioles and callus-like structures on leaf surfaces (*arrows*). (**d–f**) Close up of developing blades on petioles showing the presence of trichomes (*arrows*). (**g–i**) Close up of callus-like structures formed over the surface of leaves. (**j–l**) Adventitious root development of *Arabidopsis* seedlings grown for 18 days on 0 (**j**) or 56 μM *N*-isobutyl decanamide (**k, l**). Note the proliferation of adventitious roots on the stem region of *N*-isobutyl decanamide-treated plants. (**m–p**) Development of the root apical zone of 12-day-old

without producing blades or other organs (Ha et al. 2003). Two of the clearest effects of *N*-isobutyl decanamide on shoot development were the ectopic induction of outgrowths along the leaf petioles and the formation of callus-like structures on blades (Fig. 11.1b, c). The outgrowths formed on petioles resembled a leaf blade in that they showed the presence of trichomes, a class of differentiated epidermal cells commonly present in blades but not in petioles (Fig. 11.1d–f). In addition, *N*-isobutyl decanamide treatments were found to induce the production of callus-like structures on leaves (Fig. 11.1g–i). These structures sustained growth as plant development progressed and expressed genes indicative of proliferative activity such as cyclin B1, which is involved in cell cycle transitions (López-Bucio et al. 2007). These results suggest that, in plants treated with the alkamide, leaf cells do not exit from the cell cycle with normal developmental timing, resulting in ectopic cell divisions. The fact that petiole cells do not undergo correct developmental specification and are diverted towards other developmental fates, such as blade formation, indicates that the alkamide is also capable of reprogramming petiole cells to initiate the de novo formation of organs in differentiated cells.

11.4.3 Root Development

Roots perform the essential activities of providing water, nutrients and physical support to the plant. The primary root originates in the embryo and produces many lateral roots during the lifetime of a plant, and each of these produces more lateral roots. The quantity and placement of these determine the architecture of the root system; in turn, this plays a major role in determining whether a plant will survive in a particular climate or environment (Malamy and Benfey 1997; Casimiro et al. 2003). During the post-embryonic development of plants, new axes of growth emerge from shoot tissues through adventitious organogenesis. This is particularly important in crops, such as maize, in which adventitious root formation provides a flexible way for plants to alter their form and resource allocation in response to environmental changes or injury. While lateral roots typically form from the primary root pericycle, adventitious roots form naturally from stem tissue. Lateral and adventitious root formation is a complex process affected by multiple endogenous factors, including phytohormones such as auxin, and environmental factors such as light and wounding (Casimiro et al. 2003).

Different reports indicate that NAEs and alkamides may play an important role in regulating root architecture, with stimulating or repressing effects in biomass production depending on the compound, the concentration in the medium and

Fig. 11.1 (continued) *Arabidopsis* seedlings grown on 0, 14, 28 and 56 µM *N*-isobutyl decanamide respectively. The shorter distance between the root tip and the root hair zone is an indication of a faster differentiation process (compare **m** and **n**) and an increase in root hair density. Note the formation of lateral roots close to the root tip (**o**) and the conversion of the primary root tip and lateral roots into callus-like structures (**p**). Scale bars (**a–l**)=5 mm, (**m–p**)=500 µm

conditions of culture. Micromolar concentrations of NAE 12:0 and NAE 18:2 supplied to *A. thaliana* seedlings grown in agar plates inhibited primary root elongation and disrupted normal cell growth in a dose-dependent and selective manner (Blancaflor et al. 2003; Motes et al. 2005; Wang et al. 2006). Kanbe et al. (1993) showed that amidenin, a non-substituted alkamide isolated from the actino-mycete *Amycolatopsis* sp., promoted the growth of rice plants at concentrations of 0.6 and 1.8×10^{-5} M and inhibited growth at concentration of 6×10^{-5} M. Similarly, the alkamides affinin and *N*-isobutyl decanamide showed a dose-dependent effect on biomass production in *A. thaliana*, which correlated with primary root growth inhibition and enhanced lateral root proliferation (Ramírez-Chávez et al. 2004).

As a first step in exploring the structure–activity relationships of NAEs and alkamides, López-Bucio and co-workers (2007) quantified the root growth response of *Arabidopsis* seedlings to natural and synthetic compounds. From a group of similar chain length NAEs and alkamides, they identified *N*-isobutyl decanamide, a C10 saturated alkamide that is naturally produced in *Acmella radicans* (Ríos-Chávez et al. 2003) and *Cissampelos glaberrima* (Laurerio-Rosario et al. 1996), as the most active compound in inhibiting primary root growth and stimulating lateral root formation. The plant regenerative properties of *N*-isobutyl decanamide were tested further in *Arabidopsis* explants that were treated with various concen-trations of this compound. Cultivation of stem explants that harboured the shoot apical meristem on a nutrient-rich medium lacking alkamides resulted in the forma-tion of plants with fully developed shoot and root systems. By contrast, explants obtained from stems or primary roots resulted in the development of adventitious roots in stem explants or of lateral roots in explants from primary roots (Campos-Cuevas et al. 2008). *N*-isobutyl decanamide treatments were found to induce adventitious root formation both in explants or intact *Arabidopsis* seedlings (Fig. 11.1j–l). This effect was accompanied by the differentiation of the primary root meristem, followed by root hair and lateral root formation close to the primary root tip (Fig. 11.1m–p). The stimulation of root branching by *N*-isobutyl decana-mide was related to a general plant growth-promoting effect of this compound in plants regenerated from explants, indicating a potential use of alkamides in the propagation of plants and/or explants by tissue culture means (Campos-Cuevas et al. 2008).

11.5   Signals Interacting with NAEs and Alkamides

11.5.1   Auxins

The effects of NAEs and alkamides in root and shoot development suggest that phytohormones could be involved in the responses of plants to these compounds. Auxins are involved in altering primary root growth and in promoting lateral and adventitious root formation (Woodward and Bartel 2005). Several lines of evidence indicate that the effects of alkamides on root system architecture are independent of auxin signalling. First, the examination of primary root growth of auxin-resistant mutants, *aux1-7*, *eir1* and *axr4-2*, in response to high affinin concentrations revealed a primary root growth inhibition similar to wild-type plants (Ramírez-Chávez et al. 2004). Second, affinin and *N*-isobutyl decanamide were able to induce high numbers of adventitious roots in shoot explants from auxin-resistant mutants (Campos-Cuevas et al. 2008), indicating a normal cell proliferating response. Third, affinin and *N*-isobutyl decanamide failed to activate the expression of the auxin-inducible gene markers *DR5:uidA* and *BA3:uidA* in primary roots and during adventitious root formation in explants (Ramírez-Chávez et al. 2004; Campos-Cuevas et al. 2008). This information suggests that alkamides regulate plant development by an auxin-independent signalling mechanism.

11.5.2 Cytokinins

Cytokinins are purine derivatives that promote and maintain plant cell division in cultures and are also involved in various differentiation processes, including shoot formation, primary root growth and callus formation (Howell et al. 2003). Three different cytokinin receptors have been described, which activate gene expression in a cytokinin-dependent manner. These receptors are sensor histidine kinases, encoded by the *cre1/ahk4/wol*, *ahk2* and *ahk3* genes in *Arabidopsis* (Kakimoto 2003).

The possibility that alkamides could regulate organ development in roots and shoots interacting with cytokinin signalling was investigated recently. This was achieved by evaluating the activation of cytokinin gene expression markers in response to *N*-isobutyl decanamide and testing the responses of single, double and triple mutant combinations for the three known cytokinin receptors to this alkamide (López-Bucio et al. 2007). Interestingly, the ectopic formation of blades on petioles and callus-like structures on leaves was related to an enhanced expression of *ARR5: uidA*, a cyokinin-inducible gene marker in shoots. The triple cytokinin receptor mutant *cre1-12/ahk2-2/ahk3-3* was insensitive to *N*-isobutyl decanamide treatment, showing absence of blades on petioles and callus-like structures in leaves under elevated concentrations of this alkamide. These results suggest that alkamides interact with cytokinin signalling to control cell division and differentiation processes during plant development. The molecular mechanisms underlying this interaction are currently under investigation.

11.5.3 Nitric Oxide

Nitric oxide (NO) is a critical signalling molecule in several vital processes in both mammals and plants. Because of its gaseous nature, it is a highly permeable molecule that freely diffuses thorough biological membranes. NO is synthesized from the amino acid L-arginine. In plant systems, it can be produced principally by two routes: (1) from L-arginine by a nitric oxide synthase-like protein or (2) from nitrite via nitrate reductase and nitrite reductase enzyme action. Once synthesized and released, NO acts as mediator of developmental processes (Neill et al. 2003). Experimental data indicate the signalling roles for NO in processes such as xylem differentiation, programmed cell death, and lateral and adventitious root formation (Lamattina et al. 2003).

The relationship between NO and alkamides was investigated recently in *A. thaliana* shoot explants (Campos-Cuevas et al. 2008). *N*-isobutyl decanamide treatment was found to induce NO accumulation in different stages of plant development. NO was detected by confocal microscopic analysis at the sites of adventitious root formation, and its concentration increased with alkamide treatment. Whether NO mediates the adventitious root response and other morphogenetic responses of plants to alkamides remains to be investigated.

11.6 Cellular Alterations Underlying Plant Responses to NAEs and Alkamides: Cell Cycle Progression and Microtubule Stability

The molecular mechanisms underlying NAE and alkamide responses comprise parallel inhibition of cell elongation and localized stimulation of cell division. These processes are likely to function as focal points in their regulation of plant architecture (Blancaflor et al. 2003; Ramírez-Chávez et al. 2004).

11.6.1 Cell Cycle Progression

Cell division activity in plants is localized in small groups of cells, called meristems, which are already present in the embryo and are active during most of the cell cycle of the plant. The cell cycle that occurs in dividing cells consists of the alternating phases of DNA replication (S phase) and chromosome separation (mitosis, or M phase), interrupted by gaps known as G1 (interval between M and S phases) and G2 (interval between S and M phases). Important controls operate at the transition points as cells move from G1 into S phase, and from G2 into M phase (Beemster et al. 2003). Both the G1–S and G2–M phase transitions can be controlled in plant cells in response to phytohormones, such as auxins and cytokinins. An example of G2 control is found in the development of lateral root primordia, which are derived from pericycle cells arrested in G2. These cells move into M phase upon auxin stimulation and then continue to proliferate, producing a lateral root primordium that eventually emerges from the side of the primary root (Himanen et al. 2002).

Ramírez-Chávez et al. (2004) evaluated the effects of affinin on cell division by using the G2–M specific marker *CycB1:uidA*, which is expressed only in dividing cells. In response to micromolar concentrations of affinin, a dose–response reduction in the number of cells expressing *CycB1:uidA* was observed in root meristems. Meristem cells of *Arabidopsis* roots grown in the presence of NAE 12:0 also displayed abnormal cell division patterns characterized by oblique wall formation and the appearance of numerous irregularly shaped cells that varied in size (Blancaflor et al. 2003). Interestingly, the opposite effect was observed in leaves exposed to *N*-isobutyl decanamide, in which the formation of blades on petioles and callus-like structures could be observed (López-Bucio et al. 2007). These results suggest that localized effects of alkamides are important for cell proliferating activity of these compounds.

11.6.2 Microtubule Stability

Cellular elongation in roots has been shown to be dependent on the cortical microtubule cytoskeleton. Drugs that either depolymerize or stabilize microtubules in roots cause a significant reduction in root growth rate that eventually leads to radial expansion (Baskin et al. 1994). In primary roots, cortical microtubules in cells of the elongation zone are typically arranged along to the longitudinal axis of the root and uniformly distributed (Fig. 11.2a). This orientation shifts to oblique or longitudinal arrays as the cells make their transition into the maturation zone. Continuous exposure to 50 μM NAE 12:0 (Blancaflor et al. 2003) or 36 μM *N*-isobutyl decanamide (Méndez-Bravo et al., unpublished data) caused radial swelling in roots of *Arabidopsis* seedlings (Fig. 11.2a, b). Instead of the typical cylindrical cell shape observed in the differentiation zone of untreated roots, cells of alkamide-treated roots were shorter and wider. Microtubules in cells with altered shape were oriented in random directions and appeared to be fragmented and disorganized (Fig. 11.2c, d). This change in microtubule orientation and size, and cell elongation creates a link between acylamides and root architecture.

11.7 AHLs: Inter-Kingdom Signals for Plant–Bacterial Interactions

Bacterial cells communicate with each other using chemical signals. Specifically, they release, detect, and respond to the accumulation of compounds that allow bacteria to coordinate their gene expression in responses to changes in the population density, a process commonly referred to as quorum-sensing (Waters and Bassler 2005). Many processes in bacteria are regulated by quorum-sensing, including symbiosis, virulence, antibiotic production and biofilm formation. Gram-negative bacteria use *N*-acyl-homoserine lactones (AHLs) to communicate and regulate their quorum-sensing; these compounds contain a conserved homoserine lactone (HL)

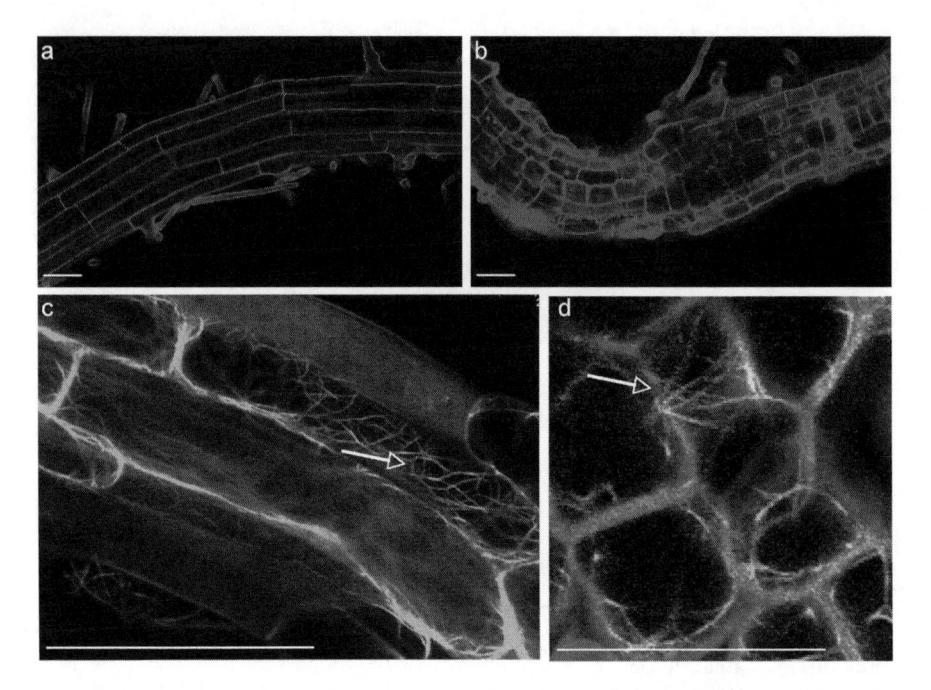

Fig. 11.2 Effects of *N*-isobutyl decanamide on cell elongation and microtubule stability. (**a, b**) Confocal microscope images of (respectively) 0 and 56 μM *N*-isobutyl decanamide-treated root epidermal cells of WT *Arabidopsis thaliana* seedlings. Note the shorter size of alkamide-treated cells. (**c, d**) Organization of epidermal microtubules in root cells of 12-day-old *Arabidopsis* transgenic seedlings expressing a fimbrin::GFP construct. **c** Microtubules in the epidermal cells of control roots and **d** roots treated with 56 μM *N*-isobutyl decanamide. Note the fragmented and randomly oriented microtubules in alkamide-treated cells. Scale bars = 50 μm

ring and an amide (*N*)-linked acyl side chain (Table 11.1). The acyl groups of naturally occurring AHLs range from 4 to 18 carbons in length; they can be saturated or unsaturated and with or without a C-3 substituent (Reading and Sperandio 2006). Recent evidence shows, however, that quorum-sensing signalling is not restricted to bacterial cell-to-cell communication, but also allows communication between plants and their prokaryote partners (Hughes and Sperandio 2008). The presence of AHL-producing bacteria in the rhizosphere of tomato induced the salicylic acid and ethylene-dependent defence response that plays an important role in the activation of systemic resistance in plants, and conferred resistance to the fungal pathogen *Alternaria alternata* (Schuhegger et al. 2006). In addition, certain *Rhizobium* mutants that fail to produce or sense AHLs were unable to nodulate legume plants, suggesting that AHLs may play a role in nodulation (Zheng et al. 2006). Our research aiming to clarify the role of AHLs in plant development revealed that these compounds exert strong and specific cellular responses similar to NAEs and alkamides, including arrested cell proliferation in the primary root meristem and enhanced lateral root formation (Ortíz-Castro et al. 2008).

11.8 Concluding Remarks

Most plant hormones are pleiotropic rather than specific, that is, each has more than one effect on the growth and development of plants. Auxin, for instance, stimulates the rate of cell elongation at low concentrations with inhibitory effects at high concentrations, it causes shoots to grow upwards and roots downwards, and it promotes the formation and growth of lateral roots and shoot branches. Auxin also causes the plant to produce a second hormone, ethylene, to elicit a plethora of additional responses. Ethylene, in turn, regulates auxin biosynthesis and alters auxin-responsive gene expression (Stephanova et al. 2007). The other well-known plant hormones gibberellins, abscisic acid and cytokinins have a similarly complex array of functions. Such interactions may also exist for NAEs and alkamides. These compounds are part of a vast array of amino compound-containing lipids ubiquitous in organisms from bacteria to mammals. Lipids are integral components of the cell membrane, the prime and essential limit of cells with its environment. Small signalling NAEs can be released from the membrane by PLD enzymes in response to biotic or abiotic stimuli (Bargmann and Munnik 2006). The presented information suggests that acylamides can be perceived by plants. NAEs and alkamides may actually function by activating other signalling pathways already important for plant development, such as cytokinins and abscisic acid, or by inducing accumulation of NO, a highly active messenger for cellular responses. This information suggests that these lipids are not merely structural components, but also important players involved in the orchestrated adjustment of anatomical characteristics. Such adjustment may account for the morphological adaptations to limit exposure to unfavourable environmental conditions.

NAEs and alkamides are believed to act, at least in part, by an endocannabinoid-like signalling mechanism similar to that described in animals (López-Bucio et al. 2006; Morquecho-Contreras and López-Bucio 2007). Recent findings add further complexity for plant responses to these lipids. In a search for alkamide-related compounds from bacteria that could modify root growth, we identified the *N*-acyl homoserine lactones, a class of bacterial quorum-sensing signals, as important regulators of plant morphogenesis (Ortíz-Castro et al. 2008). Thus, the possibility is open that amino compound-containing lipids could represent a novel class of signals for plant–bacterial communication. Elucidating how alkamides interact with classic and novel signals to regulate plant development remains a major challenge. The adoption of an integrated approach that combines genetics, molecular biology, cell biology and analytical chemistry should increase our knowledge on the signalling pathways involved in small lipid perception in plants.

References

Anaya AL, Macías-Rubalcava M, Cruz-Ortega R, García-Santana C, Sánchez-Monterrubio PN, Hernández-Bautista BE, Mata R (2005) Allelochemicals from *Stauranthus perforatus*, a rutaceous three of the Yucatán Peninsula, México. Phytochemistry 66:487–494

Bargmann BO, Munnik T (2006) The role of phospholipase D in plant stress responses. Curr Opin Plant Biol 9:515–522

Baskin TI, Wilson JE, Cork A, Williamson RE (1994) Morphology and microtubule organization in Arabidopsis roots exposed to orizalyn or taxol. Plant Cell Physiol 35:935–942

Bauer R, Reminger P (1989) TLC and HPLC analysis of alkamides in *Echinacea* drugs. Planta Med 55:367–371

Beemster GTS, Fiorani F, Inzé D (2003) Cell cycle: the key to plant growth control? Trends Plant Sci 8:154–158

Blancaflor EB, Hou G, Chapman KD (2003) Elevated levels of N-lauroylethanolamine, an endogenous constituent of desiccated seeds, disrupt normal root development in *Arabidopsis thaliana* seedlings. Planta 217:206–217

Campos-Cuevas JC, Pelagio-Flores R, Raya-González J, Méndez-Bravo A, Ortiz-Castro R, López-Bucio J (2008) Tissue culture of *Arabidopsis thaliana* explants reveals a stimulatory effect of alkamides on adventitious root formation and nitric oxide accumulation. Plant Sci 174:165–173

Caramelo JJ, Florin-Christensen J, Delfino JM (2003) Phospholipase activity on N-acyl phosphatidylethanolamines is critically dependent on the N-acyl chain length. Biochem J 374:109–115

Casimiro I, Beekman T, Graham N, Bhalerao R, Zhang H, Casero P, Sandberg G, Bennett M (2003) Dissecting *Arabidopsis* lateral root development. Trends Plant Sci 8:165–171

Chapman KD (2004) Occurrence, metabolism, and prospective functions of N-acylethanolamines in plants. Prog Lipid Res 43:309–327

Christensen L, Lam J (1991) Acetylenes and related compounds in Heliantheae. Phytochemistry 30:11–49

Finch-Savage WE, Leubner-Metzger G (2006) Seed dormancy and the control of germination. New Phytol 171:501–523

Ha CM, Kin GT, Kim BC, Jun JH, Soh MS, Ueno Y, Machida D, Tsukaya H, Nam HG (2003) The *BLADE-ON-PETIOLE 1* gene controls leaf pattern formation through the modulation of meristematic activity in *Arabidopsis*. Development 130:161–172

Hansen H, Moesgaard B, Petersen G, Hansen H (2002) Putative neuroprotective actions of N-acylethanolamines. Pharmacol Ther 95:119–127

Herms DA, Mattson WJ (1992) The dilemma of plants: to grow or defend. Quat Rev Biol 67:283–319

Himanen K, Boucheron E, Vaneste S, Almeida-Engler J, Inzé D, Beeckamn T (2002) Auxin-mediated cell cycle activation during early lateral root initiation. Plant Cell 14:2339–2351

Howell SH, Lall S, Che P (2003) Cytokinins and shoot development. Trends Plant Sci 9:453–459

Hughes DT, Sperandio V (2008) Inter-kingdom signaling: communication between bacteria and their hosts. Nature Rev Microbiol 6:111–120

Kakimoto T (2003) Perception and signal transduction of cytokinins. Annu Rev Plant Biol 54:605–627

Kanbe K, Naganawa H, Okamura M, Sasaki T, Hamada M, Okami Y, Takeuchi T (1993) Amidenin, a new plant growth regulating substance isolated from *Amycolatopsis* sp. Biosci Biotechnol Biochem 57:1261–1263

Kashiwada Y, Ito C, Katagiri H, Mase I, Komatsu K, Namba T, Ikeshiro Y (1997) Amides of the fruit of *Zanthoxylum* spp. Phytochemistry 44:1125–1127

Kunos G, Jarai Z, Batkai S, Isaac EJ, Liu J, Wagner JA (2000) Endocannabinoids as vascular modulators. Chem Phys Lipids 108:159–168

Lafrance D, Marion D, Pezolet M (1990) Study of the structure of N-acyl dihexadecanoyl phosphatidylethanolamines in aqueous dispersion by infrared and Raman spectroscopies. Biochemistry 29:4592–4599

Lamattina L, García-Mata C, Graciano M, Pagnussat G (2003) Nitric oxide: the versatility of an extensive signal molecule. Annu Rev Plant Biol 54:109–136

Laurerio-Rosario S, Silva A, Parente J (1996) Alkamides from *Cissampelos glaberrima*. Planta Med 62:376–377

Laux T, Mayer KFX (1998) Cell fate regulation in the shoot meristem. Sem Cell Dev Biol 9:195–200

Letchamo W, Livesey J, Arnason TJ, Bergeron C, Krutilina VS (1999) Cichoric acid and isobutylamide content in *Echinacea purpurea* as influenced by flower developmental stages. In: Janick J (ed) Perspectives on new crops and new uses. ASHS Press, Alexandria, VA, pp 494–498

Liu J, Wang L, Harvey-White J, Osei-Hyiaman D, Razdan R, Gong Q, Chan AC, Zhou Z, Huang BX, Kim HY, Kunos G (2006) A biosynthetic pathway for anandamide. Proc Natl Acad Sci USA 103:13345–13350

López-Bucio J, Acevedo-Hernández G, Ramírez-Chávez E, Molina-Torres E, Herrera-Estrella L (2006) Novel signals for plant development. Curr Opin Plant Biol 6:280–287

López-Bucio J, Millán-Godínez M, Méndez-Bravo A, Morquecho-Contreras A, Ramírez-Chávez E, Molina-Torres J, Pérez-Torres A, Higuchi M, Kakimoto T, Herrera-Estrella L (2007) Cytokinin receptors are envolved in alkamide regulation of root and shoot development in *Arabidopsis*. Plant Physiol 145:1703–1713

LoVerme J, Russo R, La Rana G, Fu J, Farthing J, Raso G, Meli R, Hohmann A, Calignano A, Piomelli D (2006) Rapid broad-spectrum analgesia through activation of peroxisome proliferator-activated receptor alpha. J Pharmacol Exp Ther 319:1051–1061

Malamy J, Benfey P (1997) Down and out in *Arabidopsis*: the formation of lateral roots. Trends Plant Sci 2:390–401

Molina-Torres J, Salgado-Garciglia R, Ramírez-Chávez E, del Río R (1996) Purely oleofinic alkamides in *Heliopsis longipes* and *Acmella* (*Spillanthes*) *oppositifolia*. Biochem System Ecol 24:43–47

Morquecho-Contreras A, López-Bucio J (2007) Cannabinoid-like signaling and other new developmental pathways in plants. Int J Plant Dev Biol 1:34–41

Motes CM, Pechter P, Min-Yoo C, Yuh-Shu W, Chapman KD, Blancaflor E (2005) Differential effects of two phospholipase D inhibitors, 1-butanol and *N*-acylethanolamine, on in vivo cytoskeletal organization and *Arabidopsis* seedling growth. Protoplasma 226:109–123

Movahed P, Jonsson BA, Birnir B, Wingstrand JA, Jorgensen TD, Ermund A, Sterner O, Zygmunt PM, Hogestatt ED (2005) Endogenous unsaturated C18 *N*-acylethanolamines are vanilloid receptor (TRPV1) agonists. J Biol Chem 280:38496–38504

Neill SJ, Desikan R, Hancock JT (2003) Nitric oxide signaling in plants. New Phytol 159:11–35

Okamoto Y, Morishita J, Tsuboi K, Tonai T, Ueda N (2004) Molecular characterization of a phospholipase D generating anandamide and its congeners. J Biol Chem 279:5298–5305

Ortíz-Castro R, Martínez-Trujillo M, López-Bucio J (2008) *N*-acyl-L-homoserine lactones: a class of bacterial quórum-sensing signals alter post-embryonic root development in *Arabidopsis thaliana*. Plant Cell Environ 31:1497–1509

Oz M, Alptekin A, Tchugunova Y, Dinc M (2005) Effects of saturated long-chain *N*-acylethanolamines on voltage-dependent Ca^{2+} fluxes in rabbit T-tubule membranes. Arch Biochem Biophys 434:344–351

Pappan K, Austin-Brown S, Chapman K, Wang X (1998) Substrate selectivities and lipid modulation of plant phospholipase Dα, -β, and -γ. Arch Biochem Biophys 353:131–140

Paria BC, Dey SK (2000) Ligand-receptor signaling with endocannabinoids in preimplantation embryo development and implantation. Chem Phys Lipids 108:211–220

Potters G, Pasternak TP, Guisez Y, Palme KJ, Jansen MAK (2007) Stress-induced morphogenic responses: growing out of trouble? Trends Plant Sci 12:98–105

Qu L, Chen Y, Wang X, Scalzo R, Davis JM (2005) Patterns of variation in alkamides and cichoric acid in roots and above ground parts of *Echinacea purpurea* (L.) Moench. HortScience 40:1239–1242

Ramírez-Chávez E, López-Bucio J, Herrera-Estrella L, Molina-Torres J (2004) Alkamides isolated from plants promote growth and alter root development in *Arabidopsis*. Plant Physiol 134:1058–1068

Razem FA, Baron K, Hill RD (2006) Turning on gibberellin and abscisic acid signaling. Curr Opin Plant Biol 9:454–459

Reading NC, Sperandio V (2006) Quorum-sensing: the many languages of bacteria. FEMS Microbiol Lett 254:1–11

Ríos-Chávez P, Ramírez-Chávez E, Armenta-Salinas C, Molina-Torres J (2003) *Acmella radicans* var. *radicans*: in vitro culture stablisment and alkamide content. In Vitro Cell Dev Biol-Plant 39:37–41

Schuhegger R, Ihring A, Gantner S, Bahnweg G, Knappe C, Vogg G, Hutzler P, Schmid M, Van Breusegem F, Eber L, Hartmann A, Langebartels C (2006) Induction of systemic resistance in tomato by *N*-acyl-L-homoserine lactone-producing rhizosphere bacteria. Plant Cell Environ 29:909–918

Shresta R, Dixon RA, Chapman KD (2003) Molecular identification of a functional homologue of the mammalian fatty acid amide hydrolase in *Arabidopsis thaliana*. J Biol Chem 278: 34990–34997

Shresta R, Kim SC, Dyer JM, Dixon RA, Chapman KD (2006) Plant fatty acid (ethanol) amide hydrolases. Biochem Biophys Acta 176:324–334

Simon GM, Cravatt BF (2006) Endocannabinoid biosynthesis proceeding through glycerophospho-*N*-acylethanolamine and a role for alpha/beta-hydrolase in this pathway. J Biol Chem 281:26465–26472

Stephanova AN, Jeonga Y, Likhacheva AV, Alonso JM (2007) Multilevel interactions between ethylene and auxin in *Arabidopsis* roots. Plant Cell 19:2169–2185

Swamy MJ, Ramakrishnan M, Angerstein B, Marsh D (2000) Spin-label electron spin resonance studies on the mode of anchoring and vertical location of the *N*-acyl chain in *N*-acyl phosphatidylethanolamines. Biochemistry 39:12476–12484

Teaster ND, Motes C, Tang Y, Wiant W, Cotter MQ, Wang YH, Kilaru A, Venables BJ, Hasenstein KH, González G, Blancaflor E, Chapman KD (2007) *N*-acylethanolamine metabolism interacts with abscisic acid signaling in *Arabidopsis thaliana* seedlings. Plant Cell 19:2454–2469

Tripathy S, Venables B, Chapman K (1999) *N*-acylethanolamines in elicitor signal transduction and activation of defense gene expression. Plant Physiol 121:1299–1308

Tsuboi K, Sun YX, Okamoto Y, Araki N, Tonai T, Ueda N (2005) Molecular characterization of *N*-acylethanolamine-hydrolyzing acid amidase, a novel member of the choloylglycine hydrolase family with structural and functional similarity to acid ceramidase. J Biol Chem 280:11082–11092

Venables BJ, Waggoner CA, Chapman KD (2005) *N*-acylethanolamines in selected legumes. Phytochemistry 66:1913–1918

Wang X (2004) Lipid signaling. Curr Opin Plant Biol 7:329–336

Wang YS, Shresta R, Kilaru A, Wiant W, Venables BJ, Chapman KD, Blancaflor E (2006) Manipulation of *Arabidopsis* fatty acid amide hydrolase expression modifies plant growth and sensitivity to *N*-acylethanolamines. Proc Natl Acad Sci USA 103:12197–12202

Waters CM, Bassler BL (2005) Quorum-sensing: cell-to-cell communication in bacteria. Annu Rev Cell Dev Biol 21:319–346

Weyers JDB, Paterson NW (2001) Plant hormones and the control of physiological processes. New Phytol 152:375–407

Wilson RI, Nicoll RA (2002) Endocannabinoid signaling in the brain. Science 296:678–682

Woodward AW, Bartel B (2005) Auxin: regulation, action and interaction. Ann Bot 95:707–735

Worrall D, Ng CKY, Hetherington AM (2003) Sphingolipids, new players in plant signaling. Trends Plant Sci 8:317–320

Wymann MP, Schneiter R (2008) Lipid signaling in disease. Nature Rev Mol Cell Biol 9:163–176

Zheng H, Zhong Z, Lai X, Chen WX, Li S, Zhu J (2006) A *luxR*/*luxI*-type quorum sensing system in a plant bacterium *Mesorhizobium tianshanense*, controls symbiotic nodulation. J Bacteriol 188:1943–1949

Chapter 12
The Roles of *YUCCA* Genes in Local Auxin Biosynthesis and Plant Development

Y. Zhao

12.1 Introduction

Auxin is the first identified plant hormone and, arguably, has the broadest physiological roles in plant growth and development compared to other plant hormones. Despite its essential roles in plant development, dissection of auxin biosynthesis in plants has proven to be very difficult. Early physiological and analytic biochemical studies established that indole-3-acetic acid (IAA), the main auxin in plants, can be synthesized from both tryptophan (Trp)-dependent and Trp-independent pathways (Bartel 1997). It is only recently, however, that molecular genetic studies in Arabidopsis have begun to identify key auxin biosynthetic genes.

12.2 Identification of YUCCA Flavin Monooxygenases as Key Enzymes in Auxin Biosynthesis

Early genetic studies on auxin biosynthesis in plants were conducted mainly in Trp auxotroph mutants on the basis of the hypothesis that Trp was a main precursor for IAA biosynthesis (Last et al. 1991; Wright et al. 1991). Surprisingly, none of the Trp biosynthesis mutants analyzed were auxin-deficient. In contrast, some of the Arabidopsis and maize *trp* mutants had elevated concentrations of IAA conjugates (Wright et al. 1991). Further isotope labelling and feeding experiments demonstrated that IAA is also synthesized from a Trp-independent pathway (Wright et al. 1991). Trp auxotroph mutants are probably not the optimum starting point for investigating auxin biosynthesis, for several reasons. First, the *trp* mutants used in

Y. Zhao
Section of Cell and Developmental Biology, University of California San Diego, 9500 Gilman Drive, La Jolla, CA 92093-0116, USA
e-mail: y3zhao@ucsd.edu

E-C. Pua and M.R. Davey (eds.),
Plant Developmental Biology – Biotechnological Perspectives: Volume 2,
DOI 10.1007/978-3-642-04670-4_12, © Springer-Verlag Berlin Heidelberg 2010

auxin research were not completely Trp-deficient, and the residual Trp synthesis capacity complicates interpretations of feeding/labelling experiments. Second, Trp is essential for protein synthesis, and it is difficult to differentiate phenotypes caused by defects in auxin biosynthesis and Trp deficiency.

Perhaps due to the lack of knowledge on developmental defects that auxin deficiency may cause, no systematic forward genetic screens have ever been conducted to isolate auxin-deficient mutants. A lack of auxin-deficient mutants in the literature suggests that auxin is either essential for plant growth or is synthesized by redundant pathways/genes. However, investigations of auxin biosynthesis pathways in plant pathogenic bacteria provided well-defined phenotypes for plants that overproduce auxin (Comai and Kosuge 1982, 1983). For example, expression of the bacterial auxin biosynthesis gene *iaaM* in Arabidopsis under the strong cauliflower mosaic virus (CaMV) 35S promoter leads to auxin overproduction (Romano et al. 1995). Light-grown *iaaM* overexpressing in Arabidopsis seedlings exhibited long hypocotyls and epinastic cotyledons (Fig. 12.1; Romano et al. 1995). Auxin-overproduction phenotypes are also observed in two recessive Arabidopsis mutants, namely *superroot1* (*sur1*; Boerjan et al. 1995) and *superroot2* (*sur2*; Delarue et al. 1998). In addition to producing extensive adventitious roots, the *sur* mutants possessed long hypocotyls and epinastic cotyledons (Boerjan et al. 1995; Delarue et al. 1998). Both SUR1 and SUR2 are involved in biosynthesis of indolic glucosinolate from indole-3-acetaldoxime, which is also an intermediate for IAA biosynthesis in Arabidopsis (Barlier et al. 2000; Mikkelsen et al. 2004). Inactivation of either SUR1 or SUR2 diverts more indole-3-acetaldoxime to IAA synthesis, thereby causing auxin overproduction. Furthermore, Arabidopsis seedlings have long hypocotyls and epinastic cotyledons when grown at higher temperatures, due to auxin overproduction (Gray et al. 1998). Taken together, it appears that light-grown Arabidopsis seedlings have elongated hypocotyls and epinastic cotyledons when auxin concentrations are elevated.

The fact that auxin overproduction causes characteristic developmental phenotypes suggests that such phenotypes may be used for identifying key auxin biosynthesis genes through a gain-of-function screen for auxin-overproduction mutants.

Fig. 12.1 Auxin overproduction causes distinct phenotypes in Arabidopsis. Overexpression of the bacterial auxin biosynthesis gene *iaaM* (*middle*) or *YUC* flavin monooxygenase (*right*) leads to auxin overproduction. The overexpression lines have long hypocotyls and epinastic cotyledons compared to wild-type Arabidopsis (*left*). Bar = 1 mm

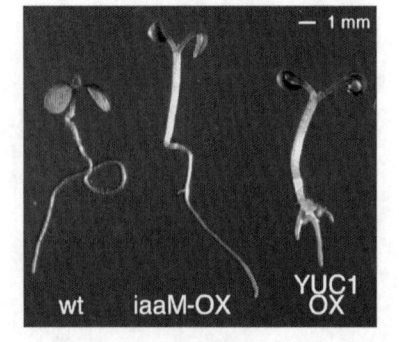

The underlying hypothesis is that constitutive activation of a rate-limiting auxin biosynthesis enzyme may lead to auxin overproduction. The Arabidopsis *yucca* (later renamed as *yuc1D*) mutant was identified as a dominant, gain-of-function, long hypocotyl mutant in an activation-tagging screen (Fig. 12.1; Zhao et al. 2001). In addition, *yuc1D* has short primary roots, more root hairs, and increased apical dominance (Zhao et al. 2001). The observed phenotypes of *yuc1D* are indicative of elevated IAA levels. Furthermore, the auxin reporter DR5-GUS is up-regulated in *yuc1D* (Zhao et al. 2001), and *yuc1D* phenotypes are partially suppressed by over-expression of the *iaaL* gene, which presumably inactivates free IAA by conjugating IAA with the amino acid lysine (Zhao et al. 2001). Direct auxin analysis shows that *yuc1D* has 50% more free IAA than do wild-type plants (Zhao et al. 2001). Moreover, explants of *yuc1D* can propagate in auxin-free media, but not the wild-type Arabidopsis, further demonstrating that the *yuc1D* has elevated auxin concentrations (Zhao et al. 2001). The physiological and genetic analyses clearly demonstrate that *yuc1D* is an auxin-overproduction mutant.

The phenotypes of *yuc1D* are conferred by overexpression of the *YUC1* flavin monooxygenase gene caused by the insertion of four copies of the CaMV 35S enhancers near the *YUC1* gene. Overexpression of *YUC1* not only increases auxin concentrations, but also renders the plants resistant to 5-methyl Trp, a toxic Trp analog, indicating that the increased auxin concentrations in *yuc1D* are probably synthesized through a Trp-dependent auxin biosynthesis pathway (Zhao et al. 2001). Furthermore, biochemical analysis suggests that YUC1 catalyzes a rate-limiting step in auxin biosynthesis by converting tryptamine to N-hydroxyl tryptamine (Zhao et al. 2001).

YUC1 belongs to a large gene family with 11 members in the Arabidopsis genome (Zhao et al. 2001; Cheng et al. 2006, 2007a). Inactivation of a single *YUC* gene does not cause obvious developmental defects, suggesting that members of the *YUC* genes have overlapping functions. Systematic analysis of loss-of-function *yuc* mutant combinations has revealed that *YUC* genes play essential roles in many developmental processes, including embryogenesis, seedling growth, vascular formation and flower development. For example, flowers of the *yuc1 yuc4* double mutants have fewer floral organs and are completely sterile (Fig. 12.2; Zhao et al. 2001; Cheng et al. 2006, 2007a). The quadruple mutants *yuc1 yuc4 yuc10 yuc11* fail to form the basal part of the embryo, while *yuc1 yuc2 yuc4 yuc6* plants have strong vascular defects (Cheng et al. 2006, 2007a). The observed developmental defects of *yuc* mutant combinations are indicative of partial auxin deficiency. Further experimentation demonstrates that the developmental defects in the *yuc* mutants are, indeed, caused by partial auxin deficiency (Cheng et al. 2007a). The first piece of evidence is that *yuc* mutants display synergistic genetic interactions with known auxin mutants, such as *pin1*, *pid* and *aux1*, further supporting the view that *yuc* mutants are partially auxin-deficient (Cheng et al. 2007a). For instance, *yuc1 yuc4 pin1* plants fail to produce true leaves. *yuc1 yuc4 pid* triple mutants do not have cotyledons, whereas *pin1*, *pid* or *yuc1 yuc4* alone do not cause such developmental defects (Cheng et al. 2007a). In addition, the expression levels of the auxin reporter DR5-GUS are decreased in the *yuc* mutants (Cheng et al.

Fig. 12.2 *YUC* genes are essential for normal Arabidopsis development. Wild-type Arabidopsis flower buds (*left*) are closed by sepals, while flower buds of *yuc1 yuc4* double mutants (*right*) are open and have fewer floral organs. Bar = 1 mm

2007a). Finally, the developmental defects of *yuc1 yuc4* double mutants are rescued by the expression of the bacterial auxin biosynthesis gene *iaaM* under the control of the *YUC1* promoter, and the *yuc1 yuc2 yuc6* triple mutants are rescued by iaaM driven by the *YUC6* promoter (Cheng et al. 2006).

Genetic, biochemical and physiological analyses of both gain-of-function and loss-of-function *yuc* mutants demonstrate convincingly that YUC flavin monooxygenases catalyze a rate-limiting step in Trp-dependent auxin biosynthesis. Overexpression of *YUC* genes leads to auxin-overproduction phenotypes, while disruption of *YUC* genes causes auxin-deficient phenotypes, which can be rescued by in vivo auxin production in a tissue-specific manner.

12.3   *YUC* Genes Have Dynamic Expression Patterns

The identification of *YUC* genes as key auxin biosynthetic components provides molecular markers for defining the sites of auxin biosynthesis. Interestingly, *YUC* genes in Arabidopsis are not expressed ubiquitously. Rather, their expression is restricted to discrete groups of cells (Cheng et al. 2006, 2007a). For example, *YUC1* is expressed only in the apical meristem in a mature embryo (Cheng et al. 2006, 2007a). The specific expression patterns of *YUC* genes raise an interesting question, namely what is the physiological role of restricting auxin synthesis to specific cells? There is no apparent answer to this question at present. However, there is evidence that the sites of auxin synthesis by the *YUC* genes are important for normal plant development. All the 11 *YUC* genes in Arabidopsis show unique expression patterns (Cheng et al., unpublished data). Consequently, different combinations of the Arabidopsis *yuc* mutants display different developmental phenotypes, indicating that the differences in phenotypes are probably caused by differences in expression patterns. For example, *yuc1 yuc2 yuc4 yuc6* quadruple mutants appear to be normal in embryogenesis, but later show severe defects in flowers and vascular forms. Conversely, the *yuc1 yuc4 yuc10 yuc11* quadruple mutants fail to produce the basal part of the embryo. Another observation is that inactivation of *YUC* genes leads to the inhibition of DR5-GUS expression only in tissues where *YUC* genes are expressed. For example, *YUC1* and *YUC4* are expressed in leaf promordia and young leaves. DR5-GUS expression is dramatically decreased in young leaves, but there is no obvious change in roots of *yuc1 yuc4* double mutants. Because auxin biosynthesis, mediated by the *YUC* genes, appears to be limited to discrete groups of cells, it is warranted to investigate the relationship between local auxin synthesis and polar transport. Genetic interactions between *yuc1 yuc4* and *pin1* and *aux1* suggest that complex mechanisms exist that coordinate the two processes (Cheng et al. 2006, 2007a). Identification of the transcription factors that specify the expression patterns of the *YUC* genes will help to understand how the patterns are generated. At present, only STY/SHI genes and LEC2 have been implicated in regulating *YUC* expression. Mutations in these genes can cause developmental defects related to auxin biosynthesis pathways (Sohlberg et al. 2006; Stone et al. 2008).

12.4 *YUC* Genes Are Conserved in the Plant Kingdom

YUC-like flavin monooxygenases are widely distributed throughout the plant kingdom. Database searches show that *YUC* genes exist in all of the sequenced plant genomes, including those of moss, rice and Arabidopsis. Functional analysis of the rice *YUC* genes demonstrates that *YUC* genes also conduct auxin biosynthesis in rice, and play essential roles in many processes of plant growth and development (Woo et al. 2007; Yamamoto et al. 2007; Fujino et al. 2008). Functional characterization of the *YUC1* gene (FLOOZY) in petunia not only demonstrates the critical roles of *YUC* genes in auxin biosynthesis and plant development, but also shows that *YUC* genes function non-autonomously (Tobena-Santamaria et al. 2002). Recently, the role of *YUC* genes in auxin biosynthesis and plant development has been reported in maize (Gallavotti et al. 2008) and tomato (Expósito-Rodríguez et al. 2007). Therefore, it is likely that *YUC*-mediated auxin biosynthesis is conserved amongst plant species, based on the sequence homology and functional analyses of the *YUC* genes in several plant species. This is in contrast to other auxin biosynthesis genes, such as CYP79B2/B3, which, so far, have been found only in *Brassica*, and thus do not appear to be widely distributed (Zhao et al. 2002).

12.5 Dissection of Auxin Action Mechanisms on the Basis of Auxin Biosynthesis

Previous genetic screens for Arabidopsis mutants defective in auxin signalling were mainly conducted at seedling stages in the presence of excess exogenous auxin (Lincoln et al. 1990). The main phenotypic readout was the primary root elongation, because exogenous auxin inhibits primary root elongation. A caveat of the previous genetic screens is that some of the key auxin genes may not be identified if the genes are not expressed in the root, or are essential for root development. For example, *PIN1* (Galweiler et al. 1998) and *PID* (Christensen et al. 2000) play an essential role in auxin-mediated flower development, but inactivation of PIN1 or PID does not lead to auxin resistance in roots, indicating that some components in auxin pathways may have been missed from the auxin-resistant mutant screens.

The identification of *yuc* mutants, which are partially auxin-deficient, provides an opportunity to genetically dissect auxin action mechanisms from a different approach, for several reasons. First, the developmental defects of *yuc1 yuc4* double mutants occur mainly in flowers and vascular systems (Cheng et al. 2006), thus allowing for screening genes important for auxin action in a non-rooting system. Second, previous screens for auxin mutants were conducted in the presence of excess exogenous auxin. In contrast, the *yuc* mutants allow genetic screens to be conducted in low endogenous auxin background. Third, exogenous auxin may not perturb the in vivo auxin gradients that may be important for plant development, whereas the *yuc* mutants may alter auxin gradients due to the restricted expression patterns of the *YUC* genes (Cheng et al. 2006). Therefore, genetic screens in the *yuc* mutant background complement previous genetic screens, and may help to identify novel components in auxin pathways.

Recent identification and molecular analyses of the *yuc1 yuc4* enhancer, *npy1* (*naked pins in yuc mutants 1*), demonstrate the power of such genetic approaches using *yuc* mutants as starting materials (Cheng et al. 2007b, 2008). Flowers of *yuc1 yuc4* are defective, but the double mutants still make flowers. When the *NPY1* gene is inactivated in the *yuc1 yuc4* background, the resulting triple mutants form pin-like inflorescences (Fig. 12.3), a phenotype that is observed also in the known auxin

Fig. 12.3 The auxin biosynthetic mutants *yuc1 yuc4* are used to identify the *NPY1* gene. Mutations in *NPY1* in the *yuc1 yuc4* background abolish the formation of floral organs (*right*), whereas *npy1* alone is similar to wild-type (*left*). Bar = 30 mm

mutants *pin1*, *pid* and *mp*. Inactivation of *NPY1* alone does not cause obvious defects in flower development.

NPY1 encodes a protein containing a BTB domain at its N-terminus and an NPH3 (non-phototropic hypocotyl 3) domain in the middle (Cheng et al. 2007b). *NPY1* belongs to a gene family with 32 members in the Arabidopsis genome. NPH3, the founding member of this family, was identified as a key component for phototropic response (Motchoulski and Liscum 1999). NPH3 physically interacts with the photoreceptor PHOT1 (Liscum and Briggs 1995; Motchoulski and Liscum 1999), which is homologous to PID and is a member of the AGC kinase family (Cheng et al. 2007b). In response to directional blue light, PHOT1 is activated and NPH3 is de-phosphorylated, causing plants to grow towards the light (Liscum and Briggs 1995; Motchoulski and Liscum 1999; Pedmale and Liscum 2007). Because NPY1 is homologous to NPH3, and PID is homologous to PHOT1, it appears that auxin-mediated organogenesis and phototropic response use analogous mechanisms. Another similarity between these two pathways is that both require the participation of an auxin response factor. Inactivation of *ARF7/NPH4* leads to the abolishment of phototropic response (Harper et al. 2000), while disruption of *ARF5/MONOPTEROS* causes the formation of pin-like inflorescences (Przemeck et al. 1996). Further genetic analysis of NPY1 homologs and PID homologs has established that the NPY and PID families play essential roles in auxin-mediated organogenesis (Cheng et al. 2008). The genetic interactions among YUCs, PIDs and NPYs have led to the discovery of a linear pathway that controls auxin action in Arabidopsis organogenesis (Cheng et al. 2008).

12.6   Conclusions

In summary, YUC flavin monooxygenases are key enzymes of auxin biosynthesis. These enzymes play essential roles in many aspects of plant growth and development. The identification of *YUC* genes assists in clarifying auxin transport, movements and dynamics, because the location of the *YUC* flavin monooxygenases defines the starting points for auxin transport. Moreover, the *yuc* mutants provide a sensitized background for isolating genes important for auxin-mediated development, as demonstrated by the identification of *NPY1*. Further genetic analyses of additional *yuc1 yuc4* enhancers/suppressors will help filling in the gaps in understanding auxin action in plant growth and development.

Acknowledgements Research in my lab is supported by the NIH grant #R01GM68631 and the NSF Plant Genome grant DBI-0820729. The author thanks members of the lab for their comments.

References

Barlier I, Kowalczyk M, Marchant A, Ljung K, Bhalerao R, Bennett M, Sandberg G, Bellini C (2000) The *SUR2* gene of *Arabidopsis thaliana* encodes the cytochrome P450 CYP83B1, a modulator of auxin homeostasis. Proc Natl Acad Sci USA 97:14819–14824

Bartel B (1997) Auxin biosynthesis. Annu Rev Plant Physiol Plant Mol Biol 48:51–66

Boerjan W, Cervera MT, Delarue M, Beeckman T, Dewitte W, Bellini C, Caboche M, Van Onckelen H, Van Montagu M, Inze D (1995) *superroot*, a recessive mutation in Arabidopsis, confers auxin overproduction. Plant Cell 7:1405–1419

Cheng Y, Dai X, Zhao Y (2006) Auxin biosynthesis by the YUCCA flavin monooxygenases controls the formation of floral organs and vascular tissues in *Arabidopsis*. Genes Dev 20:1790–1799

Cheng Y, Dai X, Zhao Y (2007a) Auxin synthesized by the YUCCA flavin monooxygenases is essential for embryogenesis and leaf formation in *Arabidopsis*. Plant Cell 19:2430–2439

Cheng Y, Qin G, Dai X, Zhao Y (2007b) NPY1, a BTB-NPH3-like protein, plays a critical role in auxin-regulated organogenesis in *Arabidopsis*. Proc Natl Acad Sci USA 104:18825–18829

Cheng Y, Qin G, Dai X, Zhao Y (2008) *NPY* genes and AGC kinases define two key steps in auxin-mediated organogenesis in *Arabidopsis*. Proc Natl Acad Sci USA 105:21017–21022

Christensen SK, Dagenais N, Chory J, Weigel D (2000) Regulation of auxin response by the protein kinase PINOID. Cell 100:469–478

Comai L, Kosuge T (1982) Cloning characterization of *iaaM*, a virulence determinant of *Pseudomonas savastanoi*. J Bacteriol 149:40–46

Comai L, Kosuge T (1983) Transposable element that causes mutations in a plant pathogenic *Pseudomonas* sp. J Bacteriol 154:1162–1167

Delarue M, Prinsen E, Onckelen HV, Caboche M, Bellini C (1998) *Sur2* mutations of *Arabidopsis thaliana* define a new locus involved in the control of auxin homeostasis. Plant J 14:603–611

Expósito-Rodríguez M, Borges AA, Borges-Pérez AB, Hernández M, Pérez JA (2007) Cloning and biochemical characterization of *ToFZY*, a tomato gene encoding a flavin monooxygenase involved in a tryptophan-dependent auxin biosynthesis pathway. J Plant Growth Regul 26: 329–340

Fujino K, Matsuda Y, Ozawa K, Nishimura T, Koshiba T, Fraaije MW, Sekiguchi H (2008) *NARROW LEAF 7* controls leaf shape mediated by auxin in rice. Mol Genet Genomics 279:499–507

Gallavotti A, Barazesh S, Malcomber S, Hall D, Jackson D, Schmidt RJ, McSteen P (2008) *sparse inflorescence1* encodes a monocot-specific *YUCCA*-like gene required for vegetative and reproductive development in maize. Proc Natl Acad Sci USA 105:15196–15201

Galweiler L, Guan C, Muller A, Wisman E, Mendgen K, Yephremov A, Palme K (1998) Regulation of polar auxin transport by AtPIN1 in *Arabidopsis* vascular tissue. Science 282:2226–2230

Gray WM, Ostin A, Sandberg G, Romano CP, Estelle M (1998) High temperature promotes auxin-mediated hypocotyl elongation in *Arabidopsis*. Proc Natl Acad Sci USA 95:7197–7202

Harper RM, Stowe-Evans EL, Luesse DR, Muto H, Tatematsu K, Watahiki MK, Yamamoto K, Liscum E (2000) The *NPH4* locus encodes the auxin response factor ARF7, a conditional regulator of differential growth in aerial Arabidopsis tissue. Plant Cell 12:757–770

Last RL, Bissinger PH, Mahoney DJ, Radwanski ER, Fink GR (1991) Tryptophan mutants in *Arabidopsis*: the consequences of duplicated tryptophan synthase β genes. Plant Cell 3:345–358

Lincoln C, Britton JH, Estelle M (1990) Growth and development of the *axr1* mutants of *Arabidopsis*. Plant Cell 2:1071–1080

Liscum E, Briggs WR (1995) Mutations in the *NPH1* locus of Arabidopsis disrupt the perception of phototropic stimuli. Plant Cell 7:473–485

Mikkelsen MD, Naur P, Halkier BA (2004) Arabidopsis mutants in the C-S lyase of glucosinolate biosynthesis establish a critical role for indole-3-acetaldoxime in auxin homeostasis. Plant J 37:770–777

Motchoulski A, Liscum E (1999) *Arabidopsis* NPH3: A NPH1 photoreceptor-interacting protein essential for phototropism. Science 286:961–964

Pedmale UV, Liscum E (2007) Regulation of phototropic signaling in *Arabidopsis* via phosphorylation state changes in the phototropin 1-interacting protein NPH3. J Biol Chem 282: 19992–20001

Przemeck GK, Mattsson J, Hardtke CS, Sung ZR, Berleth T (1996) Studies on the role of the *Arabidopsis* gene *MONOPTEROS* in vascular development and plant cell axialization. Planta 200:229–237

Romano CP, Robson PR, Smith H, Estelle M, Klee H (1995) Transgene-mediated auxin overproduction in Arabidopsis: hypocotyl elongation phenotype and interactions with the hy6-1 hypocotyl elongation and *axr1* auxin-resistant mutants. Plant Mol Biol 27:1071–1083

Sohlberg JJ, Myrenas M, Kuusk S, Lagercrantz U, Kowalczyk M, Sandberg G, Sundberg E (2006) *STY1* regulates auxin homeostasis and affects apical-basal patterning of the Arabidopsis gynoecium. Plant J 47:112–123

Stone S L, Braybrook SA, Paula SL, Kwong LW, Meuser J, Pelletier J, Hsieh TF, Fischer RL, Goldberg RB, Harada JJ (2008) *Arabidopsis LEAFY COTYLEDON2* induces maturation traits and auxin activity: Implications for somatic embryogenesis. Proc Natl Acad Sci USA 105:3151–3156

Tobena-Santamaria R, Bliek M, Ljung K, Sandberg G, Mol JN, Souer E, Koes R (2002) FLOOZY of petunia is a flavin mono-oxygenase-like protein required for the specification of leaf and flower architecture. Genes Dev 16:753–763

Woo YM, Park HJ, Su'udi M, Yang JI, Park JJ, Back K, Park YM, An G (2007) Constitutively wilted 1, a member of the rice YUCCA gene family, is required for maintaining water homeostasis and an appropriate root to shoot ratio. Plant Mol Biol 65:125–136

Wright AD, Sampson MB, Neuffer MG, Michalczuk L, Slovin JP, Cohen JD (1991) Indole-3-acetic acid biosynthesis in the mutant maize *orange pericarp*, a tryptophan auxotroph. Science 254:998–1000

Yamamoto Y, Kamiya N, Morinaka Y, Matsuoka M, Sazuka T (2007) Auxin biosynthesis by the *YUCCA* genes in rice. Plant Physiol 143:1362–1371

Zhao Y, Christensen SK, Fankhauser C, Cashman JR, Cohen JD, Weigel D, Chory J (2001) A role for flavin monooxygenase-like enzymes in auxin biosynthesis. Science 291:306–309

Zhao Y, Hull AK, Gupta NR, Goss KA, Alonso J, Ecker JR, Normanly J, Chory J, Celenza JL (2002) Trp-dependent auxin biosynthesis in *Arabidopsis*: involvement of cytochrome P450s CYP79B2 and CYP79B3. Genes Dev 16:3100–3112

Chapter 13
Role of Cytokinin in the Regulation of Plant Development

T. Kiba and H. Sakakibara

13.1 Introduction

Cytokinins were discovered in a search for the identity of a cell division factor in plants. Kinetin, though it is not a naturally occurring compound, was the first substance defined as a cytokinin (Miller et al. 1955), while *trans*-zeatin (tZ; Fig. 13.1, structure A) was the first natural cytokinin isolated from living plants (Letham 1963). Although discovered as a cell division factor, cytokinins have been found to influence diverse aspects of plant growth and development when applied exogenously, including senescence, apical dominance, and nutrient signaling (Mok 1994; Mok and Mok 2001). Since the 1950s, considerable information has accumulated concerning cytokinin chemistry and physiology. However, cytokinins have been the least studied plant hormone with regard to the molecular mechanisms of biosynthesis and signaling. Only recently have genes and mutants involved in these pathways begun to be identified and characterized. Major breakthroughs, such as identification of the genes for plant adenosine phosphate-isopentenyltransferase (IPT) (Kakimoto 2001; Takei et al. 2001) and cytokinin receptor (Inoue et al. 2001; Yamada et al. 2001), have greatly increased our understanding of cytokinin function in plant growth and development. This chapter summarizes the current understanding of cytokinin biosynthesis and signaling pathways, and considers emerging knowledge on how these pathways influence plant development. For additional information, readers are directed to a number of reviews, book chapters, and historical overviews on cytokinin (Mok 1994; Mok and Mok 2001; Schmülling et al. 2003; Sakakibara 2004, 2006; Mizuno 2005; Müller and Sheen 2007; To and Kieber 2008; Frugier et al. 2008).

T. Kiba and H. Sakakibara
RIKEN Plant Science Center, 1-7-22 Suehiro, Tsurumi Yokohama 230-0045, Japan
e-mail: sakaki@riken.jp

E-C. Pua and M.R. Davey (eds.),
Plant Developmental Biology – Biotechnological Perspectives: Volume 2,
DOI 10.1007/978-3-642-04670-4_13, © Springer-Verlag Berlin Heidelberg 2010

13.2 Cytokinin Biosynthesis and Metabolism

13.2.1 Chemical Structure and Activity of Cytokinins

Naturally occurring cytokinins are adenine derivatives with either an N^6-isoprenoid or an N^6-aromatic side chain (Shaw 1994; Mok and Mok 2001). In addition, many diverse structures have been isolated from plants varying in side chain and/or in sugar and sugar phosphate conjugation (Shaw 1994). Currently, cytokinin bases, including tZ (Fig. 13.1a) and N^6-(Δ^2-isopentenyl)adenine (iP; Fig. 13.1b), are considered to be the bioactive forms based on their binding activity to plant cytokinin receptors (Yamada et al. 2001; Yonekura-Sakakibara et al. 2004; Romanov et al. 2006). Interestingly, the receptors are different in terms of ligand preferences (Yamada et al. 2001; Yonekura-Sakakibara et al. 2004; Romanov et al. 2006), suggesting that variation in cytokinin side-chain structure may be important for their physiological functions. This notion constitutes an interesting hypothesis that is yet to be tested.

According to the current model for cytokinin biosynthesis (Fig. 13.2), active cytokinin concentrations could be controlled by the following three steps: de novo synthesis, activation, and degradation. Recent identification of several genes involved in these steps sheds light on the elaborate mechanisms to control the amount, molecular species, and activity of cytokinins.

13.2.2 De Novo Synthesis

The initial step of cytokinin biosynthesis in higher plants is N-prenylation of adenosine $5'$-phosphates (AMP, ADP, or ATP) with dimethylallyl diphosphate (DMAPP) to form iP riboside $5'$-(mono, di, or tri) phosphate (iPRMP, iPRDP, or iPRTP). This reaction is catalyzed by IPT (Fig. 13.2). The first gene encoding *IPT* was found in *Agrobacterium tumefaciens* (Akiyoshi et al. 1984). Characterization of the DNA sequence for the *A. tumefaciens* gene has led to the identification of *IPT* genes in *Arabidopsis thaliana* (Kakimoto 2001; Takei et al. 2001). In this model plant, there are seven *IPT* genes (*AtIPT1* and *AtIPT3–AtIPT8*) involved in cytokinin

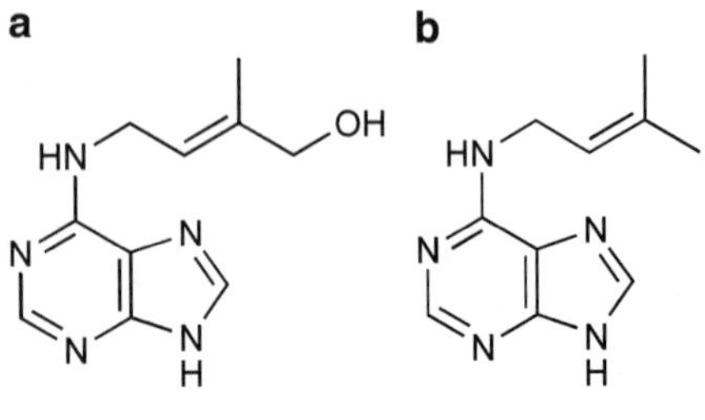

Fig. 13.1 Chemical structures of the bioactive cytokinins *trans*-zeatin (**a**), and N^6-(Δ^2-isopentenyl) adenine (**b**)

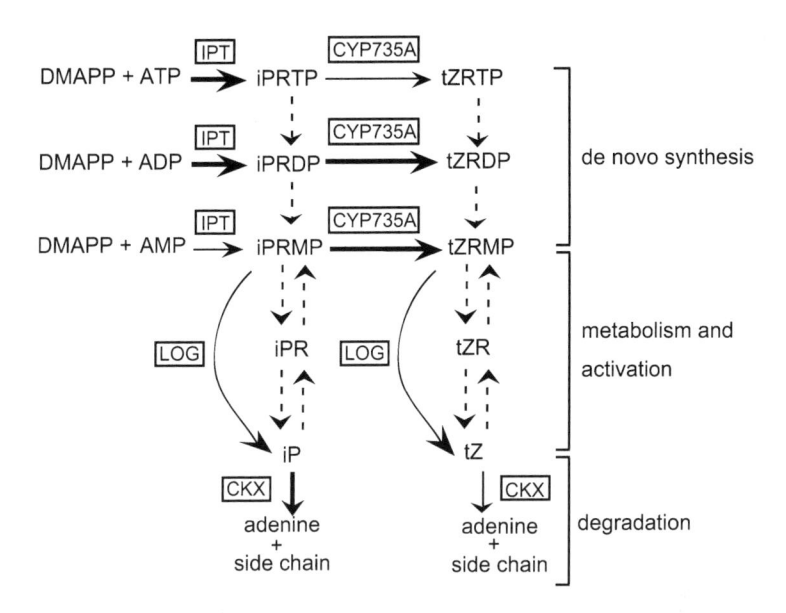

Fig. 13.2 A current scheme of major pathways of cytokinin de novo synthesis, metabolism, and degradation. The initial step of cytokinin biosynthesis is catalyzed by the adenosine phosphate-isopentenyltransferase (IPT) to form N^6-(Δ^2-isopentenyl)adenine riboside 5′-phosphates (iPRMP, iPRDP, or iPRTP). The iP nucleotides are converted to the corresponding *trans*-zeatin (tZ) nucleotides by the cytochrome P450 monooxygenase (CYP735A). LOG catalyzes iPRMP and tZRMP to yield bioactive cytokinin species, such as iP and tZ. Finally, the active cytokinins are degraded by cytokinin oxidase/dehydrogenase (CKX). *Solid arrows* Reactions catalyzed by known gene products, *dashed arrows* reactions with unidentified gene products. For further details, also refer to Sakakibara (2006) and Hirose et al. (2008)

biosynthesis (Kakimoto 2001; Takei et al. 2001). The higher-order *atipt* mutants display a severe reduction in cytokinin levels and growth retardation, demonstrating that *IPTs* are required for biosynthesis of cytokinin in plants (Miyawaki et al. 2006). Plant IPTs are distinct from bacterial IPTs in their substrate preferences, favoring ADP or ATP over AMP (Kakimoto 2001).

13.2.3 Activation

The iP nucleotides produced by IPT are metabolized to cytokinin bases, such as iP and tZ, to become bioactive. Recently, two key enzymes involved in such metabolic reactions were identified. They are the cytochrome P450 monooxygenase (CYP735A) in *Arabidopsis* (Fig. 13.2; Takei et al. 2004a) and cytokinin nucleoside 5′-monophosphate phosphoribohydrolase (LONELY GUY) in rice (Fig. 13.2; Kurakawa et al. 2007). CYP735A, encoded by the two genes *CYP735A1* and *CYP735A2* in *Arabidopsis*, catalyzes *trans*-hydroxylation of an iP nucleotide, but not of an iP nucleoside or iP base, to form a tZ nucleotide (Fig. 13.2). CYP735A utilizes iPRDP and iPRMP preferentially over iPRTP (Takei et al. 2004a). LONELY GUY (LOG) acts in a novel step of cytokinin activation that directly converts a cytokinin nucleotide to a cytokinin base. LOG exclusively hydrolyzes cytokinin nucleoside $5'$-monophosphates, such as iPRMP and tZRMP, but not the di- or triphosphates (Fig. 13.2). There are ten LOG family genes in rice with distinct patterns of expression (Kurakawa et al. 2007). The rice *log* mutants display severe developmental defects (Kurakawa et al. 2007), indicating that this activation step is of great relevance to plant development.

Characterizations of partially purified enzymes and feeding experiments have suggested the existence of stepwise activation pathways catalyzed by nucleotidases and nucleosidases (Chen 1997). However, genes encoding these enzymes, as well as the physiological significance of these pathways remain to be determined.

13.2.4 Degradation

One of the major pathways for cytokinin inactivation is oxidative degradation (Armstrong 1994), and the best-characterized component in this pathway is cytokinin oxidase/dehydrogenase (CKX). CKX cleaves an isoprenoid side chain from isoprenoid cytokinin bases and nucleosides, thus irreversibly inactivating cytokinins (Fig. 13.2). Biochemical characterizations of recombinant CKXs expressed in bacteria demonstrated that these enzymes favor iP and iP riboside over tZ and tZ riboside as substrates (Werner et al. 2006 and references therein). The Arabidopsis and rice genomes contain seven and 11 *CKX* genes (*AtCKX1–AtCKX7*, *OsCKX1–OSCKX11*), respectively, which exhibit overlapping yet distinct patterns of expression (Ashikari et al. 2005; Werner et al. 2006).

13.3 Cytokinin Signaling

During the past decade, multistep histidine-to-aspartate (His-to-Asp) phosphorelay has emerged as a mechanism plants employ to perceive and transmit the cytokinin signal. Amongst the potential His-Asp phosphorelay components in *Arabidopsis* implicated in cytokinin signaling are three histidine kinases (HKs), six histidine-containing phosphotransfer factors (HPs), and 23 response regulators (RRs). The Arabidopsis *RR* (*ARR*) gene family members are further classified into three subtypes based on the structural designs and functions of encoded proteins, specifically, ten type-A *RR*s, 11 type-B *RR*s, and two type-C *RR*s (Mizuno 2005; Müller and Sheen 2007; To and Kieber 2008).

The current model of cytokinin signal transduction by His-Asp phosphorelay involves four major steps (Fig. 13.3), namely, (1) cytokinin sensing and phosphorylation of HPs by HKs, (2) nuclear translocation of phosphorylated HPs, followed

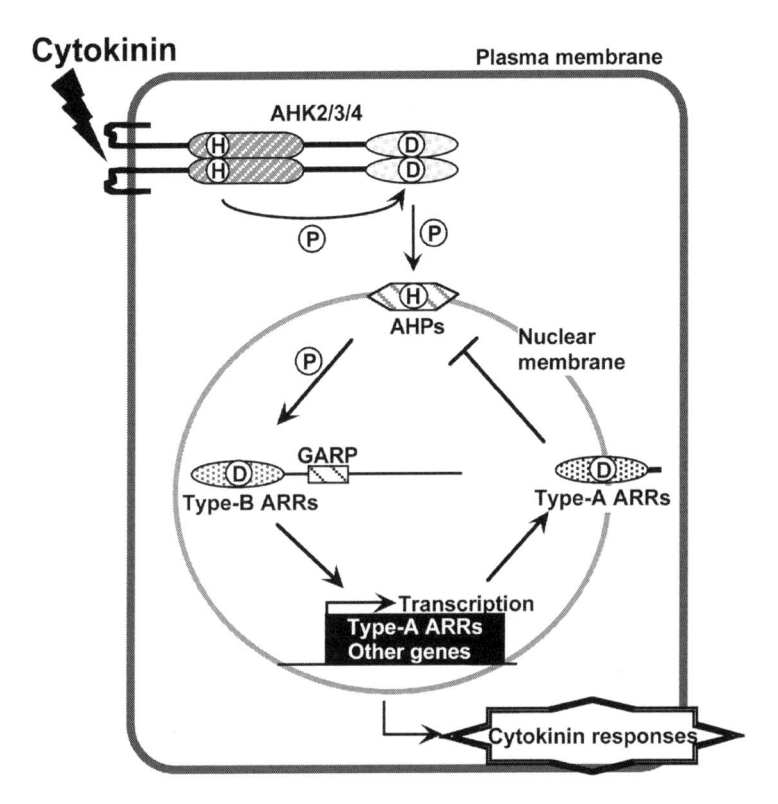

Fig. 13.3 The current framework of cytokinin perception and signal transduction by His-Asp phosphorelay in *Arabidopsis*. There are four steps mediated by four components, cytokinin-sensor histidine kinases (HKs), histidine-containing phosphotransfer factors (HPs), type-B response regulators (RRs), and type-A RRs: (1) Arabidopsis HKs (AHK2/3/4) sense cytokinins and phosphorylate Arabidopsis HPs (AHPs), (2) the phosphorylated AHPs move into the nucleus and donate a phosphoryl group to type-B Arabidopsis RRs (ARR), (3) the phosphorylated type-B ARRs activate transcription of target genes, including type-A *ARR* genes, (4) the type-A ARRs act as negative regulators to form a feedback loop. *Arrows with "P"* indicate phosphorelays between histidine (H)- and aspartate (D)-residues; the other arrows and the "dead end" line indicate positive and negative interactions, respectively. Also refer to Mizuno (2005) and To and Kieber (2008) for more information

by donation of a phosphoryl group to type-B RRs, (3) transcriptional activation of target genes, including type-A *RR* genes, by phosphorylated type-B RRs, and (4) negative regulation by type-A RRs to form a feedback loop (Mizuno 2005; Müller and Sheen 2007; To and Kieber 2008).

13.3.1   HKs Act as Cytokinin Sensors

The very first cytokinin receptor, CRE1/AHK4/WOL (AHK4), was identified in a genetic screening for cytokinin response mutants in a tissue culture assay (Inoue et al. 2001). In *Arabidopsis*, there are two AHK4 homologs (AHK2 and AHK3) that also serve as cytokinin receptors (Higuchi et al. 2004; Nishimura et al. 2004). These receptors share three domains, including a transmembrane domain, a HK domain, and a receiver domain. Several studies demonstrated that AHKs directly bind cytokinins (Yamada et al. 2001; Romanov et al. 2006) and exhibit cytokinin-dependent HK activity on HPs (Hwang and Sheen 2001; Inoue et al. 2001; Yamada et al. 2001). An *ank2,3,4* triple mutant is almost insensitive to cytokinin and displays severe growth and developmental defects (Higuchi et al. 2004; Nishimura et al. 2004). Together, these results suggest that AHKs act as cytokinin-sensing HKs and play a central role in cytokinin perception in plants (Fig. 13.3). Interestingly, AHKs show partially overlapping yet distinct patterns of expression and binding affinities for cytokinin species (Yamada et al. 2001; Higuchi et al. 2004; Yonekura-Sakakibara et al. 2004; Romanov et al. 2006). Furthermore, AHK4, but not AHK2 or AHK3, mediates bidirectional phosphorelay (Mähönen et al. 2006a), suggesting that each AHK may possess a different physiological role.

13.3.2 HPs Mediate the Cytokinin Signal

HPs have been implicated in cytokinin signaling by the findings that (1) HPs interact with both cytokinin receptors (HKs) and RRs, and function as phosphorelay intermediates between them (Urao et al. 2000; Dortay et al. 2006), (2) HPs move into the nucleus after cytokinin treatment (Hwang and Sheen 2001; Yamada et al. 2004), and (3) the *ahp1,2,3,4,5* quintuple mutant shows morphological and cytokinin sensitivity defects reminiscent of an *ank2,3,4* triple mutant (Hutchison et al. 2006). Thus, current models propose that HPs act to shuttle cytokinin signals between the cytoplasm and nucleus, most possibly between HKs and RRs (Fig. 13.3).

Recently, a pseudo-HP (AHP6) that lacks a conserved His-residue necessary for phosphorelay was identified and proposed to impede cytokinin signaling by interfering with the phosphorelay between HKs and RRs (Mähönen et al. 2006b; see Sect. 13.4.3).

13.3.3 Type-B RRs Are Transcription Factors that Positively Regulate Cytokinin Responses

In addition to an N-terminal receiver domain, type-B RR proteins contain long C-terminal extensions, in which a transcription activation domain and a DNA-binding domain (GARP-domain) reside. Various studies in vitro, in heterologous systems, and in planta demonstrated that type-B RRs bind to DNA in a sequence-specific manner with a core sequence (A/G)GAT(T/C), and act as transcriptional activators (Sakai et al. 2001; Hosoda et al. 2002). Overexpression of type-B *RRs* generally results in cytokinin hypersensitivity or a constitutive-response phenotype (Sakai et al. 2001; Imamura et al. 2003). In contrast, the knockout mutants of some type-B *RRs* show reduced sensitivity to cytokinins (Sakai et al. 2001; Argyros et al. 2008), indicating that type-B RRs positively regulate cytokinin signaling. Furthermore, the phenotype of the *arr1,10,12* triple mutant is quite analogous to an *ahk2,3,4* triple mutant (Ishida et al. 2008). These findings suggest that type-B RRs are required to output the cytokinin signal perceived by HKs (Fig. 13.3).

13.3.4 Type-A RRs Act as Negative Regulators of Cytokinin Signaling

Originally identified as cytokinin primary response genes in maize (Sakakibara et al. 1998) and *Arabidopsis* (Brandstatter and Kieber 1998; Taniguchi et al. 1998), type-A RRs are now implicated in cytokinin signaling. Type-A RRs consist almost solely of a receiver domain, and do not contain any apparent output domain. Overexpression of type-A RRs generally results in cytokinin hyposensitivity (Kiba et al. 2003; Lee et al. 2007), whereas higher-order mutants display an increased sensitivity to cytokinin (To et al. 2004), indicating that type-A RRs are negative regulators of cytokinin signaling. In addition to these findings, the following observations led to a model in which type-A RRs constitute a negative feedback loop of the cytokinin signaling pathway (Fig. 13.3). Transcription of type-A *RRs* is rapidly induced by type-B RRs in response to cytokinin (Hwang and Sheen 2001; Taniguchi et al. 2007). Type-A RRs can interact and phosphorelay with HPs (Imamura et al. 1998; Urao et al. 2000; Dortay et al. 2006). Type-A RRs possess inherent phosphatase activity (Imamura et al. 1998). A conserved Asp-residue in type-A RRs is necessary for the phosphorelay, phosphatase activity, and negative regulator function (Imamura et al. 1998; To et al. 2007; Lee et al. 2008).

13.3.5 Downstream Targets of His-Asp Phosphorelay

Considerable efforts have been made to determine the target genes for type-B RRs. Genome-wide expression profiling analyses utilizing mutants and transgenic plants overexpressing engineered proteins, such as the constitutive active ARR1ΔDDT (Taniguchi et al. 2007) and dominant negative ARR1-SRDX (Heyl et al. 2008), have identified many target genes including the type-A *ARRs* and *CYTOKININ RESPONSE FACTORs* (*CRFs*; Brenner et al. 2005; Kiba et al. 2005; Rashotte et al. 2006; Taniguchi et al. 2007; Heyl et al. 2008). Among them, only type-A *ARRs* and *CRFs* have so far been characterized in detail (Rashotte et al. 2006).

The CRFs are plant-specific transcription factors belonging to the APETALA2 superfamily. Cytokinin not only regulates the transcription of *CRFs*, but also control nuclear translocation of CRF proteins. Furthermore, expression profiling revealed that CRFs share a large fraction of target genes with type-B ARRs, suggesting that CRFs function in tandem with type-B ARRs to regulate cytokinin responses (Rashotte et al. 2006).

13.4 Molecular Mechanisms of Cytokinin Action in Plant Development

13.4.1 Maintenance of Vegetative Shoot Apical Meristems

Plant meristems are made up of pluripotent stem cells the proliferation of which generates all post-embryonically formed organs. The functional maintenance of the shoot apical meristem (SAM) is achieved through a balance between the production of organ primordia formed at the periphery, and the growth of indeterminate cells at the center (Fletcher 2002). At the molecular level, SAM function is maintained by several groups of regulators, including two classes of homeodomain transcription factors: Class-1 KNOTTED1-like homeobox (KNOX) proteins that are expressed in the indeterminate cells, and a WUSCHEL (WUS) protein that is expressed in a defined group of cells (referred to as the organizing center) below the indeterminate cells (Fletcher 2002).

Cytokinins have been implicated in SAM function, based on observations that exogenous application of cytokinin or overexpression of the bacterial *IPT* gene enhance cell division, SAM activity, and formation of meristem-like tissues (Mok 1994; Rupp et al. 1999). However, compelling evidence for these observations has been provided only recently by analyzing mutants and transgenic plants impaired in cytokinin homeostasis and signaling. Mutants and transgenic plants with diminished cytokinin concentrations, such as in higher-order *atipt* mutants (Miyawaki et al. 2006) and transgenic plants overexpressing *CKX* (Werner et al. 2003), or decreased cytokinin sensitivity such as in higher-order *ahk* (Higuchi et al. 2004), *ahp* (Hutchison et al. 2006), and type-B *arr* mutants (Ishida et al. 2008), display reduced SAM activity and size, consolidating the view that cytokinin is required for proper SAM function.

Recently, molecular links between SAM function and cytokinins have emerged. KNOX proteins were found to activate a subset of *IPT* genes in *Arabidopsis* and in rice (Fig. 13.4a; Jasinski et al. 2005; Yanai et al. 2005; Sakamoto et al. 2006). Overexpression of *KNOX* genes, including SHOOT MERISTEMLESS (STM), results in a rapid increase in mRNA levels of a subset of *IPT* genes, and in a concomitant accumulation of cytokinins (Yanai et al. 2005; Sakamoto et al. 2006). Either application of cytokinin or expression of bacterial *IPT* in the SAM partially

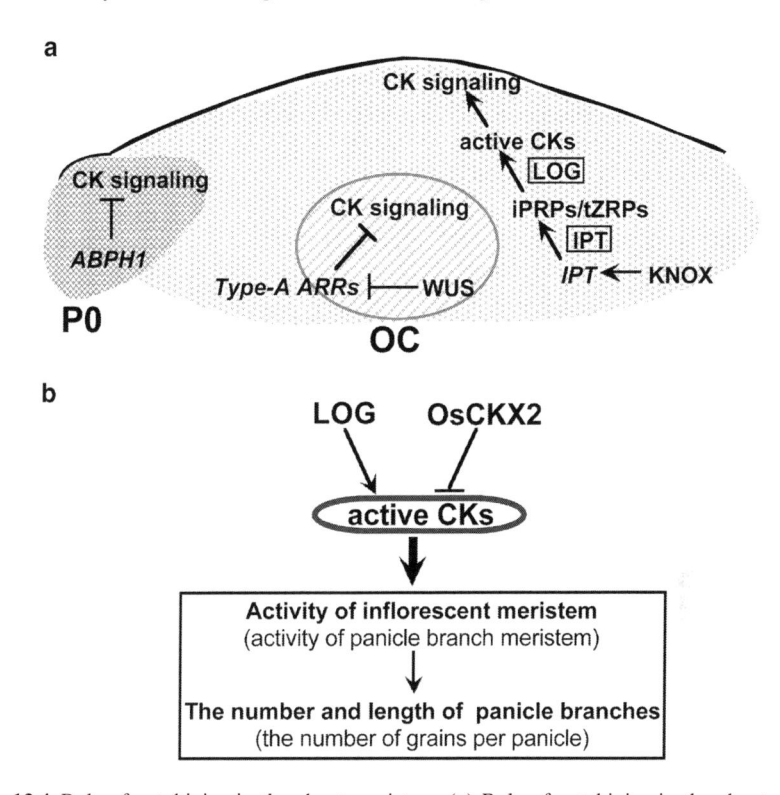

Fig. 13.4 Role of cytokinins in the shoot meristem. (**a**) Role of cytokinins in the shoot apical meristem (SAM). The dome-shaped structure of the SAM is drawn schematically. The class-1 KNOTTED1-like homeobox (KNOX) proteins, which are expressed throughout the SAM, induce de novo synthesis of cytokinins (CK) by activating isopentenyltransferase (*IPT*) gene expression. The resultant cytokinins (iPRPs and tZRPs) are converted to bioactive cytokinins by LONELY GUY (LOG). WUSCHEL (WUS), which is specifically expressed in the organizing center (OC), represses the transcription of type-A *ARR* genes, thus up-regulating cytokinin signal transduction in OC. At the site of an incipient primordium (P0), *ABPH1* (a maize type-A RR) restricts the growth of the meristem by inhibiting cytokinin signalling. (**b**) Role of cytokinin in the inflorescence meristem of rice. In the inflorescence meristem, LOG and cytokinin oxidase/dehydrogenase (OsCKX2) act antagonistically to regulate active cytokinin levels. An accumulation of active cytokinins in the inflorescence meristem enhances the activity of the meristem (and/or panicle branch meristem), eventually resulting in an increase in the number and length of panicle branches. *Arrows and "dead end" lines* Positive and inhibitory interactions, respectively

rescues the *stm* mutant (Yanai et al. 2005), whereas a mutation in AHK4 (*wol*) enhances a weak *stm* allele (Jasinski et al. 2005), indicating that cytokinin is required to mediate KNOX function in the SAM. Consistently, a loss-of-function mutation in the rice *LOG* gene, which is expressed specifically in shoot meristem tips, causes a reduction in SAM size (Kurakawa et al. 2007), thereby corroborating the requirement for cytokinin biosynthesis in the SAM. Together, these results demonstrate that regulation of local cytokinin biosynthesis plays an integral part in SAM maintenance (Fig. 13.4a).

Another homeobox protein, WUS, has been shown to repress the transcription of type-A *RR* genes in *Arabidopsis* (Leibfried et al. 2005). An increase in *WUS* expression, either by an inducible *WUS* transgene or by a *clavata3* (*clv3*) mutation, results in the complete disappearance of *ARR7* (type-A *ARR*) mRNA from the SAM. Conversely, suppression of *WUS* by *CLV3* induction causes an expansion of the area in which *ARR7* is expressed. Binding assays in vivo and in vitro demonstrated that WUS binds directly to the *ARR7* promoter sequence. These results indicate that the role of WUS in SAM maintenance involves direct repression of type-A *ARR* genes, thus up-regulating cytokinin signal transduction in the organizing center (Fig. 13.4a; Leibfried et al. 2005). Interestingly, a maize type-A *RR* gene was identified as the causative gene of *abphyll* (*abph1*) mutants that show an enlarged SAM and an altered phyllotaxy (Giulini et al. 2004). *ABPH1* is expressed in a small region of the SAM at the site of the incipient primordium. Thus, it is proposed that *ABPH1* restricts the growth of indeterminate cells by down-regulating cytokinin signal transduction at the site of the incipient primordium (Fig. 13.4a). Taken together, these results suggest that the regulation of local cytokinin signaling is also required for correct SAM function.

Recently, Müller and Sheen (2008) reported that auxin suppresses cytokinin signaling by activating the transcription of a subset of type-A *ARRs* in root stem-cell specification during early embryogenesis. Controlling type-A RRs may represent an efficient way to fine-tune cytokinin signaling throughout plant development.

13.4.2   Inflorescence Meristem Activity

Upon the transition from vegetative to reproductive growth, a vegetative SAM is converted to a reproductive SAM that produces inflorescences instead of leaves (Wang and Li 2008). During the conversion in *Arabidopsis*, a vegetative SAM first adopts inflorescence meristem identity, and then the inflorescence meristem produces floral meristems. In rice, the inflorescence meristem generates branch and spikelet meristems before producing floral meristems (Wang and Li 2008). The activity and size of these meristems are the major determinants of the inflorescence patterning, and ultimately of grain yield (Takeda and Matsuoka 2008). To avoid confusion, these meristems will collectively be referred to as inflorescence meristems in this section.

Several lines of recent evidence have shown that cytokinins are closely relevant to inflorescence meristem activity. A quantitative trait locus *Grain number1a* (*Gn1a*) of rice that controls inflorescence branching pattern and grain number has been cloned and shown to encode a cytokinin oxidase/dehydrogenase, *OsCKX2* (Ashikari et al. 2005). *OsCKX2* is expressed in inflorescence meristems, and reduced expression of *OsCKX2* results in cytokinin accumulation in the meristem (Ashikari et al. 2005). Transgenic plants harboring two copies of *OsCKX2* generate an inflorescence with fewer grains, whereas transgenic plants expressing reduced levels of *OsCKX2* develop an inflorescence with more grains (Ashikari et al. 2005), indicating that the *Gn1a* trait is controlled by cytokinin. Consistently, the rice *log* mutants produce a small inflorescence with fewer branches and grains (Kurakawa et al. 2007), suggesting that cytokinin concentrations in the inflorescence meristem are regulated by the antagonistic action of LOG and OsCKX2 (Fig. 13.4b). In addition, a small inflorescence phenotype was also observed in transgenic rice plants that overexpress a type-A *RR* gene, *OsRR6* (Hirose et al. 2007). Taken together, these results demonstrate that cytokinin positively regulates inflorescence meristem activity, and also highlight cytokinin as one of the major determinants of grain yield in rice.

13.4.3   Root Meristem Maintenance and Root Vascular Development

Cytokinin has been shown to play opposite roles in regulating root and shoot growth. Although the inhibitory effect of cytokinin on root growth was first described as early as 1957 (Skoog and Miller 1957), only recently have mechanisms of cytokinin action in the root meristem been elucidated. Applying exogenous cytokinins or increasing endogenous cytokinin concentrations results in a decrease in root proximal meristem size and growth; mutants and transgenic plants with low cytokinin concentrations or sensitivity generally display enhanced proximal meristem size and growth (Dello Ioio et al. 2007). Altering cytokinin concentrations at the transition zone (TZ), where root proximal meristem cells initiate differentiation, is sufficient to affect the meristem, indicating that cytokinin controls cell differentiation at the TZ. Furthermore, the analysis of cytokinin signaling mutants revealed that AHK3, ARR1, and ARR12 mediate this effect (Dello Ioio et al. 2007). Together, these observations suggest that cytokinin promotes cell differentiation in the root proximal meristem, thereby reducing proximal meristem size and growth. However, the higher-order cytokinin signaling mutants *ahk2, 3, 4* (Higuchi et al. 2004; Nishimura et al. 2004), *ahp1, 2, 3, 4, 5* (Hutchison et al. 2006), and *arr1, 10, 12* (Ishida et al. 2008) fail to maintain root meristem, indicating that a basal level of cytokinin signaling is necessary for root meristem maintenance.

In addition to defects in root meristem maintenance, these higher-order cytokinin signaling mutants develop a primary root with aberrant vasculature consisting exclusively of protoxylem (Higuchi et al. 2004; Nishimura et al. 2004; Hutchison et al. 2006; Mähönen et al. 2006a; Ishida et al. 2008). Similarly, ectopic expression of *CKX* in vascular cells results in a phenocopy of these mutants (Mähönen et al. 2006a, b), indicating that cytokinin is required for correct vascular development. Two recent studies substantiated the requirement for cytokinin in vascular development, and further provided a mechanistic explanation for the exclusive xylem phenotype (Fig. 13.5a; Mähönen et al. 2006a, b). Mähönen et al. (2006b) in *Arabidopsis* performed a suppressor screen for the determinate root growth phenotype of the *wooden leg* (*wol*) mutant, which is caused by a missense mutation in *AHK4*, and identified a mutation at the *AHP6* locus. A mutation in *AHP6* results in a reduction of protoxylem cell files in both the *wol* mutant and wild-type backgrounds, indicating that AHP6 functions to promote protoxylem differentiation. Consistently, *AHP6* is expressed in developing protoxylem and adjoining pericycle cell files. Biochemical assays in vitro demonstrated that AHP6 has the potential to inhibit the phosphorelay mediating cytokinin signal. The *ahp6* mutant plants display an increased sensitivity to cytokinin inhibition of protoxylem differentiation, confirming that AHP6 functions as a negative regulator of cytokinin signaling. Collectively, these data suggest that the function of cytokinins in vascular development is to inhibit protoxylem differentiation, and AHP6 specifies the spatial domain of protoxylem differentiation by interfering with cytokinin signaling (Fig. 13.5a). Hence, cytokinins are required for correct vascular development.

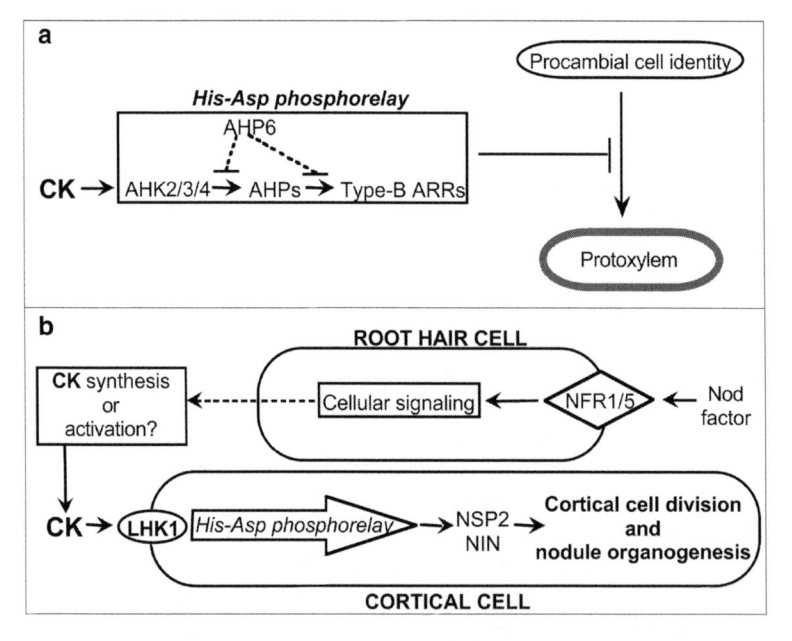

Fig. 13.5 Role of cytokinins in the regulation of **a** root vascular development, and **b** nodule organogenesis. (**a**) In root vascular development, cytokinin (CK) signals mediated by the His-Asp phosphorelay inhibit procambial cells to differentiate into protoxylem. AHP6 impedes cytokinin signaling by interfering with phosphorelay, most probably at multiple steps, thereby specifying the spatial domain of protoxylem differentiation. (**b**) Nod factor is perceived by receptors (NFR1 and NFR5) at the plasma membrane of the root hair cell, provoking cellular signaling in the root hair cell and presumably an accumulation of cytokinin. The Lotus His-kinase 1 (LHK1) perceives cytokinins and initiates the phosphorelay in the cortical cell. At least two factors, the NODULE INCEPTION (NIN) and NODULATION SIGNALING PATHWAY 2 (NSP2), act downstream of LHK1 to induce cortical cell division and subsequently nodule organogenesis. *Arrows and "dead end" lines* Positive and inhibitory interactions, respectively; *solid lines* defined interactions, *dashed lines* presumed interactions

13.4.4   Nodule Organogenesis

Legumes and rhizobia interact symbiotically to form nodules. Nodule formation is initiated by the binding of bacterial nodulation (Nod) factors to plant kinase-like receptors (NFR1 and NFR5). The perception of Nod-factor provokes coordinated responses in root hair cells and cortical cells, including root hair curling and thread initiation in root hair cells, and dedifferentiation and activation of cell division in cortical cells that give rise to nodule primordia. The release of bacteria, and subsequent pattern formation and cell differentiation lead to the formation of mature nodules (Oldroyd and Downie 2008).

Many investigations have shown a positive correlation between cytokinin and nodule formation. Overexpression of the *CKX* gene (Lohar et al. 2004), or suppression of a cytokinin receptor gene expression (Gonzalez-Rizzo et al. 2006), results in reduced nodule numbers. Conversely, exogenous cytokinin application induces responses similar to Nod factors (or rhizobia), including cortical cell division, amyloplast deposition, and expression of early nodulin gene, although cytokinin treatment does not lead to the formation of complete nodules (Lorteau et al. 2001). Despite these evidences, data have remained elusive whether cytokinins are essential for nodule formation. Recently, this question was elucidated by complementary analyses of loss-of-function and gain-of-function mutations in the cytokinin receptor *Lotus histidine kinase 1* (*LHK1*) locus (Murray et al. 2007; Tirichine et al. 2007). A loss-of-function mutant of *LHK1* (*hit1*) fails to induce a subset of nodulin genes and cortical cell division, as a result producing only a few nodules (Murray et al. 2007). A gain-of-function mutation in *LHK1* (*snf2*) leads to the formation of spontaneous nodules in the absence of rhizobia (Tirichine et al. 2007). In heterologous assay systems, LHK1 functions as a cytokinin sensor. These assays also showed that the *hit1* mutation disrupts LHK1 function, and the *snf2* mutation locks LHK1 in a constitutively active status (Murray et al. 2007; Tirichine et al. 2007). Consistently, *hit1* mutants exhibit cytokinin hyposensitivity (Murray et al. 2007), whereas *snf2* mutants display a constitutive cytokinin response and increased sensitivity to cytokinins (Tirichine et al. 2007). These experiments show unequivocally that cytokinin signaling is necessary and sufficient for nodule organogenesis (Fig. 13.5b).

Analyses of the genetic interaction between *LHK1* and major nodulation-related genes placed *LHK1* downstream of the Nod factor receptor genes (*NFR1* and *NFR5*; Radutoiu et al. 2003) and a cellular signaling component *CCaMK* gene (Tirichine et al. 2006), both of which act in the root hair cell (Fig. 13.5b; Tirichine et al. 2007). However, how Nod factor perception and signaling at the root hair cell brings about LHK1 activation at the cortical cell is still an open question. One attractive hypothesis is that the de novo synthesis and/or activation of cytokinins occur locally in response to Nod factor perception. This hypothesis remains to be investigated in the future.

13.4.5   Other Developmental Events

Recent studies have provided invaluable insights into the underlying mechanisms of cytokinin function in leaf senescence (Kim et al. 2006), apical dominance (Tanaka et al. 2006), and nutrient signaling (Takei et al. 2004b). Analysis of a gain-of-function mutant of the *AHK3* locus revealed that cytokinin signaling via AHK3 and ARR2 is required for cytokinin inhibition of leaf senescence (Kim et al. 2006). Tanaka et al. (2006) reported that auxin causes a reduction in local cytokinin concentrations in *Pisum sativum* by repressing a subset of *IPT* genes (*PsIPT1* and *PsIPT2*) in the nodal stem, thereby inhibiting outgrowth of axillary buds. Characterization of *AtIPT3* demonstrated that *AtIPT3* is a key determinant of cytokinin biosynthesis in response to nitrate availability (Takei et al. 2004b). Although these are compelling indications that cytokinins are involved in these developmental events, further research is required to determine the precise mode of cytokinin function. Several other reviews provide a more detailed summary on these topics (Sakakibara et al. 2006; Ongaro and Leyser 2008; Hirose et al. 2008).

13.5   Perspectives

Thanks to the rapid accumulation of knowledge and materials over the past decade on cytokinin metabolism and signaling, tremendous progress has been achieved in understanding the molecular mechanisms that underlie cytokinin functions in plants. Nevertheless, several questions remain to be addressed in the future. What is the physiological significance of variants in cytokinin structure—for example, side-chain variations of isoprenoid cytokinins? Since there is great potential for redundancy in the components of cytokinin metabolism and signaling, how is specificity achieved, and what are the physiological and developmental consequences of such specificity? How do cytokinins bring about distinct developmental changes in different cell types, tissues, and organs? Now that mutants and genomic information useful to address these questions are available, it will not take long until the answers to these intriguing questions are known.

References

Akiyoshi DE, Klee H, Amasino RM, Nester EW, Gordon MP (1984) T-DNA of *Agrobacterium-tumefaciens* encodes an enzyme of cytokinin biosynthesis. Proc Natl Acad Sci USA 81: 5994–5998

Argyros RD, Mathews DE, Chiang YH, Palmer CM, Thibault DM, Etheridge N, Argyros DA, Mason MG, Kieber JJ, Schaller GE (2008) Type B response regulators of *Arabidopsis* play key roles in cytokinin signaling and plant development. Plant Cell 20:2102–2116

Armstrong DJ (1994) Cytokinin oxidase and the regulation of cytokinin degradation. In: Mok DWS, Mok MC (eds) Cytokinins. Chemistry, activity, and function, CRC Press, Boca Raton, FL, pp 139–154

Ashikari M, Sakakibara H, Lin S, Yamamoto T, Takashi T, Nishimura A, Angeles ER, Qian Q, Kitano H, Matsuoka M (2005) Cytokinin oxidase regulates rice grain production. Science 309:741–745

Brandstatter I, Kieber JJ (1998) Two genes with similarity to bacterial response regulators are rapidly and specifically induced by cytokinin in *Arabidopsis*. Plant Cell 10:1009–1019

Brenner WG, Romanov GA, Kollmer I, Burkle L, Schmülling T (2005) Immediate-early and delayed cytokinin response genes of *Arabidopsis thaliana* identified by genome-wide expression profiling reveal novel cytokinin-sensitive processes and suggest cytokinin action through transcriptional cascades. Plant J 44:314–333

Chen C-M (1997) Cytokinin biosynthesis and interconversion. Physiol Plant 101:665–673

Dello Ioio R, Linhares FS, Scacchi E, Casamitjana-Martinez E, Heidstra R, Costantino P, Sabatini S (2007) Cytokinins determine Arabidopsis root-meristem size by controlling cell differentiation. Curr Biol 17:678–682

Dortay H, Mehnert N, Burkle L, Schmülling T, Heyl A (2006) Analysis of protein interactions within the cytokinin-signaling pathway of *Arabidopsis thaliana*. FEBS J 273:4631–4644

Fletcher JC (2002) Shoot and floral meristem maintenance in *Arabidopsis*. Annu Rev Plant Biol 53:45–66

Frugier F, Kosuta S, Murray JD, Crespi M, Szczyglowski K (2008) Cytokinin: secret agent of symbiosis. Trends Plant Sci 13:115–120

Giulini A, Wang J, Jackson D (2004) Control of phyllotaxy by the cytokinin-inducible response regulator homologue *ABPHYL1*. Nature 430:1031–1034

Gonzalez-Rizzo S, Crespi M, Frugier F (2006) The *Medicago truncatula* CRE1 cytokinin receptor regulates lateral root development and early symbiotic interaction with *Sinorhizobium meliloti*. Plant Cell 18:2680–2693

Heyl A, Ramireddy E, Brenner WG, Riefler M, Allemeersch J, Schmülling T (2008) The transcriptional repressor ARR1-SRDX suppresses pleiotropic cytokinin activities in *Arabidopsis*. Plant Physiol 147:1380–1395

Higuchi M, Pischke MS, Mähönen AP, Miyawaki K, Hashimoto Y, Seki M, Kobayashi M, Shinozaki K, Kato T, Tabata S, Helariutta Y, Sussman MR, Kakimoto T (2004) In planta functions of the Arabidopsis cytokinin receptor family. Proc Natl Acad Sci USA 101: 8821–8826

Hirose N, Makita N, Kojima M, Kamada-Nobusada T, Sakakibara H (2007) Overexpression of a type-A response regulator alters rice morphology and cytokinin metabolism. Plant Cell Physiol 48:523–539

Hirose N, Takei K, Kuroha T, Kamada-Nobusada T, Hayashi H, Sakakibara H (2008) Regulation of cytokinin biosynthesis, compartmentalization and translocation. J Exp Bot 59:75–83

Hosoda K, Imamura A, Katoh E, Hatta T, Tachiki M, Yamada H, Mizuno T, Yamazaki T (2002) Molecular structure of the GARP family of plant Myb-related DNA binding motifs of the Arabidopsis response regulators. Plant Cell 14:2015–2029

Hutchison CE, Li J, Argueso C, Gonzalez M, Lee E, Lewis MW, Maxwell BB, Perdue TD, Schaller GE, Alonso JM, Ecker JR, Kieber JJ (2006) The Arabidopsis histidine phosphotransfer proteins are redundant positive regulators of cytokinin signaling. Plant Cell 18:3073–3087

Hwang I, Sheen J (2001) Two-component circuitry in *Arabidopsis* cytokinin signal transduction. Nature 413:383–389

Imamura A, Hanaki N, Umeda H, Nakamura A, Suzuki T, Ueguchi C, Mizuno T (1998) Response regulators implicated in His-to-Asp phosphotransfer signaling in *Arabidopsis*. Proc Natl Acad Sci USA 95:2691–2696

Imamura A, Kiba T, Tajima Y, Yamashino T, Mizuno T (2003) *In vivo* and *in vitro* characterization of the ARR11 response regulator implicated in the His-to-Asp phosphorelay signal transduction in *Arabidopsis thaliana*. Plant Cell Physiol 44:122–131

Inoue T, Higuchi M, Hashimoto Y, Seki M, Kobayashi M, Kato T, Tabata S, Shinozaki K, Kakimoto T (2001) Identification of CRE1 as a cytokinin receptor from *Arabidopsis*. Nature 409:1060–1063

Ishida K, Yamashino T, Yokoyama A, Mizuno T (2008) Three type-B response regulators, ARR1, ARR10 and ARR12, play essential but redundant roles in cytokinin signal transduction throughout the life cycle of *Arabidopsis thaliana*. Plant Cell Physiol 49:47–57

Jasinski S, Piazza P, Craft J, Hay A, Woolley L, Rieu I, Phillips A, Hedden P, Tsiantis M (2005) KNOX action in *Arabidopsis* is mediated by coordinate regulation of cytokinin and gibberellin activities. Curr Biol 15:1560–1565

Kakimoto T (2001) Identification of plant cytokinin biosynthetic enzymes as dimethylallyl diphosphate: ATP/ADP isopentenyltransferases. Plant Cell Physiol 42:677–685

Kiba T, Yamada H, Sato S, Kato T, Tabata S, Yamashino T, Mizuno T (2003) The type-A response regulator, ARR15, acts as a negative regulator in the cytokinin-mediated signal transduction in *Arabidopsis thaliana*. Plant Cell Physiol 44:868–874

Kiba T, Naitou T, Koizumi N, Yamashino T, Sakakibara H, Mizuno T (2005) Combinatorial microarray analysis revealing Arabidopsis genes implicated in cytokinin responses through the His->Asp phosphorelay circuitry. Plant Cell Physiol 46:339–355

Kim HJ, Ryu H, Hong SH, Woo HR, Lim PO, Lee IC, Sheen J, Nam HG, Hwang I (2006) Cytokinin-mediated control of leaf longevity by AHK3 through phosphorylation of ARR2 in *Arabidopsis*. Proc Natl Acad Sci USA 103:814–819

Kurakawa T, Ueda N, Maekawa M, Kobayashi K, Kojima M, Nagato Y, Sakakibara H, Kyozuka J (2007) Direct control of shoot meristem activity by a cytokinin-activating enzyme. Nature 445:652–655

Lee DJ, Park JY, Ku SJ, Ha YM, Kim S, Kim MD, Oh MH, Kim J (2007) Genome-wide expression profiling of *ARABIDOPSIS RESPONSE REGULATOR 7(ARR7)* overexpression in cytokinin response. Mol Genet Genomics 277:115–137

Lee DJ, Kim S, Ha YM, Kim J (2008) Phosphorylation of *Arabidopsis* response regulator 7 (ARR7) at the putative phospho-accepting site is required for ARR7 to act as a negative regulator of cytokinin signaling. Planta 227:577–587

Leibfried A, To JP, Busch W, Stehling S, Kehle A, Demar M, Kieber JJ, Lohmann JU (2005) WUSCHEL controls meristem function by direct regulation of cytokinin-inducible response regulators. Nature 438:1172–1175

Letham DS (1963) Zeatin, a factor Inducing cell division isolated from *Zea mays*. Life Sci: 569–573

Lohar DP, Schaff JE, Laskey JG, Kieber JJ, Bilyeu KD, Bird DM (2004) Cytokinins play opposite roles in lateral root formation, and nematode and *Rhizobial* symbioses. Plant J 38:203–214

Lorteau MA, Ferguson BJ, Guinel FC (2001) Effects of cytokinin on ethylene production and nodulation in pea (*Pisum sativum*) cv. Sparkle. Physiol Plant 112:421–428

Mähönen AP, Higuchi M, Törmäkangas K, Miyawaki K, Pischke MS, Sussman MR, Helariutta Y, Kakimoto T (2006a) Cytokinins regulate a bidirectional phosphorelay network in *Arabidopsis*. Curr Biol 16:1116–1122

Mähönen AP, Bishopp A, Higuchi M, Nieminen KM, Kinoshita K, Törmäkangas K, Ikeda Y, Oka A, Kakimoto T, Helariutta Y (2006b) Cytokinin signaling and its inhibitor AHP6 regulate cell fate during vascular development. Science 311:94–98

Miller CO, Skoog F, Von Saltza MH, Strong F (1955) Kinetin, a cell division factor from deoxyribonucleic acid. J Am Chem Soc 77:1392–1393

Miyawaki K, Tarkowski P, Matsumoto-Kitano M, Kato T, Sato S, Tarkowska D, Tabata S, Sandberg G, Kakimoto T (2006) Roles of Arabidopsis ATP/ADP isopentenyltransferases and tRNA isopentenyltransferases in cytokinin biosynthesis. Proc Natl Acad Sci USA 103:16598–16603

Mizuno T (2005) Two-component phosphorelay signal transduction systems in plants: from hormone responses to circadian rhythms. Biosci Biotechnol Biochem 69:2263–2276

Mok MC (1994) Cytokinins and plant development - An overview. In: Mok DWS, Mok MC (eds) Cytokinins. Chemistry, activity, and function, CRC Press, Boca Raton, FL, pp 155–166

Mok DW, Mok MC (2001) Cytokinin metabolism and action. Annu Rev Plant Physiol Plant Mol Biol 52:89–118

Müller B, Sheen J (2007) Advances in cytokinin signaling. Science 318:68–69

Müller B, Sheen J (2008) Cytokinin and auxin interaction in root stem-cell specification during early embryogenesis. Nature 453:1094–1098

Murray JD, Karas BJ, Sato S, Tabata S, Amyot L, Szczyglowski K (2007) A cytokinin perception mutant colonized by *Rhizobium* in the absence of nodule organogenesis. Science 315:101–104

Nishimura C, Ohashi Y, Sato S, Kato T, Tabata S, Ueguchi C (2004) Histidine kinase homologs that act as cytokinin receptors possess overlapping functions in the regulation of shoot and root growth in *Arabidopsis*. Plant Cell 16:1365–1377

Oldroyd GE, Downie JA (2008) Coordinating nodule morphogenesis with rhizobial infection in legumes. Annu Rev Plant Biol 59:519–546

Ongaro V, Leyser O (2008) Hormonal control of shoot branching. J Exp Bot 59:67–74

Radutoiu S, Madsen LH, Madsen EB, Felle HH, Umehara Y, Gronlund M, Sato S, Nakamura Y, Tabata S, Sandal N, Stougaard J (2003) Plant recognition of symbiotic bacteria requires two LysM receptor-like kinases. Nature 425:585–592

Rashotte AM, Mason MG, Hutchison CE, Ferreira FJ, Schaller GE, Kieber JJ (2006) A subset of Arabidopsis AP2 transcription factors mediates cytokinin responses in concert with a two-component pathway. Proc Natl Acad Sci USA 103:11081–11085

Romanov GA, Lomin SN, Schmülling T (2006) Biochemical characteristics and ligand-binding properties of *Arabidopsis* cytokinin receptor AHK3 compared to CRE1/AHK4 as revealed by a direct binding assay. J Exp Bot 57:4051–4058

Rupp HM, Frank M, Werner T, Strnad M, Schmülling T (1999) Increased steady state mRNA levels of the *STM* and *KNAT1* homeobox genes in cytokinin overproducing *Arabidopsis thaliana* indicate a role for cytokinins in the shoot apical meristem. Plant J 18:557–563

Sakai H, Honma T, Aoyama T, Sato S, Kato T, Tabata S, Oka A (2001) ARR1, a transcription factor for genes immediately responsive to cytokinins. Science 294:1519–1521

Sakakibara H (2004) Cytokinin biosynthesis and metabolism. In: Davies PJ (ed) Plant hormones. Biosynthesis, signal transduction, action! Springer, Dordrecht, pp 95–114

Sakakibara H (2006) Cytokinins: activity, biosynthesis, and translocation. Annu Rev Plant Biol 57:431–449

Sakakibara H, Suzuki M, Takei K, Deji A, Taniguchi M, Sugiyama T (1998) A response-regulator homologue possibly involved in nitrogen signal transduction mediated by cytokinin in maize. Plant J 14:337–344

Sakakibara H, Takei K, Hirose N (2006) Interactions between nitrogen and cytokinin in the regulation of metabolism and development. Trends Plant Sci 11:440–448

Sakamoto T, Sakakibara H, Kojima M, Yamamoto Y, Nagasaki H, Inukai Y, Sato Y, Matsuoka M (2006) Ectopic expression of KNOTTED1-like homeobox protein induces expression of cytokinin biosynthesis genes in rice. Plant Physiol 142:54–62

Schmülling T, Werner T, Riefler M, Krupková E, Manns IBY (2003) Structure and function of cytokinin oxidase/dehydrogenase genes of maize, rice, *Arabidopsis* and other species. J Plant Res 116:241–252

Shaw G (1994) Chemistry of adenine cytokinins. In: Mok DWS, Mok MC (eds) Cytokinins. Chemistry, activity, and function, CRC Press, Boca Raton, FL, pp 15–34

Skoog F, Miller CO (1957) Chemical regulation of growth and organ formation in plant tissue cultured *in vitro*. Symp Soc Exp Biol 11:118–131

Takeda S, Matsuoka M (2008) Genetic approaches to crop improvement: responding to environmental and population changes. Nature Rev Genet 9:444–457

Takei K, Sakakibara H, Sugiyama T (2001) Identification of genes encoding adenylate isopentenyltransferase, a cytokinin biosynthesis enzyme, in *Arabidopsis thaliana*. J Biol Chem 276:26405–26410

Takei K, Yamaya T, Sakakibara H (2004a) Arabidopsis *CYP735A1* and *CYP735A2* encode cytokinin hydroxylases that catalyze the biosynthesis of *trans*-zeatin. J Biol Chem 279:41866–41872

Takei K, Ueda N, Aoki K, Kuromori T, Hirayama T, Shinozaki K, Yamaya T, Sakakibara H (2004b) *AtIPT3* is a key determinant of nitrate-dependent cytokinin biosynthesis in *Arabidopsis*. Plant Cell Physiol 45:1053–1062

Tanaka M, Takei K, Kojima M, Sakakibara H, Mori H (2006) Auxin controls local cytokinin biosynthesis in the nodal stem in apical dominance. Plant J 45:1028–1036

Taniguchi M, Kiba T, Sakakibara H, Ueguchi C, Mizuno T, Sugiyama T (1998) Expression of Arabidopsis response regulator homologs is induced by cytokinins and nitrate. FEBS Lett 429:259–262

Taniguchi M, Sasaki N, Tsuge T, Aoyama T, Oka A (2007) ARR1 directly activates cytokinin response genes that encode proteins with diverse regulatory functions. Plant Cell Physiol 48:263–277

Tirichine L, Imaizumi-Anraku H, Yoshida S, Murakami Y, Madsen LH, Miwa H, Nakagawa T, Sandal N, Albrektsen AS, Kawaguchi M, Downie A, Sato S, Tabata S, Kouchi H, Parniske M, Kawasaki S, Stougaard J (2006) Deregulation of a Ca^{2+}/calmodulin-dependent kinase leads to spontaneous nodule development. Nature 441:1153–1156

Tirichine L, Sandal N, Madsen LH, Radutoiu S, Albrektsen AS, Sato S, Asamizu E, Tabata S, Stougaard J (2007) A gain-of-function mutation in a cytokinin receptor triggers spontaneous root nodule organogenesis. Science 315:104–107

To JPC, Kieber JJ (2008) Cytokinin signaling: two-components and more. Trends Plant Sci 13:85–92

To JPC, Haberer G, Ferreira FJ, Deruere J, Mason MG, Schaller GE, Alonso JM, Ecker JR, Kieber JJ (2004) Type-A Arabidopsis response regulators are partially redundant negative regulators of cytokinin signaling. Plant Cell 16:658–671

To JPC, Deruere J, Maxwell BB, Morris VF, Hutchison CE, Ferreira FJ, Schaller GE, Kieber JJ (2007) Cytokinin regulates type-A Arabidopsis Response Regulator activity and protein stability via two-component phosphorelay. Plant Cell 19:3901–3914

Urao T, Miyata S, Yamaguchi-Shinozaki K, Shinozaki K (2000) Possible His to Asp phosphorelay signaling in an Arabidopsis two-component system. FEBS Lett 478:227–232

Wang Y, Li J (2008) Molecular basis of plant architecture. Annu Rev Plant Biol 59:253–279

Werner T, Motyka V, Laucou V, Smets R, Van Onckelen H, Schmülling T (2003) Cytokinin-deficient transgenic Arabidopsis plants show multiple developmental alterations indicating opposite functions of cytokinins in the regulation of shoot and root meristem activity. Plant Cell 15:2532–2550

Werner T, Kollmer I, Bartrina I, Holst K, Schmülling T (2006) New insights into the biology of cytokinin degradation. Plant Biol 8:371–381

Yamada H, Suzuki T, Terada K, Takei K, Ishikawa K, Miwa K, Yamashino T, Mizuno T (2001) The Arabidopsis AHK4 histidine kinase is a cytokinin-binding receptor that transduces cytokinin signals across the membrane. Plant Cell Physiol 42:1017–1023

Yamada H, Koizumi N, Nakamichi N, Kiba T, Yamashino T, Mizuno T (2004) Rapid response of Arabidopsis T87 cultured cells to cytokinin through His-to-Asp phosphorelay signal transduction. Biosci Biotechnol Biochem 68:1966–1976

Yanai O, Shani E, Dolezal K, Tarkowski P, Sablowski R, Sandberg G, Samach A, Ori N (2005) Arabidopsis KNOXI proteins activate cytokinin biosynthesis. Curr Biol 15:1566–1571

Yonekura-Sakakibara K, Kojima M, Yamaya T, Sakakibara H (2004) Molecular characterization of cytokinin-responsive histidine kinases in maize. Differential ligand preferences and response to *cis*-zeatin. Plant Physiol 134:1654–1661

Chapter 14
Light Signalling in Plant Developmental Regulation

A. Galstyan and J.F. Martínez-García

14.1 Introduction

During their life cycles, plants pass through several developmental stages, such as germination, seedling establishment, flowering induction and seed production, followed by senescence. All stages are shaped by the combination of endogenous (genotype) and exogenous (environmental conditions) factors. Due to their sessile nature, plants cannot change their "behaviour" (e.g. moving to a new location) to adapt and/or avoid unfavourable conditions. Therefore, plants have developed precise sensing mechanisms (i.e. receptors) to monitor their living surroundings, displaying a great plasticity in their development. In this respect, some authors have argued that the plasticity of development plays a role in plant survival analogous to that of behaviour in animals.

Amongst many environmental factors influencing plant growth (e.g. temperature, gravity, wind, humidity, water and nutrient availability), light is one of the most important factors because of its role as an informative and positional signal. The main focus of this chapter is to review our current understanding of the role of light signalling in the regulation of plant development.

A. Galstyan
Centre for Research in Agricultural Genomics (CSIC-IRTA-UAB), c. Jordi Girona 18-26, 08034 Barcelona, Spain

J.F. Martínez-García
Institució Catalana de Recerca i Estudis Avançats (ICREA), Passeig Luís Companys 23, 08010 Barcelona, Spain
Centre for Research in Agricultural Genomics (CSIC-IRTA-UAB), c. Jordi Girona 18-26, 08034 Barcelona, Spain
e-mail: jmggmg@cid.csic.es

E-C. Pua and M.R. Davey (eds.),
Plant Developmental Biology – Biotechnological Perspectives: Volume 2,
DOI 10.1007/978-3-642-04670-4_14, © Springer-Verlag Berlin Heidelberg 2010

14.2 Plant Photomorphogenesis: Various Responses to a Complex Stimulus

The information content of a light signal controls every aspect of the plant, including its architecture, structural development and morphogenesis. The control of morphogenesis by light is called *photomorphogenesis*. Light characteristics, such as intensity, wavelength, direction and duration, provide plants with important cues of, for example, the proximity of other vegetation and the season of the year. Highly sensitive plant photoreceptors monitor the light environment and transduce it into a cellular signal, which affects endogenous mechanisms of growth control and differentiation. As a consequence, light modulates a variety of processes in a plant's life, such as germination, seedling de-etiolation, shade avoidance responses and flowering induction, collectively defined as photomorphogenesis.

14.3 Sensing Changes in Light Conditions: Multiple Photoreceptors Continuously Monitor the Light Environment

In the control of photomorphogenesis, classic photobiological studies have shown that plants are most sensitive to UV-B, UV-A/blue (B), red (R), and far-red (FR) light. It has been well established that there are at least four different types of photoreceptors identified in *Arabidopsis*, which currently is the best characterized plant system. These include the phytochromes, cryptochromes, phototropins and a newly recognized set of B light photoreceptors, the zeitlupes (Fig. 14.1a). The action of these photoreceptors controls plant photomorphogenesis through the complex transcriptional regulatory networks upon perception of B, R or FR light.

As light sensors, photoreceptor activity is tightly controlled by light in multiple ways. Prominently, light affects the photoreceptor physicochemical properties, transforming the light signal into a biochemical cellular signal. In addition, light can influence photoreceptor gene expression and photoreceptor function mainly at the post-transcriptional level, by regulating protein abundance, altering their subcellular localization and/or their ability to interact with other proteins (Bae and Choi 2008).

14.3.1 Phytochromes

Phytochromes, the first family of plant photoreceptors to be discovered, are most sensitive to the R and FR region of the light spectrum. In *Arabidopsis*, phytochromes are encoded by a small gene family of five members (*PHYA* to *PHYE*), which are responsible for regulating various R and FR light responses, including seed germination, seedling de-etiolation (Fig. 14.2a), shade avoidance and flowering.

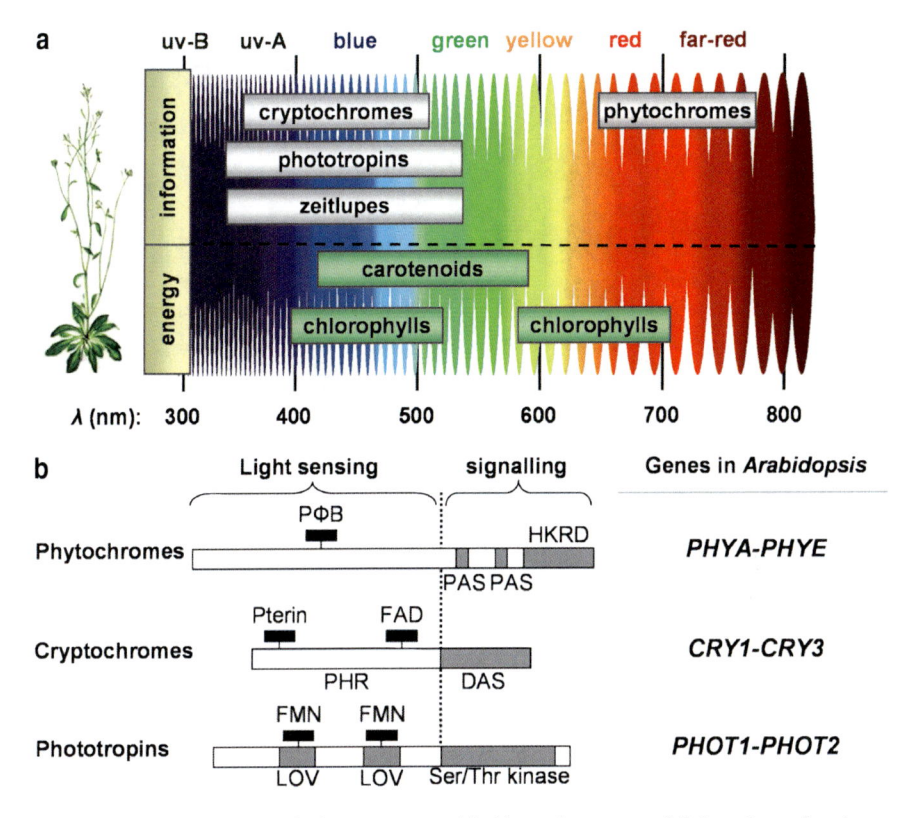

Fig. 14.1 Light spectrum and photoreceptors. (**a**) Absorption range of light colours for photomorphogenic and photosynthetic photoreceptors. (**b**) Schematic representation of the main structural features of major *Arabidopsis* photoreceptors

PHY genes were also identified in all other plant species (e.g. monocotyledons, gymnosperms, ferns, mosses and algae), in cyanobacteria, fungi and even in non-photosynthetic bacteria (Montgomery and Lagarias 2002).

Plant phytochromes are typically homodimeric proteins. Each subunit can be divided into an N-terminal photosensory domain and a C-terminal domain (Fig. 14.1b). The N-terminal domain covalently binds a single bilin chromophore (PΦB), which confers the ability to absorb R and FR light. The C-terminal domain, which contains two PAS domains followed by a two-component His kinase-related domain (HKRD), confers dimerization ability and a nuclear localization signal. Based on their light stability, phytochromes are classified as either photolabile (type I) or photostable (type II). In *Arabidopsis*, phyA is a type I and phyB-phyE are type II phytochromes. Phytochromes are synthesized in the inactive Pr form (λ_{max} of absorbance, 666 nm). In darkness, phytochromes are located in the cytoplasm; upon light perception, the Pr form is rapidly converted into the biologically active Pfr form (λ_{max} of absorbance, 730 nm) and translocated to the nucleus (Fig. 14.3a). The photoconversion between Pr and Pfr upon R or FR absorption is

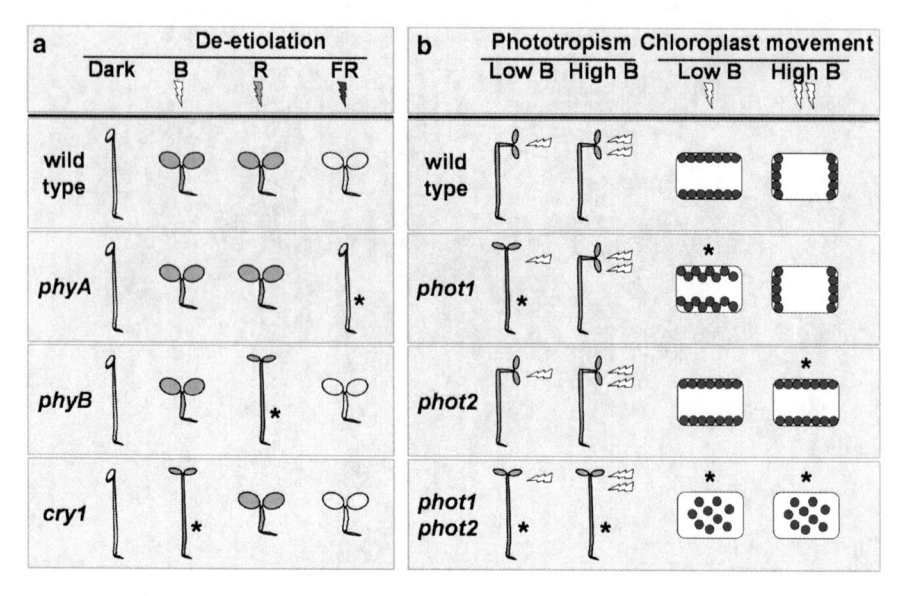

Fig. 14.2 Schematic summary of phenotypes in different photoreceptor mutants. (**a**) Seedling phenotypes of wild type, *phyA*, *phyB* and *cry1* mutants growing in the dark or monochromatic B, R or FR light. (**b**) Seedling phototropic responses to directional B light and chloroplast migration in leaf mesophyll cells of wild type, *phot1*, *phot2* and *phot1phot2* mutants. Low and high B refer to light intensity. Asterisks indicate differential phenotypes compared to the wild type

reversible, allowing phytochromes to act as a molecular switch. Upon return to darkness, Pfr is converted back to Pr, which is called dark reversion (Chen et al. 2004; Bae and Choi 2008).

Phytochrome responses have been classified based on the amount of radiation energy required to obtain the response. High irradiance responses (HIRs) require prolonged or high-frequency intermittent illumination; low fluence responses (LFRs) are the classical R/FR reversible responses, and very low fluence responses (VLFRs) are not reversible and are sensitive to a broad spectrum of light from 300–780 nm. Genetic analyses of *Arabidopsis* phytochrome mutants showed that phyA is responsible for the VLFR and FR-HIR, and that phyB is the prominent phytochrome responsible for the LFR and R-HIR during photomorphogenesis. Different phytochromes have distinct, redundant, antagonistic and synergistic roles in regulating plant responses (Bae and Choi 2008).

14.3.2 Cryptochromes

Cryptochromes are UV-A (~320–400 nm) and B (~400–500 nm) light receptors that play an important function in a number of B light-regulated responses, such as seedling de-etiolation, photoperiod-dependent flowering and resetting of the circadian oscillator. In *Arabidopsis*, cryptochromes are encoded by at least three genes, *CRY1*,

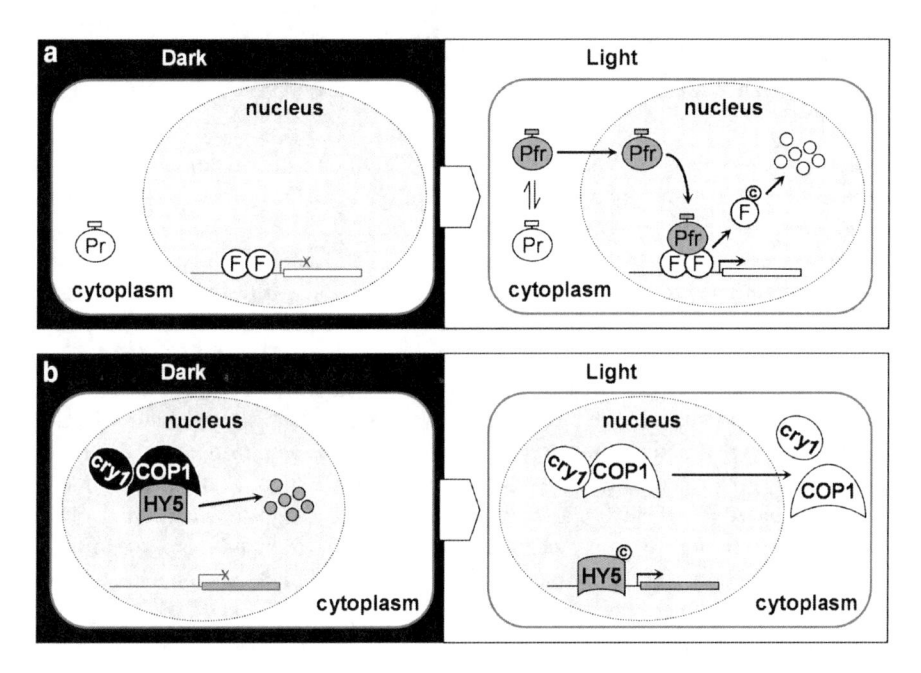

Fig. 14.3 Simplified model summarizing different mechanisms involved in (**a**) phytochrome and (**b**) cryptochrome signalling. F refers to PIF transcription factors. Light-induced conformational changes of cry1 and COP1 are represented by a light shading. © refers to phosphorylation of PIF and HY5 factors. For simplicity, only phytochrome and cryptochrome activated genes are shown

CRY2 and *CRY3*. Cry1 plays the prevalent role in the de-etiolation response to high B light intensities (Fig. 14.2b), while cry2 is mostly important in this response to low intensities. Cryptochromes possess two domains, a conserved photosensory N-terminal photolyase-related (PHR) domain and a divergent, intrinsically unstructured C-terminal DAS domain, which is absent in cry3. The PHR domain non-covalently binds two chromophores, a flavin adenine dinucleotide (FAD) and a pterin or deazaflavin (Fig. 14.1b), which provides the light-sensing capacity. The C-terminal DAS domain has the ability to transduce the signals perceived by the PHR domain and is important for nuclear/cytosol trafficking and protein–protein interactions. Similar to phytochromes, distinct cryptochromes have different light stability: cry1 is photostable (like phyB-phyE), while cry2 is photolabile (like phyA). This is interesting for the coordination of cryptochromes and phytochromes in the regulation of light responses in plants. Cry2 is constitutively nuclear-localized, whereas cry1 is nuclear in the dark but largely cytoplasmic in the light. In contrast, the cry3 C-terminal domain is replaced with a transient peptide sequence targeting it to both chloroplasts and mitochondria (Lin and Shalitin 2003).

The PHR domain shares sequence similarity to DNA photolyases. Based on this similarity, it has been suggested that cryptochromes can bind directly to chromatin, but this has been demonstrated only for cry3. Nevertheless, there is no report of cryptochromes possessing DNA repair activity upon UV-B damages.

14.3.3 Phototropins, Other LOV Domain-Containing Proteins and UV-B Receptors

Phototropins are plant-specific photoreceptors identified as regulators of multiple B light responses, including phototropism, rapid inhibition of hypocotyl growth, stomatal aperture and chloroplast movements (see Sect. 14.4.4). In *Arabidopsis*, phototropins are encoded by the two genes *PHOT1* and *PHOT2* that work together to control the abovementioned responses. Phot1 is specialized for low B light intensities, while phot2 is more important for high intensities (Fig. 14.2b). Phototropins have two distinct domains, a C-terminal Ser/Thr kinase domain and an N-terminal region, which encodes two light-sensory LOV (light, oxygen and voltage) subdomains. These LOV domains non-covalently bind the chromophore flavin mononucleotide (FMN; Fig. 14.1b) and can undergo B light-dependent autophosphorylation. Phot1 and phot2 are plasma membrane-associated. Following light activation, a fraction of phot1 is released into the cytoplasm. *PHOT2* expression is light-induced, while phot1 protein concentrations are decreased in seedlings upon extended exposure to light. These effects might account for the greater importance of phot1 and phot2 under low and high light-mediated processes respectively (Briggs and Christie 2002).

In addition to phototropins, a few other *Arabidopsis* proteins have LOV domains. They have a similar structural organization, with an N-terminal LOV domain followed by an F-box and several Kelch repeats. They are known as ZTL (ZEITLUPE), LKP2 (LOV, KELCH REPEAT PROTEIN 2) and FKF1 (FLAVIN-BINDING, KELCH REPEAT, F-BOX 1). Genetic analyses in *Arabidopsis* indicated their function in sustaining normal circadian clock function and photoperiod-dependent flowering. Despite there being no direct functional evidence, they can be classified structurally as a new class of photoreceptors. Although UV-B radiation (280–315 nm) regulates photomorphogenic responses, no receptors have been identified (Chen et al. 2004; Jenkins 2009).

14.4 Physiological Responses During Photomorphogenesis: Roles of Photoreceptors in Plant Development

Analyses of mutants deficient in one or more photoreceptors allow the identification of a specific photoreceptor for a particular aspect of light-regulated development. Some of the physiological responses analyzed to understand light signalling in *Arabidopsis* are summarized below.

14.4.1 Germination

In unfavourable environmental conditions, seeds might remain dormant in a dry state for extended periods, germinating only when conditions improve. In *Arabidopsis*, seed dormancy is controlled by abiotic factors, such as temperature, time of storage, available nutrients and light (colour and intensity). It has been well established that light-dependent germination is mediated entirely by phytochromes, mainly by phyA and phyB. Low quantity of R or FR light (VLFR) or continuous FR light (HIR) induce germination mediated through phyA, whereas germination induced by R/FR pulses (LFR) is largely controlled by phyB. Mutant analyses also demonstrated the contribution of phyE to this response (Jiao et al. 2007).

14.4.2 De-Etiolation

After germination, seedlings activate two different developmental programs, depending on whether they grow in darkness (skotomorphogenesis/etiolation) or in light (photomorphogenesis/de-etiolation). De-etiolation refers to the transition of a dark-grown seedling to a photoautotrophically competent seedling induced by its first exposure to light. This response involves a switch from the skotomorphogenic pattern of seedling growth (characterized by rapid hypocotyl elongation, small cotyledons forming the apical hook and pale yellow colour) to the photomorphogenic pattern (characterized by inhibited hypocotyl elongation, apical hook opening, expansion of cotyledons/leaves and synthesis of anthocyanins and chlorophylls). Several photoreceptors regulate these responses (Fig. 14.2a). The regulation of de-etiolation involves not only phytochromes and cryptochromes, but also the circadian clock and phytohormones (Quail 2002).

14.4.3 Phototropism

Phototropism refers to plant movement in a direction determined by a light source. Within a plant, shoots normally grow towards light (positive phototropism), while roots grow away from light (negative phototropism). In *Arabidopsis*, phototropic responses are controlled mainly by phototropins (Fig. 14.2b), although phytochromes and cryptochromes also contribute to their regulation (Jiao et al. 2007).

14.4.4 Chloroplast Movement

Chloroplast movement in *Arabidopsis* is regulated by B light. It can be classified in two separate responses depending on light intensity, namely an *accumulation response* to low intensities that helps to maximize light capturing for photosynthesis, and an *avoidance response* to high intensities that ameliorates the potential damaging effects of excess light energy. Phototropins are the main photoreceptors controlling these responses (Fig. 14.2b), while phytochromes also participate. In particular, phot2 has been shown to regulate the chloroplast avoidance responses under the high intensity of B light (Briggs and Christie 2002).

14.4.5   Shade Avoidance Syndrome

Under high plant density conditions, competition for light requires developmental adaptation to either tolerate or avoid shading by nearby competitors. The shade avoidance syndrome (SAS) refers to a set of plant responses aimed to adapt growth and development to environments of high plant density, like those found in both natural (e.g. forests) and agricultural (e.g. orchards) communities. SAS includes responses such as inhibition of seed germination, induction of hypocotyl/stem/internode elongation, inhibition of leaf expansion and induction of flowering time. The presence of neighbouring plants results in a reduction in the R to FR ratio (R:FR) caused by a specific enrichment in FR light reflected from the surface of neighbouring leaves. The R:FR changes are perceived by the phytochromes, directly affecting the equilibrium between the Pr (inactive) and Pfr (active) forms. Although phyB is the major phytochrome controlling SAS, genetic and physiological analyses have shown that other phytochromes act redundantly with phyB in the control of some aspects of SAS development, like flowering time (phyD, phyE), petiole elongation (phyD, phyE) and internode elongation between rosette leaves (phyE). In addition, phyA inhibits excessive SAS responses under low R:FR ratio light, whereas phyB inhibits SAS responses under a high R:FR ratio light (Bae and Choi 2008).

14.4.6   Photoperiodic Responses

The ability to measure the daylength (photoperiod) allows plants to adapt to seasonal changes. One of the best known responses to photoperiod is flowering. Based on this response, plants can be classified as long-day, short-day and day-neutral. Several models have been developed to explain how plants measure daylength to regulate photoperiodic responses. The external coincidence model is currently the most widely accepted. The model proposes that daylength measurement relies on a circadian oscillator that controls the levels of a regulatory molecule the activity of which is modulated by light. The photoperiodic signal is generated only when light overlaps with a phase of the circadian cycle in which the levels of the regulatory molecule is above a certain threshold. This model proposes a dual role for light, by entraining the circadian clock and by directly affecting the production of the signal.

The molecular genetic dissection of flowering time in *Arabidopsis*, which is a facultative long-day plant, has shown that several photoreceptors, light signalling and circadian clock components are essential for adequate photoperiodic responses.

Putative photoreceptors ZTL and other LOV domain-containing proteins regulate flowering time and circadian rhythms. However, phyA and cry2 seem to be the main photoperiodic photoreceptors. Genetic analyses also resulted in the identification of *CO* (*CONSTANS*) and *FT* (*FLOWERING LOCUS T*) genes as specifically involved in the perception of the photoperiod signal. CO is an important regulator of *FT* expression. *CO* expression cycles (regardless of daylength) with a peak about 16–20 h after dawn. In contrast, *FT* is induced only when light is perceived at a time when *CO* is highly expressed. This effect points to post-transcriptional regulation of CO activity by light. Indeed, it was demonstrated that FR and B light, which are perceived by phyA and cry2 respectively, regulate the stability of CO protein through inhibition of proteasome-dependent CO degradation. Thus, CO protein is a light-sensitive flowering regulator that was proposed to be produced with a circadian rhythm, which is in agreement with the essence of the external coincidence model. In addition, evidence indicates that FT protein also promotes flowering (Kobayashi and Weigel 2007; Mas 2008).

14.5 Photoreceptor Signal Transduction

Signalling or signal transduction refers to those events occurring from perception of a physical or biological signal to the first change in gene expression elicited by the signal. In plants, these terms may also refer to the set of events between the perception of a signal and the appearance of a measurable growth or developmental change (Quail 2002). Therefore, four research areas related to the mechanisms by which light signalling is transduced have been summarized below, these being genetic analyses of light signalling, protein kinase activity, proteasome-based protein degradation and light-regulated transcriptional networks.

14.5.1 Genetic Analyses: Identification of Key Players

To define novel components of the different photoreceptor signalling networks, several laboratories have performed genetic screens based on the seedling de-etiolation process in *Arabidopsis*. These screens identified two major classes of mutants, namely (1) the *cop* (*constitutive photomorphogenic*)/*det* (*de-etiolated*)/*fus* (*fusca*) class, where seedlings de-etiolate in complete darkness as if they had perceived a light signal, and (2) a class that develops normally in darkness, but has altered responsiveness to light signals. Because the *cop/det/fus* mutants are recessive and act more or less pleiotropically, these components have generally been postulated to function downstream of the convergence of the cryptochrome and phytochrome pathways, and to encode negative regulators of photomorphogenesis. In the other class, the mutants have been classified according to the colour of light they are unable to perceive, namely B, R and/or FR. They include components that appear to be specific to either phyA or phyB, or to phytochrome versus cryptochrome, signalling. These mutants suggest that early steps in each pathway involve intermediates dedicated to the individual photoreceptors and that the separate pathways converge downstream in a "signal integration" process that drives later common events in the de-etiolation response. However, there are also mutants within this group that show a general reduction in all types of light signalling, such as *hy5* (*long hypocotyl 5*). *HY5* encodes a bZIP transcription factor the action of which is thought to be repressed by its COP1-mediated degradation (see Sect. 14.5.3). As a component involved in B-, R- and FR-induced responses, HY5 is currently considered to act downstream of convergence points between signals from multiple photoreceptors and other developmental signals, together with COP/DET/FUS (Quail 2002).

Mutants specifically impaired in de-etiolation in FR are considered phyA-signalling mutants, as phyA is the only photoreceptor acting under these light conditions (Fig. 14.2a). Similarly, mutants specifically impaired in de-etiolation in R are generally considered phyB-signalling components (Fig. 14.2a). For example, the specific phyA-signalling mutants are expected to identify components that act close to phyA action and upstream of any branch point in a general photomorphogenic signalling pathway. This is the case of *FHY1* (*FAR-RED ELONGATED HYPOCOTYL 1*), its paralogue *FHL* (*FHY1-LIKE*), *FHY3* and *FAR1* (*FAR-RED IMPAIRED RESPONSE 1*). Functional analyses of these mutants indicated a role for FHY1 and FHL in the nuclear localization of active phyA (PfrA). Upon light activation, PfrA interacts with FHY1 and FHL in the cytoplasm. The nuclear localization signal of FHY1 then allows nuclear entry of the complex. FHY3 and FAR1 also control phyA nuclear accumulation by directly regulating the expression of *FHY1* and *FHL*. In addition, other phyA-signalling components encoding putative transcription factors have been genetically identified, such as *LAF1* (*LONG AFTER FAR-RED LIGHT 1*), encoding a MYB protein, or *HFR1* (*LONG HYPOCOTYL IN FAR-RED 1*) that encodes an atypical bHLH (basic-helix-loop-helix) protein highly similar to PIF3 (PHYTOCHROME-INTERACTING FACTOR 3; see below; Bae and Choi 2008).

Yeast two-hybrid screens combined with reverse genetic approaches have been performed to uncover early phytochrome signalling events. The best understood component of this group is probably PIF3, a predicted bHLH transcription factor. Additional PIF and PIL (PIF3-LIKE) factors have been identified based on sequence similarity. PIF3, the founder member of the group, is a DNA-binding protein that binds preferentially to PfrA and PfrB. PfrB can interact with PIF3 bound to a DNA sequence in vitro, leading to the provocative model for the direct regulation of gene expression by phytochromes in response to light (Fig. 14.3a). Although conclusive evidence in favour of this or any other model is still lacking, chromatin immunoprecipitation (ChIP) assays have led to the identification of direct target genes of PIF3 (Martinez-Garcia et al. 2000; Shin et al. 2007). The PIF3 protein is degraded rapidly in response to light (Sect. 14.5.2), which led to the conclusion that this protein functions transiently in light signalling pathways. Subsequent studies have shown that PIF3 re-accumulates in the dark during recurring light–dark cycles, suggesting a role in fine-tuning photomorphogenic development throughout the plant life cycle. Genetic data also show that PIF3 functions as a negative regulator of morphological phenotypes under R (i.e. shorter hypocotyls), and as a positive regulator in chloroplast development (greening processes) and anthocyanin accumulation. A study of PIF3 function indicates that it has an early role in regulating specific gene expression and a long-term role inhibiting hypocotyl elongation under continuous irradiation. These two functions seem to require different molecular activities. In early signalling, PIF3 acts positively as a transcription factor, exclusively requiring its DNA-binding capacity. PIF3 seems to function as a constitutive co-activator in this process without the need for phytochrome binding. This possibility implies the existence of additional factors photoactivated by phytochromes, which function in concert with PIF3. During long-term irradiations, PIF3 acts exclusively through its phyB-interacting capacity to control hypocotyl elongation by decreasing phyB protein abundance (and thereby R light sensitivity), apparently without participating directly in the transduction chain as a signalling intermediate. The mechanism by which PIF3 modulates phyB levels is unknown (Al-Sady et al. 2008).

A limited list of cryptochrome-signalling components has also been genetically identified. These include *HYH* (*HY5 HOMOLOG*), *SUB1* (*SHORT UNDER BLUE LIGHT 1*) and *HFR1*, also identified as a phyA-signalling component. All these components are negative regulators of cryptochrome signalling. In addition, *PP7* (*PROTEIN PHOSPHATASE 7*) was shown to be the only positive regulator of cryptochrome action (Li and Yang 2007).

Phototropine-mediated signalling is less understood. At least two components of a hypothetical phot1-complex have been identified, NPH3 (NON-PHOTOTROPIC HYPOCOTYL 3) and RPT2 (ROOT PHOTOTROPISM 2) that belong to a novel plant-specific gene family. NPH3 and RPT2 directly interact with phot1, co-localizing in the plasma membrane. Phot1/NPH3 and phot1/RPT2 interactions are proposed to regulate phototropin-dependent hypocotyl inhibition and stomata opening respectively. This finding indicates early branching during phototropin-mediated signal transduction.

UV-B light triggers the inhibition of *Arabidopsis* hypocotyl elongation and transcriptional regulation of gene expression. COP1 and HY5 act together in the nucleus in UV-B responses. In addition, two signalling components were described in the UV-B perception pathway: *ULI3* (*UV-B LIGHT INSENSITIVE 3*), encoding a protein of unknown function, and *UVR8* (*UV RESISTANT LOCUS 8*), which encodes a protein associated with chromatin that is required for the UV-B induction of transcription (Chen et al. 2004; Jenkins 2009).

14.5.2 Phosphorylation/Dephosphorylation

The central biochemical events that constitute the signal transfer from the photo-activated photoreceptors to their primary signalling partners are subjects of research in many laboratories. Phosphorylation/dephosphorylation of photoreceptors and their signalling intermediates might play an important role in regulating plant growth responses to perceived light. Structural and evolutionary studies indicate that plant phytochromes have evolved from prokaryotic bacteriophytochromes, which resemble two-component His kinases. Whereas crytochromes have no resemblance to kinases, phototropins contain a C-terminal Ser/Thr kinase domain. All phytochromes, cryptochromes and phototropins possess autophophorylation activity.

Because bacteriophytochromes have been demonstrated to posses light-regulated His kinase activity, it was concluded that protein kinase activity is an early and important event in bacteriophytochrome signalling. However, plant phytochromes are most likely not His kinases, since key conserved residues in the HKRD are missing. Oat phyA possesses Ser/Thr protein kinase activity in vitro, with Pfr being more active. The kinase activity of phyA was confirmed after identification of several substrates, such as PSK1 (PHYTOCHROME SUBSTRATE KINASE 1), Aux/IAAs and cryptochromes. Phosphorylation of phyA modulates photoresponses in several ways, by controlling the subcellular localization of phyA, its stability and its affinity towards downstream signal transducers. Several components were identified to regulate phytochrome activity both via phosphorylation (NDPK2, NUCLEOSIDE DIPHOSPHATE KINASE 2) and dephosphorylation (FYPP, FLOWER-SPECIFIC PHYTOCHROME-ASSOCIATED PROTEIN PHOSPHATASE; PAPP5, PHYTO-CHROME-ASSOCIATED PROTEIN PHOSPHATASE 5), which activate and deactivate phyA activity respectively. Little is known about a possible kinase activity of the other plant phytochromes (Bae and Choi 2008).

B light-dependent autophosphorylation of cry1 and cry2 is also important for their function. In the case of cry2, phosphorylation is associated with proteolytic degradation. The C-terminal domain of cry2 fused to GUS results in constitutive signalling activity and constitutive phosphorylation. It has been suggested that light activation of N-terminus of cry1 induces a conformational changes in its C-terminus that allows its autophosphorylation and dimerization, and possible interaction with downstream partner proteins. There are also reports that phyA can phosphorylate cryptochromes in vitro, although the functional relevance of this interaction is not clear (Chen et al. 2004).

Phototropins are perhaps the only photoreceptors with well-established kinase activity. Since recombinant phot1 undergoes light-dependent autophosphorylation in the absence of any other plant proteins, it can be concluded that phot1 is necessary and sufficient for light-regulated protein kinase activity. Recombinant phot2 has similar spectral and protein kinase properties. Structural studies suggest that upon light absorption there are large light-driven structural rearrangements that liberate the protein kinase domain and presumably allow protein kinase activity (Chen et al. 2004).

Light signal transduction is highly regulated by phosphorylation also at the level of photoreceptor direct interaction partner proteins. For instance, rapid degradation of nuclear-localized PIF3 protein requires phyA, phyB and phyD. Although the mechanism of light-induced PIF degradation is not clear, the data suggest that the first step in the light-induced degradation of PIF3 is its phosphorylation after direct physical interaction with phytochromes. Phosphorylated PIF3 may then be degraded by the 26S proteasome (see Sect. 14.5.3). Similar processes were described for additional components of the phytochrome signalling pathway, such as other PIFs/PILs, HFR1 and HY5. In summary, the effect of protein phosphorylation in light signalling differs from stabilization of the protein (e.g. HY5) to targeting it to proteasome-mediated degradation (e.g. HFR1, PIF3; Jiao et al. 2007).

14.5.3   Ubiquitination/Proteasome-Mediated Proteolysis

Regulation of protein degradation is a fundamental part of light signalling, especially through the 26S proteasome pathway. In order for a protein to be degraded by this system, it is polyubiquitinated through the sequential action of three enzymatic activities, namely an E1 ubiquitin (Ub)-activating enzyme, an E2 Ub-conjugating enzyme and an E3 Ub ligase, which recognizes and recruits both Ub-charged E2 and the target protein to catalyze Ub binding to a specific lysine in the target. From the several *cop/det/fus* mutants identified, nine have been cloned. Their gene products organize into three complexes in vivo, namely the COP1 complex, COP9 signalosome (CSN) and CDD complex. These three multi-protein complexes are directly involved in the Ub/proteasome-mediated protein degradation pathway. The best characterized component is COP1, a RING finger-type E3 Ub ligase. COP1 is required for the light-dependent degradation of several transcription factors involved in the light-regulated transcriptional network. In the dark, COP1 localizes in the nucleus and is able to bind transcription factors, such as HY5, LAF1, HFR1, and then ubiquinate and target them for degradation by the 26S proteasome. Upon light perception, COP1 is translocated to the cytoplasm, allowing the accumulation of those transcriptional factors necessary for photomorphogenesis (Fig. 14.3b). COP1 is also required for the accumulation of both PIF3 and FHY1 in the dark, although the mechanism of this regulation remains unclear.

Light-induced change in COP1 nucleocytoplasmic localization is modulated via interactions with CSN. The direct interaction of COP1 with phyA, phyB, cry1 and cry2 indicates an additional light-dependent regulation of COP1 activity, for example, targeting photolabile photoreceptors (phyA and cry2) for light-dependent degradation. Moreover, it was suggested that, upon light exposure, cry1 conformation rapidly changes and antagonizes the COP1 effect on HY5 and other transcription factors necessary for photomorphogenesis (Fig. 14.3b). Integrating different light signals, COP1 provides a point of convergence for photoreceptor-mediated signal transduction pathways, acting as a key regulator of light responsive gene expression (Yi and Deng 2005).

14.5.4 Light-Regulated Transcriptional Networks: Changes in Gene Expression

Transcriptional networks have been implicated in mediating photomorphogenesis. R/FR-sensing phytochromes convert perceived light information into absolute concentrations and/or concentration gradients of the Pfr active form (see Sect. 14.3.1). Based on the hypothesis that Pfr can rapidly modulate gene expression by interacting with different PIFs (see Sect. 14.5.1), it was proposed that upon light perception phytochromes initiate complex transcriptional networks instrumental for the implementation of photomorphogenic responses. In agreement with this hypothesis, global gene expression analyses have shown that phytochrome photoperception is associated with massive alterations in gene expression. Consistently, some studies showed that several PIFs, such as PIF1 and PIF3, have a role in controlling gene expression as transcription factors in a phytochrome-dependent manner. Also, direct target genes of phytochrome action, such as *HFR1*, *PIL1*, *PAR1* (*PHYTOCHROME RAPIDLY REGULATED 1*) and *ATHB4* (*ARABIDOPSIS THALIANA HOMEOBOX 4*), are specifically involved in the implementation of SAS responses (Jiao et al. 2007; Roig-Villanova et al. 2007).

Gene expression regulation is also a major signalling mechanism underlying cryptochrome action, as demonstrated by transcriptome analysis. It is not clear as to which of the cryptochrome-regulated genes is directly involved in B light de-etiolation responses and how cryptochromes generally regulate gene expression. The genetic identification of components of cryptochrome signalling suggests that important events might occur not only in the nucleus (e.g. HY5, HYH, HFR1 and PP7) but also in the cytoplasm (e.g. SUB1; Li and Yang 2007).

14.6 Light Interaction with Endogenous Networks

Since many plant hormones regulate the same cell division and expansion growth responses modulated by photomorphogenesis, it is expected that light signals converge at the endogenous mechanisms of growth, providing a means to integrate light environment information with endogenous developmental programs, such as those controlled by phytohormones and the circadian clock.

14.6.1 Hormone Connections

Auxins mediate a diverse range of responses including cell division, expansion and differentiation. In addition, auxins coordinate plant development (affecting organ patterning, tropic responses and the architecture of shoot and root) as signals transmitted from cell to cell or from one organ to another. Asymmetric distribution of specific influx and efflux carriers, encoded by the *AUX* and *PIN* genes and located at specific positions in the cell membrane within a cell and organ, provides a simple but effective means to establish polar (directional) transport of auxin. Therefore, light might interact with auxins at different levels, including auxin biosynthesis, transport and responses.

Early work showed a role for phytochromes in regulating auxin levels in corn and oat coleoptiles. Recently, low R:FR perception was shown to increase the concentrations of endogenous auxins, involving the action of *TAA1* (*TRYPTO-PHAN AMINOTRANSFERASE OF ARABIDOPSIS 1*) that encodes an auxin bio-synthetic enzyme required for full induction of SAS responses (Tao et al. 2008). Applications of auxin transport inhibitors, combined with genetic analyses, have also involved phytochromes in light control of auxin transport. Phototropins were also suggested to alter the activity or localization of auxin transport carriers. Upon directional light perception, an auxin gradient is rapidly established by the action of auxin efflux carriers such as PIN proteins that transport the hormone out of the cell. It seems highly probable that phototropin signalling leads to an alteration in the activity or localization of auxin transport facilitators, modifying polar movement of the hormone and resulting in phototropic growth in response to light.

At the molecular level, auxin rapidly activates the transcription of three gene families, namely *Aux/IAA*, *SAUR* and *GH3* genes. Work from several laboratories evidenced that light appears to regulate the expression of members of the *Aux/IAA*, *SAUR* and *GH3* families, some particularly by phyA and phyB action. Recently, the SAS regulator PAR1 was described as a direct repressor of *SAUR* genes, rapidly connecting shade- and auxin-regulated transcriptional networks. Proteolysis may represent another convergence point between auxin and light signalling. Aux/IAAs proteins operate by binding to ARF (AUXIN RESPONSIVE FACTOR) transcription factors to regulate their action negatively, providing a mechanism by which auxin modulates gene expression. Auxin controls Aux/IAA protein levels via its degradation by the 26S proteasome. Genetic analyses showed that mutations that stabilize IAA3, IAA7 and IAA17 result in mild de-etiolated phenotypes in the dark. These observations suggest that normal turnover of Aux/IAAs is impor-tant to repress photomorphogenesis in the dark (Halliday and Fankhauser 2003; Roig-Villanova et al. 2007).

Brassinosteroids (BRs) are powerful growth promoters that act synergistically with auxins. The first evidence of possible interaction between light and BR signalling was the identification of the *det2* mutant. *DET2* encodes an enzyme involved in BR biosynthesis. This and most other BR biosynthesis mutants identi-fied are dwarfed, dark-green and display a de-etiolated seedling phenotype in the dark. Mutants affected in BR biosynthesis show an enhanced phyA-mediated VLFR and reduced HIR and LFR responses, suggesting the involvement of BRs in phytochrome signalling. BRs regulate the expression of some light-signalling genes such as *PIF3* and *CIP1* (*COP1 INTERACTING PROTEIN 1*), although the biological relevance of this interaction is unknown. *BAS1* (*PHYB-4 ACTIVATION-TAGGED SUPPRESSOR 1*) gene, which encodes a BR inactivating enzyme, was proposed as a modulator of photomorphogenesis based on the increased hypocotyl response of *bas1* mutants to exogenous BRs, specifically in white and FR light. Recent work has proposed a role for the SAS modulator ATHB4 in BR signalling, integrating shade perception and BR-mediated growth (Halliday and Fankhauser 2003; Sorin et al. 2009).

Gibberellins (GAs) control multiple aspects of plant development, including induction of germination, leaf expansion, stem elongation and flowering. Phytochromes regulate GA levels and biosynthesis at several stages of plant development. During germination, the expression of genes encoding GA biosynthetic enzymes is regulated by phyB and by the antagonistic action of the bHLH transcription factors SPT (SPATULA) and PIF1/PIL5. During elongation growth, it was demonstrated that end-of-day FR light treatment increases bioactive GA concentrations, specifically in the elongating stem of cowpea, although genetic studies of *phyB ga1* double mutants (*GA1* encodes an enzyme acting in GA biosynthesis) suggest that phytochrome control also affects GA signalling. Recently, it was shown that some *pif* mutants have an altered sensitivity to GAs. The molecular basis for this phenotype is dependent on the activity of DELLA proteins, negative regulators of GA signalling. Notably, DELLA proteins physically interact with several members of the PIF family, an interaction that inhibits the ability of PIF proteins to bind to and regulate their target promoters. Therefore, PIF proteins represent another integration point between these two pathways (Alabadi and Blazquez 2009).

Cytokinins (CKs) are hormones involved in cell division that affect a wide range of developmental responses. Photomorphogenic development can be mimicked by exogenous application of high concentrations of CKs to dark-grown seedlings. Consistently, the *amp* mutant that contains elevated CK concentrations displays some aspects of de-etiolation when grown in the dark. *ARR4*, an early CK-responsive gene that encodes a response regulator, is also induced by R light via phyB perception. ARR4 also interacts with the N-terminus of phyB, stabilizing the active PfrB form, suggesting that CK signalling might enhance phyB signalling by altering the PrB/PfrB ratio. Consistently, plants overexpressing *ARR4* show hypersentivity to R light. However, this interaction does not explain how CKs promote some aspects of de-etiolation in the dark.

Ethylene is a gaseous hormone that regulates many aspects of plant development, such as the triple response in etiolated seedlings, fruit ripening and senescence. ACC oxidase, a key enzyme in ethylene biosynthesis, is light-regulated in a phytochrome-dependent manner in both *Arabidopsis* and *Sorghum*. The maintenance of the apical hook in etiolated seedling requires normal ethylene perception and signalling, while the unhooking response to light requires functional photoreceptors. Experiments in *Arabidopsis* suggest that perception of ethylene, rather than its production, is light-modulated during de-etiolation (Halliday and Fankhauser 2003). Furthermore, mutation in *BIG* confers aberrant photomorphogenic phenotypes and altered responses to ethylene, CKs, GAs and auxins. The study of *BIG* gene suggests a complex, non-linear interaction between photomorphogenesis and multiple hormones (Halliday and Fankhauser 2003).

14.6.2   Light–Clock Signal Integration

In plants, the circadian clock controls daily changes in gene expression, growth, photosynthetic activity and seasonal flowering. This rhythmic mechanism synchronizes internal signalling processes with external light cues, driving a vast array of metabolic and developmental responses. Clock components and photoreceptors have an intimate relationship; light signals transduced by the phytochromes and cryptochromes ensure that the clock is in tune with daily light/dark cycles. This process, known as photoentrainment, is achieved by adjusting the phase and the period of the oscillator relative to the prevailing photoperiod. Analyses of *CAB:: LUC* (*CHLOROPHYLL A/B BINDING PROTEIN::LUCIFERASE*) transgenic lines, which display circadian luciferase activity in vivo, in different photoreceptor null mutants indicated a role for phytochromes and cryptochromes in controlling photoentrainment under different light conditions.

Genetic and molecular analyses have identified *elf3* (*early flowering 3*) and *tic* (*time for coffee*) mutants, because light induces high levels of *CAB::LUC* expression during the dark period, a time when this response is suppressed in the wild type. This indicates that ELF3 and TIC participate in the differential regulation of day and night time sensitivity to light, a mechanism known as circadian gating that ensures correct entrainment of the clock to changing dawn and dusk signals. In the case of ELF3, moderation of the phytochrome signal may be direct, since ELF3, which is localized in the nucleus, interacts with phyB in vitro. Photoentrainment is also controlled by ZTL, a putative photoreceptor (see Sect. 14.3.3). ZTL has been shown to control protein levels of TOC1 (TIMING OF CAB 1), a component of the biological clock, via ubiquitination and subsequent proteolysis by the 26S proteasome (Mas 2008).

14.7   Applied Aspects of Photomorphogenic Research

Many traits selected for plant productivity in modern agriculture are greatly influenced by photoreceptors. Because of these observations, photoperception modification has become an appealing target for crop improvement.

14.7.1   What Is Fit Under Natural Conditions Might Be Inadequate for Agriculture

In natural environments, competition for light is central for plant success and survival to reproductive maturity, and activation of SAS developmental programs aims to overgrow or survive putative plant competitors even before the plant is actually shaded. However, activation of SAS responses in most crop species is disadvantageous for plant productivity in modern agriculture because it results in resource allocation to stem growth, at the cost of leaf growth and the development of storage and reproductive structures. As this example illustrates, modern agricultural practices place different constraints on plant growth that have not been necessarily selected for during plant evolution, although the natural environment selects for certain traits. Traditionally, breeding efforts have been focused on optimizing grain yields by modulating those characteristics that affect these traits, such as *plant height*, *branching* and *time of flowering*. With the recent interest in lignocellulosic-based biofuels, however, a new breeding paradigm may emerge to optimize biomass at the expense of grain yield. Therefore, an understanding of light signalling might lead to the judicial manipulation of photoreceptor signalling pathways to improve the characteristics of these new traits.

14.7.2   Classical Breeding for the Development of Agronomical Varieties Has Selected Light-Regulated Traits

14.7.2.1   Plant Height

Dwarfing is a method utilized for increasing yield by reducing the resources allocated to structural growth. Dwarfing also increases yield in cereal crops by increasing the resistance to mechanical flattening ("lodging") by wind, rain or heavy yield production. Indeed, selection of semi-dwarf varieties of wheat and rice has contributed to the so-called green revolution and has become the choice of most growers. Molecular analyses of these cereal dwarf varieties have shown that they carry mutant alleles that affect gibberellin pathways. An alternative way to control this trait in newly developed crops is to reduce stem elongation by specifically suppressing SAS responses, rather than hormone pathways. This approach would suppress stem elongation only under high plant density conditions typical of crop monocultures (Kebrom and Brutnell 2007).

14.7.2.2   Branching

In cereal grasses, plant breeders have selected to attenuate some but not all SAS responses within modern crop varieties. One of the most dramatic effects of shade in the grasses is the production and proliferation of basal auxiliary meristems that develop into tillers (lateral branches). In general, high planting densities (low R/FR) result in increased apical dominance at the expense of tiller development. In species or varieties in which tillering is a component of yield, the tillering component of the SAS has been tempered to permit tiller production in crop monocultures; for example, selection of rice was for high tillering varieties that are more productive than low tillering varieties under long growing seasons. However, tiller proliferation in maize is often a negative component of yield, and genetic variation has been selected to enhance or maintain suppression of tillers (Kebrom and Brutnell 2007).

14.7.2.3 Flowering Time

Many agriculturally important plant species have critical daylength requirements for flowering and fruiting. The optimization of photoperiod has become a standard practice, where crops are sown only in the season with the appropriate daylength, by supplementing with artificial light to extend photoperiod or by covering the plants with black cloths to shorten it. The knowledge from physiological studies of phytochrome action has allowed the substitution of constant light by single pulses in the night period to control flowering in some species. Such practices permit considerable cost savings to growers and have had dramatic impacts on the flower industry. The understanding of the molecular basis of daylength perception and how this is coupled to the corresponding evocation response has allowed the identification of key photoperiod regulators, such as CO. In that context, it has been shown that overexpression of *Arabidopsis CO* in potato results in delayed tuberization, suggesting a conserved function for CONSTANS activity in unrelated photoperiod-modulated responses (Kobayashi and Weigel 2007). However, there are no reports of biotechnological applications of this information.

14.8 Is There a General Strategy to Modulate Photomorphogenic Traits for Crop Improvement?

For rapid transfer of knowledge from basic understanding to crop improvement, it would be desirable to find a general strategy to modulate photomorphogenic traits. However, all the photomorphogenic traits for biotechnological improvement are also affected by other exogenous factors (e.g. cold) and endogenous pathways (e.g. hormones), complicating design strategies to improve some traits without undesirable side effects. Altogether, current experience suggests that the knowledge generated by research in photomorphogenesis will be implemented slowly into the important task of crop improvement. In that respect, one fundamental aspect that needs to be better understood is the mechanisms of communication between organs when responding to light signals (Bou-Torrent et al. 2008).

Acknowledgements We thank laboratory members and Dr. M. Phillips for comments on the manuscript.

References

Alabadi D, Blazquez MA (2009) Molecular interactions between light and hormone signaling to control plant growth. Plant Mol Biol 69:409–417

Al-Sady B, Kikis EA, Monte E, Quail PH (2008) Mechanistic duality of transcription factor function in phytochrome signaling. Proc Natl Acad Sci USA 105:2232–2237

Bae G, Choi G (2008) Decoding of light signals by plant phytochromes and their interacting proteins. Annu Rev Plant Biol 59:281–311

Bou-Torrent J, Roig-Villanova I, Martinez-Garcia JF (2008) Light signaling: back to space. Trend Plant Sci 13:108–114

Briggs WR, Christie JM (2002) Phototropins 1 and 2: versatile plant blue-light receptors. Trends Plant Sci 7:204–210

Chen M, Chory J, Fankhauser C (2004) Light signal transduction in higher plants. Annu Rev Genet 38:87–117

Halliday KJ, Fankhauser C (2003) Phytochrome-hormonal signalling networks. New Phytol 157:449–463

Jenkins GI (2009) Signal transduction in responses to UV-B radiation. Annu Rev Plant Biol 60:407–431

Jiao Y, Lau OS, Deng XW (2007) Light-regulated transcriptional networks in higher plants. Nature Rev Genet 8:217–230

Kebrom TH, Brutnell TP (2007) The molecular analysis of the shade avoidance syndrome in the grasses has begun. J Exp Bot 58:3079–3089

Kobayashi Y, Weigel D (2007) Move on up, it's time for change—mobile signals controlling photoperiod-dependent flowering. Genes Dev 21:2371–2384

Li QH, Yang HQ (2007) Cryptochrome signaling in plants. Photochem Photobiol 83:94–101

Lin C, Shalitin D (2003) Cryptochrome structure and signal transduction. Annu Rev Plant Biol 54:469–496

Martinez-Garcia JF, Huq E, Quail PH (2000) Direct targeting of light signals to a promoter element-bound transcription factor. Science 288:859–863

Mas P (2008) Circadian clock function in *Arabidopsis thaliana*: time beyond transcription. Trends Cell Biol 18:273–281

Montgomery BL, Lagarias JC (2002) Phytochrome ancestry: sensors of bilins and light. Trends Plant Sci 7:357–366

Quail PH (2002) Phytochrome photosensory signalling networks. Nature Rev Mol Cell Biol 3:85–93

Roig-Villanova I, Bou-Torrent J, Galstyan A, Carretero-Paulet L, Portoles S, Rodriguez-Concepcion M, Martinez-Garcia JF (2007) Interaction of shade avoidance and auxin responses: a role for two novel atypical bHLH proteins. EMBO J 26:4756–4767

Shin J, Park E, Choi G (2007) PIF3 regulates anthocyanin biosynthesis in an HY5-dependent manner with both factors directly binding anthocyanin biosynthetic gene promoters in Arabidopsis. Plant J 49:981–994

Sorin C, Salla-Martret M, Bou-Torrent J, Roig-Villanova I, Martinez-Garcia JF (2009) ATHB4, a regulator of shade avoidance, modulates hormone response in Arabidopsis seedlings. Plant J 59:266–277

Tao Y, Ferrer JL, Ljung K, Pojer F, Hong F, Long JA, Li L, Moreno JE, Bowman ME, Ivans LJ, Cheng Y, Lim J, Zhao Y, Ballare CL, Sandberg G, Noel JP, Chory J (2008) Rapid synthesis of auxin via a new tryptophan-dependent pathway is required for shade avoidance in plants. Cell 133:164–176

Yi C, Deng XW (2005) COP1 - from plant photomorphogenesis to mammalian tumorigenesis. Trends Cell Biol 15:618–625

Part IV
Molecular Genetics of Developmental Regulation

Chapter 15
RNA Silencing in Plants

A. Eamens, S.J. Curtin, and P.M. Waterhouse

15.1 Introduction

RNA or gene silencing is a broad term used to describe the vast array of related processes involving RNA–RNA, RNA–DNA, RNA–protein or protein–protein interactions that ultimately result in the repression of gene expression (Wang and Metzlaff 2005). RNA silencing-related mechanisms have been described in almost all living organisms, with the notable exception of bacteria and the yeast *Sacchromyces cerevisiae*, and is termed transcriptional (TGS) or posttranscriptional gene silencing (PTGS) in plants, quelling in fungi and RNA interference (RNAi) in animals, insects and nematodes. In plants, RNA silencing pathways have evolved to extraordinary levels of complexity and diversity, playing crucial roles in providing protection against invading nucleic acids derived from viruses or replicating transposons, controlling chromatin modifications as well as regulating endogenous gene expression to ensure normal plant growth and development. RNA silencing mechanisms regulate gene expression through the suppression of mRNA transcription or by the initiation of sequence-specific mRNA degradation. In animals, RNA silencing also regulates gene expression by inhibiting translation. However, this mechanism of silencing appears to occur less frequently in plants. Due to the fact that RNA silencing-related mechanisms have been observed in almost all eukaryotes, RNA silencing is considered to have ancient evolutionary origins. An increasing volume of evidence suggests that RNA silencing was a major

A. Eamens and P.M. Waterhouse
CSIRO Plant Industry, 1600, Canberra, ACT 2601, Australia
School of Molecular and Microbial Biosciences, University of Sydney, Sydney, NSW 2006, Australia
e-mail: a.eamens@usyd.edu.au

S.J. Curtin
CSIRO Plant Industry, 1600, Canberra, ACT 2601, Australia
School of Wine & Food Science, Charles Sturt University, Wagga Wagga, NSW2678, Australia
University of Minnesota, St. Paul, Minnesota 55108–6026, USA

E-C. Pua and M.R. Davey (eds.),
Plant Developmental Biology – Biotechnological Perspectives: Volume 2,
DOI 10.1007/978-3-642-04670-4_15, © Springer-Verlag Berlin Heidelberg 2010

contributing factor in the evolution of multicellular, eukaryotic organisms from prokaryotic progenitors (Sharp 2001; Margis et al. 2006).

15.2   History of RNA Silencing in Plants

Observations of gene silencing were first made in plants in the early 1990s and were initially thought to be a side effect of introducing transgenes into the plant genome via *Agrobacterium tumefaciens*-mediated transformation. In an early study of tobacco (*Nicotiana tobaccum*) plants sequentially transformed with two T-DNA vectors encoding different selectable marker genes, it was observed that the expression of the marker gene encoded by the first T-DNA vector became inactive upon introduction of the second T-DNA (Matzke et al. 1989). The observed transgene inactivation was correlated with methylation of the promoter sequences driving the expression of the selectable marker genes encoded by the two vectors. The authors went on to show that the initiation of DNA methylation and inactivation of the initial selectable marker gene was dependent on the genomic integration of the second T-DNA. They suggested that the substantial homology shared by the two vectors, including two copies of the *nopaline synthase* promoter per T-DNA insert, may have initiated methylation of the first vector. In addition, when petunia (*Petunia hybrida*) plants were engineered to carry additional copies of the flower pigmentation gene, chalcone synthase (chsA), with the aim of intensifying the purple colouration of flowers, an unexpected result was observed. Instead of producing intensely purple flowers, these transformed petunia lines expressed a dramatic range of flower pigmentation, including flowers of dark purple, patterns of purple and white colouration and flowers that were completely white. Experimental analyses revealed that both the endogenous and transgene-encoded versions of chsA were silenced to differing degrees depending on the colouration phenotype expressed by individual plant lines. Nuclear run-off transcription analysis demonstrated that the loss of *chsA* mRNA was not associated with reduced transcription of either the introduced or endogenous form of the chsA gene. The authors referred to this phenomenon as "cosuppression" (Napoli et al. 1990; van der Krol et al. 1990).

Cosuppression-like observations were also made in plant lines engineered to express viral-encoded sequences, namely the viral coat protein (CP) or a segment of viral replicase, and such plant lines generally conferred resistance to the virus from which these introduced sequences were derived. In one such study, Lindbo et al. (1993a) produced transgenic tobacco plants expressing a non-translatable CP sequence from *Tobacco etch virus* (TEV) and showed that these plant lines were resistant to TEV infection, but not to the unrelated *Potato virus X* (PVX). Plants infected with TEV returned to a 'recovered', healthy non-infected state approximately 3 to 5 weeks after the initial inoculation event, and recovered leaves did not support subsequent inoculations with TEV. The authors went on to show that TEV resistance was associated with a decreased steady-state level of transgene mRNA, but not with reduced rates of transgene transcription, and concluded that the sequence-specific RNA degradation was induced by excessive levels of transgene- and virus-derived CP mRNA. The authors also postulated that a host-encoded RNA-dependent RNA polymerase (RDR) was involved in this RNA-mediated resistance mechanism (Lindbo et al. 1993b).

Continued focus on over-expression of the chsA gene in petunia also suggested that transgene-derived RNA was acting as the trigger to induce the sequence specificity of cosuppression. Purple-coloured flowers were shown to express the *chsA* transcript at high levels. However, *chsA* expression in white flowers actively undergoing cosuppression was dramatically reduced. Molecular analyses of cosuppressed flowers revealed not only that the expression of full-length *chsA* transcript was reduced, but also that truncated versions of *chsA* accumulated to high levels in plants with white flowers. In addition, a percentage of these truncated mRNAs were shown to be poly(A)$^-$ molecules that expressed extensive secondary structure. Metzlaff et al. (1997) suggested that the over-expression of the chsA transgene resulted in the formation of aberrant RNA species in the nucleus that formed extensive secondary structures leading to their degradation. The initiation of chsA silencing was subsequently associated with the presence of multiple T-DNA insertions, integrated into the plant genome at a single locus in an inverted-repeat (IR) orientation. Furthermore, PTGS of chsA was shown to be exclusive to transformant lines harbouring such IR multi-copy insertions and that PTGS failed to initiate in any of the single-copy T-DNA insertion lines analysed (Stam et al. 1997).

Correlation of the presence of IR T-DNA insertions with the initiation of cosuppression in petunia led to the development of the first IR transgene for gene silencing in plants. The IR vector developed by Waterhouse et al. (1998) was designed to express hairpin RNA (hpRNA) against the β-glucuronidase (gus or uidA) reporter gene in rice (*Oryza sativa*). The authors showed that expression of the hpRNA vector, from which a double-stranded RNA (dsRNA) molecule is transcribed, resulted in very high levels of silencing, with GUS expression silenced in 90% of transformant lines when the hpRNA construct was superimposed on the endogenous GUS activity of rice callus via *Agrobacterium*-mediated transformation. The hpRNA vector was also shown to induce much more potent silencing than did conventional sense or antisense versions of the *gus* transgene. In the same study, the authors also showed that tobacco plants transformed with both sense and antisense transgenes encoding a *Potato virus Y* (PVY) protease protein were highly resistant to PVY infection. From these results, Waterhouse et al. (1998) proposed a model for a plant surveillance system that is induced by a dsRNA trigger to direct the sequence specificity of PTGS; Fig. 15.1 presents a simplified version of their original model. In the same year, Fire et al. (1998) showed that the introduction of dsRNA into the nematode *Caenorhabditis elegans* resulted in significant silencing of the mRNA in question, compared to conventional silencing strategies based on the introduction of a sense or antisense transcript alone. Of considerable interest was their finding that only a small amount of dsRNA was required to induce efficient RNA silencing, suggesting that a catalytic or amplification step was required in the RNAi process.

Following the 1998 discovery that dsRNA triggers RNA silencing in plants (Waterhouse et al. 1998), nematodes (Fire et al. 1998), protozoa (Ngo et al. 1998)

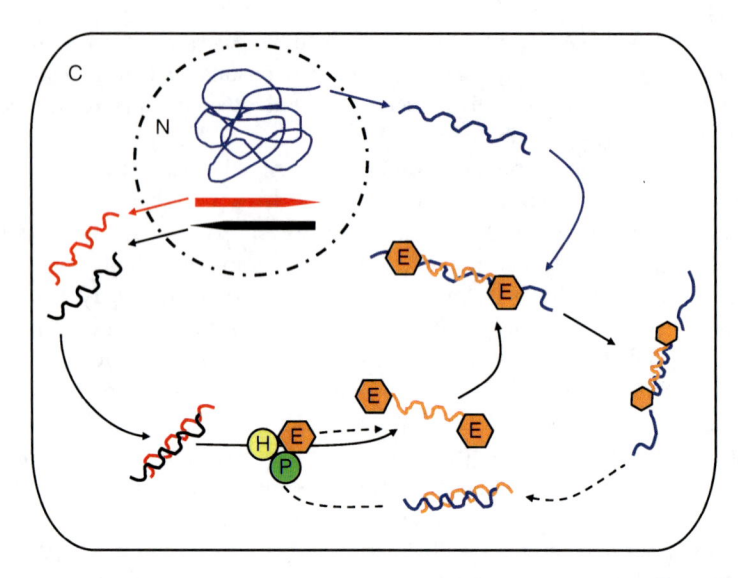

Fig. 15.1 The model proposed by Waterhouse et al. (1998) to explain dsRNA-directed PTGS. Both sense (*black pointed-line*) and antisense (*red pointed-line*) RNA is transcribed from the introduced transgenes in the nucleus (N), and is exported to the cytoplasm (C) where these single-stranded RNA (ssRNA) molecules hybridise to form dsRNA (*black and red duplex*). The dsRNA is recognised by a protein complex, consisting of a helicase (H, *yellow circle*), a dsRNA-dependent RNA polymerase (P, *green circle*) and an RNase L-like nuclease (E, *orange hexagon*). This complex transcribes a complementary RNA (cRNA, *orange squiggle*) transcript to which the RNase L-like nuclease is attached. The cRNA–RNase L complex searches for, and attaches to homologous endogene- or viral-derived transcripts (*blue squiggle*), and cleaves the ssRNA surrounding the dsRNA duplex (*orange and blue duplex*). The cleaved ssRNAs are then degraded by other plant nucleases, either silencing the expression of an endogenous gene or to provide virus resistance. The intact RNA duplex is resistant to degradation and can act as a template for the synthesis of additional cRNA–RNase L complexes (*dashed black arrows*). Once the cycle is initiated, there is no further requirement for the nuclear-encoded dsRNA trigger

and insects (Kennerdell and Carthew 1998), the second major discovery in the field of RNA silencing was made in plants that following year. Hamilton and Baulcombe (1999) identified and associated small RNA (sRNA) molecules with plants actively undergoing PTGS. The authors screened for sRNAs in four different silencing backgrounds, these being (1) tomato (*Lycopersicon esculentum*) plants undergoing cosuppression of a endogenous gene, following the insertion of a homologous transgene, (2) tobacco plants undergoing PTGS of the gus reporter gene, (3) *Nicotiana benthamiana* plants in which the introduced green fluorescent protein (gfp) gene was systemically silenced following inoculation with *Agrobacterium* harbouring a gfp transgene, and (4) *N. benthamiana* plants inoculated with PVX. The authors showed that the silencing of the three targeted genes and the infecting virus was associated with the accumulation of sRNA species ∼21 to 25 nucleotides (nt) in length, and that these sRNAs were of both the sense and antisense polarity. In addition, the authors revealed a correlation between the level of sRNA accumulation and the silencing efficiency conferred by each of the analysed silencing systems.

These sRNA species are now collectively referred to as small-interfering RNAs (siRNAs). Subsequent studies in fly (*Drosophila melanogaster*) in vitro systems revealed the biochemical features of siRNAs; they generally have 2-nt $3'$ overhangs with respect to the complementary strand of the dsRNA duplex from which they are processed, and both strands of the siRNA duplex carry a $5'$ phosphate and $3'$ hydroxyl group (Zamore et al. 2000). Additional experiments in *Drosophila* embryos identified the RNase III-like endonuclease Dicer as being the endonuclease responsible for processing siRNA duplexes from longer molecules of dsRNA (Bernstein et al. 2001). *Carpel factory* (*caf*), an Arabidopsis (*Arabidopsis thaliana*) mutant line characterised by floral meristem determinacy defects and other whole-plant organ morphogenesis abnormalities, was originally identified in a developmental screen by Jacobsen et al. (1999) and has since been identified as one of the four Dicer-like (DCL) proteins encoded for by the Arabidopsis genome. Molecular dissection of the *caf* mutant revealed that the CAF gene encodes an endonuclease protein with similarities to both DExH/DEAD-box type RNA helicases and RNase III proteins (Jacobsen et al. 1999). We now know that the floral defects associated with the *caf* phenotype result from the mutant's inability to process a class of endogenous sRNAs from their precursor molecules of dsRNA (Golden et al. 2002; Park et al. 2002). The CAF gene is now referred to as DCL1, the founding member of the plant-specific RNase III-like endonuclease family of proteins that are responsible for dsRNA cleavage to produce species of sRNAs.

In an earlier developmental screen, a mutant line termed *Argonaute1* (*ago1*) was identified. The Arabidopsis *ago1* mutant expressed unexpanded pointed cotyledons, very narrow rosette leaves, and a single thickened and partially fasciated inflorescence (Bohmert et al. 1998). Homology searches not only revealed that the Arabidopsis genome encodes nine additional AGO proteins, but also showed that AGO1 expressed substantial homology to the *C. elegans* RNAi-deficient 1 (RDE-1) protein. RDE-1 deficient worms have been shown to be unable to mount an efficient RNAi response when exposed to a dsRNA trigger (Tabara et al. 1999). AGO1 has since been shown to play a central role in RNA silencing in plants, with AGO1 using single-stranded sRNA guide strands processed from sRNA duplexes to direct cleavage of complementary mRNA transcripts (Baumberger and Baulcombe 2005). The molecular characterisation of AGO1 formed the final piece of the simplest form of the RNA silencing puzzle in plants, that is, following the formation of a dsRNA molecule, a DCL protein recognises its specific dsRNA substrate and processes this molecule into shorter ~21- to 25-nt duplexes of dsRNA, from which a single strand—the sRNA guide strand—is released to an AGO protein to direct a specific RNA silencing-mediated response.

15.3 The Parallel Gene Silencing Pathways of Plants

As mentioned above, it is generally believed that RNA silencing is an evolutionarily conserved process with the basic components of RNA silencing pathways possibly arising before the divergence of plants and animals (Sharp 2001; Margis et al. 2006). The fact that divergent organisms share similar pathway components suggests that RNA silencing is a fundamental biological process and a universal endogenous tool for the regulation of gene expression (Cogoni and Macino 2000). In plants, the parallel gene silencing pathways have been shown to be involved in controlling the expression of developmentally regulated genes, repressing the mobility of endogenous transposable elements and defending against invading viral-derived nucleic acids. In Arabidopsis, four classes of endogenous sRNA species have been documented: the microRNA (miRNA; Lee and Ambros 2001), *trans*-acting siRNA (tasiRNA; Dunoyer et al. 2005; Xie et al. 2005), natural-antisense siRNA (natsiRNA; Borsani et al. 2005) and repeat-associated siRNA (rasiRNA; Meister and Tuschl 2004) classes. The RNA silencing pathway(s) in which each of these classes of sRNA directs a specific RNA silencing mechanism are discussed below and a schematic representation of each pathway is provided in Fig. 15.2.

15.3.1   The MicroRNA Pathway

MicroRNAs are encoded by endogenous genes (MIR genes), from which a primary non-protein coding message is transcribed (pri-miRNA). The pri-miRNA sequence contains an imperfect hairpin stem-loop structure allowing the molecule to fold-back onto itself to form dsRNA. The first miRNA, lin-4, was isolated in *C. elegans* (Lee et al. 1993) and, to date, hundreds of unique miRNAs have been identified in Arabidopsis (Millar and Waterhouse 2005). In plants, the pri-miRNA transcript is cleaved by DCL1 in the nucleus, with the combined action of the double-stranded RNA binding protein (dsRBP), Hyponastic Leaves1 (HYL1), to produce a shorter precursor-miRNA (pre-miRNA) molecule of dsRNA. The first DCL1-catalysed cleavage step in the miRNA biogenesis pathway is made just below the miRNA duplex region, and the duplex itself is then liberated from the stem-loop structure of the pre-miRNA transcript by the second DCL1-catalysed cleavage step of this pathway. As for the first dsRNA cleavage step preformed by DCL1, efficient pre-miRNA processing by DCL1 requires HYL1. Recently, it has been shown that the efficient processing of some classes of pri-miRNA also requires the action of the ethylene zinc finger protein Serrate (SE). However, the exact role of SE in miRNA processing remains to be determined (Yang et al. 2006). Once liberated from the pre-miRNA transcript, the 2-nt 3' overhangs of miRNA duplexes are methylated by the sRNA-specific methyltransferase Hua Enhancer1 (HEN1). Small-interfering RNA duplexes are also methylated by HEN1, a process that is thought to protect all sRNA species from polyuridylation and subsequent degradation (Chen et al. 2002; Yu et al. 2005). The methylated miRNA duplex is then exported to the cytoplasm by the *Drosophila* Exportin-5 (Exp-5) homolog Hasty (HST), and as outlined for SE, the exact role specified by HST in miRNA biogenesis remains to be determined because certain families of miRNA are transported from the nucleus to the cytoplasm via a HST-independent mechanism (Park et al. 2005). In the

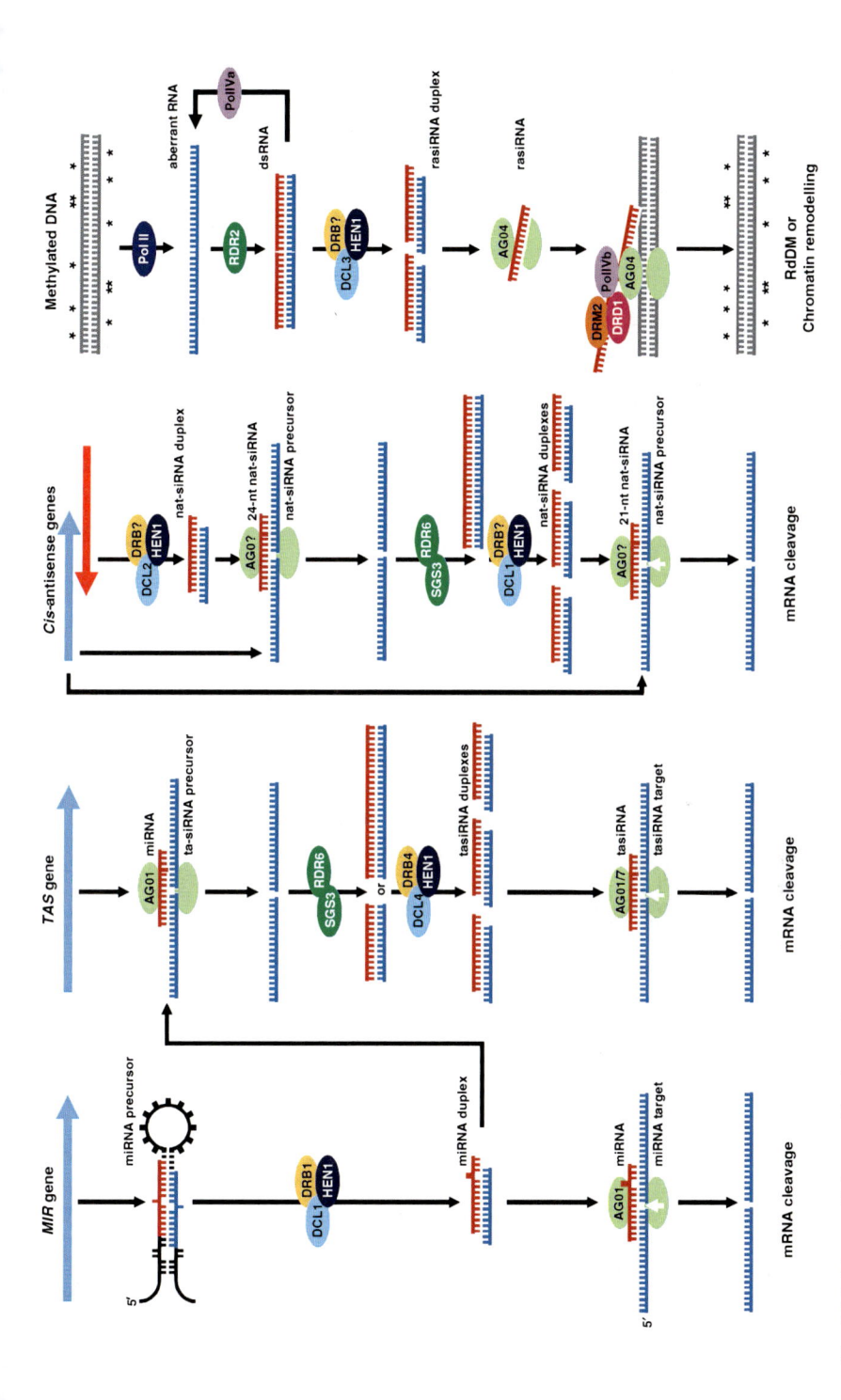

Fig. 15.2 The parallel gene silencing pathways of the model dicotyledonous plant species, *Arabidopsis thaliana*: schematic representation of the parallel RNA silencing pathways, including the miRNA, tasiRNA, natsiRNA and rasiRNA/RdDM pathways. The individual step, or steps directed by each of the RNA silencing-associated proteins discussed in the text of this chapter are outlined

cytoplasm, the guide strand from the miRNA duplex is loaded onto AGO1, the catalytic centre of the plant RNA-induced silencing complex (RISC). The miRNA guide strand is used by RISC to direct silencing of complementary mRNAs, mediated by the 'Slicer' endonuclease activity of AGO1 for transcript cleavage (Baumberger and Baulcombe 2005), and the majority of plant miRNAs appear to repress gene expression via transcript cleavage. This is in direct contrast to miRNA-mediated repression of gene expression in animals and insects, where the predominant mode of action is translational repression, mediated by the binding of the miRNA to its target mRNA (Mallory and Vaucheret 2006). Interestingly, DCL1 and HYL1 have recently been shown to co-localise to distinct nuclear bodies, termed nuclear dicing bodies or 'D-bodies' (Fang and Spector 2007). Furthermore, the miRNA biogenesis machinery HEN1, SE and AGO1 were also shown to localise to the nucleus, specifically to regions surrounding these nuclear bodies, suggesting that D-bodies may act as small specialised factories for miRNA production in plants. The nuclear localisation of AGO1 also raises the possibility that miRNA duplexes are directly loaded onto AGO1, prior to this protein's nuclear exportation to the cytoplasm, where AGO1 containing RISC would detect cognate mRNAs for cleavage.

15.3.2 The trans-*Acting siRNA Pathway*

The tasiRNA class of endogenous sRNAs are generated via an overlapping, yet distinct silencing pathway from the miRNA biogenesis pathway. In plants, two miRNAs, miR173 and miR390, have been shown to induce an additional level of complexity to the control of gene expression for normal development (Axtell et al. 2007). These two miRNAs target five tasiRNA transcripts (*TAS*) in a DCL1/HYL1-mediated process to initiate the 21-nt phasing of tasiRNA production. However, instead of being identified for silencing, these miRNA-cleaved non-protein coding *TAS* transcripts become templates for dsRNA synthesis by the RDR RDR6. RDR6 requires the assistance of the coiled-coil protein Suppressor of Gene Silencing3 (SGS3) to produce the dsRNA molecule that is processed into phased 21-nt tasiRNAs by DCL4, in a sequential process initiated at the miRNA cleavage site (Vazquez et al. 2004; Yoshikawa et al. 2005; Xu et al. 2006). As shown for DCL1/HYL1-directed processing of MIR genes, DCL4 requires the coordinated action of the dsRBP DRB4 to efficiently process *TAS* dsRNA into tasiRNAs. The tasiRNAs then target their own specific transcripts for degradation via AGO-catalysed mRNA cleavage; nearly one third of the 23 characterised auxin response factor (ARF) genes in Arabidopsis have been shown to be tasiRNA targets. The ARF genes encode transcription factors that induce auxin signalling during plant development (Jones-Rhoades and Bartel 2004; Allen et al. 2005). Thus, the tasiRNA biogenesis pathway is directly linked to the miRNA pathway and, like miRNAs, this specific class of sRNA plays a pivotal role in regulating normal plant development.

15.3.3 The Natural-Antisense siRNA Pathway

More than 2,000 pairs of natural-antisense genes have been identified in Arabidopsis. These *cis*-antisense gene pairs are transcribed from different DNA strands to produce a dsRNA molecule via annealing of the overlapping $3'$ ends of the two individual transcripts (Jin et al. 2008). Annealing of the complementary $3'$ ends provides a substrate for DCL2-mediated dsRNA cleavage and the generation of a single 24-nt natural-antisense siRNA (natsiRNA). The single natsiRNA then guides the cleavage of one transcript from the *cis*-antisense gene pair, to produce a cleaved RNA transcript that is converted into dsRNA by RDR6 and SGS3. This RDR6/SGS3-generated dsRNA molecule is converted into phased 21-nt natsiRNA duplexes by DCL1. The phased 21-nt natsiRNAs, like tasiRNAs, are in turn used to direct the targeting of cognate mRNAs for degradation. For example, Borsani et al. (2005) showed one such natural-antisense gene pair being induced by salt stress in Arabidopsis. One of the *cis*-antisense genes, designated SRO5, encodes a protein of unknown function, while the other gene of this pair encodes Δ^1-pyroline-5-carboxylate dehydrogenase (P5CDH). Expression of the SRO5 gene is induced by salt stress, resulting in DCL2-mediated generation of the single 24-nt natsiRNA that in turn results in DCL1-mediated cleavage of the *P5CDH* transcript. Downregulation of *P5CDH* expression results in the accumulation of proline, leading to the plant's increased tolerance to salt stress.

15.3.4 The Repeat-Associated siRNA/RNA-Directed DNA Methylation Pathways

Transcriptional gene silencing (TGS) is an epigenetic mechanism, resulting in the silencing of an introduced transgene or an endogenous sequence through the inactivation of their promoter sequences via DNA methylation. DNA methylation is essential for normal plant and animal development. In plants, the majority of DNA methylation is associated with repetitive sequences, such as transposons and retroelements, and DNA methylation of such repeat sequences is thought to occur as a natural suppressor to control their expression (Wassenegger 2005). In Arabidopsis, repeat sequences have also been shown to be an extremely rich source of siRNAs. These repeat-associated siRNAs (rasiRNAs) differ to miRNAs and tasiRNAs as they are of the 24-nt size class. The rasiRNAs are thought to direct DNA methylation to transcriptionally silence repetitive sequences in the plant genome to maintain genome integrity (Chan et al. 2005). In addition, mutations to genes encoding various proteins involved in the parallel gene silencing pathways of Arabidopsis also cause disruption to heterochromatic structures. The loss of DNA methylation from regions of repetitive DNA suggests that this unique size class of endogenous siRNA plays an important role in regulating normal plant development (Chan et al. 2004; Onodera et al. 2005).

Wassenegger et al. (1994) were the first to show that viroid replication could induce RNA-directed DNA methylation (RdDM) of a homologous transgene. Subsequently, Jones et al. (1998) demonstrated that nuclear DNA sequences, homologous to viral sequences, became methylated upon infection with a cytoplasmically replicating virus. In addition, Wang et al. (2001) reported a cytoplasmically replicating viral satellite RNA inducing high levels of DNA methylation of a homologous satellite transgene. The authors of all three studies speculated that a sequence-specific RNA signal was able to enter the nucleus to direct RdDM. Wang et al. (2001) went on to show that the sequence-specific DNA methylation observed in their study was associated with the accumulation of satellite-specific siRNAs, providing strong evidence that siRNAs also act to trigger RdDM and TGS. From earlier studies, it was postulated that siRNAs guide DNA methyltransferases to homologous sequences throughout the plant genome, and subsequent studies of DNA methylation in RNA silencing mutants identified several silencing proteins as being involved in RdDM, including DCL3 and AGO4 (Zilberman et al. 2003; Chan et al. 2004).

Following these initial findings, a number of investigations have concentrated on identifying RdDM mutants in Arabidopsis to identify all RNA silencing machinery involved in the RdDM pathway (Mette et al. 2000; Kanno et al. 2004, 2005). Results of these studies have begun to form a detailed picture of the RdDM pathway in Arabidopsis. In brief, methylated DNA is thought to act as a template for the transcription of an aberrant mRNA molecule, by a protein expressing RNA polymerase activity, be it RNA polymerase II (PolII), RDR2, or the recently identified plant-specific PolIVa. Alternatively, it has been suggested that PolIVa could

Table 15.1 The RNA silencing-associated proteins of *Arabidopsis thaliana*

Protein	Locus	Protein class and function
AGO1	At1g48410	RNA slicer - core component of plant RISC
AGO4	At2g27040	RNA slicer - involved in the establishment phase of RdDM
AGO6	At2g32940	RNA slicer - rasiRNA-directed heterochromatin formation
CMT3	At1g69770	Methyltransferase - maintenance phase of RdDM
DCL1	At1g01040	RNase III - miRNA, natsiRNA and tasiRNA biogenesis
DCL2	At3g03300	RNase III - natsiRNA biogenesis and viral defence
DCL3	At3g43920	RNase III - rasiRNA biogenesis and establishment phase of RdDM
DCL4	At5g20320	RNase III - tasiRNA biogenesis and viral defence
DRM2	At3g17310	Methyltransferase - establishment phase of RdDM
HYL1	At1g09700	dsRBP - miRNA and tasiRNA biogenesis
DRB4	At3g62800	dsRBP - miRNA and tasiRNA biogenesis
DRD1	At2g16390	SNF2-like chromatin remodelling factor - establishment phase of RdDM
HEN1	At4g29160	sRNA-specific methyltransferase - sRNA biogenesis
HST	At3g05040	Exportin-5 ortholog - miRNA exportation from nucleus
MET1	At5g49160	Methyltransferase - maintenance phase of RdDM methylation
PolIVa	At1g63020	DNA-dependent RNA polymerase - establishment phase of RdDM
PolIVb	At2g40030	DNA-dependent RNA polymerase - establishment phase of RdDM
RDR2	At4g11130	RNA-dependent RNA polymerase - rasiRNA biogenesis
RDR6	At3g49500	RNA-dependent RNA polymerase - tasiRNA and natsiRNA biogenesis
SGS3	At5g23570	Coiled-coil protein - tasiRNA and natsiRNA biogenesis

recognise RDR2-generated dsRNA to transcribe additional aberrant molecules of RNA in a self-perpetuating loop (Vaucheret 2005). The dsRNA molecule is subsequently processed into 24-nt siRNAs by DCL3, and these siRNAs are methylated by HEN1 and used by AGO4 to direct the actual sequence-specific DNA methylation step of RdDM. The methylation step of RdDM is mediated by the combined actions of the SNF2-like chromatin remodelling protein Defective in RNA-directed DNA Methylation1 (DRD1), the alternate form of the plant-specific PolIV, PolIVb, and the primary de novo DNA methyltransferase Domains Rearranged Methylase2 (DRM2). Once established, the maintenance DNA methyltransferases Methyltransferase1 (MET1) and Chromomethylase3 (CMT3) would maintain DNA methylation to allow the methylated DNA to act as template for further rounds of RdDM (Cao et al. 2003; Zilberman et al. 2003; Kanno et al. 2004, 2005). Table 15.1 presents each of the abovementioned proteins identified as functioning in the parallel gene silencing pathways of Arabidopsis.

15.4 RNA Silencing as an Antiviral Defence Mechanism

As mentioned above, plants transformed with viral-derived sequences show an increased resistance to the virus from which the introduced sequence was derived, and numerous related observations suggest that RNA silencing is a natural antiviral defence mechanism that has evolved to combat or silence the expression of such foreign invading nucleic acids. All organisms from bacteria to humans can be infected by viruses but, unlike most animals, plants do not have an immune system with which to defend themselves. Instead, plants have expanded the number of genes encoding components of the various RNA silencing pathways to develop a powerful defence system of their own (Waterhouse 2006). For example, Arabidopsis encodes four DCL genes and rice has six, compared to animals that encode only a single Dicer protein. Most plant viruses have single-strand RNA genomes that, upon infection, are transcribed into dsRNA either by a virally encoded RNA polymerase or by the plant-encoded RDR RDR6, as part of the plant's own defence-response mechanism (Waterhouse and Fusaro 2006). Virally derived dsRNA is recognised by at least two of the plant's DCL enzymes to produce different size classes of siRNA. The size class of viral siRNA produced depends not only on which DCL protein processes the dsRNA, but also on the strain of the infecting virus. For example, when Arabidopsis is infected with *Tobacco rattle virus* (TRV), 21- and 24-nt siRNAs are produced but, when infected by the unrelated virus *Turnip crinkle virus* (TCV), only 22-nt viral-derived siRNA is generated. These siRNA accumulation profiles change when viruses are used to infect plant lines in which the expression of one of the DCL proteins has been eliminated, revealing the hierarchal action of the DCL gene family to ensure the infected plant has the greatest chance possible of mounting an efficient antiviral response to protect its genome integrity (Deleris et al. 2006). In Arabidopsis, DCL4 appears to be the first line of defence against viral infection. In the absence of DCL4, however, DCL2 activity comes to the fore to process viral-derived dsRNA to initiate a siRNA-directed PTGS response (Waterhouse and Fusaro 2006).

The redundant actions of RNA silencing proteins in plants are thought to be required for negating the effects of viral infection, as almost all plant viruses encode multifunctional proteins that can suppress the plant's own PTGS-mediated defence mechanisms. The first viral silencing suppressors to be identified were the helper component protease (P1/HC-Pro) from Potyviruses and the 2b protein from Cucumoviruses (Anandalakshmi et al. 1998; Soards et al. 2002). Following the initial discovery of Potyvirus HC-Pro and Cucumovirus 2b suppressor proteins, numerous analogous proteins were shown to be encoded by other plant RNA and DNA viruses, including Tombus virus-encoded p19, Closterovirus p21, Potexvirus p25 and the p38 suppressor from Carmoviruses. Viral suppressors of gene silencing enhance the virus's infectivity either by interfering with the action of RNA silencing machinery directly, or by binding to dsRNA or siRNA species to sequester their subsequent use in directing RNA silencing (reviewed in Moissard and Vionnet 2004; Voinnet 2005; Li and Ding 2005). Not surprisingly, viral suppressor proteins such as p19 and p21 also have the ability to bind to endogenous sRNA duplexes, be they miRNA or siRNA, affecting the accumulation and functions of the endogenous RNA silencing pathways directed by these specific sRNA classes, and ultimately resulting in alterations to plant development (Chapman et al. 2004; Dunoyer et al. 2004). For example, HC-Pro does not appear to interact directly with miRNAs, but it does enhance their accumulation in plants, possibly by preventing separation of the miRNA duplex, thereby indirectly influencing plant development (Chapman et al. 2004). Recent studies have shown that some classes of plant viruses have also evolved additional survival mechanisms that minimise their interference with the endogenous siRNA and miRNA biogenesis pathways to limit their detrimental effects to plant development upon infection.

The discovery of RNA silencing also serving as a natural antiviral defence mechanism in plants has provided the plant scientific community with a new avenue for the development of antiviral strategies. The expression of transgenes encoding hpRNA has been shown to form a powerful tool for engineering virus resistance in plants. Wang et al. (2000) have been able to induce complete resistance to *Barley yellow dwarf virus* (BYDV) in barley (*Hordeum vulgare*) plants transformed with such a hpRNA transgene. Similar results were achieved in tobacco plants expressing a hpRNA construct against PVY (Smith et al. 2000). More recently, an alternate approach has been employed to control virus infection in plants. Artificial miRNA (amiRNA) vectors consisting of a sequence of the *Arabidopsis* miR159a pre-miRNA transcript in which the endogenous miRNA sequence had been replaced with an amiRNA specific to the viral suppressor genes p69 and HC-Pro were shown to provide resistance against *Turnip yellow mosaic virus* (TYMV) and *Turnip mosaic virus* (TMV) infection respectively. In addition, a dimeric vector that expressed both of the abovementioned amiRNAs simultaneously, and targeting both viral suppressors, conferred resistance to TYMV and TMV infection in inoculated Arabidopsis plants (Niu et al. 2006). In a separate study, an amiRNA vector targeting the 2b viral suppressor of the *Cucumber mosaic virus* (CMV), a silencing suppressor that interacts with and blocks the Slicer activity of AGO1, was also shown to confer resistance to CMV in transformed tobacco plants. Qu et al. (2007) went on to show a strong correlation between viral resistance and the expression level of the 2b-specific amiRNA for individual plant lines.

15.5 Current Applications of RNA Silencing Strategies to Alter Plant Development

RNA silencing at both the transcriptional and posttranscriptional level has been used as a research tool for almost a decade. We are now starting to see these technologies being used not only to validate gene function, but also to develop commercially focused applications in plant science. In addition to developing plant lines that are resistant to viral attack, RNA silencing-related approaches are currently being applied to a diverse spectrum of plant development issues. These include studying the individual roles specified by unique miRNA species in plant development, the modification of plant metabolic pathways for the production of medical therapeutics and other consumer-friendly plant-derived products, and even focusing on the specialised field of producing designer flowers.

An alternate amiRNA-like approach has recently been developed in plants, not to artificially over-express a particular endogenous miRNA of interest, but to antagonise the endogenous miRNA's ability to bind and cleave its specific target (s), providing the plant research community with a new functional analysis tool to study miRNA action. This novel approach has been termed 'target-mimicry' and relies on the expression of a small, non-protein coding RNA that contains a complementary miRNA cleavage site within its sequence. A 23-nt motif is engineered into the non-coding mRNA to express critical mismatched sequences to the miRNA under analysis, most notably a mismatched bugle directly opposite the miRNA cleavage position for plant miRNAs at positions 9–11 of the miRNA. Franco-Zorrilla et al. (2007) used this approach to study the effects of 'knocking-out' the expression of *Arabidopsis* miRNAs miR156 and miR319, to produce transformant lines that expressed marked developmental phenotypes. The authors demonstrated that target-mimicry consisting of a non-cleavable mRNA that forms a non-productive interaction with the corresponding miRNA provides a new approach to study miRNA function in plant development.

Another widely utilised RNA silencing approach has been the alteration of metabolic pathways in plants. For example, RNA silencing technologies have been used to alter cottonseed oil to make it healthier for human consumption. Cottonseed oil has high levels of palmitic acid, which provides the oil with stability at the high temperatures required for deep-frying but also gives the oil its low-density lipoprotein (LDL) cholesterol-raising properties in humans. Liu et al. (2002) used hpRNA-encoded transgenes to silence the $\Delta9$ and $\Delta12$ desaturases that catalyse the biosynthesis of the fatty acids oleic acid and stearic acid, to produce plants with seed oil that is much more suitable for human consumption. These fatty acids give the oil a greater thermostability, as required for its industrial applications, but without the unwanted LDL cholesterol-raising properties. Similarly, the starch composition of wheat destined for human consumption in developed countries has been modified via transformation of wheat with a hpRNA vector targeting an isoform of the starch-branching enzyme. Altering the wheat starch amylose/amylopectin ratio with the use of a hpRNA vector targeting the starch-branching enzyme isoform, Regina et al. (2006) produced a high-amylose wheat that, if widely adopted in Western countries, could have significant public health benefits.

RNA silencing technologies also have many important non-food applications to specifically alter certain aspects of plant development. These include altering the photosynthetic pathway in algae to increase bioreactor performance (Mussgnug et al. 2007), and metabolically altering the morphine biosynthesis pathway in poppies to increase the yield of pharmaceutically significant compounds (Allen et al. 2004). However, the next application of an RNA silencing technology for the release of a commercially viable plant-developed product is one that will deliver aesthetic, rather than nutritional, environmental or medical benefits. RNA silencing has been used in the generation of the blue rose, the development of which has been a long-running quest for plant breeders. Rose plants naturally lack an enzyme involved in the biosynthesis of dihydromyricetin, an intermediate compound in the production of dephinidin-based anthocyanins, involving the major pigmentation gene for violet- and blue-coloured flowers. Initially, when the gene encoding this enzyme in *Viola* was introduced into roses, the transformed plants expressed flowers with purple-coloured flowers, as one of the rose-encoded genes involved in the conversion of dihydromyricetin into delphinidin is also responsible for the conversion of other intermediate compounds into red and yellow pigments. However, Katsumoto et al. (2007) silenced this gene through the introduction of a hpRNA-encoding construct and, subsequently, introduced a homologous gene from *Iris* to produce rose plants with 'true' blue-coloured flowers.

15.6 Concluding Remarks

RNA silencing is one of the most fundamental and conserved mechanisms for the control of gene expression and the regulation of development in eukaryotic organisms. In the early 1990s, plant biologists could not have envisioned at the time just how crucial their pioneering work on cosuppression and other RNA silencing-related mechanisms would become to modern science today. In plants, two of the most important findings in the field of RNA silencing were (1) the identification of dsRNA as the trigger for inducing RNA silencing mechanisms (Waterhouse et al. 1998) and (2) the association of sRNA species generated from dsRNA molecules (Hamilton and Baulcombe 1999). We now know that these sRNAs processed from longer molecules of dsRNA are used as guides to direct the specific functions of the individual RNA silencing pathways. Due to the additional elegant research conducted by individual researchers all over the globe, we today have a much greater understanding of how the individual endogenous RNA silencing pathways of plants work to regulate development. The continuing efforts of the plant biology community to identify the remaining pieces of the puzzle, that is, the RNA silencing pathways of plants, will not only increase our understanding of the interactions between these complex biological pathways, but also allow for the continuing development of additional RNA silencing technologies for future applications in the regulation of plant development.

References

Allen RS, Millgate AG, Chitty JA, Thisleton J, Miller JA, Fist AJ, Gerlach WL, Larkin PJ (2004) RNAi-mediated replacement of morphine with the nonnarcotic alkaloid reticuline in opium poppy. Nature Biotechnol 22:1559–1566

Allen E, Xie Z, Gustafson AM, Carrington JC (2005) MicroRNA-directed phasing during trans-acting siRNA biogenesis in plants. Cell 121:207–221

Anandalakshmi R, Pruss GJ, Ge X, Marathe R, Mallory AC, Smith TH, Vance VB (1998) A viral suppressor of gene silencing in plants. Proc Natl Acad Sci USA 95:13079–13084

Axtell MJ, Synder JA, Bartel DP (2007) Common functions for diverse small RNAs of land plants. Plant Cell 19:1750–1769

Baumberger N, Baulcombe DC (2005) Arabidopsis ARGONAUTE1 is an RNA Slicer that selectively recruits microRNAs and short interfering RNAs. Proc Natl Acad Sci USA 102:11928–11933

Bernstein E, Caudy AA, Hammond SM, Hannon GJ (2001) Role for a bidentate ribonuclease in the initiation step of RNA interference. Nature 409:363–366

Bohmert K, Camus I, Bellini C, Bouchez D, Caboche M, Benning C (1998) AGO1 defines a novel locus of Arabidopsis controlling leaf development. EMBO J 17:170–180

Borsani O, Zhu J, Verslues PE, Sunkar R, Zhu J-K (2005) Endogenous siRNAs derived from a pair of natural cis-antisense transcripts regulate salt tolerance in Arabidopsis. Cell 123:1279–1291

Cao X, Aufsatz W, Zilberman D, Mette MF, Huang MS, Matzke M, Jacobsen SE (2003) Role of the DRM and CMT3 methyltransferases in RNA-directed DNA methylation. Curr Biol 13:2212–2217

Chan SW, Zilberman D, Xie Z, Johansen LK, Carrington JC, Jacobsen SE (2004) RNA silencing genes control de novo DNA methylation. Science 303:1336

Chan SW, Henderson IR, Jacobsen SE (2005) Gardening the genome: DNA methylation in Arabidopsis thaliana. Nature Rev Genet 6:351–360

Chapman EJ, Prokhnevsky AI, Gopinath K, Dolja VV, Carrington JC (2004) Viral RNA silencing suppressors inhibit the microRNA pathway at an intermediate step. Genes Dev 18:1179–1186

Chen X, Liu J, Cheng Y, Jia D (2002) HEN1 functions pleiotropically in Arabidopsis development and acts in C function in the flower. Development 129:1085–1094

Cogoni C, Macino G (2000) Post-transcriptional gene silencing across kingdoms. Curr Opin Genet Dev 10:638–643

Deleris A, Gallego-Bartolome J, Bao J, Kasschau KD, Carrington JC, Voinnet O (2006) Hierarchical action and inhibition of plant Dicer-like proteins in antiviral defense. Science 313:68–71

Dunoyer P, Lecellier CH, Parizotto EA, Himber C, Voinnet O (2004) Probing the microRNA and small interfering RNA pathways with virus-encoded suppressors of RNA silencing. Plant Cell 16:1235–1250

Dunoyer P, Himber C, Voinnet O (2005) DICER-LIKE4 is required for RNA interference and produces the 21-nucleotide small interfering RNA component of the plant cell-to-cell silencing signal. Nature Genet 37:1356–1360

Fang Y, Spector DL (2007) Identification of nuclear dicing bodies containing proteins for micro-RNA biogenesis in living *Arabidopsis* plants. Curr Biol 17:818–823

Fire A, Xu S, Montgomery MK, Kostas SA, Driver SE, Mello CC (1998) Potent and specific genetic interference by double-stranded RNA in *Caenorhabditis elegans*. Nature 391:806–811

Franco-Zorrilla JM, Valli A, Todesco M, Mateos I, Puga MI, Rubio-Somoza I, Leyva A, Weigel D, García JA, Paz-Ares J (2007) Target mimicry provides a new mechanism for regulation of microRNA activity. Nature Genet 39:1033–1037

Golden TA, Schauer SE, Lang JD, Pien S, Mushegian AR, Grossniklaus U, Meinke DW, Ray A (2002) *SHORT INTEGUMENTS1/SUSPENSOR1/CARPEL FACTORY*, a Dicer homolog, is a maternal effect gene required for embryo development in Arabidopsis. Plant Physiol 130:808–822

Hamilton AJ, Baulcombe DC (1999) A species of small antisense RNA in posttranscriptional gene silencing in plants. Science 286:950–952

Jacobsen SE, Running MP, Meyerowitz EM (1999) Disruption of an RNA helicase/RNase III gene in *Arabidopsis* causes unregulated cell division in floral meristems. Development 126:5231–5243

Jin H, Vacic V, Girke T, Lonardi S, Zhu JK (2008) Small RNAs and the regulation of *cis*-antisense transcripts in *Arabidopsis*. BMC Mol Biol 9:6

Jones AL, Thomas CL, Maule AJ (1998) *De novo* methylation and co-suppression induced by a cytoplasmically replicating plant RNA virus. EMBO J 17:6385–6393

Jones-Rhoades MW, Bartel DP (2004) Computational identification of plant microRNAs and their targets, including stress-induced miRNA. Mol Cell 14:787–799

Kanno T, Mette MF, Kreil DP, Aufsatz W, Matzke M, Matzke AJM (2004) Involvement of a putative SNF2 chromatin remodeling protein DRD1 in RNA-directed DNA methylation. Curr Biol 14:801–805

Kanno T, Huettel B, Mette MF, Aufsatz W, Jaligot E, Daxinger L, Kriel DP, Matzke M, Matzke AJM (2005) Atypical RNA polymerase subunits required for RNA-directed DNA methylation. Nature Genet 37:761–765

Katsumoto Y, Fukuchi-Mizutani M, Fukui Y, Brugliera F, Holton TA, Karan M, Nakamura N, Yonekura-Sakakibara K, Togami J, Pigeaire A, Tao GQ, Nehra NS, Lu CY, Dyson BK, Tsuda S, Ashikari T, Kusumi T, Mason JG, Tanaka Y (2007) Engineering of the rose flavonoid biosynthetic pathway successfully generated blue-hued flowers accumulating delphinidin. Plant Cell Physiol 48:1589–1600

Kennerdell JR, Carthew RW (1998) Use of dsRNA-mediated genetic interference to demonstrate that *frizzled* and *frizzled 2* act in the wingless pathway. Cell 95:1017–1026

Lee RC, Ambros V (2001) An extensive class of small RNAs in *Caenorhabditis elegans*. Science 294:862–864

Lee RC, Feinbaum RL, Ambros V (1993) The *C. elegans* heterochronic gene *lin-4* encodes small RNAs with antisense complementarity to *lin-4*. Cell 75:843–854

Li HW, Ding SW (2005) Antiviral silencing in animals. FEBS Lett 579:5965–5973

Lindbo JA, Silva-Rosales L, Proebsting WM, Dougherty WG (1993a) Induction of highly specific antiviral state in transgenic plants: implications for regulation of gene expression and virus resistance. Plant Cell 5:1749–1759

Lindbo JA, Silva-Rosales L, Dougherty WG (1993b) Pathogen derived resistance to potyviruses: working, but why? Sem Virol 4:369–379

Liu Q, Singh SP, Green AG (2002) High-stearic and high-oleic cottonseed oils produced by hairpin RNA-mediated post-transcriptional gene silencing. Plant Physiol 129:1732–1743

Mallory AC, Vaucheret H (2006) Functions of microRNAs and related small RNAs in plants. Nature Genet 38:S31–S36

Margis R, Fusaro AF, Smith NA, Curtin SJ, Watson JM, Finnegan EJ, Waterhouse PM (2006) The evolution and diversification of Dicers in plants. FEBS Lett 580:2442–2450

Matzke MA, Priming M, Trnovsky J, Matzke AJM (1989) Reversible methylation and inactivation of marker genes in sequentially transformed tobacco plants. EMBO J 8:643–649

Meister G, Tuschl T (2004) Mechanisms of gene silencing by double-stranded RNA. Nature 431:343–349

Mette MF, Aufsatz W, van der Winden J, Matzke MA, Matzke AJ (2000) Transcriptional silencing and promoter methylation triggered by double-stranded RNA. EMBO J 19:5194–5201

Metzlaff M, O'Dell M, Cluster PD, Flavell RB (1997) RNA-mediated RNA degradation and chalcone synthase A silencing in petunia. Cell 88:845–854

Millar AA, Waterhouse PM (2005) Plant and animal microRNAs: similarities and differences. Funct Integrat Genomics 5:129–135

Moissard G, Vionnet O (2004) Viral suppression of RNA silencing in plants. Mol Plant Pathol 5:71–82

Mussgnug JH, Thomas-Hall S, Rupprecht J, Foo A, Klassen V, McDowall A, Schenk PM, Kruse O, Hankamer B (2007) Engineering photosynthetic light capture: impacts on improved solar energy to biomass conversion. Plant Biotechnol J 5:802–814

Napoli C, Lemieux C, Jorgensen R (1990) Introduction of a chimeric chalcone synthase gene into petunia results in reversible co-suppression of homologous genes in trans. Plant Cell 2:279–289

Ngo H, Tschudi C, Gull K, Ullu E (1998) Double-stranded RNA induces mRNA degradation in Trypanosoma brucei. Proc Natl Acad Sci USA 95:14687–14692

Niu QW, Lin SS, Reyes JL, Chen KC, Wu HW, Yeh SD, Chua NH (2006) Expression of artificial microRNAs in transgenic Arabidopsis thaliana confers virus resistance. Nature Biotechnol 24:1420–1428

Onodera Y, Haag JR, Ream T, Nunes PC, Pontes O, Pikaard CS (2005) Plant nuclear RNA polymerase IV mediates siRNA and DNA methylation-dependent heterochromatin formation. Cell 120:613–622

Park W, Li J, Song R, Messing J, Chen X (2002) CARPEL FACTORY, a Dicer homolog, and HEN1, a novel protein, act in microRNA metabolism in Arabidopsis thaliana. Curr Biol 12:1484–1495

Park MY, Wu G, Gonzalez-Sulser A, Vaucheret H, Poethig RS (2005) Nuclear processing and export of microRNAs in Arabidopsis. Proc Natl Acad Sci USA 102:3691–3696

Qu J, Ye J, Fang RX (2007) Artificial microRNA-mediated virus resistance in plants. J Virol 81:6690–6699

Regina A, Bird A, Topping D, Bowden S, Freeman J, Barsby T, Kosar-Hashemi B, Li Z, Rahman S, Morell M (2006) High-amylose wheat generated by RNA interference improves indices of large-bowel health in rats. Proc Natl Acad Sci USA 103:3546–3551

Sharp PA (2001) RNA interference – 2001. Genes Dev 15:485–490

Smith NA, Singh SP, Wang M-B, Stoutjesdijik PA, Green AG, Waterhouse PM (2000) Gene expression: total silencing by intron-spliced hairpin RNAs. Nature 407:319–320

Soards AJ, Murphy AM, Palukaitis P, Carr JP (2002) Virulence and differential local and systemic spread of cucumber mosaic virus in tobacco are affected by the CMV 2b protein. Mol Plant Microbe Interact 7:647–653

Stam M, de Bruin R, Kenter S, van der Hoorn RAL, van Blokland R, Mol JNM, Kooter JM (1997) Post-transcriptional silencing of chalcone synthase in Petunia by inverted transgene repeats. Plant J 12:63–82

Tabara H, Sarkissian M, Kelly WG, Fleenor J, Grishok A, Timmons L, Fire A, Mello CC (1999) The rde-1 gene, RNA interference, and transposon silencing in C. elegans. Cell 99:123–132

van der Krol AR, Mur LA, Beld M, Mol JNM, Stuitje AR (1990) Flavonoid genes in petunia: addition of a limited number of gene copies may lead to a suppression of gene expression. Plant Cell 2:291–299

Vaucheret H (2005) RNA polymerase IV and transcriptional silencing. Nature Genet 37:659–660

Vazquez F, Vaucheret H, Rajagopalan R, Lepers C, Gasciolli V, Mallory AC, Hilbert J-L, Bartel, DP, Crete P (2004) Endogenous trans-acting siRNAs regulate the accumulation of Arabidopsis mRNAs. Mol Cell 16:69–79

Voinnet O (2005) Non-cell autonomous RNA silencing. FEBS Lett 579:5858–5871

Wang M-B, Metzlaff M (2005) RNA silencing and antiviral defense in plants. Curr Opin Plant Biol 8:216–222

Wang M-B, Abbott DC, Waterhouse PM (2000) A single copy of a virus-derived transgene encoding hairpin RNA gives immunity to barley yellow dwarf virus. Mol Plant Pathol 1:347–356

Wang M-B, Wesley SV, Finnegan EJ, Smith NA, Waterhouse PM (2001) Replicating satellite RNA induces sequence-specific DNA methylation and truncated transcripts in plants. RNA 7:16–28

Wassenegger M (2005) The role of the RNAi machinery in heterochromatin formation. Cell 122:13–16

Wassenegger M, Hiemes S, Riedel L, Sänger HL (1994) RNA-directed *de novo* methylation of genomic sequences in plants. Cell 76:567–576

Waterhouse PM (2006) Defence and counterdefence in the plant world. Nature Genet 2:138–139

Waterhouse PM, Fusaro AF (2006) Viruses face double defence by plant small RNAs. Science 313:54–55

Waterhouse PM, Graham MW, Wang M-B (1998) Virus resistance and gene silencing in plants can be induced by simultaneous expression of sense and antisense RNA. Proc Natl Acad Sci USA 95:13959–13964

Xie Z, Allen E, Wilken A, Carrington JC (2005) DICER-LIKE4 functions in trans-acting siRNA biogenesis and vegetative phase change in plants. Proc Natl Acad Sci USA 102:12984–12989

Xu L, Yang L, Pi L, Liu Q, Ling Q, Yang H, Poethig RS, Huang H (2006) Genetic interactions between the *AS1-AS2* and *RDR6-SGS3-AGO7* pathway for leaf morphogenesis. Plant Cell Physiol 47:853–863

Yang L, Liu Z, Lu F, Dang A, Huang H (2006) SERRATE is a novel nuclear regulator in primary miRNA processing in Arabidopsis. Plant J 47:841–850

Yoshikawa M, Peragine A, Park MY, Poethig RS (2005) A pathway for the biogenesis of *trans*-acting siRNAs in *Arabidopsis*. Genes Dev 19:2164–2175

Yu L, Yu X, Shen R, He Y (2005) *HYL1* gene maintains venation and polarity of leaves. Planta 221:231–242

Zamore PD, Tuschl T, Sharp PA, Bartel DP (2000) RNAi: double-stranded RNA directs the ATP-dependent cleavage of mRNA at 21 to 23 nucleotide intervals. Cell 101:25–33

Zilberman D, Cao X, Jacobsen SE (2003) *ARGONAUTE4* control of locus-specific siRNA accumulation and DNA and histone methylation. Science 299:716–719

Chapter 16
DNA Methylation: a Dynamic Regulator of Genome Organization and Gene Expression in Plants

E.J. Finnegan

16.1 Introduction

The role of DNA methylation in biological processes has been debated since 1975 when it was proposed that DNA methylation could regulate gene expression in a heritable manner (Holliday and Pugh 1975; Riggs 1975). There can be little dispute that DNA methylation plays a role in genome defence; it provides a means for discriminating between self and non-self DNA in bacteria and inactivates mobile elements in higher eukaryotes (McClelland 1981; Yoder et al. 1997). The role of DNA in regulating gene expression in development has been more controversial (Bird 1995, 1997; Yoder et al. 1997). The long-held view that DNA methylation is a static mark on DNA, locking segments of the genome in a quiescent state, argues against a role in gene regulation. New data suggest that DNA methylation is a dynamic process, with the pattern of methylation being a balance between the activities of DNA methyltransferases and demethylases (Zhu et al. 2007). The dynamic nature of cytosine methylation raises the possibility that it may regulate gene expression in response to developmental or environmental cues, and it has been shown recently that gene expression can be regulated by dynamic changes in DNA methylation in mammalian cells (Kangaspeska et al. 2008; Metivier et al. 2008). The tools are now in hand to determine whether methylcytosine (^{m}C) plays a similar role in regulating plant gene expression during development.

Many of the recent advances in studying DNA methylation have been made in *Arabidopsis thaliana* because of the availability of the genome sequence, as well as the biochemical and genetic resources to support these studies. This review discusses the recent advances in technologies for mapping ^{m}C across the whole genome, the machinery involved in adding and removing methylation, as well as

E.J. Finnegan
CSIRO, Plant Industry, GPO Box 1600, Canberra, ACT 2601, Australia
e-mail: jean.finnegan@csiro.au

E-C. Pua and M.R. Davey (eds.),
Plant Developmental Biology – Biotechnological Perspectives: Volume 2,
DOI 10.1007/978-3-642-04670-4_16, © Springer-Verlag Berlin Heidelberg 2010

Box 16.1

In the nucleus, DNA is wrapped around nucleosomes consisting of two copies each of the four histone proteins H2A, H2B, H3 and H4, to form a complex know as chromatin. Chemical modification of these histones modulates gene expression. Transcribed genes are associated with high levels of acetylation of lysine residues in the amino-terminal tails of H3 and H4, as well as H3 di- and trimethyl-lysine 4 (H3K4me2 and H3K4me3). In plants, inactive, densely methylated DNA is generally associated with chromatin enriched in H3K9me2 (Gendrel et al. 2002); heterochromatin in plants is characterized by high levels of both methylcytosine and H3K9me2. In contrast to animals, H3K9me3 does not co-localize to heterochromatic DNA and is not generally associated with methylated regions (Turck et al. 2007). H3K9me3 is a relatively rare modification in plant chromatin and its influence on gene expression is not fully understood. The repressive chromatin mark H3K27me3 is added to chromatin by polycomb group proteins and in plants does not co-localize with methylated DNA (Turck et al. 2007).

the interplay between DNA methylation and associated chromatin (Box 16.1). The roles of ^{m}C in regulating genome stability, plant development and response to the environment are explored.

16.2 Mapping DNA Methylation

16.2.1 Technological Advances in Mapping Methylated Cytosine

In order to understand the role of DNA methylation in plant processes, it is first necessary to determine exactly which cytosines are methylated in particular tissues at different stages of development or in response to environmental challenges. Mapping ^{m}C requires a means to discriminate between methylated and non-methylated DNA, followed by a detection system to identify those regions containing ^{m}C. Various methods to discriminate between methylated and non-methylated DNA have been developed, but until recently the detection was limited to short regions of the genome.

Early studies relied on differential cleavage of DNA by methylation-sensitive restriction enzymes. When combined with Southern hybridization or polymerase chain reaction (PCR) using primers that flank the restriction site, this assayed methylation in a known sequence but only where ^{m}C occurs within the recognition site of the enzyme used (Bird and Southern 1978; Singer-Sam et al. 1990). Subsequent developments allowed the methylation status of more cytosines to be assessed. Methylation-sensitive amplification polymorphism, an adaptation of amplified fragment length polymorphism in which methylation-sensitive restriction enzymes were used to cleave DNA prior to amplification, allowed methylation changes in anonymous sequences across the genome to be identified (Reyna-Lopez et al. 1997). The restriction enzyme McrBC, which cleaves DNA between two ^mC residues separated by 40 to 3,000 bp, was used to identify densely methylated regions. Cleavage by McrBC occurs preferentially when ^mC is preceded by a purine, but is independent of the base 3′ to ^mC (Sutherland et al. 1992; Stewart and Raleigh 1998). Other assays are based on enrichment strategies such as immunoprecipitation using an antibody against ^mC (Weber et al. 2005) or affinity purification of ^mCG using the methyl-binding domain of human MeCP2 (methyl CpG binding protein 2) bound to agarose beads (Cross et al. 1994) to identify methylated DNA. Treatment of DNA with bisulfite, which converts cytosine to uracil but does not modify ^mC, followed by PCR and sequencing was a vast improvement over other technologies as it allowed the methylation status of every cytosine within a region of interest to be measured (Frommer et al. 1992).

Technological advances now allow the mapping of DNA methylation at the genome-wide level, with the most sophisticated providing single-base resolution of ^mC (reviewed in Zilberman and Henikoff 2007; Beck and Rakyan 2008). Using the same means to discriminate between methylated and non-methylated DNA, these advances depend on the development of whole-genome arrays (Borevitz and Ecker 2004) and on high-throughput sequencing, so-called deep sequencing (Mardis 2008). Needless to say, these approaches are limited to organisms for which the genome has been sequenced.

There are several variations of these genome-wide approaches and it is worth considering the merits and limitations of each (Zhang et al. 2006, 2008; Zilberman et al. 2006; Cokus et al. 2008; Lister et al. 2008; Reinders et al. 2008). The resolution of array-based hybridization is limited by the length of the DNA probes on the array; for example, the Affymetrix array for the Arabidopsis genome provides resolution at the level of 35 base pairs (Dalma-Weiszhausz et al. 2006). The resolution of approaches that employ affinity purification or immunoprecipitation is correlated with the length of fragments enriched; the longer the fragment, the greater the chance of creating a false positive through the enrichment of non-methylated sequences that are flanked by methylated regions. Bisulfite treatment, followed by hybridization to a genome array, increases the resolution of array-based technologies as the hybridization intensity at each probe relies on the presence of ^mC (i.e. whether or not the sequence is converted) in the region that hybridizes to the probe itself, rather than in the flanking sequences (Reinders et al. 2008). Only the DNA treated with bisulfite, followed by deep sequencing, permits single-base resolution of ^mC (Cokus et al. 2008; Lister et al. 2008). The amount of DNA required for the hybridization-based assays varies, with more sample DNA being required for affinity purification and immunoprecipitation than for bisulfite treatment (2–20 μg, compared with 100 ng).

Of these different techniques, bisulfite treatment followed by deep sequencing unquestionably provides the most information. It is not only the most expensive of the techniques but also requires the most sophisticated algorithms to interpret the data. In bisulfite-treated DNA, unmethylated cytosine is replaced by thymine in the final sequence, and so a region of DNA with both C and ^{m}C will be a perfect match to neither the original reference genome, nor to an in silico conversion that simulates the bisulfite treatment. The choice of technique will depend on the resolution required, the amount of tissue (and ultimately DNA) available and budgetary constraints.

16.2.2 High-Resolution Maps of DNA Methylation in the Arabidopsis Genome

The DNA methylome of Arabidopsis has now been mapped by several laboratories using young seedlings (Tran et al. 2005a; Zhang et al. 2006, 2008; Zilberman et al. 2006; Cokus et al. 2008; Reinders et al. 2008) or floral tissue (Lister et al. 2008). Similar approaches have been taken to map the methylome of 16,000 promoter sequences in primary human somatic cells and in mature sperm (Weber et al. 2007).

The higher resolution of these new approaches to mapping ^{m}C has confirmed earlier reports that DNA methylation in Arabidopsis is concentrated in the transposon-rich pericentric DNA on each chromosome as well as at the heterochromatic knob on chromosome 4 (Zhang et al. 2006, 2008; Zilberman et al. 2006; Vaughn et al. 2007; Cokus et al. 2008; Reinders et al. 2008). A more surprising finding is that there is considerable ^{m}C within the transcribed region of 20–30% of annotated genes, whereas the promoter and 3′ flanking regions of genes are relatively hypomethylated (Tran et al. 2005a; Zhang et al. 2006, 2008; Zilberman et al. 2006; Cokus et al. 2008). Methylation within genes is largely in a CG context and is not associated with corresponding small RNA sequences (Tran et al. 2005a; Vaughn et al. 2007; Zhang et al. 2008). Promoter methylation was observed in only about 5% of genes. These genes are expressed at a lower level and tend to show tissue-specific expression (Zhang et al. 2006). Methylation within the body of genes also correlated with the level of transcription, with genes transcribed at moderate levels being more highly methylated than genes expressed at low or high levels (Zilberman et al. 2006; Zhang et al. 2008). Transcription of body methylated genes increased in strong *met1* mutants, suggesting that DNA methylation may interfere with transcription elongation (Zilberman et al. 2006; Lister et al. 2008). This idea is supported by the observation that extensive CG methylation within the coding region of the *PHYA* gene was associated with transcriptional silencing (Chawla et al. 2007).

Pseudogenes and non-expressed genes show a much higher level of methylation than do expressed genes. This methylation is evenly distributed across the whole sequence. DNA methylation of transposable elements is also uniformly distributed across the entire sequence, and occurs in all sequence contexts. Loss of CG methylation in a *met1* mutant was correlated with increased transcription of promoter methylated genes as well as pseudogenes and some transposable elements (Zhang et al. 2006; Lister et al. 2008). Loss of non-CG methylation was also associated with increased expression of some transposons as well as a number of genes (Lindroth et al. 2001; Cao and Jacobsen 2002; Zhang et al. 2006; Lister et al. 2008), suggesting that both symmetric and non-symmetric methylation are associated with repression of transcription.

In addition to determining the frequency of ^{m}C in each sequence context (55% CG, 23% CHG and 22% CHH, where H represents A, G or T), genome-wide bisulfite sequencing has revealed that DNA methyltransferases show strong sequence preference over and above the CG, CHG and CHH contexts (Cokus et al. 2008; Lister et al. 2008). For example, CG dinucleotides within the sequence ACGT are methylated at a lower frequency than CCCG, whereas CAG and CTG were more highly methylated than CCG. Highly methylated asymmetric cytosines occur in the sequence CTA. ^{m}C occurred about 900-fold more frequently in this context than in the sequence CCC. As demonstrated previously for mammalian DNA, CG dinucleotides were generally symmetrically methylated or remained unmodified, and CHG sites also showed a strong tendency to be methylated symmetrically (Cokus et al. 2008).

16.3 Methylation Patterns Are the Balance Between Methyltransferase and Demethylase Activities

16.3.1 DNA Methyltransferases

There are three well characterized families of DNA methyltransferases in plants, two of which have counterparts in other eukaryotes, while the third is found only in plants (Table 16.1; reviewed in Goll and Bestor 2005; Chan et al. 2005; Pavlopoulou and Kossida 2007). Methylation of cytosine in all sequence contexts is established by enzymes of the DOMAINS REARRANGED METHYLTRANSERASE (DRM) family that are similar to the Dnmt3 enzymes in mammals (Okano et al. 1998; Cao et al. 2000). As their name suggests, the conserved motifs of the catalytic domain in plant enzymes are arranged in a novel order. Establishment of methylation by DRMs occurs through the RNA-directed DNA methylation pathway (see Sect. 16.4).

The CHROMOMETHYLASE (CMT) family, which is unique to plants, is characterized by the insertion of a chromodomain between conserved motifs II and IV of the catalytic domain of the protein (Henikoff and Comai 1998). CMT enzymes maintain methylation of cytosines in a CHG context and play a redundant role with DRMs in the "maintenance" of methylation of asymmetrically located cytosines (Bartee et al. 2001; Lindroth et al. 2001; Papa et al. 2001; Cao and Jacobsen 2002). The activity of CMT3 is disrupted by mutation of kryptonite (KYP), which dimethylates histone 3 lysine 9 (H3K9; Jackson et al. 2002), suggesting that CMT may only methylate DNA associated with repressive chromatin (Text box 1). Under in vitro condition, CMT3 binds the histone H3 amino tail that is trimethylated on both lysine 9 and lysine 27 (Lindroth et al. 2004). If the in vitro binding reflects what happens in vivo, then chromatin domains enriched in

Table 16.1 RNA-dependent DNA methylation pathway: plant genes required for addition, removal or maintenance of DNA methylation patterns (*nt* nucleotide)

Gene name	Abbr.	Arabidopsis gene number	Function
DICER-LIKE 3	DCL3	At3g43920	Cleavage of dsRNA to 24-nt small RNAs
ARGONAUTE 4	AGO4	At2g27040	Binds 24-nt siRNAs, de novo DNA methylation, RNA cleavage
ARGONAUTE 6	AGO6	At2g32940	Accumulation of heterochromatin associated siRNAs, DNA methylation
RNA-DEPENDENT RNA-POLYMERASE 2	RDR2	At4g11130	Converts ssRNA from heterochromatic template to dsRNA
DOMAINS REARRANGED METHYLTRANSFERASE 2	DRM2	At5g14620	De novo DNA methylation of CG, CHG and CHH sites
DEFECTIVE in RNA-DIRECTED DNA METHYLATION 1	DRD1	At2g16390	SNF2-like chromatin remodelling protein required for de novo DNA methylation with POL V
DEFECTIVE IN MERISTEM SILENCING	DMS3	At3g49250	Structural-maintenance-of-chromosomes hinge domain protein required for de novo methylation; involved in siRNA interactions with nucleic acid binding partner?
RNA POLYMERASE IV	POLIV	Several subunits	Plant-specific polymerase with novel largest (NRPD1a, At1g63020) and second-largest subunits (At3g23780); template unknown; required for accumulation of siRNAs from methylated DNA
RNA POLYMERASE V	POLV	Several subunits	Plant-specific polymerase with novel largest (NRPD1b, A2g40030) and second-largest subunits (At3g23780); short POL V generated transcripts are required for de novo DNA methylation of overlapping or adjacent genes
HISTONE DEACETYLASE 6	HDA6	At5g63110	Putative histone deacetylase; removal of histone acetylation and maintenance of CG methylation
KRYPTONITE	KYP	At5g13960	SRA domain protein that binds methylcytosine in CHG, CHH sequences; dimethylation of histone H3 lysine 9 (H3K9me2); maintenance of non-CG methylation (also known as SUVH4)
SUPPRESSOR OF VARIEGATION H5	SUVH5	At2g35160	H3K9 methyltransferase; maintenance of non-CG methylation
SUPPRESSOR OF VARIEGATION H6	SUVH6	At2g22740	SRA domain protein that binds methylcytosine in CHG, CHH sequences; H3K9 methyltransferase; maintenance of non-CG methylation
CHROMOMETHYLASE 3	CMT3	At1g69770	DNA methyltransferase; maintenance of non-CG methylation

(*continued*)

Table 16.1 (continued)

Gene name	Abbr.	Arabidopsis gene number	Function
METHYLTRANSFERASE 1	*MET1*	At5g49160	DNA methyltransferase; maintenance of CG methylation, weak de novo CG methylation
CLASSY 1	*CLSY1*	At3g42670	SNF2-like chromatin remodelling protein required for accumulation of siRNAs with POL IV
Accessory proteins required for DNA methylation			
DECREASED DNA METHYLATION 1	*DDM1*	At5g66750	SNF2-like chromatin remodelling protein required for DNA methylation in all sequence contexts, particularly in repeated sequences
REQUIRED TO MAINTAIN REPRESSION 1	*RMR1*		Maize SNF2-like chromatin remodelling protein required for accumulation of siRNAs and maintenance of DNA methylation and repression of paramutated Pl-Roades
Removal of DNA methylation; blocking the spread of DNA methylation			
DEMETER	*DME*	At5g04560	DNA glycosylase/lyase; demethylation of imprinted genes in the central cell
REPRESSOR OF SILENCING 1	*ROS1*	At2g36490	DNA glycosylase/lyase; demethylation of DNA in somatic tissue
DEMETER-LIKE 2	*DML2*	At3g10010	DNA glycosylase/lyase; demethylation of DNA in somatic tissue
DEMETER-LIKE 3	*DML3*	At4g34060	DNA glycosylase/lyase; demethylation of DNA in somatic tissue
INCREASED IN BONSAI METHYLATION 1	*IBM1*	At3g07610	Jumonji domain C protein; putative H3K9me2 demethylase
DNA methyl-binding proteins			
METHYL-BINDING PROTEIN 5	*MBD5*	At3g46580	Methyl-binding domain protein; binds methylcytosine in CG and CHH sequences
METHYL-BINDING PROTEIN 6	*MBD6*	At5g59380	Methyl-binding domain protein; binds methylcytosine in CG dinucleotides
METHYL-BINDING PROTEIN 7	*MBD7*	At5g59800	Methyl-binding domain protein; binds methylcytosine in CG dinucleotides
VARIANT IN METHYLATION 1	*VIM1*	At1g57820	SRA, PHD and RING domain protein; binds methylcytosine in CG and CHG sequences; essential for the maintenance of DNA methylation in the centromeric repeat

H3K9me3 and H3K27me3 should co-localize with ^{m}C. This is not the case, suggesting that in vivo CMT3 does not bind either H3K9me3 or H3K27me3 (Turck et al. 2007). The link between KYP and CMT3 activities suggests that CMT3 is more likely to bind the repressive mark, H3K9me2, in vivo. This would be consistent with the observation that CMT3 is important for the maintenance, rather than the establishment of asymmetric methylation.

The remaining class of plant methyltransferase is the METHYLTRANSFERASE (MET) family; its main function being the maintenance of CG methylation (Finnegan and Dennis 1993; Kishimoto et al. 2001). The MET enzymes are analogous to the Dnmt1 (DNA methyltransferase 1) family of enzymes in mammals (Bestor et al. 1988). The Arabidopsis METI protein has been shown to have some de novo methylation activity at CG dinucleotides at transgene targets of RNA directed DNA methylation (RdDM; Aufsatz et al. 2004). Although DRM2 and CMT3 are the only known methyltransferases that target CHH sites, about 20% of cytosines in this context remain methylated in the Arabidopsis triple *drm1 drm2 cmt3* mutant (Cokus et al. 2008; Lister et al. 2008). Perhaps DRM1, CMT2 or a member of the MET family has a low level of activity at non-CG sites.

16.3.2 Chromatin-Modifying Proteins Are Essential for DNA Methylation

Mutant screens for plants with reduced levels of DNA methylation have identified two proteins that belong to different clades of the SNF2 ATP-dependent chromatin remodelling protein family (Jeddeloh et al. 1999; Kanno et al. 2004). DECREASED DNA METHYLATION 1 (DDM1), which was identified in a screen for mutants with reduced methylation of repeated sequences (Table 16.1; Vongs et al. 1993), is required for methylation of cytosines in all sequence contexts, particularly in repeated sequences. Loss of DNA methylation of single-copy sequences was observed only after several generations of inbreeding (Kakutani et al. 1996). The mode of action of DDM1 is still unclear, but it may allow access of methyltransferases to DNA. A homologue of DDM1, Lsh1, is essential for normal genome-wide patterns of methylation in mammalian cells (Dennis et al. 2001), suggesting that DDM1-like proteins are a common component of DNA methylation pathways across kingdoms.

DEFECTIVE in RNA-DIRECTED DNA METHYLATION (DRD1), which belongs to a plant-specific group of SNF2 chromatin remodelling proteins, was identified in a screen for the release of transcriptional silencing (Kanno et al. 2004). In *drd1* mutants, non-CG methylation was almost eliminated from the targets of RNA-directed methylation, but methylation of centromeric and rDNA repeats was unaffected. It is not known how DRD1 affects methylation, but it could act by displacing nucleosomes and/or unpairing DNA to allow formation of a heteroduplex between DNA and a guide RNA (Kanno et al. 2004). The specificity of DRD1 for the RNA-directed DNA methylation pathway distinguishes it from DDM1-like proteins, which are regulators of global DNA methylation. It is intriguing that DRD1 plays a role in both the addition and removal of DNA methylation (see Sect. 16.3.3).

Other SNF2 chromatin remodelling proteins that play a role in DNA methylation include the maize protein REQUIRED TO MAINTAIN REPRESSION 1 (RMR1), which belongs to a separate branch of the DRD1 clade (Hale et al. 2007), and the more distantly related mammalian protein ATRX, which has similarity to DRD1 only in the binding sites for Mg^{++} and ATP (Gibbons et al. 2000).

16.3.3 DNA Demethylases

Over the years, there have been several reports of enzymatic demethylation in mammalian systems, but few of these have been substantiated (reviewed in Kapoor et al. 2005), whereas DNA glycosylase activity has been shown to catalyse demethylation in chick cells (Jost 1993). In plants, removal of ^{m}C is catalysed by enzymes that have dual DNA glycosylase/lyase activities via a base-excision repair pathway. In Arabidopsis, there is a small family of these enzymes named DEMETER-LIKE, after the founding member, DEMETER (DME; Table 16.1; Gong et al. 2002; Kapoor et al. 2005; Gehring et al. 2006).

REPRESSOR OF SILENCING 1 (ROS1), the first member of this family with demonstrated demethylase activity, was initially shown to be required for demethylation of a transgene (Gong et al. 2002). Subsequently, ROS1 was found to demethylate endogenous Arabidopsis genes and transposable elements (Zhu et al. 2007). In *ros1* mutants, there was no change in the level of CG methylation, but methylation of CHG and CHH increased in the majority of the sequences examined. The increase in DNA methylation was associated with decreased transcription of these loci. Even sequences like transposons that are normally maintained in a silent state became more densely methylated in a *ros1* mutant (Zhu et al. 2007). These data suggest that methylation is a dynamic process, with the final pattern being an equilibrium between the addition and removal of ^{m}C. Further evidence that DNA methylation is curtailed by the activity of demethylases was obtained by inbreeding *ros1*. After three or more generations of selfing, *ros1* plants showed a range of developmental abnormalities including shorter, early flowering plants with abnormal flowers, short siliques and a reduced seed set (Gong et al. 2002). These phenotypes have not yet been associated with changes in methylation of specific genes, although hypermethylation of several endogenous genes has been observed in *ros1* plants (Kapoor et al. 2005). These observations support the idea that ROS1 activity is essential for protecting some genes from becoming methylated and silenced.

There was no change in global DNA methylation in the genome of a plant carrying mutations in the three demethylases ROS1 and its related enzymes DEMETER-LIKE 2 (DML2) and DML3 (Penterman et al. 2007a). Hypermethylation was observed at about 180 loci, almost 80% of which lie in protein coding genes. In contrast to the genic methylation seen in wild-type plants, methylation at these loci was higher in the promoter and/or 3' ends of genes, suggesting that DML enzymes normally protect these regions from becoming methylated (Penterman et al. 2007a; Lister et al. 2008). Hypermethylation occurred in all sequence contexts, consistent with the capacity of these enzymes to excise ^{m}C, independent of context. Like *ROS1*, *DML2* and *DML3* are ubiquitously expressed and comparison of methylation levels in the single and triple mutants showed that these enzymes act redundantly at many loci. Methylation at some loci is controlled by a single demethylase; for example, the *SUPERMAN* locus, which becomes hypermethylated in plants with low levels of DNA methylation, is primarily demethylated by DML2 (Penterman et al. 2007a). Single, double and triple mutants were phenotypically normal at least in the first generation of homozygosity, and there was little change in expression of the hypermethylated loci (Penterman et al. 2007a). As for the *ros1* mutant, accumulation and/or spread of DNA methylation associated with loss of activity of the three DML enzymes may be cumulative, with several generations being required before detrimental effects on gene expression occur.

What determines whether or not a locus is susceptible to hypermethylation in the absence of DML activity and, conversely, what targets DML enzymes to particular loci? These questions may be answered, in part, by the finding that loci demethylated by DML enzymes commonly overlap repeated sequences and/or transposons, and are enriched in small RNA populations (Fig. 16.1; Penterman et al. 2007b). Almost 80% of loci that are demethylated by DMLs have homology to small RNAs, 24 nucleotides in length, suggesting that hypermethylation occurs through the RrDM pathway (see Sect. 16.4; Figs. 16.1, 16.2). Consistent with this, hypermethylation of loci in a *ros1* mutant requires the pathway of RNA POLYMERASE IV (POL IV)/RNA-dependent RNA polymerase (RDR2)/DICER-LIKE 3 (DCL3)/ARGONAUTE 4 (AGO4; Penterman et al. 2007b). One intriguing possibility is that DML enzymes operate in a complex that targets specific genes for demethylation by interacting with the same small RNAs that direct DNA methylation (Kapoor et al. 2005; Fig. 16.2). In support of this, ROS1 demethylation is dependent on

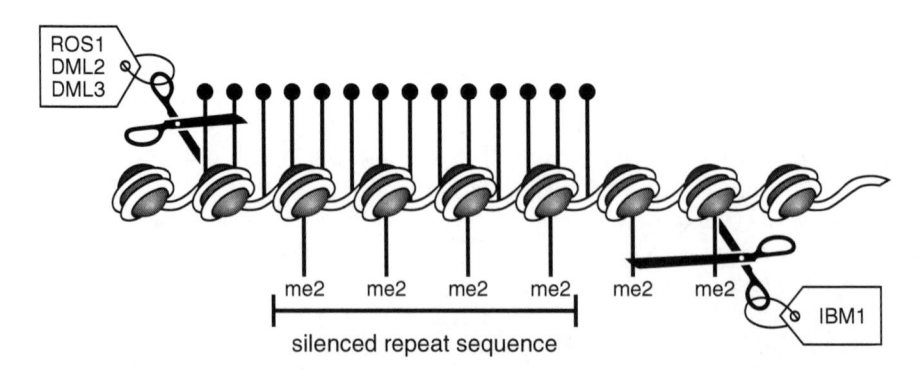

Fig. 16.1 Mechanisms to prevent the spread of heterochromatic marks into flanking sequences: DNA demethylases ROS1, DML2 and DML3 remove methylcytosine (*filled lollipops*) in DNA adjoining repeat sequences, while the putative histone H3 lysine 9 demethylase IBM1 prevents the spread of both DNA methylation and H3K9(me2), indicated by me2

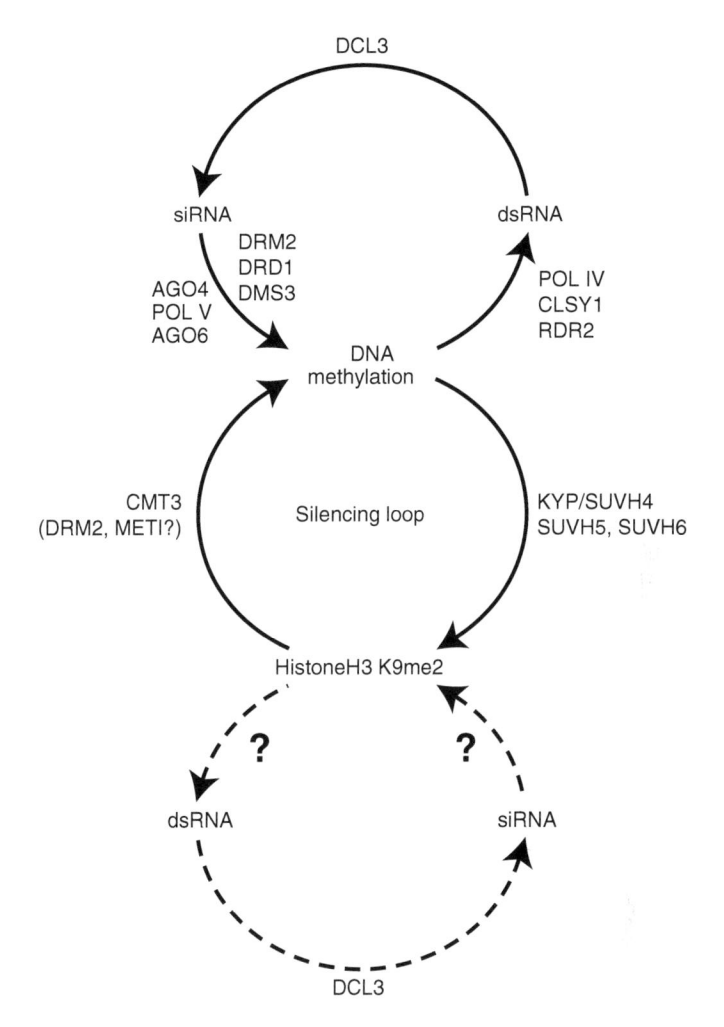

Fig. 16.2 DNA methylation and histone methylation form a self-reinforcing silencing loop. The *central circle* indicates the interdependence of DNA methylation and dimethylation of histone H3 lysine 9. Histone methyltransferases KYP, SUVH6 (and probably SUVH5) are recruited to methylated DNA as the SRA domain in these proteins bind methylcytosine in non-CG contexts, leading to dimethylation of H3K9. This in turn recruits CMT3, a DNA methyltransferase that maintains methylation in non-CG contexts. Methylation of DNA is required for the accumulation of corresponding siRNAs; this process is shown in the *upper circle*. POL IV, CLSY1 and RDR2 are required for the production of double-stranded RNA, which is processed to 24-nucleotide small RNAs by DCL3. AGO4, AGO6, POL V, DRD1, DRM2 and DMS3 play a role in RNA-directed DNA methylation. It is not known whether siRNAs direct H3K9 methylation in plants (*dotted lines in lower circle*), which in turn could lead to the accumulation of more siRNAs. This process certainly occurs in *S. pombe*, which lacks DNA methylation

RDR2 activity (Penterman et al. 2007b), which is also required for the accumulation of small interference RNAs (siRNAs) from at least some loci. Furthermore, DRD1, another component of the RdDM pathway, facilitates demethylation of CG dinucleotides in targets of this pathway (Huettel et al. 2006).

In addition to its role in regulating RdDM, ROS1 also antagonises methylation at some sites for which there are no corresponding small RNAs, suggesting a broader housekeeping function (Penterman et al. 2007b). ROS1 expression is down-regulated in both *met1* and *drm2* mutants as well as *rdr2*, *drd1* and *pol iv* mutants, suggesting a possible feedback loop regulating genomic methylation (Huettel et al. 2006; Penterman et al. 2007b).

While ROS1, DML2 and DML3 protect genes from potentially deleterious DNA methylation in somatic tissues, DME, the remaining member of this family, is active only in the central cell of the female gametophyte (Choi et al. 2002). The maternal allele of *DME* is essential for seed viability. It is expressed primarily in the central cell where it activates expression of three imprinted genes, *MEDEA (MEA)*, *FLOWERING WAGININGEN (FWA)* and *FERTILISATION INDEPENDENT SEED 2 (FIS2*; Choi et al. 2002; Kinoshita et al. 2004; Gehring et al. 2006; Jullien et al. 2006; Fig. 16.3). The maternal alleles of these three genes are expressed in the developing endosperm, the fertilization product of the central cell and one of the two sperm cells. Neither maternal nor paternal alleles are expressed in the developing embryo, and the paternal allele remains inactive in the endosperm. The paternal allele of *DME* plays no role in the development of either embryo or endosperm.

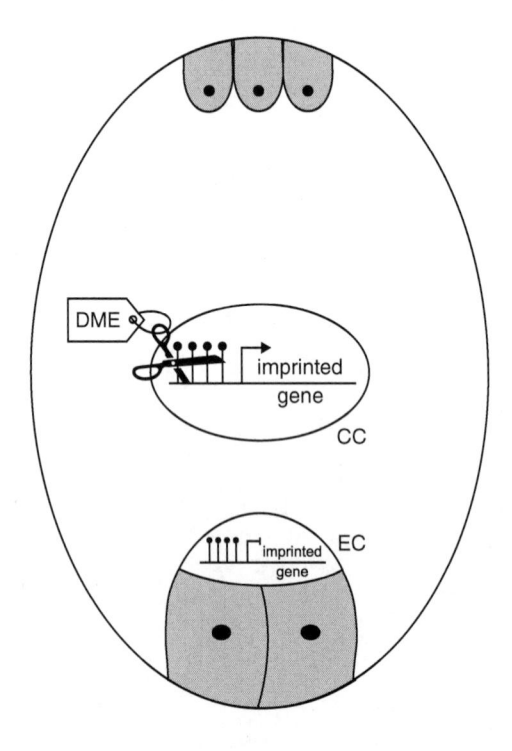

Fig. 16.3 Demethylation and activation of imprinted genes occurs in the central cell of the female gametophyte. DNA demethylase, DME, is expressed primarily in the central cell (CC) of the female gametophyte. DME removes methylcytosine (*filled lollipops*) from the critical region of the imprinted genes *MEA*, *FWA* and *FIS2*, leading to their activation and continued expression in the central cell and developing endosperm after fertilization. In the egg cell (EC), the lack of DME activity ensues that imprinted genes remain silenced, before and after fertilization

16.4 Targeting DNA Methylation

About 5–7% of cytosines (24% cytosines in CG dinucleotides, 6.7% in CHG and 1.7% in CHH) in the Arabidopsis genome are methylated, indicating that there must be mechanisms to target methylation to particular sequences. One such mechanism is the well characterized RdDM pathway, which results in dense methylation, particularly at cytosines in CHG and CHH contexts (reviewed in Chapman and Carrington 2007; Huettel et al. 2007; Matzke et al. 2007). The main targets of RdDM are repeated sequences located around the centromeres of each of the Arabidopsis chromosomes, the heterochromatic knob on chromosome 4 and transposable elements at dispersed sites throughout the genome. Consistent with this, deep sequencing has demonstrated that these sequences are enriched in populations of small RNAs, particularly those 24 nucleotides in length (Rajagopalan et al. 2006; Kasschau et al. 2007; Lister et al. 2008). Transgenes that have the capacity to form double-stranded RNA, due either to their repeat nature or to high expression levels, are also targets of RdDM.

Many of the components of the RdDM pathway have been identified through genetic screens for relief of RNA-directed transcriptional gene silencing (Table 16.1). Components of this pathway include DCL3, AGO4, AGO6, RDR2 and DRM2, as well as plant-specific proteins, including DRD1, DEFECTIVE IN MERISTEM SILENCING (DMS3) and two novel RNA polymerases, POL IV and POL V (formerly known as POL IVa and POL IVb respectively; Zilberman et al. 2003; Chan et al. 2004; Herr et al. 2005; Kanno et al. 2005, 2008; Onodera et al. 2005; Zheng et al. 2007). Other members of the DCL, AGO and RDR families operate in the microRNA (miRNA), trans-acting siRNA (tasiRNA) and/or the viral defence pathways in Arabidopsis (reviewed in Vaucheret 2006; Chapman and Carrington 2007). Unlike these small RNA pathways that act in the cytoplasm, much of the RdDM pathway occurs within the nucleus or its compartments, including the nucleolus, Cajal bodies and AB bodies (Pontes et al. 2006; Fang and Spector 2007; Li et al. 2008).

The RdDM pathway begins with the generation of double-stranded RNA derived from endogenous repeats. This can occur by readthrough transcription of an inverted repeat or the production of aberrant RNA that is made double-stranded through the action of RDR2. DCL3 cleaves this double-stranded RNA into 24-nucleotide siRNAs, which are bound by AGO4. An RNA–protein complex that includes AGO4 and POL V works with an SNF2 chromatin remodelling protein, DRD1 and DRM2 to methylate the homologous target DNA (Kanno et al. 2005; Huettel et al. 2006; El-Shami et al. 2007). POL V derived transcripts serve as scaffolds for guide siRNAs that target heterochromatin formation; guide siRNAs could interact either with these transcripts or with the complementary target DNA (Wierzbicki et al. 2008). The most recently identified component of the RdDM pathway, a structural-maintenance-of-chromosomes hinge domain-containing protein known as DMS3, may provide some clues (Kanno et al. 2008). Authentic structural-maintenance-of-chromosome proteins, such as cohesin and condensin, are large ATPases that play a role in chromosome segregation, higher-order chromosome structure, DNA repair and long-range gene regulation (Losada and Hirano 2005). Recently, a mammalian protein, SmcHD1, that has homology to the SMC-hinge domain of DMS3, has been shown to be involved in maintaining X chromosome inactivation (Blewitt et al. 2008). It has been suggested that DMS3, which lacks an ATPase domain, stabilizes siRNA–DNA or siRNA–RNA duplexes, facilitating the methylation of target DNA (Kanno et al. 2008). Whatever the nature of this interaction, methylation is directed to the strand opposite to that which could hybridize to the siRNA (Lister et al. 2008).

Once methylation has been established at RdDM targets, other proteins including the histone deacetylase HDA6, histone methyltransferases KYP (also know as SUPPRESSOR OF VARIEGATION H4, SUVH4), SUVH5 and SUVH6, as well as DNA methyltransfereases DRM2, CMT3 and MET1 play a role in maintaining methylation (Bartee et al. 2001; Cao and Jacobsen 2002; Ausfatz et al. 2002, 2004; Cao et al. 2003; Ebbs et al. 2005; Ebbs and Bender 2006). Maintenance of methylation is frequently associated with accumulation of siRNAs corresponding to the methylated target (Fig. 16.2). The process leading to accumulation of siRNAs is poorly understood, but seems to require transcription of the heterochromatinized repeat sequences (Lister et al. 2008), an apparent contradiction to the concept that these sequences are inactive. This paradox may be explained by the observation that transcription of heterochromatic DNA in *Schizosaccharomyces pombe* occurs only during early S phase, coincident with a loss of H3K9me2. siRNAs are generated from the double-stranded RNA, followed by rapid restoration of H3K9me2 by late S phase (Kloc et al. 2008; Fig. 16.2). In plants, POL IV and another member of the SNF2 family of chromatin remodelling proteins, CLSY1, are required for the production of siRNAs from a methylated target, perhaps by transcription of the methylated template, or of an aberrant RNA produced by Pol II (Vaughn and Martiennsen 2005; Pontes et al. 2006; Smith et al. 2007). Not all repeat elements require the same suite of enzymes to ensure maintenance of methylation and siRNA accumulation, suggesting that there are several means to maintain heterochromatin (Lippman et al. 2003; Zilberman et al. 2004; Chan et al. 2006; Mosher et al. 2008).

Differential accumulation of siRNAs in Arabidopsis ecotypes Columbia (Col) and Landsberg *erecta* (L*er*) has been associated with DNA methylation polymorphisms. In all, 68 loci that accumulated a higher level of siRNAs in L*er* than in Col were identified and, for most loci examined, the corresponding DNA was methylated in L*er* but not Col (Zhai et al. 2008). It is not clear how the differential accumulation of siRNAs arises between ecotypes. Only about a third of all methylation in Arabidopsis occurs in regions that are represented in the small RNA population (Lister et al. 2008), suggesting that other mechanisms can establish methylation at sequences not associated with small RNAs.

One such example of methylation that is not associated with small RNAs is the CG methylation seen in the coding regions of transcribed genes. Methylation within genic regions is also highly polymorphic between ecotypes (Vaughn et al. 2007). In a survey of 96 accessions, genic methylation was highly polymorphic with about half of all methylated genes being methylated in only one or two ecotypes. It has been proposed that genic methylation could be established by the RdDM pathway triggered by the production of double-stranded RNAs, perhaps arising from converging sense and (rare) antisense transcripts (Zilberman et al. 2006). In the absence of continued production of the RNA trigger, CpG methylation may then be maintained by METI. Alternatively, other unknown mechanisms may establish CpG methylation.

16.5 Interplay Between DNA Methylation and Chromatin Modifications

In a cell, DNA is associated with histone proteins in a complex known as chromatin (Text box 1). While chromatin was originally thought to provide an inert packaging matrix, it is now clear that modification of chromatin proteins influences gene expression (Text box 1). There are several lines of evidence pointing to a cross talk between the methylation status of DNA and modifications of histone H3, particularly in heterochromatin.

16.5.1 Heterochromatin Exists in a Self-Reinforcing Silencing Loop

In *metI* and *ddm1* mutants, DNA methylation is lost from the heterochromatin associated with each centromere. In addition to DNA methylation, H3K9me2 is also depleted at centromeric heterochromatin (Gendrel et al. 2002; Soppe et al. 2002; Tariq et al. 2003). Conversely, mutation of the H3K9 methyltransferase KYP leads to a reduction in the levels of both H3K9me2 and DNA methylation in repeated sequences and in targets of RdDM (Jackson et al. 2002; Johnson et al. 2002). While KYP is the main H3K9 methyltransferase in Arabidopsis, the two related proteins SUVH5 and SUVH6 are required to maintain non-CpG methylation at particular loci (Ebbs et al. 2005; Ebbs and Bender 2006). These observations suggest that DNA methylation and H3K9me2 form a self-reinforcing pair of modifications to maintain DNA in repressive chromatin (Fig. 16.2).

The molecular basis for the self-perpetuation of this epigenetic silencing loop is becoming clear. The SRA (SET and RING associated) domain of KYP and SUVH6 binds methylated DNA, with stronger binding to ^mC in CHG and CHH contexts (Johnson et al. 2007). Loss-of-function *kyp* mutants that have missense mutations in the SRA domain show reduced binding to methylated DNA. Thus, methylation of DNA recruits histone methyltransferases, leading to H3K9 dimethylation. On the other hand, targets of the KYP histone methyltransferase are also substrates for CMT DNA methylation (Tran et al. 2005b; Chan et al. 2006), suggesting that K9me2 recruits CMT3 to repressive chromatin, reinforcing the silencing loop.

16.5.2 A Putative Histone Demethylase Prevents the Spread of DNA Methylation

In general, single-copy sequences that are adjacent to densely methylated repeats escape DNA methylation. As already discussed, one mechanism preventing the spread of methylation to DNA flanking repeats is active demethylation (Penterman et al. 2007b). A second means to prevent the spread of DNA methylation was revealed through a genetic screen for mutants showing ectopic methylation of a genic region (Saze et al. 2008). One such mutant, *increase in bonsai methylation 1* (*ibm1*), resulted in hypermethylation of the *BONSAI* (*BNS*) gene. *BNS* lies adjacent to a LINE element, the DNA of which is methylated and associated with chromatin enriched in K9me2 (Saze and Kakutani 2007). Mutation of *IBM1* led to a rapid and extensive spread of both K9me2 and ^mC from the LINE1 element into the adjacent gene *BNS*, leading to transcriptional silencing within the first generation of homozygosity. IBM1 encodes a Jumonji C (jmjC) domain-containing protein with similarity to H3K9 demethylases (Yamane et al. 2006; Klose et al. 2007; Saze et al. 2008), suggesting that IBM1 may block the spread of K9me2 into sequences flanking heterochromatin (Fig. 16.1). Mutation of *kyp* suppressed the *ibm1* hypermethylation phenotype, supporting the idea that spread of K9me2 leads to spread of DNA methylation. Not surprisingly, this phenotype was also suppressed by mutation of *cmt3*.

   BONSAI also became stably hypermethylated in a *ddm1* mutant (to produce the stable *bns* epi-allele), but only after several generations of inbreeding (Saze and Kakutani 2007). Mutation of *ibm1* did not produce a precise phenocopy of the *bns* phenotype; *ibm1* was pleiotropic, suggesting that a number of genes, perhaps those adjacent to densely methylated repeated sequences, were transcriptionally silenced (Saze et al. 2008). All the *ibm1* phenotypes were suppressed in the double *ibm1 kyp* mutant, supporting this idea.

16.5.3 DNA Methyl-Binding Proteins Mediate the Interplay Between DNA Methylation and Chromatin Modification

The MBD (methyl-CpG-binding domain) class of ^mC-binding proteins, of which the mouse protein MeCP2 is the founding member, share homology in their MBDs (Nan et al. 1993). In mammals, MBD proteins act as transcriptional repressors by recruiting histone deacetylases to chromatin. The Arabidopsis genome encodes 13 MBD-containing proteins, only three of which, AtMBD5, AtMBD6 and AtMBD7, have been demonstrated to bind ^mC in one or more sequence contexts in vitro (Zemach and Grafi 2003; Ito et al. 2003; Scebba et al. 2003). AtMBD5, 6 and 7 localize to the densely methylated chromocentres, and this localization is disrupted in *met1* and *ddm1* mutants, suggesting that they also bind methylated DNA in vivo (Zemach and Grafi 2003; Zemach et al. 2005). AtMBD6 may play a role in silencing of ribosomal RNA genes, as it interacts with protein complexes that have histone deacetylase activity and co-localizes with 18S rDNA (Zemach and Grafi 2003; Zemach et al. 2005).

AtMBD7 interacts in a yeast two-hybrid assay with PRMT11 (At4g29510), a protein arginine methyltransferase that methylates histones, as well as its interacting partner, AtMBD7 (Scebba et al. 2007). These findings suggest that AtMBD7 mediates a cross talk between DNA methylation and histone arginine methylation, by recruiting AtPRMT11 to methylated DNA. AtMBD7 has two functional MBDs, raising the possibility that binding of AtMBD7 may bring together ^{m}C from distant sites on the same chromosome, or even from different chromosomes, leading to chromatin compaction (Grafi et al. 2007).

VARIANT IN METHYLATION 1 (VIM1), like KYP, binds methylated DNA through an extended SRA domain. In contrast to KYP, VIM1 binds ^{m}C in CG and CHG but not in the context CHH (Johnson et al. 2007; Woo et al. 2007). VIM1 co-localizes with methylated DNA at the chromocentres and is essential for methylation of the centromeric repeat DNA. *VIM1* was identified in a screen for natural variation in cytosine methylation (Woo et al. 2007). The *vim1* mutant shows hypomethylation and decondensation of the centromeric repeat DNA in interphase, and an altered distribution of the centromere-specific histone H3 variant. In contrast to other mutants that affect DNA methylation, *vim1* showed no loss of methylation in pericentric repeat sequences or at the *FWA* locus, and has no major phenotypic abnormalities. VIM1 has similarities with the mammalian protein UHRF1 (ubiquitin-like, containing PHD and RING finger domains 1). Both proteins contain plant homeo- and ring finger-domains, in addition to the SRA methyl-binding domain (Bostick et al. 2007; Woo et al. 2007). UHRF1 plays a role in maintaining global DNA methylation in mammalian cells, probably through its interaction with the maintenance methyltransferase DNMT1. UHRF1 shows a strong preference for hemimethylated CG dinucleotides, the preferred substrate of DNMT1, and co-localizes with DNMT1 throughout S phase (Bostick et al. 2007). It has been suggested that UHRF1 tethers DNMT1 to DNA, facilitating maintenance methylation following DNA replication. Perhaps VIM1 plays a similar role to maintain methylation at the centromeric repeat.

16.5.4   Ubiquitination of Histone H2B Prevents DNA Methylation

Histone modifications other than H3K9 methylation can influence the methylation status of DNA. *UBP26*, which encodes a thiol protease that cleaves ubiquitin from mono-ubiquitinated H2B (ubH2B), was identified in a screen for the release of transcriptional silencing (Sridhar et al. 2007). In the *ubp26-1* mutant, the level of histone H3K9me2 decreased, and RNA-directed DNA methylation of transgenes and transposons was suppressed. Transposons *AtMULE1*, *AtGP1* and *AtLINE1*, which are transcriptionally active in the mutant, were associated with elevated levels of UbH2B and H3K4me3 (Sridhar et al 2007). In yeast and animals, H2B ubiquitination is required for methylation of H3K4 and, consistent with this, there is an increase in the global level of H3K4me3 in the *ubp26-1* mutant (Sridhar et al. 2007).

16.6 Genome Stability Is Mediated by CpG Methylation

Arabidopsis plants with reduced levels of CG methylation, due to the presence of a *METI* antisense construct, have a pleiotropic phenotype that becomes more severe over several generations of inbreeding (Finnegan et al. 1996; Ronemus et al. 1996). While ectopic expression of genes and reactivation of transposons that are normally silenced by methylation were both anticipated and observed (Kakutani 1997; Soppe et al. 2000; Lippman et al. 2003), it was surprising to discover that some endogenous genes became hypermethylated and were transcriptionally silenced (Jacobsen and Meyerowitz 1997; Jacobsen et al. 2000). It was proposed that CG methylation maintained the genome in a state of organization that was compatible with the program of gene expression associated with normal development (Finnegan et al. 1998). Two recent studies showed that loss of CG methylation is associated with genome-wide activation of alternative epigenetic regulatory mechanisms, which provides support for this idea (Mathieu et al. 2007; Lister et al. 2008).

Loss of CG methylation triggers de novo non-CG methylation in both genic and heterochromatic loci across the genome, as has been observed at the *SUPERMAN* and *AGAMOUS* loci in *METI* antisense plants (Jacobsen et al. 2000). De novo methylation was stochastic, and results from the mistargeting of RNA-directed DNA methylation as well as the down-regulation of genes encoding demethylase activities (Mathieu et al. 2007). Mistargeting of RdDM may not be completely random. The frequent occurrence of *sup* epi-alleles in plants with low DNA methylation suggests that some sequences may be more susceptible to hypermethylation than others (Finnegan et al. 1996; Jacobsen and Meyerowitz 1997). Early-generation *met1-3* mutants showed a reduction in H3K9me2 at loci that are normally packaged into heterochromatin. After further inbreeding, the level of H3K9me2 associated with repetitive DNA increased, even though some sequences (180 bp, 45S rDNA and TSI) remained unmethylated. This suggests that H3K9me2 may accumulate in the absence of DNA methylation.

The biological consequences of the aberrant non-CG methylation and accumulation of K9me2 were tested by combining the *met1-3* with mutants in *drm2*, a de novo methyltransferase, or the lysine 9 methyltransferase, *kyp*. Homozygous double-mutant plants had more severe developmental abnormalities in the first generation than their *met1-3* single-mutant siblings (Matthieu et al. 2007). These observations suggest that stochastic changes in methylation of DNA and histone H3K9 offer some protection to the plant in the absence of CG methylation.

Other epigenetic mechanisms, such as polycomb-mediated repression, may also be activated in response to loss of CG methylation. The distribution of H3 trimethyllysine 27 (K27me3), the modification imposed by Polycomb complexes, is relocated in a *met1-3* mutant (Mathieu et al. 2005). Alterations in the activity of Polycomb complexes may account for the down-regulation of the floral repressor *FLOWER-ING LOCUS C* (*FLC*) in *met1* antisense plants (Sheldon et al. 1999). Down-regulation of *FLC* is not associated with hypermethylation of the *FLC* locus (Finnegan et al. 2005) but, as a natural target of Polycomb regulation (Bastow et al. 2004), could be associated with elevated K27me3 caused by the activation of Polycomb complexes. The activation of epigenetic mechanisms may also explain the repression of the seed abortion phenotype of the FIS class of genes, when fertilized with pollen from plants with reduced levels of CG methylation (Luo et al. 1999).

16.7 DNA Methylation Regulates Genes During Development and in Response to External Stimuli

DNA methylation clearly plays an important role in maintaining transposable elements in a quiescent state, but is there any evidence to support the idea that methylation of DNA regulates plant development? Methylation of cytosine residues in DNA increases the complexity of the primary DNA sequence and can block the binding of transcription factors and other proteins when present in their binding sites (McClelland 1981; Banks et al. 1988; Gierl et al. 1988; Iguchi-Argia and Schaffner 1989; Staiger et al. 1989; Comb and Goodman 1990; Inamdar et al. 1991). If DNA methylation does regulate transcription, then one would predict that disrupting DNA methylation should affect gene expression. Loss of DNA methylation in all contexts leads to the activation of many genes (Zhang et al. 2006; Lister et al. 2008). Despite the changes in gene expression seen in methylation-deficient mutants, however, the evidence that DNA methylation regulates plant gene activity in a developmentally relevant manner, with the exception of imprinted genes, is largely lacking. The capacity to map ^{m}C at single-base resolution will allow this question to be addressed in a systematic manner by monitoring tissue-specific changes in DNA methylation and correlating these with changes in gene expression.

16.7.1 Imprinting

DNA methylation is important for the regulation of imprinted gene expression in animals, where there is differential methylation of maternal and paternal alleles of imprinted genes in most or all cells of the organism (Paulsen and Ferguson-Smith 2001). In plants, imprinting was first observed in maize endosperm, where the maternal copy of certain alleles of zein and tubulin genes is hypomethylated and more highly expressed than the paternal allele (Lund et al. 1995a, b; Alleman and Doctor 2000). Preferential expression of the maternally inherited alleles of other endosperm-specific genes, including the *R* locus involved in aleurone pigmentation, *drz1* and *no-apical meristem related protein1*, has also been observed, suggesting that imprinting may be quite common in plants (Kermicle 1970; Chaudhuri and Messing 1994). A mechanism that allows differential methylation and expression of the maternal allele of imprinted genes has been elucidated in Arabidopsis.

Maternal allele-specific expression of the Arabidopsis genes *FWA*, *MEA* and *FIS2* is dependent on the activity of DME, which is expressed in the central cell of the female gametophyte (Fig. 16.3). Demethylation of critical regions in the *cis*-regulatory elements of these genes is essential for expression in both the central cell and developing endosperm (Choi et al. 2002; Kinoshita et al. 2004; Gehring et al. 2006; Jullien et al. 2006). The paternal alleles of *FWA* and *FIS2* remain inactive throughout endosperm development. This silenced state is dependent on the maintenance methyltransferase MET1 (Jullien et al. 2006). In contrast, repression of the paternal *MEA* allele requires the activity of the FIS–PRC2 complex that includes the MEA protein transcribed from the maternal allele (Gehring et al. 2006). In vegetative tissues, both alleles of *FWA* and *MEA* are maintained in the inactive state through methylation of *cis*-regulatory sequences. Loss of methylation in *met1* mutants or ectopic expression of *DME* leads to ectopic expression of these genes (Kinoshita et al. 1999, 2004; Choi et al. 2002). *FIS2* remains inactive in vegetative tissues even when the critical sequences become demethylated in a *met1* mutant, suggesting that the transcription factors required to activate *FIS2* are absent (Jullien et al. 2006). The maternal allele of the maize gene *Fie1* is hypomethylated in the central cell relative to sperm and egg cells, suggesting that a DME orthologue may regulate maternal allele expression in maize endosperm (Hermon et al. 2007).

16.7.2 Response to the Environment

As sessile organisms, plants are exposed to a variety of biotic and abiotic stresses during their lifetime, and need to adapt or respond to these stresses to ensure their survival and reproductive success. There is increasing evidence that plants use epigenetic mechanisms to modulate their response to biotic and abiotic stresses. Changes in histone modification and/or small RNA populations have been associated with changing gene expression in response to different environmental challenges (Sung and Amasino 2004; Sunkar and Zhu 2004; Bastow et al. 2004; Borsani et al. 2005; Tsuji et al. 2006; Katiyar-Agarwal et al. 2007). Changes in DNA methylation, both at the global level and at that of individual genes, have also been observed in response to environmental stress (Wada et al. 2004; Choi and Sano 2007; Boyko et al. 2007; Rozhon et al. 2008).

Transposons are hypomethylated and reactivated in response to the extreme (and unnatural) stress of tissue culture and also to climatic stress or fungal elicitors (Brettell and Dennis 1991; Hirochika et al. 1996; Takeda et al. 1999; Kalendar et al. 2000; Hashida et al. 2003, 2006). Transposition of stress-reactivated transposons can lead to heritable changes in gene expression or in genome organization

(McClintock 1984; Jiang et al. 2003). Retrotransposons are abundant in plant genomes. Some reside in densely methylated repetitive DNA that is heterochromatic, whereas others exist as isolated elements that are embedded within euchromatin (Peterson-Burch et al. 2004). Reactivation of retroelements, particularly those located within euchromatic DNA, can lead to the activation or silencing of flanking genes (Kashkush et al. 2003; Lippman et al. 2004). Although the DNA of isolated retroelements is frequently methylated, the associated histone modifications are a combination of those normally seen in hetero- and euchromatin. These elements are more readily activated by mutation in the POL V/DRD1 pathway (Huettel et al. 2006). If RdDM and demethylation are targeted by the same small RNAs, then changing the balance of activity of these two players in response to stress could lead to reversible demethylation of these isolated retroelements. This in turn could modulate the expression of genes flanking these elements, allowing for stress-inducible changes in gene expression.

Changes in DNA methylation in response to stress offer the potential to change methylation patterns in the progeny of a stressed plant, perhaps allowing for rapid adaptation to environmental conditions. The plant germline is established late in development. Any changes in DNA methylation that occur in response to environmental stresses can accumulate in the meristem and be transmitted to the progeny, as there is no evidence for the resetting of DNA methylation patterns during plant gamete formation (Finnegan et al. 1996; Kakutani et al. 1996). In this way, stress responses can affect gene expression in both the generation that is exposed to stress and future generations.

16.8 Conclusions

Research into the role of DNA methylation has entered an exciting phase, thanks to the development of genome-wide technologies for mapping the epigenome. The ability to correlate information on the DNA methylome, the transcriptome and the small RNA'ome has created a new opportunity for exploring the role of DNA methylation in plant processes (Lister et al. 2008). Only when these technologies become affordable for high-throughput analyses of the methylome in individual tissues, at different developmental times or in response to different environmental challenges will a definitive assessment of the role of DNA methylation in regulating expression gene by gene be possible. Finally, as more plant genome sequences become available, it will become clear whether the "rules" learned in Arabidopsis are applicable to other plant species, particularly those with larger genomes, where genes are more commonly flanked by transposable elements (Zhang 2008).

Acknowledgements I would like to thank Liz Dennis, Ian Greaves, Estelle Jaligot and Alain Rival for their critical reading of the manuscript. Thanks too to Xiaofeng Cao, Steve Jacobsen, Tetsuji Kakutani, Marjori Matzke and Jurek Paskowski who provided me with their data prior to publication.

References

Alleman M, Doctor J (2000) Genomic imprinting in plants: observations and evolutionary implications. Plant Mol Biol 43:147–161

Aufsatz W, Mette MF, van der Winden J, Matzke M, Matzke AJM (2002) HDA6, a putative histone deacetylase needed to enhance DNA methylation induced by double stranded RNA. EMBO J 21:6832–6841

Aufsatz W, Mette MF, Matzke AJM, Matzke M (2004) The role of METI in RNA-directed *de novo* and maintenance methylation of CG dinucleotides. Plant Mol Biol 54:793–804

Banks JA, Masson P, Fedoroff N (1988) Molecular mechanisms in the developmental regulation of the maize *Suppressor-mutator* transposable element. Genes Dev 2:1364–1380

Bartee L, Malagnac F, Bender J (2001) *Arabidopsis cmt3* chromomethylase mutations block non-CG methylation and silencing of an endogenous locus. Genes Dev 15:1753–1758

Bastow R, Mylne JS, Lister C, Lippman Z, Martienssen RA, Dean C (2004) Vernalization requires epigenetic silencing of *FLC* by histone methylation. Nature 427:164–167

Beck S, Rakyan VK (2008) The methylome: approaches for global DNA methylation profiling. Cell (in press)

Bestor T, Laudano A, Mattaliano R, Ingram V (1988) Cloning and sequencing of a cDNA encoding DNA methyltransferase of mouse cells. J Mol Biol 203:971–983

Bird AP (1995) Gene number, noise reduction and biological complexity. Trends Genet 11:94–100

Bird AP (1997) Does DNA methylation control transposition of selfish elements in the germline? Trends Genet 13:469–470

Bird AP, Southern EM (1978) Use of restriction enzymes to study eukaryote DNA methylation I. The methylation pattern of ribosomal DNA from *Xenopus laevis*. J Mol Biol 118:22–47

Blewitt ME, Gendrel A-V, Pang Z, Sparrow DB, Whitelaw N, Craig JM, Apedaile A, Hilton DJ, Dunwoodie SL, Brockdorff N, Kay GF, Whitelaw E (2008) SmcHD1, containing a structural-maintenance-of-chromosomes hinge domain, has a critical role in X inactivation. Nature Genet 40:663–669

Borevitz JO, Ecker JR (2004) Plant genomics: the third wave. Annu Rev Genomics Hum Genet 5:443–477

Borsani O, Zhu J, Verslues PE, Sunkar R, Zhu J-K (2005) Endogenous siRNAs derived from a pair of natural cis-antisense transcripts regulate salt tolerance in *Arabidopsis*. Cell 123:1279–1291

Bostick M, Kim JK, Eteve P-O, Clark A, Pradhan S, Jacobsen SE (2007) UHRF1 plays a role in maintaining DNA methylation in mammalian cells. Science 317:1760–1764

Boyko A, Kathira P, Zemp FJ, Yao Y, Pogribny I, Kovalchuk I (2007) Transgenerational changes in the genome stability and methylation in pathogen-infected plants. Nucleic Acids Res 35:1714–1725

Brettell RIS, Dennis ES (1991) Reactivation of a silent *Ac* following tissue culture is associated with heritable changes in its methylation pattern. Mol Gen Genet 229:365–372

Cao X, Jacobsen SE (2002) Locus-specific control of asymmetric and CpNpG methylation by the *DRM* and *CMT3* methyltransferase genes. Proc Natl Acad Sci USA 99:16491–16498

Cao X, Springer NM, Muszynski MG, Phillips RL, Kaepplar S, Jacobsen SE (2000) Conserved plant genes with similarity to mammalian *de novo* DNA methyltransferases. Proc Natl Acad Sci USA 97:4979–4984

Cao X, Ausfatz W, Zilberman D, Mette MF, Huang MS, Matzke M, Jacobsen SE (2003) Role of the *DRM* and *CMT3* methyltransferases in RNA-directed DNA methylation. Curr Biol 13:2212–2217

Chan SW-L, Zilberman D, Xie Z, Johnson LM, Carrington JC, Jacobsen SE (2004) RNA silencing genes control de novo DNA methylation. Science 303:1336

Chan SW, Henderson IR, Jacobsen SE (2005) Gardening the genome: DNA methylation in *Arabidopsis thaliana*. Nature Rev Genet 6:351–360

Chan SW-L, Henderson IR, Zhang X, Shah G, Chien JS-C, Jacobsen SE (2006) RNAi, DRD1 and histone methylation actively target developmentally important non-CG DNA methylation in *Arabidopsis*. PLoS Genet 2:e83. doi:10.1371/journal.pgen.0020083

Chapman EJ, Carrington JC (2007) Specialization and evolution of endogenous small RNA pathways. Nature Rev Genet 8:884–896

Chaudhuri S, Messing J (1994) Allele-specific paternal imprinting of *dzr1*, a post-transcriptional regulator of zein accumulation. Proc Natl Acad Sci USA 91:4867–4871

Chawla R, Nicholson SJ, Folta KM, Srivastava V (2007) Transgene-induced silencing of Arabidopsis phytochrome A gene via exonic methylation. Plant J 52:1105–1118

Choi C-S, Sano H (2007) Abiotic-stress induces demethylation and transcriptional activation of a gene encoding a glycerolphosphodiesterase-like protein in tobacco plants. Mol Gen Genomics 277:589–600

Choi Y, Gehring M, Johnson L, Hannon M, Harada JJ, Goldberg RB, Jacobsen SE, Fischer RL (2002) DEMETER, a DNA glycosylase domain protein, is required for endosperm gene imprinting and seed viability in *Arabidopsis*. Cell 110:33–42

Cokus SJ, Feng SH, Zhang X, Chen Z, Merriman B, Haudenschild CD, Pradhan S, Nelson SF, Pellegrini M, Jacobsen SE (2008) Shotgun bisulfite sequencing of the *Arabidopsis* genome reveals DNA methylation patterning. Nature 452:215–219

Comb M, Goodman HM (1990) CpG methylation inhibits proenkephalin gene expression and binding of the transcription factor AP-2. Nucleic Acids Res 18:3975–3982

Cross SH, Charlton JA, Nan X, Bird AP (1994) Purification of CpG islands using a methylated DNA binding column. Nature Genet 6:236–244

Dalma-Weishausz DD, Warrington J, Tanimoto EY, Miyada CG (2006) The affymetrix GeneChip platform: an overview. Methods Enzymol 410:3–28

Dennis K, Fan T, Geiman T, Yan Q, Muegge K (2001) Lsh, a member of the SNF2 family, is required for genome-wide methylation. Genes Dev 15:2940–2944

Ebbs ML, Bender J (2006) Locus-specific control of DNA methylation by the *Arabidopsis* SUVH5 histone methyltransferase. Plant Cell 18:1166–1176

Ebbs ML, Bartee L, Bender J (2005) H3 lysine 9 is maintained on a transcribed inverted repeat by combined action of SUVH6 and SUVH4 methyltransferases. Mol Cell Biol 25:10507–10515

El-Shami M, Pontier D, Lahmy S, Braun L, Picart C, Vega D, Hakimi MA, Jacobsen SE, Cooke R, Lagrange T (2007) Reiterated WG/GW motifs form functionally and evolutionarily conserved ARGONAUTE-binding platforms in RNAi-related components. Genes Dev 21:2539–2544

Fang Y, Spector DL (2007) Identification of nuclear dicing bodies containing proteins for micro-RNA biogenesis in living *Arabidopsis* plants. Curr Biol 17:818–823

Finnegan EJ, Dennis ES (1993) Isolation and identification by sequence homology of a putative cytosine methyltransferase from *Arabidopsis thaliana*. Nucleic Acids Res 21:2383–2388

Finnegan EJ, Peacock WJ, Dennis ES (1996) Reduced DNA methylation in *Arabidopsis thaliana* results in abnormal plant development. Proc Natl Acad Sci USA 93:8449–8454

Finnegan EJ, Genger RK, Peacock WJ, Dennis ES (1998) DNA methylation in plants. Annu Rev Plant Physiol Plant Mol Biol 49:223–248

Finnegan EJ, Kovac KA, Jaligot E, Sheldon CC, Peacock WJ, Dennis ES (2005) The down-regulation of *FLOWERING LOCUS C* (*FLC*) expression in plants with low levels of DNA methylation and by vernalization occurs by distinct mechanisms. Plant J 44:420–432

Frommer M, Mcdonald LE, Millar DS, Collis CM, Watt F, Grigg GW, Molloy PL, Paul CL (1992) A genomic sequencing protocol that yields a positive display of 5-methylcytosine residues in individual DNA strands. Proc Natl Acad Sci USA 89:1827–1831

Gehring M, Huh JH, Hsieh T-F, Penterman J, Choi Y, Harada JJ, Goldberg RB, Fischer RL (2006) DEMETER DNA glycosylase establishes *MEDEA* Polycomb gene self-imprinting by allele-specific demethylation. Cell 124:495–506

Gendrel A-V, Lippman Z, Yordan C, Colot V, Martienssen R (2002) Dependence of heterochromatic histone H3 methylation patterns on the *Arabidopsis* gene *DDM1*. Science 297:1871–1873

Gibbons RJ, McDowell TL, Raman S, O'Rouke DM, Garrick D, Ayyub H, Higgs DR (2000) Mutations in *ATRX*, encoding a SWI/SNF-like protein, causes diverse changes in the pattern of DNA methylation. Nature Genet 24:368–371

Gierl A, Lutticke S, Saedler H (1988) *TnpA* product encoded by the transposable element En-1 of *Zea mays* is a DNA binding protein. EMBO J 7:4045–4053

Goll MG, Bestor TH (2005) Eukaryote cytosine methyltransferases. Annu Rev Biochem 74:481–514

Gong Z, Morales-Ruiz T, Ariza RR, Roldan-Arjona T, David L, Zhu JK (2002) *ROS1*, a repressor of transcriptional gene silencing in *Arabidopsis*, encodes a DNA glycosylase/lyase. Cell 111:803–814

Grafi G, Zemach A, Pitto L (2007) Methyl-CpG-binding domain (MBD) proteins in plants. Biochim Biophys Acta 1769:287–294

Hale CJ, Stonaker JL, Gross SM, Hollick JB (2007) A novel Snf2 protein maintains *trans*-generational regulatory states established by paramutation in maize. PLoS Biol 5:e275. doi:10.1371/journal.pbio.0050275

Hashida SN, Kitamura K, Mikami T, Kishima Y (2003) Temperature shift coordinately changes the activity and methylation state of transposon *TAM3* in *Antirrhinum majus*. Plant Physiol 132:1207–1216

Hashida SN, Uchiyama T, Martin C, Kishima Y, Sano Y, Mikami T (2006) The temperature-dependent change in methylation of the *Antirrhinum* transposon *Tam3* is controlled by the activity of its transposase. Plant Cell 18:104–118

Henikoff S, Comai L (1998) A DNA methyltransferase homolog with a chromodomain exists in multiple polymorphic forms in *Arabidopsis*. Genetics 149:307–318

Hermon P, Srilunchang K, Zou J, Dresselhaus T, Danilevskaya ON (2007) Activation of the imprinted Polycomb group gene *Fie1* gene in maize endosperm requires demethylation of the maternal allele. Plant Mol Biol 64:387–395

Herr AJ, Jensen MB, Dalmay T, Baulcombe DC (2005) RNA polymerase IV directs silencing of endogenous DNA. Science 308:118–120

Hirochika H, Sugimoto K, Otsuki Y, Tsugawa H, Kanda M (1996) Retrotransposons of rice involved in mutations induced by tissue culture. Proc Natl Acad Sci USA 93:7783–7788

Holliday R, Pugh JE (1975) DNA modification mechanisms and gene activity during development. Science 187:226–232

Huettel B, Kanno T, Daxinger L, Ausfatz W, Matzke M, Matzke AJM (2006) Endogenous targets of RNA-directed DNA methylation and Pol IV in *Arabidopsis*. EMBO J 25:2828–2836

Huettel B, Kanno T, Daxinger L, Bucher E, van der Winden J, Matzke M, Matzke AJM (2007) RNA-directed DNA methylation mediated by DRD1 and Pol IVb: a versatile pathway for transcriptional gene silencing in plants. Biochim Biophys Acta 1769:358–374

Iguchi-Ariga SMM, Schaffner W (1989) CpG methylation of the cAMP responsive enhancer/promoter sequence TGACGTCA abolishes specific factor binding as well as transcriptional activation. Genes Dev 3:612–619

Inamdar NM, Ehrlich KC, Ehrlich M (1991) CpG methylation inhibits binding of several sequence-specific DNA-binding proteins from pea, wheat, soybean and cauliflower. Plant Mol Biol 17:111–123

Ito M, Koike A, Koizumi N, Sano H (2003) Methylated DNA-binding proteins from *Arabidopsis*. Plant Physiol 133:1747–1754

Jackson JP, Lindroth AM, Cao X, Jacobsen SE (2002) Control of CpNpG DNA methylation by the KRYPTONITE histone H3 methyltransferase. Nature 416:556–560

Jacobsen SE, Meyerowitz EM (1997) Hypermethylated *SUPERMAN* epigenetic alleles in *Arabidopsis*. Science 277:1100–1103

Jacobsen SE, Sakai H, Finnegan EJ, Cao X, Meyerowitz EM (2000) Ectopic hypermethylation of flower-specific genes in *Arabidopsis*. Curr Biol 10:179–186

Jeddeloh JA, Stokes TL, Richards EJ (1999) Maintenance of genomic methylation requires a SWI2/SNF2-like protein. Nature Genet 22:94–97

Jiang N, Bao Z, Zhang X, Hirochika H, Eddy SR, McCouch SR, Wessler SR (2003) An active DNA transposon family in rice. Nature 421:163–167

Johnson LM, Cao X, Jacobsen SE (2002) Interplay between two epigenetic marks: DNA methylation and histone H3 lysine 9 methylation. Curr Biol 12:1–20

Johnson LM, Bostick M, Zhang X, Kraft E, Henderson I, Callis J, Jacobsen SE (2007) The SRA domain methyl-cytosine-binding domain links DNA and histone methylation. Curr Biol 17:379–384

Jost J-P (1993) Nuclear extracts of chicken embryos promote an active demethylation of DNA by excision repair of 5-methyldeoxycytidine. Proc Natl Acad Sci USA 90:4684–4688

Jullien PE, Kinoshita T, Ohad N, Berger F (2006) Maintenance of DNA methylation during the *Arabidopsis* life cycle is essential for parental imprinting. Plant Cell 18:1360–1372

Kakutani T (1997) Genetic characterization of late-flowering traits induced by DNA hypomethylation mutation in *Arabidopsis thaliana*. Plant J 12:1447–1451

Kakutani T, Jeddeloh JA, Flowers SK, Munakata K, Richards EJ (1996) Developmental abnormalities and epimutations associated with DNA hypomethylation mutations. Proc Natl Acad Sci USA 93:12406–12411

Kalendar R, Tanskanen J, Immonen S, Nevo E, Schulman AH (2000) Genome evolution of wild barley (*Hordeum spontaneum*) by *BARE-1* retrotransposon dynamics in response to sharp microclimate divergence. Proc Natl Acad Sci USA 97:6603–6607

Kangaspeska S, Stride B, Metivier R, Polycarpou-Schwarz M, Ibbersen D, Carmouche RP, Benes V, Gannon F, Reid G (2008) Transient cyclical methylation of promoter DNA. Nature 452:112–115

Kanno T, Mette MF, Kreil DP, Ausfatz W, Matzke M, Matzke AJM (2004) Involvement of putative SNF2 chromatin remodeling protein DRD1 in RNA-directed DNA methylation. Curr Biol 14:801–805

Kanno T, Huettel B, Mette MF, Ausfatz W, Jaligot E, Daxinger L, Kreil DP, Matzke M, Matzke AJM (2005) Atypical RNA polymerase subunits required for RNA-directed DNA methylation. Nature Genet 37:761–765

Kanno T, Bucher E, Daxinger L, Huettel B, Bohmdorfer G, Gregor W, Kreil DP, Matzke M, Matzke AJM (2008) A structural-maintenance-of-chromosomes hinge domain-containing protein is required for RNA-directed DNA methylation. Nature Genet 40:670–675

Kapoor A, Agius F, Zhu J-K (2005) Preventing transcriptional gene silencing by active DNA demethylation. FEBS Lett 579:5889–5898

Kashkush K, Feldman M, Levy AA (2003) Transcriptional activation of retrotransposons alters the expression of adjacent genes in wheat. Nature Genet 33:102–106

Kasschau KD, Fahlgren N, Chapman EJ, Sullivan CM, Cumbie J, Givan SA, Carrington JC (2007) Genome-wide profiling and analysis of *Arabidopsis* siRNAs. PLoS Biol 5:e57. doi:10.1371/journal.pbio.0050057

Katiyar-Agarwal S, Gao S, Vivian-Smith A, Jin H (2007) A novel class of bacterial-induced small RNAs in *Arabidopsis*. Genes Dev 21:3123–3134

Kermicle JL (1970) Dependence of the *R*-mottled aleurone phenotype in maize on the mode of sexual transmission. Genetics 66:69–85

Kinoshita T, Yadegari R, Harada JJ, Goldberg RB, Fischer RL (1999) Imprinting of the *MEDEA* Polycomb gene in the Arabidopsis endosperm. Plant Cell 11:1945–1952

Kinoshita T, Miura A, Choi Y, Kinoshita Y, Cao X, Jacobsen SE, Fischer RL, Kakutani T (2004) One-way imprint of *FWA* imprinting in *Arabidopsis* endosperm by DNA methylation. Science 303:521–523

Kishimoto N, Sakai H, Jackson J, Jacobsen SE, Meyerowitz EM, Dennis ES, Finnegan EJ (2001) Site specificity of the *Arabidopsis* MET1 DNA methyltransferase demonstrated through hypermethylation of the *superman* locus. Plant Mol Biol 46:171–183

Kloc A, Zaratiegui M, Nora E, Martienssen R (2008) RNA interference guides histone modification during the S phase of chromosomal replication. Curr Biol 18:490–495

Klose RJ, Yan Q, Tothova Z, Yamane K, Erdjument-Bromage H, Tempst P, Gilliland DG, Kaerlin JWG (2007) The retinoblastoma binding protein RBP2 is an H3K4 demethylase. Cell 128:889–900

Li CF, Henderson IR, Song L, Fedoroff N, Lagrange T, Jacobsen SE (2008) Dynamic regulation of ARGONAUTE4 within multiple nuclear bodies in *Arabidopsis thaliana*. PLoS Genet 4:e27

Lindroth AM, Cao X, Jackson JP, Zilberman D, McCallum CM, Henikoff S, Jacobsen SE (2001) Requirement of *CHROMOMEHTYLASE3* for the maintenance of CpXpG methylation. Science 292:2077–2080

Lindroth AM, Shultis D, Jasencakova Z, Fuchs J, Johnson L, Schubert D, Patnaik D, Pradhan S, Goodrich J, Schubert I, Jenuwein T, Khorasanizadeh S, Jacobsen SE (2004) Dual histone H3 methylation marks at lysines 9 and 27 required for interaction with *CHROMOMETHYLASE3*. EMBO J 23:4286–4296

Lippman Z, May B, Yordan C, Singer T, Martienssen R (2003) Distinct mechanisms determine transposon inheritance and methylation via small interfering RNA and histone modification. PLoS Biol 1:420–428. doi:10.1371/journal/pbio.0000067

Lippman Z, Gendrel A-V, Black M, Vaughn MW, Dedhia N, McCombie WR, Lavine K, Mittal V, May B, Kasschau KD, Carrington JC, Doerge RW, Colot V, Martienssen R (2004) Role of transposable elements in heterochromatin and epigenetic control. Nature 430:471–476

Lister R, O'Malley RCO, Tonti-Fillippini J, Gregory BD, Berry CC, Millar AH, Ecker JR (2008) Highly integrated single-base resolution maps of the epigenome in *Arabidopsis*. Cell 133:1–14

Losada A, Hirano T (2005) Dynamic molecular linkersw of the genome: the first decade of SMC proteins. Genes Dev 19:1269–1287

Lund G, Ciceri P, Viotti A (1995a) Maternal-specific demethylation and expression of specific alleles of zein genes in the endosperm of *Zea mays* L. Plant J 8:571–581

Lund G, Messing J, Viotti A (1995b) Endosperm-specific demethylation and activation of specific alleles of the α-tubulin genes of *Zea mays* L. Mol Gen Genet 246:716–722

Luo M, Bilodeau P, Koltunow A, Dennis ES, Peacock WJ (1999) Genes controlling fertilization-independent seed development in *Arabidopsis thaliana*. Proc Natl Acad Sci USA 96:296–301

Mardis ER (2008) The impact of next-generation sequencing technology on genetics. Trends Genet 24:133–141

Mathieu O, Probst AV, Paszkowski J (2005) Distinct regulation of histone H3 methylation at lysines 27 and 9 by CpG methylation in *Arabidopsis*. EMBO J 24:2782–2791

Mathieu O, Reinders J, Caikovski M, Smathajitt C, Paszkowski J (2007) Transgenerational stability of the *Arabidopsis* epigenome is coordinated by CG methylation. Cell 130:851–862

Matzke M, Kanno T, Huettel B, Daxinger L, Matzke AJM (2007) Targets of RNA-directed DNA methylation. Curr Opin Plant Biol 10:512–519

McClelland M (1981) The effect of sequence specific DNA methylation on restriction endonuclease cleavage. Nucleic Acids Res 9:5859–5866

McClintock B (1984) The significance of response of the genome to challenge. Science 226:792–801

Metivier R, Gallais R, Tiffoche C, Le Peron C, Jurkowska RZ, Carmouche RP, Ibbersen D, Barath P, Demay F, Reid G, Benes V, Jeltsch A, Gannon F, Salbert G (2008) Cyclical DNA methylation of a transcriptionally active promoter. Nature 452:45–50

Mosher RA, Schwach F, Studholme D, Baulcombe DC (2008) PloIVb influences RNA-directed DNA methylation independently of its role in siRNA biogenesis. Proc Natl Acad Sci USA 105:3145–3150

Nan XS, Meehan RR, Bird A (1993) Dissection of the methyl-CPG binding domain from the chromosome protein MeCP2. Nucleic Acids Res 21:4886–4892

Okano M, Xie S, Li E (1998) Cloning and characterization of a family of novel mammalian DNA (cytosine-5) methyltransferases. Nature Genet 19:219–220

Onodera Y, Haag JR, Ream T, Nunes PC, Pontes O, Pikaard CS (2005) Plant nuclear RNA polymerase IV mediates siRNA and DNA methylation-dependent heterochromatin formation. Cell 120:613–622

Papa CM, Springer NM, Muszynski MG, Meeley R, Kaepplar SM (2001) Maize chromomethylase *Zea methyltransferase 2* is required for CpNpG methylation. Plant Cell 13:1919–1928

Paulsen M, Ferguson-Smith AC (2001) DNA methylation in genomic imprinting, development, and disease. J Pathol 195:97–110

Pavlopoulou A, Kossida S (2007) Plant cytosine-5 DNA methyltransferases: structure, function and molecular evolution. Genomics 90:530–541

Penterman J, Zilberman D, Huh JH, Ballinger T, Henikoff S, Fischer RL (2007a) DNA demethylation in the *Arabidopsis* genome. Proc Natl Acad Sci USA 104:6752–6757

Penterman J, Uzawa R, Fischer RL (2007b) Genetic interactions between DNA demethylation and methylation in Arabidopsis. Plant Physiol 145:1549–1557

Peterson-Burch BD, Nettleton D, Voytas DF (2004) Genomic neighbourhoods for *Arabidopsis* retrotransposons: a role for targeted integration in the distribution of the Metaviridae. Genome Biol 5:R78

Pontes O, Li CF, Nunes PC, Haag J, Ream T, Vitins A, Jacobsen SE, Pikaard C (2006) The *Arabidopsis* chromatin-modifying nuclear siRNA pathway involves a nucleolar RNA processing center. Cell 126:79–92

Rajagopalan R, Vaucheret H, Trejo J, Bartel DP (2006) A diverse and evolutionarily fluid set of microRNAs in *Arabidopsis thaliana*. Genes Dev 20:3407–3425

Reinders J, Vivier CD, Theiler G, Chollet D, Descombes P, Paszkowski J (2008) Genome-wide, high-resolution DNA methylation profiling using bisulfite-mediated cytosine conversion. Genome Res 18:469–476

Reyna-Lopez GE, Simpson J, Ruiz-Herrera J (1997) Differences in DNA methylation patterns are detectable during dimorphic transition of fungi by amplification of restriction polymorphisms. Mol Gen Genet 253:703–710

Riggs AD (1975) X inactivation, differentiation and DNA methylation. Cytogenet Cell Genet 14:25

Ronemus MJ, Galbiati M, Ticknor C, Chen J, Dellaporta SL (1996) Demethylation-induced developmental pleiotropy in *Arabidopsis*. Science 273:654–657

Rozhon W, Baubec T, Mayerhofer J, Mittelsten Scheid O, Jonak C (2008) Rapid quantification of global DNA methylation by isocratic cation exchange high-performance liquid chromatiography. Anal Biochem 375:354–360

Saze H, Kakutani T (2007) Heritable epigenetic mutation of a transposon-flanked *Arabidopsis* gene due to lack of the chromatin-remodeling factor DDM1. EMBO J 26:3641–3652

Saze H, Shiraishi A, Miura A, Kakutani T (2008) Control of genic DNA methylation by a jmjC domain-containing protein in *Arabidopsis thaliana*. Science 319:462–465

Scebba F, Bernacchia G, De Bastiani M, Evangelista M, Cantoni RM, Cella R, Locci MT, Pitto L (2003) *Arabidopsis* MBD proteins show different binding specificities and nuclear localizations. Plant Mol Biol 53:755–771

Scebba F, De Bastiani M, Bernacchia G, Andreucci A, Galli A, Pitto L (2007) PRMT11: a new Arabidopsis MBD7 protein partner with arginine methyltransferase activity. Plant J 52:210–222

Sheldon CC, Burn JE, Perez PP, Metzger J, Edwards JA, Peacock WJ, Dennis ES (1999) The *FLF* MADS box gene: a repressor of flowering in Arabidopsis regulated by vernalization and methylation. Plant Cell 11:445–458

Singer-Sam TP, Yang N, Mori RL, Tanguay JM, Le Bon JC, Riggs AD (1990) DNA methylation in the 5' region of the mouse PGK-1 gene and a quantitative PCR assay for methylation. In: Clawson G, Willis A, Weissback A, Jones P (eds) Nucleic acid methylation. UCLA Symposia on Molecular and Cellular Biology. Alan R Liss, New York, pp 28

Smith LM, Pontes O, Searle I, Yelina N, Yousafzai FK, Herr AJ, Pikaard CS, Baulcombe DC (2007) An SNF2 protein associated with nuclear RNA silencing and the spread of a silencing signal between cells in *Arabidopsis*. Plant Cell 19:1507–1521

Soppe WJJ, Jacobsen SE, Alonso-Blanco C, Jackson JP, Kakutani T, Koorneef M, Peeters AJM (2000) The later flowering phenotype of *fwa* mutants is caused by gain-of-function epigenetic alleles of a homeodomain gene. Mol Cell 6:791–802

Soppe WJJ, Jasencakova Z, Houben A, Kakutani T, Meister A, Huang MS, Jacobsen SE, Schubert I, Fransz PF (2002) DNA methylation controls histone H3 lysine 9 methylation and heterochromatin assembly in *Arabidopsis*. EMBO J 21:6549–6559

Sridhar VV, Kapoor A, Zhang K, Zhu J, Zhou T, Hasegawa PM, Bressan RA, Zhu J-K (2007) Control of DNA methylation and heterochromatic silencing by histone H2b deubiquitination. Nature 447:735–738

Staiger D, Kauleen H, Schell J (1989) A CACGTG motif of the *Antirrhinum majus* chalcone synthase promoter is recognized by an evolutionary conserved nuclear protein. Proc Natl Acad Sci USA 86:6930–6934

Stewart FJ, Raleigh EA (1998) Dependence of McrBC cleavage on distance between recognition elements. Biol Chem 379:611–616

Sung S, Amasino RM (2004) Vernalization in *Arabidopsis thaliana* is mediated by the PHD finger protein VIN3. Nature 427:159–164

Sunkar R, Zhu JK (2004) Novel and stress-regulated microRNAs and other small RNAs from *Arabidopsis*. Plant Cell 16:2001–2019

Sutherland E, Coe L, Raleigh EA (1992) McrBC: a multisubunit GTP-dependent restriction enzyme. J Mol Biol 225:327–348

Takeda S, Sugimoto K, Otsuki Y, Hirochika H (1999) A 13-bp *cis*-regulatory element in the LTR promoter of the tobacco retrotransposon *Tto1* is involved in responsiveness to tissue culture, wounding, methyl jasmonate and fungal elicitors. Plant J 18:383–393

Tariq M, Saze H, Probst AVL, Lichota J, Habu Y, Paskowski J (2003) Erasure of CpG methylation in *Arabidopsis* alters patterns of histone H3 methylation in heterochromatin. Proc Natl Acad Sci USA 100:8823–8827

Tran RK, Henikoff JG, Zilberman D, Ditt RF, Jacobsen SE, Henikoff S (2005a) DNA methylation profiling identifies CG methylation clusters in *Arabidopsis* genes. Curr Biol 15:154–159

Tran RK, Zilberman D, de Bustos C, Ditt RF, Henikoff JG, Lindroth AM, Delrow J, Boyle T, Kwong S, Bryson TD, Jacobsen SE, Henikoff S (2005b) Chromatin and siRNA pathways cooperate to maintain DNA methylation of small transposable elements in *Arabidopsis*. Genome Biol 6:R90. doi:10.1186/gb-2005-6-11-r90

Tsuji H, Saika H, Tsutsumi N, Hirai A, Nakazono M (2006) Dynamic and reversible changes in histone H3-Lys4 methylation and H3 acetylation occurring at submergence-inducible genes in rice. Plant Cell Physiol 47:995–1003

Turck F, Roudier F, Farrona S, Martin-Magniette M-L, Guillaume E, Buisine N, Gagnot S, Martiensen RA, Coupland G, Colot V (2007) Arabidopsis TFL2/LHP1 specifically associates with genes marked by trimethylation of histone H3 lysine 27. PLoS Genet 3:e86

Vaucheret H (2006) Post-transcriptional small RNA pathways in plants: mechanisms and regulations. Genes Dev 20:759–771

Vaughn MW, Martienssen RA (2005) Finding the right template: RNA Pol IV, a plant-specific RNA polymerase. Mol Cell 17:754–756

Vaughn MW, Tanurd IM, Lippman Z, Jiang H, Carrasquillo R, Rabinowicz PD, Dedhia N, McCombie WR, Agier N, Bulski A, Colot V, Doerge RW, Martienssen R (2007) Epigenetic natural variation in *Arabidopsis thaliana*. PLoS Biol 5:e174

Vongs A, Kakutani T, Martienssen RA, Richards EJ (1993) *Arabidopsis thaliana* DNA methylation deficient mutants. Science 260:1926–1928

Wada Y, Miyamoto K, Kusano T, Sano H (2004) Association between up-regulation of stress-responsive genes and hypomethylation of genomic DNA in tobacco plants. Mol Gen Genomics 271:658–666

Weber M, Davies JJ, Wittig D, Oakley EJ, Haase M, Lam WL, Schubler D (2005) Chromosome-wide and promoter-specific analyses identify sites of differential DNA methylation in normal and transformed human cells. Nature Genet 37:853–862

Weber M, Hellman I, Stadler MB, Ramos L, Paabo S, Rebhan M, Schubeler D (2007) Distribution, silencing potential and evolutionary impact of promoter DNA methylation in the human genome. Nature Genet 39:457–466

Wierzbicki AT, Haag JR, Pikaard CS (2008) Noncoding transcription by RNA polymerase Pol IVb/Pol V mediates transcriptional silencing of overlapping and adjacent genes. Cell 135:635–648

Woo HR, Pontes O, Pikaard CS, Richards EJ (2007) VIM1, a methylcytosine-binding protein required for centromeric heterochromatin. Genes Dev 21:267–277

Yamane K, Toumazou C, Tsukada Y, Erdjument-Bromage H, Tempst P, Wong J, Zhang Y (2006) JHDM2A, a JmjC-containing H3K9 demethylase, facilitates transcription activation by androgen receptor. Cell 125:483–495

Yoder JA, Walsh CP, Bestor TH (1997) Cytosine methylation and the ecology of intragenomic parasites. Trends Genet 13:335–340

Zemach A, Grafi G (2003) Characterization of *Arabidopsis thaliana* methyl-CpG-binding domain (MBD) proteins. Plant J 34:565–572

Zemach A, Li Y, Wayburn B, Ben-Meir H, Kiss V, Avivi Y, Kalchenko V, Jacobsen SE, Grafi G (2005) DDM1 binds Arabidopsis methyl-CpG binding domain proteins and affects their subnuclear localization. Plant Cell 17:1549–1558

Zhai J, Liu J, Liu B, Li P, Meyers BC, Chen X, Cao X (2008) Small RNA-directed epigenetic natural variation in *Arabidopsis thaliana*. PLoS Genet 4:e1000056

Zhang X (2008) The epigenetic landscape of plants. Science 320:489–492

Zhang X, Yazaki J, Sundaresan A, Cokus SJ, Chan SW-L, Chen H, Henderson IR, Shinn P, Pellegrini M, Jacobsen SE, Ecker JR (2006) Genome-wide high-resolution mapping and functional analysis of DNA methylation in *Arabidopsis*. Cell 126:1189–1201

Zhang X, Shiu S, Cal A, Borevitz JO (2008) Global analysis of genetic, epigenetic and transcriptional polymorphisms in *Arabidopsis thaliana* using whole genome tiling arrays. PLoS Genet 4:e1000032. doi:10.1371/journal.pgen.1000032

Zheng X, Zhu J, Kapoor A, Zhu J-K (2007) Role of *Arabidopsis* AGO6 in siRNA accumulation, DNA methylation and transcriptional silencing. EMBO J 26:1691–1701

Zhu J, Kapoor A, Sridhar VV, Agius F, Zhu J-K (2007) The DNA glycosylase/lyase ROS1 functions in pruning DNA methylation patterns in *Arabidopsis*. Curr Biol 17:54–59

Zilberman D, Henikoff S (2007) Genome-wide analysis of DNA methylation patterns. Development 134:3959–3965

Zilberman D, Cao X, Johansen LK, Xie Z, Carrington JC, Jacobsen SE (2003) Role of *Arabidopsis ARGONAUTE4* in RNA-directed DNA methylation triggered by inverted repeats. Curr Biol 14:1214–1220

Zilberman D, Cao X, Jacobsen SE (2004) *ARGONAUTE4* control of locus-specific siRNA accumulation and DNA and histone methylation. Science 299:716–719

Zilberman D, Gehring M, Tran RK, Ballinger T, Henikoff S (2006) Genome-wide analysis of *Arabidopsis thaliana* DNA methylation uncovers an interdependence between methylation and transcription. Nature Genet 39:61–69

Chapter 17
Molecular Mechanisms in Epigenetic Regulation of Plant Growth and Development

A. Berr and W.H. Shen

17.1 Introduction

In the eukaryotic nucleus, DNA, carrying the genetic information, is packaged into chromatin through its association with proteins. The fundamental unit of chromatin, the nucleosome, comprises roughly 147 bp of DNA wrapped around an octamer of core histone proteins (two molecules each of H2A, H2B, H3 and H4). Nucleosome and higher-order packaged chromatin fine-tune the accessibility of DNA and thus play critical roles in the regulation of genome function.

Currently, epigenetics refers to the study of heritable (mitotically and/or meiotically) changes in genome function that occur without a change in DNA sequence. Well-known epigenetic mechanisms include posttranslational covalent modifications of histone tails (e.g. acetylation, methylation and ubiquitylation), ATP-dependent chromatin remodelling and DNA methylation at cytosine residues. In this chapter, we describe diverse mechanisms of epigenetic modifications involved in several processes of plant growth and development, highlighting potential interests of epigenetic studies in both fundamental research and biotechnology applications.

17.2 Vernalization and Flowering Time

In higher plants, flowering represents the transition from vegetative to reproductive growth. Flowering at the correct time is crucial to ensure plant reproduction and seed yield. Both endogenous and environmental factors are involved in the control

A. Berr and W.H. Shen
Institut de Biologie Moléculaire des Plantes (IBMP) Centre National de la Recherche Scientifique (CNRS), Université de Strasbourg (UdS), 12 rue du Général Zimmer, 67084, Strasbourg cedex, France
e-mail: Alexandre.Berr@ibmp-ulp.u-strasbg.fr; Wen-Hui.Shen@ibmp-ulp.u-strasbg.fr

E-C. Pua and M.R. Davey (eds.),
Plant Developmental Biology – Biotechnological Perspectives: Volume 2,
DOI 10.1007/978-3-642-04670-4_17, © Springer-Verlag Berlin Heidelberg 2010

of flowering time. Prolonged incubation in cold (vernalization) during winter is memorized for blooming in spring in many plant species, including some agronomically important cereals. In the model plant *Arabidopsis*, *FLOWERING LOCUS C* (*FLC*), encoding a MADS-box transcriptional repressor, forms a converging point of autonomous and vernalization pathways in the control of flowering time (Bäurle and Dean 2006; Schmitz and Amasino 2007). Many chromatin modifiers were recently found to regulate flowering through modification of *FLC* expression.

17.2.1 Histone Methylation in FLC Activation

Forward genetic screen of mutants in the Landsberg *erecta* (L*er*) accession of *Arabidopsis* identified *early flowering in short days* (*efs*; Soppe et al. 1999). More recent molecular characterization revealed that *efs* is allelic to *SDG8* (*SET DOMAIN GROUP 8*; Kim et al. 2005). In an independent study, the authors' laboratory demonstrated that *sdg8* mutants display a reduced level of *FLC* expression and have an early flowering phenotype at all photoperiods tested (Zhao et al. 2005). Recombinant SDG8 protein showed histone methyltransferase (HMTase) activity when oligonucleosomes, but not mononucleosomes nor free histones, were used as substrates (Xu et al. 2008). *sdg8* mutants exhibit, at both global-genome and *FLC* locus-specific levels, dramatically reduced levels of dimethyl- and trimethyl- but not monomethyl-H3K36 (Zhao et al. 2005; Xu et al. 2008), indicating primary function of SDG8 in H3K36 di- and trimethylation (H3K36me2 and me3). Slightly reduced or increased levels of H3K4 methylation were also observed in the *efs*/*sdg8* mutants under the same conditions (Kim et al. 2005; Zhao et al. 2005). However, it is unclear if perturbed H3K4 methylation is a direct effect of *sdg8*, or is caused by reduction of H3K36me2 and H3K36me3. More recently, the *Arabidopsis* homolog of Trithorax *ATX1*, which encodes a H3K4-HMTase (Alvarez et al. 2003), was shown to be involved in H3K4me3 at the *FLC* locus (Pien et al. 2008). In yeast, the H3K4-HMTase SET1 and the H3K36-HMTase SET2 physically interact with the PAF1 complex during transcriptional initiation and elongation (reviewed in Shilatifard 2006). In *Arabidopsis*, mutants for the PAF1-like complex, including *vernalization independence 3* (*vip3*), *vip4*, *vip5*, *vip6*/*elf8* (*early flowering 8*) and *elf7*, have early flowering phenotypes, a low level of *FLC* expression and reduced levels of H3K4 and H3K36 methylation (He et al. 2004; Oh et al. 2004; Xu et al. 2008). Furthermore, the double mutant *sdg8vip4* has an enhanced early flowering phenotype (Zhao et al. 2005). Taken together, it is likely that H3K4 methylation by ATX1 and H3K36 methylation by SDG8 act together with PAF1 in transcriptional initiation and elongation of *FLC* (Fig. 17.1a).

Our unpublished data revealed that the double mutant of *AtUBC1* and *AtUBC2* encoding ubiquitin-conjugating enzymes (E2), as well as the single mutant of *HUB1* or *HUB2* encoding ubiquitin ligases (E3) present also a reduced level of *FLC* expression and an early flowering phenotype (Xu et al., unpublished data). HUB1/2 (Fleury et al. 2007; Liu et al. 2007) as well as AtUBC1/2 (Xu et al. 2009)

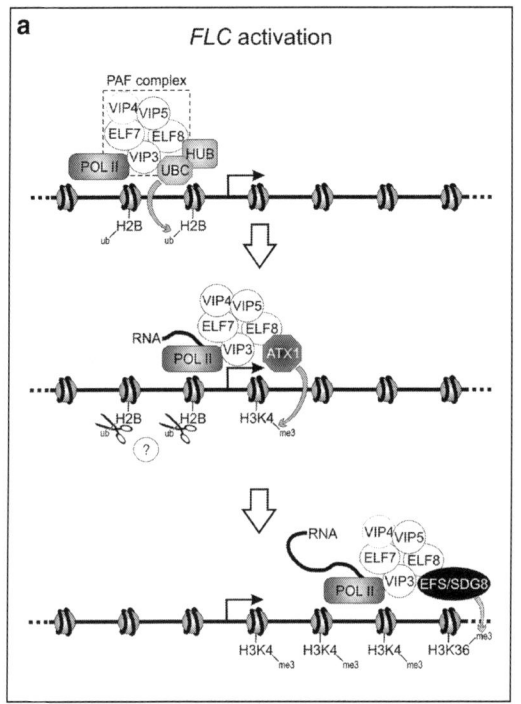
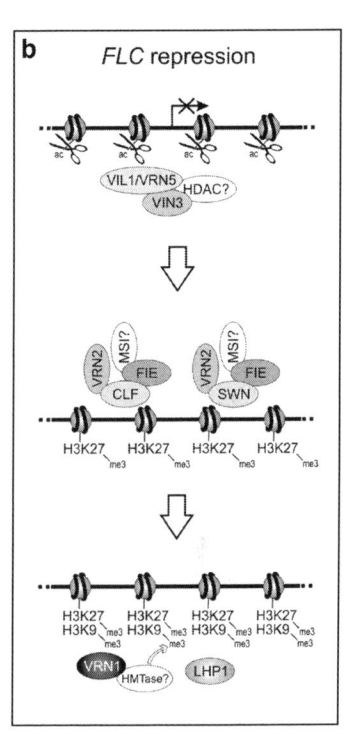

Fig. 17.1 Epigenetic regulation of *FLC* transcription. **a** *FLC* transcription through maintenance of an active chromatin state. Several complexes are required to maintain the expression of *FLC*, thus preventing inadequate flowering before spring. RNA polymerase II, PAF complex (ELF7, 8 and VIP3, 4, 5) and UBC/HUB complex are recruited to the promoter region. UBC/HUB mediates monoubiquitylation of H2B, leading to initiation of transcription. The H2B monoubiquitylation also signals for the recruitment of ATX1, which methylates H3K4, allowing transcription to enter the early elongation step. In parallel, H2B is deubiquitinated by an unknown complex. Transcription elongation requires EFS/SDG8, which methylates H3K36. **b** Vernalization induces PcG repression of *FLC*. During winter, a complex including at least *VIN3*, *VRN5* and an as yet unidentified H3 deacetylase (HDAC) is recruited at *FLC* chromatin to initiate the repression of *FLC*. The resulting hypoacetylated state creates favourable conditions for recruiting a VRN2-containing PRC2 complex, which initiates the epigenetic silencing of *FLC* via H3K27 methylation. A VRN1-containing complex is recruited and methylates H3K9 through an as yet unidentified HMTase. Together, this leads to the association of LHP1 to *FLC* chromatin and to the stable maintenance of the repressive state of *FLC* chromatin

are involved in H2B monoubiquitylation. In yeast, the AtUBC1/2 homolog RAD6 and the HUB1/2 homolog BRE1 mediate H2B monoubiquitylation, which is involved in the initiation of transcription upstream of H3K4 and H3K36 methylation (reviewed in Shilatifard 2006). Such a mechanism is likely conserved for *FLC* transcription in *Arabidopsis* (Fig. 17.1a).

Mutation in a Jumonji (Jmj)-domain protein gene, *RELATIVE OF EARLY FLOWERING 6 (REF6)*, resulted in an increased *FLC* expression and late flowering

(Noh et al. 2004). Recently, Jmj-domain proteins were shown to have histone demethylase (HDase) activity (reviewed in Shi 2007). REF6 might repress *FLC* expression by demethylating H3K4 and/or H3K36. Thus, activation of *FLC* transcription involves H3K4 and K36 methylations, which are regulated by specific HMTases and HDases.

17.2.2   Histone Methylation in FLC Repression

Upon vernalization, the active chromatin state of *FLC* is replaced progressively by a repressive state to allow plants to flower after return to a warm temperature. Characterization of *Arabidopsis* mutants defective in vernalization response revealed critical roles of H3K27 and H3K9 methylation in repression of *FLC* expression (Fig. 17.1b). Expression of *VERNALIZATION INSENSITIVE 3* (*VIN3*), which encodes a homeodomain protein, is induced by vernalization and is decreased quickly after return to warm temperatures (Sung and Amasino 2004). In wild-type plants, removal of the transcriptionally active mark H3 acetylation and acquisition of the repressive marks H3K9me3 and H3K27me3 occur on *FLC* chromatin during vernalization (Sung and Amasino 2004; Bastow et al. 2004). These events fail to occur in the *vin3* mutant, indicating a primary role of *VIN3* in the establishment of a repressive chromatin state on *FLC*. Recent studies revealed that VIN3 forms a heterodimer with its homolog VIN3-LIKE 1 (VIL1), also known as VERNALIZATION 5 (VRN5; Sung et al. 2006; Greb et al. 2007). *vil1/vrn5* mutants are also insensitive to vernalization and fail in establishing a repressive chromatin state on *FLC*.

VRN2 encodes a homolog of the *Drosophila* Polycomb group (PcG) protein, Suppressor of Zeste 12 (Su(Z)12; Gendall et al. 2001). VRN2 was found to form a PRC2 (Polycomb Repressive Complex 2) complex, together with FERTILIZATION INDEPENDENT ENDOSPERM (FIE), the *Arabidopsis* homolog of the *Drosophila* Extra Sex Combs (ESC), and with CURLY LEAF (CLF) and SWINGER (SWN), two of the three homologs (together with MEDEA) in *Arabidopsis* of the *Drosophila* Enhancer of zeste (E(z); Wood et al. 2006). Both CLF and SWN contain a signature of HMTase, the SET domain, and are likely to be responsible for the deposition of H3K27me3 on *FLC* chromatin. Several observations indicate that the PRC2 complex acts downstream of VIN3 in the maintenance of *FLC* repression. As in *vin3*, H3K9 and H3K27 methylation is impaired in *vrn2*. Unlike the case in *vin3*, vernalization-induced deacetylation of H3 still occurs in *vrn2* (Sung and Amasino 2004; Bastow et al. 2004). Also different from *VIN3*, *VRN2* expression is not responsive to vernalization. A PRC2 complex composed of CLF, FIE and EMBRYONIC FLOWER2 (EMF2) was also shown to repress the MADS-box gene *AGAMOUS-LIKE 19* (*AGL19*) encoding a flowering activator independent of the *FLC* pathway (Schönrock et al. 2006).

In *Drosophila*, PRC2-induced repression is stabilized by PRC1 that binds H3K27me3. PRC1 components could not be identified by sequence conservation in the *Arabidopsis* genome. Several observations suggest that other proteins may have PRC1-like functions in *Arabidopsis*. *VRN1* encodes a protein that binds DNA in vitro in a non-sequence-specific manner, and associates broadly and stably with chromatin throughout mitosis (Levy et al. 2002; Mylne et al. 2006). Based on the convergence of mutant phenotypes between *vrn1* and *vrn2*, VRN1 might play a PRC1-like role and could function in part analogously to *Drosophila* Psc through its DNA-binding recruiting PcG proteins to the target genes (Klymenko et al. 2006). In *vrn1*, vernalization-induced H3K27, but not H3K9, methylation still occurs, indicating particular importance of VRN1 in H3K9 methylation at *FLC* chromatin (Sung and Amasino 2004; Bastow et al. 2004). LIKE HETEROCHRO-MATIN PROTEIN 1 (LHP1) associates not only with H3K9me3, as do all the animal HP1 proteins, but also with H3K27me3, as does Pc, the H3K27me3-binding subunit of PRC1 (Turck et al. 2007; Zhang et al. 2007). LHP1 is required for maintaining *FLC* repression initiated by PRC2 during vernalization (Mylne et al. 2006; Sung et al. 2006). Future identification of VRN1 and LHP1 partners will help significantly to understand the detailed molecular mechanisms of *FLC* repression.

Arginine (R) methylation also plays repressive roles in *FLC* expression. SHK1 BINDING PROTEIN 1 (SKB1), also named PROTEIN ARGININE METHYL-TRANSFERASE 5 (AtPRMT5), was required for symmetrical dimethylation on H4R3 and *FLC* repression (Wang et al. 2007; Pei et al. 2007; Schmitz et al. 2008). *atprmt5* mutants also show attenuation of vernalization-induced H3K9me3 and H3K27me3, indicating requirement of symmetrical dimethylation on H4R3 for H3K9 and H3K27 methylation (Schmitz et al. 2008). AtPRMT10 preferentially methylates H4R3 asymmetrically and is also involved in repression of *FLC* expression (Niu et al. 2007). However, whether AtPRMT10 directly or indirectly represses *FLC* expression remains to be determined.

17.2.3 Regulation of Flowering by Histone Acetylation

Histone acetylation is broadly implicated in the transcriptional regulation of genes during *Arabidopsis* development (Tian et al. 2005). The homeostasis of histone acetylation is regulated by histone-acetyltransferases (HATs) and histone-deacetylases (HDACs). Three *Arabidopsis* HATs belonging to the subgroup of homologs of the animal p300/cAMP-responsive element-binding protein (CBP), AtHAC1, AtHAC5 and AtHAC12, were found to promote flowering (Deng et al. 2007; Han et al. 2007). Up-regulation of *FLC* expression was observed in the *hac1* mutant and, to a higher degree, in the double mutants *hac1hac5* and *hac1hac12*. However, because *FLC* activation positively associates with histone acetylation, AtHAC1, AtHAC5 and AtHAC12 might indirectly regulate *FLC* expression through an as yet unidentified upstream repressor. Despite the observation that H3 deacetylation occurs during the establishment of repressive chromatin at *FLC* (Fig. 17.1b), the identity of the HDAC involved remains unknown. A recent study on the functional characterization of HDA6, a RPD3-type HDAC in *Arabidopsis*, indicates that HDA6 is responsible for H3 deacetylation at *FLC* chromatin (Wu et al. 2008).

17.2.4 ATP-Dependent Chromatin-Remodelling Complexes in Flowering Time Control

The yeast SWR1 complex (SWR1C) remodels chromatin by replacing H2A with the variant H2AZ, associated with important regulatory functions in gene transcription (reviewed in Kamakaka and Biggins 2005). In *Arabidopsis*, mutations in putative components of SWR1C, namely *ACTIN RELATED PROTEIN 6/EARLY IN SHORT DAYS 1/SUPPRESSOR OF FRIGIDA 3* (*ARP6/ESD1/SUF3*), the SWI/ SNF ATPase *PHOTOPERIOD INDEPENDENCE1* (*PIE1*) or *SEF* (*SERRATED AND EARLY FLOWERING*)/*AtSWC6*, cause pleiotropic developmental defects including early flowering and reduction of *FLC* expression (Noh and Amasino 2003; Choi et al. 2005, 2007; Deal et al. 2005; Martin-Trillo et al. 2006; March-Diaz et al. 2007; Lázaro et al. 2008). Similarly, reduction of *FLC* expression was also observed by H2AZ knockdown (Choi et al. 2007). Together, these studies indicate that several ARP-containing chromatin-remodelling complexes are involved in *FLC* expression and control of flowering time. Mutation in *ARP4*, another homolog of components of SWR1C, leads to early flowering, specifically under long-day conditions, pointing to a specific role of ARP4 in repression of *CONSTANS* (*CO*), a flowering activator (Kandasamy et al. 2005). Mutations in *AtBRM* encoding a component of the SWI/SNF chromatin-remodelling complex lead to early flowering plants, also through the up-regulation of *CO* (Farrona et al. 2004).

17.2.5 RNAi in Flowering Time Control

Small interfering RNAs (siRNAs) can regulate gene expression by either RNA degradation or transcriptional silencing by RNA-directed DNA methylation (RdDM) and/or histone methylation (Hamilton et al. 2002; Volpe et al. 2002; Zilberman et al. 2003; Xie et al. 2004). In the *Arabidopsis* L*er* accession, *FLC* contains a transposable element inserted in an intron of the gene. siRNAs were shown to be involved in targeting *FLC* for repressive H3K9 methylation (Liu et al. 2004). In the *Arabidopsis* Columbia (Col) accession, two small RNAs corresponding to the reverse strand, just 3′ to the major poly(A) site of the mature *FLC* transcript, were identified and shown to be involved in H3K9 methylation at *FLC* in a *DICER-LIKE* (*DCL*) 2-, 3- and 4-dependent manner (Swiezewski et al. 2007). Interestingly, mutations for a single *DCL* do not trigger a strong flowering phenotype, while double *dcl1dcl3* mutants are late flowering due to an increase of *FLC* mRNA (Schmitz et al. 2007).

17.3 Parental Imprinting and Seed Development

Diploid somatic cells, arising from the fusion of male and female gametes, hold two copies of each autosomal gene, which are both classically expressed. Within placental mammals and flowering plants, a subset of genes is selectively expressed according to its maternal or paternal origin due to an epigenetic mechanism called genomic imprinting. In mammals, approximately 100–200 genes are imprinted, with many involved in embryonic and placental development (Horsthemke and Buiting 2008). In angiosperms, imprinting occurs specifically during seed development. In *Arabidopsis*, the imprinting mechanism was partially elucidated for the PcG gene *FERTILIZATION INDEPENDENT SEED2* (*FIS2*), the homeobox transcription factor gene *FLOWERING WAGENINGEN* (*FWA*), the SET-domain E(z) homolog gene *MEDEA* (*MEA*) and the MADS-box transcription factor gene *PHERES1* (*PHE1*).

17.3.1 The Maternally Expressed FWA, FIS2 and MEA Alleles

FWA, *FIS2* and *MEA* are generally silenced, except that they are expressed in the central cell and in the endosperm cells specifically from the maternal allele (Fig. 17.2). DNA CG-methylation by METHYLTRANSFERASE1 (MET1) has a crucial function in the silencing of *FWA* and *FIS2*. A pair of transposon-associated tandem repeats (*SINE*-related transposon) is present at the *FWA* promoter region, which causes RdDM via the generation of siRNAs (T. Kinoshita et al. 2004; Y. Kinoshita et al. 2007). This henceforth termed imprinting control region (ICR) has a pivotal role in the maintenance of *FWA* imprinting and is sufficient to suppress *FWA* expression. For *FIS2* silencing, DNA methylation targets a 200-bp upstream segment rich in CG-islands (Jullien et al. 2006a, b). However, it remains to be verified whether or not this region functions as an ICR in *FIS2* imprinting. During gametogenesis, *FWA* and *FIS2* are activated through DNA demethylation in female central cells (Kinoshita et al. 2004; Jullien et al. 2006b). Demethylation is achieved primarily by the DNA glycosylase *DEMETER* (*DME*) specifically expressed in the central cell. DME is a bifunctional DNA glycosylase and apurinic/apyrimidinic (AP) lyase, which initiates erasure of 5-methylcytosine in the genome through a base-excision repair pathway (Gehring et al. 2006). Expression of the demethylated maternal alleles and silencing of the methylated paternal alleles of *FWA* and *FIS2* persist in the endosperm.

DNA methylation and demethylation also occur on *MEA* (Xiao et al. 2003; Spillane et al. 2004; Gehring et al. 2006). However, because *met1* mutation with a paternally transmitted transgene fused with the *MEA* promoter does not result in transgene expression (Luo et al. 2000), it appears that DNA methylation has only subsidiary roles and that another mechanism is responsible in paternal *MEA* silencing. Histone methylation mediated by PRC2 was recently reported to be

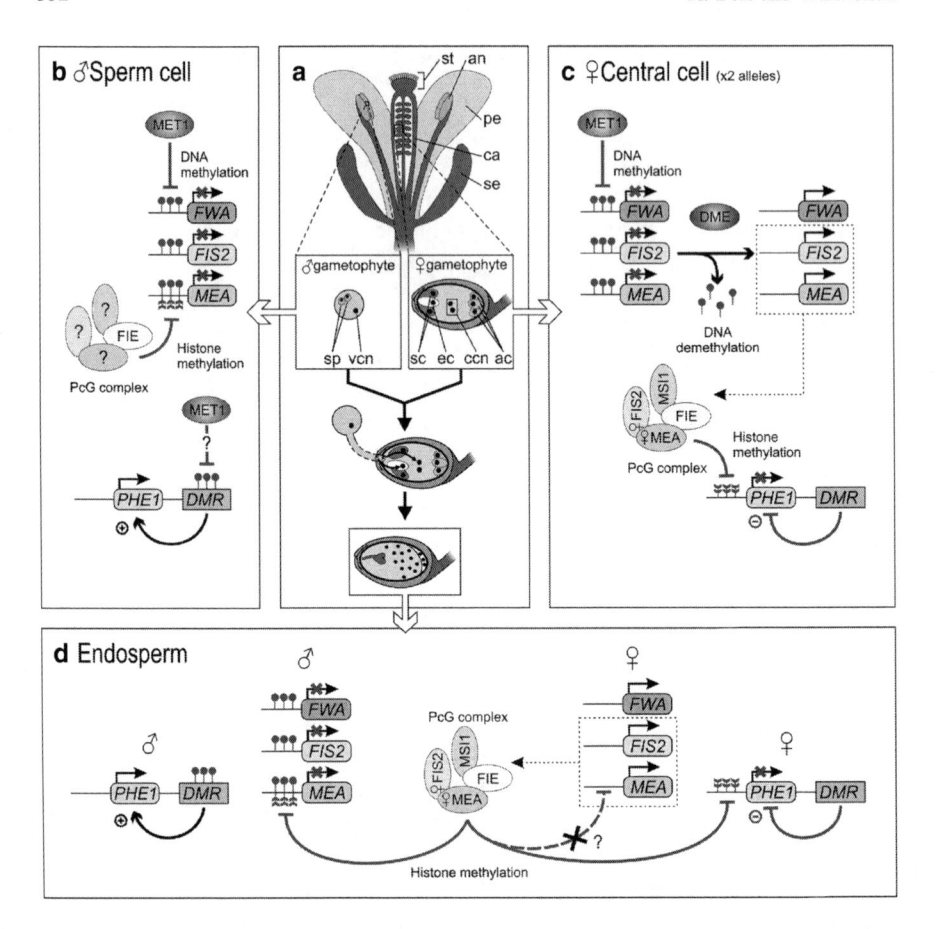

Fig. 17.2 Plant reproduction and imprinting mechanism in the endosperm. **a** Diagram of an *Arabidopsis* mature flower. The three-cell haploid male gametophyte or pollen cell produced by the anther (an) contains two haploid sperm cell (sp), together with one haploid vegetative cell nucleus (vcn). The female gametophyte formed in the carpel (ca) is composed of a central cell with two central cell nuclei (ccn), three antipodal cells (ac), two synergid cells (sc) and one haploid egg cell (ec). When the sepals (se) and petals (pe) open, the pollen is released from the anthers and, after landing on the stigma (st), germinates and produces a pollen tube containing two sperm cells. The pollen tube grows through the transmitting tract towards the ovules. When a pollen tube reaches an ovule, it penetrates the female gametophyte and releases the two sperm cells. Double fertilization occurs; one sperm cell fuses with the egg cell to initiate the zygote, while the other sperm cell fertilizes the central cell to initiate endosperm development. **b** Model for the epigenetic regulation of *FWA*, *FIS2*, *MEA* and *PHE1* in the sperm cell. *FWA* and *FIS2* are silenced as a default state through DNA methylation by MET1. Silencing of *MEA* is probably due to methylation by MET1 together with H3K27 methylation by a FIE-containing PRC2 complex. *PHE1* activation results from the methylation of the differentially methylated region (DMR), which inhibits the formation of repressive chromatin loops or blocks (as an insulator) the action of silencer elements. **c** Model for the epigenetic regulation of *FWA*, *FIS2*, *MEA* and *PHE1* in the central cell. *FWA*, *FIS2* and *MEA* are silenced in the central cell through DNA methylation by MET1. DME is expressed in the central cell and erases the MET1 methylation marks, resulting in the activation of *FWA*, *FIS2* and *MEA*. MEA and FIS2 participate in a PcG complex. The resulting

involved in *MEA* silencing. In vegetative tissues, the repressive H3K27 methylation mark was observed mostly at the 3′ ends of the *MEA* coding sequence (Jullien et al. 2006a). *MEA* ectopic expression was detected in *FIE* cosuppression plants as well as in mutants for PRC2 components (e.g. *emf2*, *vrn2*, *clf* and *swn*; Jullien et al. 2006a). In sperm cells, H3K27 methylation was observed at both the 5′ and 3′ ends of the *MEA* allele (Gehring et al. 2006; Jullien et al. 2006a). Mutation in *FIE* triggers the expression of *MEA* from both parental alleles in the endosperm (Jullien et al. 2006a). Mutants *fie*, *fis2*, *msi1* and *mea* show a similar phenotype of autonomous endosperm development without fertilization (see review by Huh et al. 2008). A PRC2 complex composed of FIE, FIS2, MSI1 and maternal-expressed MEA is proposed to silence, through H3K27 methylation, the paternal *MEA* allele in endosperm cells (Fig. 17.2). However, because *FIS2* and *MEA* are silenced in male gametes, the composition of a FIE-containing PRC2 that maintains *MEA* silencing in sperm cells needs further clarification. It is interesting to note that *MEA* is a self-imprinted gene that, from its maternal expressed allele, forms a negative feedback in silencing the paternal allele. How does the maternal *MEA* allele escape PRC2 silencing in the endosperm? The specificity is likely determined by the DNA methylation states. While the paternal *MEA* allele is methylated, the maternal *MEA* allele is demethylated by DME. Future experiments are required for understanding the crosstalk between DNA methylation and PRC2-mediated *MEA* silencing.

17.3.2 The Paternally Expressed PHE1 Allele

In contrast to the above-described genes that are maternal allele expressed, *PHE1* is preferentially expressed in endosperm from the paternal allele (Fig. 17.2). The vegetative and reproductive development of *phe1* mutants was indistinguishable from that of wild-type plants. Interestingly, the repression of the maternal *PHE1* allele is disrupted in *mea* mutant seeds with a mutated maternal *MEA* allele (Köhler et al. 2003). Current models propose that the maternal *PHE1* allele is stably silenced in the central cell through the combined action of the H3K27 methylation by the MEA-containing PRC2 complex (Köhler et al. 2005) and the unmethylated DNA state of a differentially methylated region (DMR) on *PHE1* (Makarevich et al. 2008). In sperm cells, DMR is methylated (probably by MET1) and behaves as an

---

Fig. 17.2 (continued) PRC2 complex represses the maternal *PHE1* allele together with the unmethylated DMR, which may act either by forming a repressive chromatin loop or by facilitating the long-range action of silencer regions. **d** Model of imprinting in the endosperm. Upon fertilization, maternally active *FWA*, *FIS2* and *MEA* alleles are inherited and transcribed in the endosperm, whereas paternal alleles remain silenced because inherited in an inactive state from the sperm cell. *MEA* and *FIS2* participate in a PcG complex that, in turn, represses the paternal *MEA* allele. The maternal *PHE1* allele remains unmethylated at the DMR and targeted by the PRC2 complex, whereas the methylated DMR on the paternal *PHE1* allele is not targeted by the PRC2 and thus active. Consequently, *FWA*, *FIS2* and *MEA* are maternally expressed and *PHE1* paternally expressed in the early endosperm

insulator against silencers, or inhibits the formation of repressive chromatin marks, leading to the promotion of the paternal *PHE1* allele expression. After fertilization, the maternal *PHE1* allele remains silenced and the methylated paternally inherited DMR maintains the paternal *PHE1* allele active. The paradigm between *PHE1* and *MEA* silencing by DNA methylation and PRC2-mediated H3K27 methylation is not understood.

17.3.3   Genomic Imprinting in Maize

Gene-specific imprinting was also observed for six genes in maize, including two orthologs of the *Arabidopsis* PcG gene, *FIE ZmFie1* and *ZmFie2*, the transcription factor *NO-APICAL MERISTEM (NAM) RELATED PROTEIN 1 (Nrp1)*, the homolog of the *Drosophila E(z)* gene *Mez1* and two unknown-function genes *Maternally expressed gene1 (Meg1)* and *Paternally expressed gene1 (Peg1*; Guo et al. 2003; Haun et al. 2007; Gutierrez-Marcos et al. 2004). Similar to *FWA* and *FIS2* imprinting in *Arabidopsis*, DNA methylation seems to be required for *ZmFie1* and *ZmFie2* paternal silencing. In *ZmFie1*, DNA methylation targets a putative ICR region rich in CG-islands exclusively on the paternal allele. Demethylation of the maternal *ZmFie1* allele seems to be necessary but not sufficient for *ZmFie1* activation. Indeed, demethylation of the maternal allele occurs before fertilization in the central cell and persists in the endosperm, but transcripts are detected only in the primary endosperm after fertilization (Gutierrez-Marcos et al. 2006; Hermon et al. 2007). Interestingly, the silenced *ZmFie2* paternal allele is not methylated in the central cell or sperm cell, but is hypermethylated in the endosperm. Thus, de novo methylation may occur on the silenced *ZmFie2* paternal allele in the endosperm after fertilization (Gutierrez-Marcos et al. 2006). The endosperm originates from an unusual fertilization event and has important nutritional value for seed plants and foods. There is considerable interest in obtaining further insight into epigenetic mechanisms of imprinting in endosperms.

17.4   Chromatin in Stem Cell Maintenance

In higher plants, two distinct populations of stem cells, localized in the shoot and root apical meristem (SAM and RAM), are responsible for continuous post-embryonic growth. They provide cells to maintain stem cells as well as to initiate organogenesis, leading to the development of the aerial parts and the subterranean root system of the plant respectively. Epigenetic mechanisms regulate differential gene expression whereby cells acquire specific fates.

17.4.1 SAM

In *Arabidopsis*, the Class I KNOX genes, including *STM*, *KNAT1* (also known as *BP*), *KNAT2* and *KNAT6*, are expressed in the SAM and are required for the maintenance of the indeterminate cell fate and the prevention of cell differentiation in the meristem (reviewed by Scofield and Murray 2006). A PRC2 complex with H3K27me3 HMTase activity was found to be responsible for the suppression of the Class I KNOX genes in leaves (Schubert et al. 2006; Xu et al., unpublished data). The level of H3K27me3 at *STM* was decreased in *clfswn* double mutants, whereas wild-type levels were found in *clf* and *swn* single mutants, indicating redundant roles of CLF and SWN in H3K27 methylation at *STM*. A similar redundancy between EMF2 and VRN2 was also observed for H3K27 methylation at *STM*. Thus, CLF/SWN and EMF2/VRN2 can form interchangeably different PRC2 complexes in repression of *STM* (Schubert et al. 2006). *CYCLOPHILIN 71* (*CYP71*), encoding a WD40-domain protein, was shown to be involved in the maintenance of the silenced state of several *KNOX* genes (*KNAT1*, *KNAT2* and *STM*). CYP71 protein interacts physically with histone H3, and is required for H3K27me2 and H3K27me3 at *STM* and *KNAT1* (Li et al. 2007). The MYB-domain ASYMMETRIC LEAVES1 (AS1), the LOB-domain AS2 and the WD40-domain HIRA form a protein complex involved in the repression of *KNAT1* and *KNAT2* (but not *STM* and *KNAT6*) in leaves (Byrne et al. 2000; Xu et al. 2003; Phelps-Durr et al. 2005; Guo et al. 2008). The yeast and animal HIRA proteins function as a histone chaperone in chromatin remodelling. This suggests that chromatin remodelling is involved in repression of *KNOX* genes, in addition to PRC2-mediated silencing.

The homeodomain transcription factor WUSCHEL (WUS) independently and complementarily acts with STM in the maintenance of SAM activity. In wild-type plants, *WUS* expression is restricted to the stem cell organizing centre in the SAM. Mutations in *FAS1* and *FAS2* encoding subunits of CAF1, a complex chaperoning H3 and H4 in nucleosome assembly (Polo and Almouzni 2006), result in enlarged zones of *WUS* expression and shoot fasciation (Kaya et al. 2001). *SPLAYED* (*SYD*) encoding a member of the SWI/SNF family is highly expressed in SAM, and the SYD protein was found to be recruited specifically to the *WUS* promoter in transcriptional activation (Wagner and Meyerowitz 2002; Kwon et al. 2005). It thus appears that *WUS* expression associates with chromatin remodelling.

During reproductive growth, the MADS-box gene *AGAMOUS* (*AG*) represses *WUS*, causing stem cell maintenance to terminate and thus permitting the development of floral organs. *CLF* is a critical repressor of *AG* in leaves (Goodrich et al. 1997). The *AG* locus was generally covered with H3K27me3, leading to the repression of its expression in lateral organs (Schubert et al. 2006). Interestingly, H3K4me3 (activating mark) established by *ATX1* and H3K27me3 established by the *CLF*-containing PRC2 (repressing mark) were both found to associate with silenced *AG* (Saleh et al. 2007). Moreover, ATX1 and CLF proteins were found to interact physically. Similar bivalent chromatin domains, with activating and repressing marks, are also found in a number of genes in animal stem cells. It is proposed that such genes poised for transcription may be advantaged in rapid response to a developmental switch.

17.4.2   RAM

In addition to shoot fasciation, the *fas1* and *fas2* mutants also show dysfunction of the RAM (Kaya et al. 2001; Scheres et al. 2004). *SCARECROW* (*SCR*), an essential factor in root radial patterning (Di Laurenzio et al. 1996), showed a perturbed expression pattern in *fas1* mutant plants (Kaya et al. 2001). Chromatin organization at *GLABRA2* (*GL2*), which encodes a homeodomain transcription factor and represses root hair formation (Ohashi et al. 2003), was also affected in *fas2* mutant roots (Costa and Shaw 2006). The *nrp1-1nrp2-1* mutant, containing loss of function of both *NRP1* and *NRP2* genes, has a short-root phenotype (Zhu et al. 2006). NRP1 and NRP2 proteins show highest homology with the animal SET/TAF–I/I2PP2A proteins and preferentially bind the histones H2A and H2B, suggesting their function in chaperoning these histones in nucleosome assembly. Expression of *GL2* as well as *PLETHORA2* (*PLT2*), which encodes an AP2-type transcription factor and plays crucial roles in stem cell specification and maintenance in the RAM (Aida et al. 2004), was found perturbed in *nrp1-1nrp2-1* mutant plants (Zhu et al. 2006). Together, these studies reveal crucial roles of chromatin remodelling in RAM stem cell maintenance and cell fate determination. Histone acetylation and the HDAC *HDA18* were also reported to have important functions in proper cell patterning in roots (Xu et al. 2005).

17.5   Chromatin in Plant Stress Responses

To compensate for their sessile nature, plants have evolved intricate and varied strategies to enable them to survive and adapt to abiotic stresses (e.g. excessive or inadequate temperature, water supply or salinity) and biotic stresses (e.g. pathogens or herbivores). Compared to genetic mutations, epigenetic modifications represent a rapid, heritable and reversible means to reprogram genome function in order to efficiently respond and adapt to stress.

17.5.1   Histone Acetylation in Stress Responses

Several cold-regulated (COR) genes are involved in cold acclimatization or freezing tolerance. The transcriptional activators C-repeat binding factors (CBF; e.g. CBF1, CBF2 and CBF3; Novillo et al. 2007), by binding specific promoter regulatory elements (CRT/DRE regulatory elements), stimulate expression of COR genes.

The HAT AtGCN5 in association with the transcriptional adaptor proteins ADA2a and ADA2b, forming a complex related to the yeast ADA and SAGA transcriptional activator complexes, were proposed to participate in the activation of CBF1 (Stockinger et al. 2001; Vlachonasios et al. 2003). An *Arabidopsis* cold-response transcriptome analysis identified several genes involved in RNA metabolism and chromatin remodelling (Lee et al. 2005). The flowering autonomous pathway gene *FVE*, which encodes a homolog of the mammalian retinoblastoma-associated protein, a component of a histone deacetylase (HDAC) complex, was found to delay flowering by increasing *FLC* expression under cold stress (Kim et al. 2004). More recently, the *HOS15* gene in *Arabidopsis*, encoding a WD40-repeat protein with a histone H4 deacetylase activity, was reported to be crucial for the repression of genes associated with abiotic stress tolerance (Zhu et al. 2008). Plants overexpressing *AtHD2C*, which encodes a HD2-type HDAC, are less sensitive to the stress hormone abscisic acid (ABA) and more tolerant to salt and drought stress (Sridha and Wu 2006). Similarly, the RPD3-type HDAC AtHDA19 was found to enhance, via its interaction with the transcriptional corepressor AtSin3, the level of *AtERF7* (a member of the ethylene-responsive element binding factor family) that, in turn, acts as a major transcriptional repressor in ABA responses (Song et al. 2005). Overexpression of *AtHDA19* results in an increased resistance to the pathogen *Alternaria brassicicola*, whereas silencing of *AtHDA19* results in a decreased resistance to the same pathogen (Zhou et al. 2005). Finally, another RPD3-type HDAC, *AtHDA6*, was found to interact with CORONATINE-INSENSITIVE1 (COI1), an F-box protein involved in JA-mediated plant pathogen defence responses (Devoto et al. 2002).

Several HDAC genes in rice (*Oryza sativa*) show differential expression patterns in response to different types of stress (Fu et al. 2007). The SIR2 family-related *OsSRT1* gene encoding a H3K9 HDAC is required to safeguard against genome stability and cell damage induced by oxidative stress. Down-regulation of *OsSRT1* results in an increase of H3K9 acetylation, a decrease of H3K9me2, H_2O_2 production, DNA fragmentation, cell death and lesions, mimicking plant hypersensitive responses to pathogens (Huang et al. 2007). Histone acetylation and H3K4 methylation were observed to change at the rice *ALCOHOL DEHYDROGENASE 1* (*ADH1*) and *PYRUVATE DECARBOXYLASE 1* (*PDC1*) genes following water submergence (Tsuji et al. 2006). In tomato, the HAT HAC1 is involved in activation of the thermotolerance *HsfA1a* gene (Bharti et al. 2004).

17.5.2 Chromatin-Remodelling Factors in Stress Responses

In *Arabidopsis*, the SNF2/Brahma-type *AtCHR12* was found to play a central role in mediating temporary plant growth arrest upon perceiving adverse and/or otherwise limiting environmental growth conditions (Mlynárová et al. 2007). In plants, systemic acquired resistance (SAR) provides a vital mechanism in innate immunity. SWR1C was proposed to regulate genes related to SAR response by depositing H2AZ into the controlled loci, thus coupling chromatin remodelling and innate immunity in *Arabidopsis* (March-Diaz et al. 2007). The SNF2-like *PICKLE* (*PKL*) was shown to be involved in expression of *ABI3* and *ABI5*, playing important roles in seed germination and osmotic tolerance (Perruc et al. 2007).

17.6 Perspectives

The complete genome sequence information and powerful genetic approach in *Arabidopsis* have permitted an exponential increase in the number of genes examined for their role in epigenetic regulation of plant growth and development. This will likely continue in future, and include other uncharacterized genes potentially involved in different epigenetic mechanisms. Many studies are starting to reveal essential functions of epigenetic mechanisms in, amongst others, flowering time control, seed development, cell fate maintenance and response to environmental factors. Nevertheless, molecular and functional links among different regulators or pathways remain in many cases elusive. Many complex aspects uncovered by the genetic approach need to be refined to better understand the mechanisms involved. Very importantly, other plant species, particularly crops, need more investigations. Understanding epigenetic mechanisms likely will have important impacts on biotechnology and crop breeding.

Acknowledgements The authors thank Lin Xu for helpful discussions and comments on the manuscript, and the French Agence Nationale de la Recherche (ANR) for financial support in the research project METHISTONARA. A. Berr is supported by a postdoctoral fellowship from ANR-06-Blanc-0054-01.

References

Aida M, Beis D, Heidstra R, Willemsen V, Blilou I, Galinha C, Nussaume L, Noh YS, Amasino R, Scheres B (2004) The *PLETHORA* genes mediate patterning of the *Arabidopsis* root stem cell niche. Cell 119:109–120

Alvarez-Venegas R, Pien S, Sadder M, Witmer X, Grossniklaus U, Avramova Z (2003) ATX-1, an *Arabidopsis* homolog of trithorax, activates flower homeotic genes. Curr Biol 13:627–637

Bastow R, Mylne JS, Lister C, Lippman Z, Martienssen RA, Dean C (2004) Vernalization requires epigenetic silencing of *FLC* by histone methylation. Nature 427:164–167

Bäurle I, Dean C (2006) The timing of developmental transitions in plants. Cell 125:655–664

Bharti K, Von Koskull-Döring P, Bharti S, Kumar P, Tintschl-Körbitzer A, Treuter E, Nover L (2004) Tomato heat stress transcription factor HsfB1 represents a novel type of general transcription coactivator with a histone-like motif interacting with the plant CREB binding protein ortholog HAC1. Plant Cell 16:1521–1535

Byrne ME, Barley R, Curtis M, Arroyo JM, Dunham M, Hudson A, Martienssen RA (2000) *ASYMMETRIC LEAVES1* mediates leaf patterning and stem cell function in *Arabidopsis*. Nature 408:967–971

Choi K, Kim S, Kim SY, Kim M, Hyun Y, Lee H, Choe S, Kim SG, Michaels S, Lee I (2005) *SUPPRESSOR OF FRIGIDA3* encodes a nuclear ACTIN-RELATED PROTEIN6 required for floral repression in *Arabidopsis*. Plant Cell 17:2647–2660

Choi K, Park C, Lee J, Oh M, Noh B, Lee I (2007) *Arabidopsis* homologs of components of the SWR1 complex regulate flowering and plant development. Development 134:1931–1941

Costa S, Shaw P (2006) Chromatin organization and cell fate switch respond to positional information in *Arabidopsis*. Nature 439:493–496

Deal RB, Kandasamy MK, McKinney EC, Meagher RB (2005) The nuclear actin-related protein ARP6 is a pleiotropic developmental regulator required for the maintenance of *FLOWERING LOCUS C* expression and repression of flowering in *Arabidopsis*. Plant Cell 17:2633–2646

Deng W, Liu C, Pei Y, Deng X, Niu L, Cao X (2007) Involvement of the histone acetyltransferase AtHAC1 in the regulation of flowering time *via* repression of *FLOWERING LOCUS C* in *Arabidopsis*. Plant Physiol 143:1660–1668

Devoto A, Nieto-Rostro M, Xie D, Ellis C, Harmston R, Patrick E, Davis J, Sherratt L, Coleman M, Turner JG (2002) COI1 links jasmonate signalling and fertility to the SCF ubiquitin-ligase complex in *Arabidopsis*. Plant J 32:457–466

Di Laurenzio L, Wysocka-Diller J, Malamy JE, Pysh L, Helariutta Y, Freshour G, Hahn MG, Feldmann KA, Benfey PN (1996) The *SCARECROW* gene regulates an asymmetric cell division that is essential for generating the radial organization of the *Arabidopsis* root. Cell 86:423–433

Farrona S, Hurtado L, Bowman JL, Reyes JC (2004) The *Arabidopsis thaliana* SNF2 homolog AtBRM controls shoot development and flowering. Development 131:4965–4975

Fleury D, Himanen K, Cnops G, Nelissen H, Boccardi TM, Maere S, Beemster GT, Neyt P, Anami S, Robles P, Micol JL, Inzé D, Van Lijsebettens M (2007) The *Arabidopsis thaliana* homolog of yeast *BRE1* has a function in cell cycle regulation during early leaf and root growth. Plant Cell 19:417–432

Fu W, Wu K, Duan J (2007) Sequence and expression analysis of histone deacetylases in rice. Biochem Biophys Res Comm 356:843–850

Gehring M, Huh JH, Hsieh TF, Penterman J, Choi Y, Harada JJ, Goldberg RB, Fischer RL (2006) DEMETER DNA glycosylase establishes *MEDEA* Polycomb gene self-imprinting by allele-specific demethylation. Cell 124:495–506

Gendall AR, Levy YY, Wilson A, Dean C (2001) The *VERNALIZATION 2* gene mediates the epigenetic regulation of vernalization in *Arabidopsis*. Cell 107:525–535

Goodrich J, Puangsomlee P, Martin M, Long D, Meyerowitz EM, Coupland G (1997) A Polycomb-group gene regulates homeotic gene expression in *Arabidopsis*. Nature 386:44–51

Greb T, Mylne JS, Crevillen P, Geraldo N, An H, Gendall AR, Dean C (2007) The PHD finger protein VRN5 functions in the epigenetic silencing of *Arabidopsis*. Curr Biol 17:73–78

Guo M, Rupe MA, Danilevskaya ON, Yan X, Hu Z (2003) Genome-wide mRNA profiling reveals heterochronic allelic variation and a new imprinted gene in hybrid maize endosperm. Plant J 36:30–44

Guo M, Thomas J, Collins G, Timmermans MC (2008) Direct repression of *KNOX* loci by the ASYMMETRIC LEAVES1 complex of *Arabidopsis*. Plant Cell 20:48–58

Gutierrez-Marcos JF, Costa LM, Biderre-Petit C, Khbaya B, O'Sullivan DM, Wormald M, Perez P, Dickinson HG (2004) A*maternally expressed gene1* is a novel maize endosperm transfer cell specific gene with a maternal parent-of-origin pattern of expression. Plant Cell 16:1288–1301

Gutierrez-Marcos JF, Costa LM, Dal Pra M, Scholten S, Kranz E, Perez P, Dickinson HG (2006) Epigenetic asymmetry of imprinted genes in plant gametes. Nature Genet 38:876–878

Hamilton A, Voinnet O, Chappell L, Baulcombe D (2002) Two classes of short interfering RNA in RNA silencing. EMBO J 21:4671–4679

Han SK, Song JD, Noh YS, Noh B (2007) Role of plant *CBP/p300*-like genes in the regulation of flowering time. Plant J 49:103–114

Haun WJ, Laoueille-Duprat S, O'Connell MJ, Spillane C, Grossniklaus U, Phillips AR, Kaeppler SM, Springer NM (2007) Genomic imprinting, methylation and molecular evolution of maize *enhancer of zeste* (*Mez*) homologs. Plant J 49:325–337

He Y, Doyle MR, Amasino RM (2004) PAF1-complex-mediated histone methylation of *FLOW-ERING LOCUS C* chromatin is required for the vernalization-responsive, winter-annual habit in *Arabidopsis*. Genes Dev 18:2774–2784

Hermon P, Srilunchang KO, Zou J, Dresselhaus T, Danilevskaya ON (2007) Activation of the imprinted Polycomb Group *Fie1* gene in maize endosperm requires demethylation of the maternal allele. Plant Mol Biol 64:387–395

Horsthemke B, Buiting K (2008) Genomic imprinting and imprinting defects in humans. Adv Genet 61:225–246

Huang L, Sun Q, Qin F, Li C, Zhao Y, Zhou DX (2007) Down-regulation of a *SILENT INFOR-MATION REGULATOR2*-related histone deacetylase gene, *OsSRT1*, induces DNA fragmentation and cell death in rice. Plant Physiol 144:1508–1519

Huh JH, Bauer MJ, Hsieh TF, Fischer RL (2008) Cellular programming of plant gene imprinting. Cell 132:735–744

Jullien PE, Katz A, Oliva M, Ohad N, Berger F (2006a) Polycomb group complexes self-regulate imprinting of the Polycomb group gene *MEDEA* in *Arabidopsis*. Curr Biol 16:486–492

Jullien PE, Kinoshita T, Ohad N, Berger F (2006b) Maintenance of DNA methylation during the *Arabidopsis* life cycle is essential for parental imprinting. Plant Cell 18:1360–1372

Kamakaka RT, Biggins S (2005) Histone variants: deviants? Genes Dev 19:295–316

Kandasamy MK, Deal RB, McKinney EC, Meagher RB (2005) Silencing the nuclear actin-related protein AtARP4 in *Arabidopsis* has multiple effects on plant development, including early flowering and delayed floral senescence. Plant J 41:845–858

Kaya H, Shibahara KI, Taoka KI, Iwabuchi M, Stillman B, Araki T (2001) *FASCIATA* genes for chromatin assembly factor-1 in *Arabidopsis* maintain the cellular organization of apical meristems. Cell 104:131–142

Kim HJ, Hyun Y, Park JY, Park MJ, Park MK, Kim MD, Kim HJ, Lee MH, Moon J, Lee I, Kim J (2004) A genetic link between cold responses and flowering time through *FVE* in *Arabidopsis thaliana*. Nature Genet 36:167–171

Kim SY, He Y, Jacob Y, Noh YS, Michaels S, Amasino R (2005) Establishment of the vernaliza-tion-responsive, winter-annual habit in *Arabidopsis* requires a putative histone H3 methyl transferase. Plant Cell 17:3301–3310

Kinoshita T, Miura A, Choi Y, Kinoshita Y, Cao X, Jacobsen SE, Fischer RL, Kakutani T (2004) One-way control of *FWA* imprinting in *Arabidopsis* endosperm by DNA methylation. Science 303:521–523

Kinoshita Y, Saze H, Kinoshita T, Miura A, Soppe WJ, Koornneef M, Kakutani T (2007) Control of *FWA* gene silencing in *Arabidopsis thaliana* by SINE-related direct repeats. Plant J 49:38–45

Klymenko T, Papp B, Fischle W, Köcher T, M. Schelder M, Fritsch C, Wild B, Wilm M, Müller J (2006) A Polycomb group protein complex with sequence-specific DNA-binding and selective methyl-lysine-binding activities. Genes Dev 20:1110–1122

Köhler C, Hennig L, Spillane C, Pien S, Gruissem W, Grossniklaus U (2003) The *Polycomb*-group protein MEDEA regulates seed development by controlling expression of the MADS-box gene *PHERES1*. Genes Dev 17:1540–1553

Köhler C, Page DR, Gagliardini V, Grossniklaus U (2005) The *Arabidopsis thaliana* MEDEA Polycomb group protein controls expression of PHERES1 by parental imprinting. Nature Genet 37:28–30

Kwon CS, Chen C, Wagner D (2005) *WUSCHEL* is a primary target for transcriptional regulation by SPLAYED in dynamic control of stem cell fate in *Arabidopsis*. Genes Dev 19:992–1003

Lázaro A, Gómez-Zambrano A, López-González L, Piñeiro M, Jarillo JA (2008) Mutations in the *Arabidopsis SWC6* gene, encoding a component of the SWR1 chromatin remodelling complex, accelerate flowering time and alter leaf and flower development. J Exp Bot 59:653–666

Lee B, Henderson DA, Zhu JK (2005) The *Arabidopsis* cold-responsive transcriptome and its regulation by ICE1. Plant Cell 17:3155–3175

Levy YY, Mesnage S, Mylne JS, Gendall AR, Dean C (2002) Multiple roles of *Arabidopsis VRN1* in vernalization and flowering time control. Science 297:243–246

Li H, He Z, Lu G, Lee SC, Alonso J, Ecker JR, Luan S (2007) A WD40 domain cyclophilin interacts with histone H3 and functions in gene repression and organogenesis in *Arabidopsis*. Plant Cell 19:2403–2416

Liu J, He Y, Amasino R, Chen X (2004) siRNAs targeting an intronic transposon in the regulation of natural flowering behaviour in *Arabidopsis thaliana*. Genes Dev 18:2873–2878

Liu Y, Koornneef M, Soppe WJ (2007) The absence of histone H2B monoubiquitination in the *Arabidopsis hub1* (*rdo4*) mutant reveals a role for chromatin remodeling in seed dormancy. Plant Cell 19:433–444

Luo M, Bilodeau P, Dennis ES, Peacock WJ, Chaudhury A (2000) Expression and parent-of-origin effects for *FIS2*, *MEA*, and *FIE* in the endosperm and embryo of developing *Arabidopsis* seeds. Proc Natl Acad Sci USA 97:10637–10642

Makarevich G, Villar CB, Erilova A, Köhler C (2008) Mechanism of *PHERES1* imprinting in *Arabidopsis*. J Cell Sci 121:906–912

March-Diaz R, Garcia-Dominguez M, Florencio FJ, Reyes JC (2007) SEF, a new protein required for flowering repression in *Arabidopsis*, interacts with PIE1 and ARP6. Plant Physiol 143:893–901

Martin-Trillo M, Lazaro A, Poethig RS, Gomez-Mena C, Pineiro MA, Martinez-Zapater JM, Jarillo JA (2006) *EARLY IN SHORT DAYS 1* (*ESD1*) encodes ACTIN-RELATED PROTEIN 6 (AtARP6), a putative component of chromatin remodelling complexes that positively regulates *FLC* accumulation in *Arabidopsis*. Development 133:1241–1252

Mlynárová L, Nap JP, Bisseling T (2007) The SWI/SNF chromatin-remodeling gene *AtCHR12* mediates temporary growth arrest in *Arabidopsis thaliana* upon perceiving environmental stress. Plant J 51:874–885

Mylne JS, Barrett L, Tessadori F, Mesnage S, Johnson L, Bernatavichute YV, Jacobsen SE, Fransz P, Dean C (2006) LHP1, the *Arabidopsis* homologue of HETEROCHROMATIN PROTEIN1, is required for epigenetic silencing of *FLC*. Proc Natl Acad Sci USA 103:5012–5017

Niu L, Lu F, Pei Y, Liu C, Cao X (2007) Regulation of flowering time by the protein arginine methyltransferase AtPRMT10. EMBO Rep 12:1190–1195

Noh YS, Amasino RM (2003) *PIE1*, an ISWI family gene, is required for *FLC* activation and floral repression in Arabidopsis. Plant Cell 15:1671–1682

Noh B, Lee SH, Kim HJ, Yi G, Shin EA, Lee M, Jung KJ, Doyle MR, Amasino RM, Noh YS (2004) Divergent roles of a pair of homologous jumonji/zinc-finger-class transcription factor proteins in the regulation of *Arabidopsis* flowering time. Plant Cell 16:2601–2613

Novillo F, Medina J, Salinas J (2007) *Arabidopsis* CBF1 and CBF3 have a different function than CBF2 in cold acclimation and define different gene classes in the CBF regulon. Proc Natl Acad Sci USA 104:21002–21007

Oh S, Zhang H, Ludwig P, van Nocker S (2004) A mechanism related to the yeast transcriptional regulator Paf1c is required for expression of the Arabidopsis *FLC/MAF* MADS box gene family. Plant Cell 16:2940–2953

Ohashi Y, Oka A, Rodrigues-Pousada R, Possenti M, Ruberti I, Morelli G, Aoyama T (2003) Modulation of phospholipid signaling by GLABRA2 in root-hair pattern formation. Science 300:1427–1430

Pei Y, Niu L, Lu F, Liu C, Zhai J, Kong X, Cao X (2007) Mutations in the Type II protein arginine methyltransferase AtPRMT5 result in pleiotropic developmental defects in Arabidopsis. Plant Physiol 144:1913–1923

Perruc E, Kinoshita N, Lopez-Molina L (2007) The role of chromatin-remodeling factor PKL in balancing osmotic stress responses during Arabidopsis seed germination. Plant J 52:927–936

Phelps-Durr TL, Thomas J, Vahab P, Timmermans MCP (2005) Maize ROUGH SHEATH2 and its *Arabidopsis* orthologue ASYMMETRIC LEAVES1 interact with HIRA, a predicted histone chaperone, to maintain *knox* gene silencing and determinacy during organogenesis. Plant Cell 17:2886–2898

Pien S, Fleury D, Mylne JS, Crevillen P, Inzé D, Avramova Z, Dean C, Grossniklaus U (2008) ARABIDOPSIS TRITHORAX1 dynamically regulates *FLOWERING LOCUS C* activation via histone 3 lysine 4 trimethylation. Plant Cell 20:580–588

Polo SE, Almouzni G (2006) Chromatin assembly: a basic recipe with various flavours. Curr Opin Genet Dev 16:104–111

Saleh A, Al-Abdallat A, Ndamukong I, Alvarez-Venegas R, Avramova Z (2007) The *Arabidopsis* homologs of trithorax (ATX1) and enhancer of zeste (CLF) establish 'bivalent chromatin marks' at the silent *AGAMOUS* locus. Nucleic Acids Res 35:6290–6296

Scheres B, van den Toorn H, Heidstra R (2004) Root genomics: towards digital *in situ* hybridization. Genome Biol 5:e227

Schmitz RJ, Amasino RM (2007) Vernalization: a model for investigating epigenetics and eukaryotic gene regulation in plants. Biochim Biophys Acta 1769:269–275

Schmitz RJ, Hong L, Fitzpatrick KE, Amasino RM (2007) DICER-LIKE 1 and DICER-LIKE 3 redundantly act to promote flowering *via* repression of *FLOWERING LOCUS C* in *Arabidopsis thaliana*. Genetics 176:1359–1362

Schmitz RJ, Sung S, Amasino RM (2008) Histone arginine methylation is required for vernalization-induced epigenetic silencing of *FLC* in winter-annual *Arabidopsis thaliana*. Proc Natl Acad Sci USA 105:411–416

Schönrock N, Bouveret R, Leroy O, Borghi L, Köhler C, Gruissem W, Hennig L (2006) Polycomb-group proteins repress the floral activator *AGL19* in the *FLC*-independent vernalization pathway. Genes Dev 20:1667–1678

Schubert D, Primavesi L, Bishopp A, Roberts G, Doonan J, Jenuwein T, Goodrich J (2006) Silencing by plant Polycomb-group genes requires dispersed trimethylation of histone H3 at lysine 27. EMBO J 25:4638–4649

Scofield S, Murray JAH (2006) *KNOX* gene function in plant stem cell niches. Plant Mol Biol 60:929–946

Shi Y (2007) Histone lysine demethylases: emerging roles in development, physiology and disease. Nature Rev Genet 8:829–833

Shilatifard A (2006) Chromatin modifications by methylation and ubiquitination: implications in the regulation of gene expression. Annu Rev Biochem 75:243–269

Song CP, Agarwal M, Ohta M, Guo Y, Halfter U, Wang P, Zhu JK (2005) Role of an Arabidopsis AP2/EREBP-type transcriptional repressor in abscisic acid and drought stress responses. Plant Cell 17:2384–2396

Soppe WJ, Bentsink L, Koornneef M (1999) The early-flowering mutant *efs* is involved in the autonomous promotion pathway of *Arabidopsis thaliana*. Development 126:4763–4770

Spillane C, Baroux C, Escobar-Restrepo JM, Page DR, Laoueille S, Grossniklaus U (2004) Transposons and tandem repeats are not involved in the control of genomic imprinting at the *MEDEA* locus in *Arabidopsis*. Cold Spring Harbor Symp Quant Biol 69:465–475

Sridha S, Wu K (2006) Identification of *AtHD2C* as a novel regulator of abscisic acid responses in Arabidopsis. Plant J 46:124–133

Stockinger EJ, Mao Y, Regier MK, Triezenberg SJ, Thomashow MF (2001) Transcriptional adaptor and histone acetyltransferase proteins in *Arabidopsis* and their interactions with CBF1, a transcriptional activator involved in cold-regulated gene expression. Nucleic Acids Res 29:1524–1533

Sung S, Amasino RM (2004) Vernalization in *Arabidopsis thaliana* is mediated by the PHD finger protein VIN3. Nature 427:159–164

Sung S, Schmitz RJ, Amasino RM (2006) A PHD finger protein involved in both the vernalization and photoperiod pathways in *Arabidopsis*. Genes Dev 20:3244–3248

Swiezewski S, Crevillen P, Liu F, Ecker JR, Jerzmanowski A, Dean C (2007) Small RNA-mediated chromatin silencing directed to the 3' region of the *Arabidopsis* gene encoding the developmental regulator, *FLC*. Proc Natl Acad Sci USA 104:3633–3638

Tian L, Fong MP, Wang JJ, Wei NE, Jiang H, Doerge RW, Chen ZJ (2005) Reversible histone acetylation and deacetylation mediate genome-wide, promoter-dependent and locus-specific changes in gene expression during plant development. Genetics 169:337–345

Tsuji H, Saika H, Tsutsumi N, Hirai A, Nakazono M (2006) Dynamic and reversible changes in histone H3-Lys4 methylation and H3 acetylation occurring at submergence-inducible genes in rice. Plant Cell Physiol 47:995–1003

Turck F, Roudier F, Farrona S, Martin-Magniette ML, Guillaume E, Buisine N, Gagnot S, Martienssen RA, Coupland G, Colot V (2007) Arabidopsis TFL2/LHP1 specifically associates with genes marked by trimethylation of histone H3 lysine 27. PLoS Genet 3:e86

Vlachonasios KE, Thomashow MF, Triezenberg SJ (2003) Disruption mutations of *ADA2b* and *GCN5* transcriptional adaptor genes dramatically affect Arabidopsis growth, development, and gene expression. Plant Cell 15:626–638

Volpe TA, Kidner C, Hall IM, Teng G, Grewal SI, Martienssen RA (2002) Regulation of heterochromatic silencing and histone H3 lysine-9 methylation by RNAi. Science 297: 1833–1837

Wagner D, Meyerowitz EM (2002) SPLAYED, a novel SWI/SNF ATPase homolog, controls reproductive development in *Arabidopsis*. Curr Biol 12:85–94

Wang X, Zhang Y, Ma Q, Zhang Z, Xue Y, Bao S, Chong K (2007) SKB1-mediated symmetric dimethylation of histone H4R3 controls flowering time in *Arabidopsis*. EMBO J 26:1934–1941

Wood CC, Robertson M, Tanner G, Peacock WJ, Dennis ES, Helliwell CA (2006) The *Arabidopsis thaliana* vernalization response requires a Polycomb-like protein complex that also includes VERNALIZATION INSENSITIVE 3. Proc Natl Acad Sci USA 103:14631–14636

Wu K, Zhang L, Zhou C, Yu CW, Chaikam V (2008) HDA6 is required for jasmonate response, senescence and flowering in *Arabidopsis*. J Exp Bot 59:225–234

Xiao W, Gehring M, Choi Y, Margossian L, Pu H, Harada JJ, Goldberg RB, Pennell RI, Fischer RL (2003) Imprinting of the *MEA* Polycomb gene is controlled by antagonism between MET1 methyltransferase and DME glycosylase. Dev Cell 5:891–901

Xie Z, Johansen LK, Gustafson AM, Kasschau KD, Lellis AD, Zilberman D, Jacobsen SE, Carrington JC (2004) Genetic and functional diversification of small RNA pathways in plants. PLoS Biol 2:e107

Xu L, Xu Y, Dong AW, Sun Y, Pi LM, Xu YQ, Huang H (2003) Novel *as1* and *as2* defects in leaf adaxial-abaxial polarity reveal the requirement for *ASYMMETRIC LEAVES1* and 2 and *ERECTA* functions in specifying adaxial identity. Development 130:4097–4107

Xu CR, Liu C, Wang YL, Li LC, Chen WQ, Xu ZH, Bai SN (2005) Histone acetylation affects expression of cellular patterning genes in the *Arabidopsis* root epidermis. Proc Natl Acad Sci USA 102:14469–14474

Xu L, Ménard, R, Berr A, Fuchs J, Cognat V, Meyer D, Shen WH (2009) The E2 ubiquitin-conjugating enzymes, AtUBC1 and AtUBC2, play redundant roles and are involved in activation of *FLC* expression and repression of flowering in *Arabidopsis*. Plant J 57:279–288

Xu L, Zhao Z, Dong A, Soubigou-Taconnat L, Renou JP, Steinmetz A, Shen WH (2008) Di- and tri- but not monomethylation on histone H3 lysine 36 marks active transcription of genes involved in flowering time regulation and other processes in *Arabidopsis thaliana*. Mol Cell Biol 28:1348–1360

Zhang X, Germann S, Blus BJ, Khorasanizadeh S, Gaudin V, Jacobsen SE (2007) The *Arabidopsis* LHP1 protein colocalizes with histone H3 Lys27 trimethylation. Nature Struct Mol Biol 14:869–871

Zhao Z, Yu Y, Meyer D, Wu C, Shen WH (2005) Prevention of early flowering by expression of *FLOWERING LOCUS C* requires methylation of histone H3 K36. Nature Cell Biol 7: 1156–1160

Zhou C, Zhang L, Duan J, Miki B, Wu K (2005) *HISTONE DEACETYLASE19* is involved in jasmonic acid and ethylene signaling of pathogen response in Arabidopsis. Plant Cell 17: 1196–1204

Zhu Y, Dong A, Meyer D, Pichon O, Renou JP, Cao K, Shen WH (2006) *Arabidopsis NRP1* and *NRP2* encode histone chaperones and are required for maintaining postembryonic root growth. Plant Cell 18:2879–2892

Zhu J, Jeong JC, Zhu Y, Sokolchik I, Miyazaki S, Zhu JK, Hasegawa PM, Bohnert HJ, Shi H, Yun DJ, Bressan RA (2008) Involvement of *Arabidopsis* HOS15 in histone deacetylation and cold tolerance. Proc Natl Acad Sci USA 105:4945–4950

Zilberman D, Cao X, Jacobsen SE (2003) A*ARGONAUTE4* control of locus-specific siRNA accumulation and DNA and histone methylation. Science 299:716–719

Chapter 18
Activation Tagging for Gain-of-Function Mutants

N. Marsch-Martínez and A. Pereira

18.1 Introduction

18.1.1 Importance of Mutants to Study Development

In many organisms, the study of developmental processes has made remarkable advances with the use of mutants, which are excellent tools to elucidate gene functions. In plants, mutants generated using diverse mutagens (i.e., chemical, physical or biological insertions that produce gene knockouts), or naturally occurring mutants/variants have revealed the biological role of key developmental genes. With the help of mutants, it has been possible to identify and clone numerous genes involved in many different aspects of plant development.

The first plant developmental gene to be isolated was *Knotted* (*Kn1*) from maize, by virtue of a gain-of-function dominant mutant caused by the Ds2 transposon insert (Hake et al. 1989). The *Kn1* gene was shown to encode a homeodomain-containing protein (Vollbrecht et al. 1991). Other homeotic genes isolated are those affecting flower development, the first one identified being *Deficiens* from *Antirrhinum* (Sommer et al. 1990), and subsequently from Arabidopsis, giving rise to the ABC model (Coen and Meyerowitz 1991; Weigel and Meyerowitz 1994). Since then, other developmental genes determining floral meristem identity (e.g., *LEAFY*; Weigel et al. 1992), flowering time (e.g., *CONSTANS*; Putterill et al. 1995), pollen development (e.g., *MALE STERILITY 2*; Aarts et al. 1993), and embryo development

N. Marsch-Martínez
Laboratorio Nacional de Genómica para la Biodiversidad, CINVESTAV-IPN, Campus Guanajuato, C.P. 36821, Irapuato, Gto. México

A. Pereira
Virginia Bioinformatics Institute, Virginia Polytechnic Institute and State University, Washington Street, Virginia Tech, Blacksburg, VA 24061, USA
e-mail: pereiraa@vbi.vt.edu

E-C. Pua and M.R. Davey (eds.),
Plant Developmental Biology – Biotechnological Perspectives: Volume 2,
DOI 10.1007/978-3-642-04670-4_18, © Springer-Verlag Berlin Heidelberg 2010

(e.g., *MONOPTEROS*; Hardtke and Berleth 1998), amongst many others acting in different tissues and stages, have been cloned using mutants.

The convenience of using insertion tagged mutants has facilitated the isolation of many developmental mutants that have an easily identifiable phenotype. The identification and isolation of *Kn1* from maize was achieved using several dominant gain-of-function mutants caused by endogenous transposons (Hake et al. 1989; Veit et al. 1990; Vollbrecht et al. 1991). In Arabidopsis, the first developmental gene isolated was *AGAMOUS* caused by a T-DNA insertion tag (Yanofsky et al. 1990). The utility of insertional mutagenesis in Arabidopsis using the T-DNA from *Agrobacterium* (Yanofsky et al. 1990) and heterologous transposons (Aarts et al. 1993) has aided the cloning of developmental genes using knockout mutants.

18.1.2 Phenotype Gap—the Lack of Mutant Phenotypes

In general, eukaryotic organisms have a strong genetic robustness against null mutations in numerous genes (Gu et al. 2003). Many genes of model organisms, like *Drosophila melanogaster*, *Caenorhabditis elegans*, and *Saccharomyces cerevisiae*, lack obvious loss-of-function phenotypes (Miklos and Rubin 1996; Ross-Macdonald et al. 1999). The model plant Arabidopsis, for which the whole genome is known (Arabidopsis Genome Initiative 2000), has more than 27,000 genes (TAIR release 8, www.arabidopsis.org), but for many of these genes, as in the case of *Drosophila* and *Caenorhabditis*, there is little knowledge about their exact biological role.

There are collections of plants bearing single transposon or T-DNA insertions that tag a large number of these genes. However, in most cases, knockouts do not provide informative phenotypes under "standard" conditions, as revealed by reverse genetics approaches (Bouché and Bouchez 2001). In the most detailed study of 4,000 *Ds* transposon insertions in genes, only 3.5% showed an easily identifiable mutant phenotype, mostly in developmental pathways, although conditional phenotypes might reveal other mutants (Kuromori et al. 2006). This low frequency of observed mutants can be caused by genome redundancy due to the large duplication events found in the Arabidopsis genome (Arabidopsis Genome Initiative 2000), where about two-thirds of the genome is duplicated in the form of large chromosomal segments. In addition, nearly 4,000 genes are tandemly repeated as two or more copies. Knocking out these "redundant" genes does not result in a mutant phenotype, since the loss of function of one copy can be compensated by the other copy or copies. Examples of genome redundancy and genetic analyses conducted to reveal double mutant phenotypes in Arabidopsis MADS box transcription factor developmental genes include those for, among many others, the two *SHATTER-PROOF* (Liljegren et al. 2000), as well as the *CAULIFLOWER* and *APETALA1* (Kempin et al. 1995) genes. In addition, alternative regulatory networks or metabolic pathways can also compensate the loss of function of a gene (Gu et al. 2003). Another limitation of knockout strategies is that they do not allow studies of the function of genes that act at later developmental stages, but that are essential during gamete or very early embryo development and will therefore have lethal loss-of-function phenotypes.

18.1.3 Activation Tagging for Gain-of-Function Mutants

A strategy whereby gene expression is enhanced can overcome the limitations of the knockout approaches. "Activation tagging", first proposed by Walden et al. (1994), is such a strategy. It is based on the use of enhancer, or "activating", sequences that positively influence gene expression even when placed far from the promoter (Lewin 2000). The enhancer is placed inside a tag (Fig. 18.1)—such as a transposon or T-DNA—that promotes the overexpression of adjacent endogenous genes (Neff et al. 1999; van der Graaff et al. 2000; Weigel et al. 2000), which can produce gain-of-function mutant phenotypes.

Most of the activation tagging approaches developed until now employ the well-characterized cauliflower mosaic virus (CaMV) 35S enhancer or promoter sequences as the transcriptional activator (Odell et al. 1985). In many mutants generated using the CaMV 35S enhancer, it has been shown that the enhancer increases the original expression pattern of a gene, rather than inducing ectopic or constitutive overexpression (Neff et al. 1999; van der Graaff et al. 2000; Weigel et al. 2000). However, a true constitutive promoter may have limitations for genes that have deleterious effects in early developmental stages when overexpressed.

Gain-of-function phenotypes can either directly reveal the gene function, or provide a useful clue to the pathway in which the gene is involved. Moreover, gain-of-function mutations allow visualizing the dominant phenotypes in heterozygous plants, in contrast to classic loss-of-function mutants for which the phenotype can be visualized only in homozygous plants. Moreover, enhanced expression of a gene can produce mutant phenotypes for essential genes that otherwise lead to lethal loss-of-function phenotypes. The frequency of lethal mutations, essential to the haploid gametophyte and/or the early diploid embryo, is predicted to be approximately one gene in every 100 kb (Martienssen 1998), ranging from 1,000 to 2,000 genes. For some of these genes, their functions might be revealed by

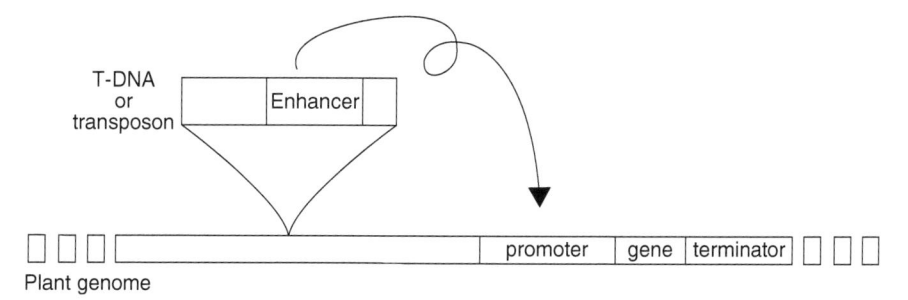

Fig. 18.1 The influence of an activation tag on a gene. The insertion of a T-DNA or transposon bearing enhancer sequences can affect the promoters of nearby genes in the genome

Table 18.1 Advantages of activation tagging over knockouts

Tagging	Advantages/disadvantages
Knockout tagging	Gene tagging generally causes recessive phenotypes. Therefore, homozygous plants are required to observe mutant phenotypes
	Knockout of genes with overlapping functions with other genes ("redundant" genes or genes belonging to the same family) might not display an evident mutant phenotype
	Gene knockout may lead to lethal phenotypes, making the study of gene functions in different developmental stages impossible
Activation tagging	Activation tagging causes dominant mutations, which can display phenotypes in heterozygous plants. However, gene knockouts can also be obtained, because in this strategy the activating sequences are in a tag that inserts in the genome
	Gene overexpression can produce phenotypes of genes with overlapping functions
	Gene overexpression might lead to viable individuals where gene knockout leads to lethal phenotypes
	Activation tagging populations are particularly suited to perform positive selection screens—for example, resistance or tolerance to chemical, physical or biological stresses
	Activation tagging can also be used to study metabolic pathways
	For biotechnological purposes, activation tagging has a high potential for the direct application of interesting genes

gain-of-function phenotypes. The main advantages of activation over the classic knockout tagging are listed in Table 18.1.

More refined activation tagging systems can employ tissue-specific or inducible promoters. Two inducible tagging systems have been reported by Matsuhara et al. (2000) and Zuo et al. (2000). Zuo and colleagues (2000) employed a chemically induced transactivation system that consists of two main components placed in a single T-DNA. The first component is a fusion of three parts, namely, a DNA-binding domain (Lex A bacterial repressor, X), an activator (VP16, V), and the regulatory region of the human estrogen receptor (E). This fusion, named VXE, is controlled by the strong constitutive promoter G10-90. The second component is an octamer of the Lex A operator fused upstream of a –46 CaMV 35S minimal promoter. This second component is located in one of the T-DNA borders, oriented outward in a way that can affect the expression of adjacent genes upon VXE activation after treatment with estrogens (Zuo et al. 2000). Matsuhara and colleagues (2000) have established a heat-shock tagging system by placing a heat-shock promoter in one of the borders of a T-DNA. In this way, a heat treatment can lead to transcription of DNA sequences downstream of the promoter, causing either sense or antisense expression. Sense transcripts can be identified using RT-PCR with specific primers.

An alternative method to random insertional-based overexpression using activation tagging is a gene-by-gene overexpression method, which was systematically used to identify the overexpression phenotypes of transcriptional factor genes in Arabidopsis by Mendel Biotechnology, amongst which a few have been published (Riechmann and Ratcliffe 2000; Haake et al. 2002; Ratcliffe et al. 2003; Broun et al. 2004). Another method to systematically overexpress Arabidopsis genes has been popularized under the term FOX hunting (Ichikawa et al. 2006), where full-length Arabidopsis cDNA from a normalized library was transformed at random, revealing 1,487 morphological mutants from 15,547 transformants with around 2.6 T-DNA loci per transformant.

Activation tagging has been effectively applied in Arabidopsis using mostly T-DNA (Kakimoto 1996; Kardailsky et al. 1999; Ito and Meyerowitz 2000; van der Fits and Memelink 2000; Borevitz et al. 2000; Lee et al. 2000; van der Graaff et al. 2000; Weigel et al. 2000; Huang et al. 2001; Zhao et al. 2001). Also, transposon insertions have been used. These include the *En/I(Spm)* transposon system (Marsch-Martínez et al. 2002; Schneider et al. 2005), and the *Ac-Ds* transposon system with the complete 35S promoter in which one of the transposon borders acts by read-out overexpression (Wilson et al. 1996). Figure 18.2 shows examples of

Fig. 18.2 Mutant phenotypes from an Arabidopsis transposon activation tagged population of Arabidopsis. Wild-type plants are indicated (WT). The other examples are developmental mutants in leaf, inflorescence and plant form. Bar = 1 cm

mutants identified in a transposon-based activation tagging population. Recently, a gene activation method was described using Arabidopsis containing *Spm* or *dSpm* transposon insertions (Sorokin et al. 2008). By overexpressing the *tnpA* transposase protein that binds to the transposon borders fused to the VP16 activation domain, *Spm/dSpm* inserted directly upstream (up to about 500 bp) of the TATA box of target genes could be activated. The plant lines displayed phenotypic variation.

The activation tagging strategy using enhancers inside a tag has also been applied in other plant species, such as barley (Ayliffe and Pryor 2007), rice (Jeong et al. 2002; Qu et al. 2008), tobacco (Suzuki et al. 2001), tomato (Mathews et al. 2003), poplar (Busov et al. 2003), and lotus (*Lotus japonicus*; Imaizumi et al. 2005). In fact, activation tagging causing dominant mutations enables it to be used in species with slower growth rates or with complex genomes, or in cell lines, as was reported in *Catharantus roseus* (van der Fits and Memelink 2000).

18.2   Genes that Modulate Development Discovered by Activation Tagging

In Arabidopsis and other plant species, activation tagging has been useful to discover genes that are involved in or modulate development, metabolism, stress resistance, and sensing of cell nutritional status. These include enzymes, components of signaling pathways, microRNAs, and transcription factors, showing that a high diversity of genes can be tagged using this strategy. Components apparently involved in modulating a single developmental pathway normally directly or indirectly affect other aspects, such as other developmental pathways and/or metabolism or stress resistance, and vice versa. Therefore, in some cases, it is difficult to define the boundaries of "gene function classification". In this chapter, the examples of activation tagging have been separated in groups, according to the first function reported for the gene. More detailed descriptions are given for genes involved in development. Metabolism, stress resistance, and nutritional status sensing genes are also discussed.

18.2.1   Genes Involved in Hormonal Biosynthesis or Signaling that Affect Development

Activation tagging has been useful for the study of hormonal pathways. There are several examples of regulators and enzymes involved in hormone signaling or biosynthesis that have been identified using this strategy.

18.2.1.1 Identification of the First Component of Cytokinin Signaling

In plants, the phytohormone cytokinin plays different physiological roles that include leaf senescence, lateral branching, and nutrient metabolism (Muller and Sheen 2007). At the cellular level, it induces cell division, chloroplast development, and shoot production, when cells are grown in medium supplemented with cytokinin (Skoog and Miller 1957). Despite the importance of this hormone for plant growth and development, knowledge regarding the cytokinin signal transduction pathway is lacking. Kakimoto (1996) used a T-DNA activation tagging vector to transform calli derived from hypocotyl segments of 50,000 Arabidopsis seedlings. The calli require the presence of cytokinin in the medium to proliferate, develop chlorophyll, and form shoots. To generate mutants in which cytokinin signaling was altered, the transformed population was screened for growth in the absence of exogenous cytokinin. Five mutants that became green and showed fast proliferation and shoot development were selected and named cytokinin-independent (*cki*). The gene that was affected by the tag in the four mutants showed similarity to the histidine kinase (HK) and receiver domains of the two component systems (Heyl and Schmülling 2003). The results of a study on gene structure and its overexpression phenotype indicate that this gene plays an important role in cytokinin recognition, perhaps at early stages of signal transduction. This was a pioneering study showing the role of HKs in cytokinin signal transduction (Muller and Sheen 2007).

Later, as the Arabidopsis genome sequence became available (Arabidopsis Genome Initiative 2000), all potential components of this phospho-relay signaling system could be identified and tested (Ferreira and Kieber 2005). In one of these tests, Hwang and Sheen (2001) showed that *CKI1* could act independently of exogenous cytokinin, suggesting that CKI1 might function in cytokinin signaling as a constitutively active component and/or could be involved in sensing endogenous signals.

18.2.1.2 First Functional Demonstration of a Cytokinin Biosynthesis Gene in Plants

The first naturally produced cytokinin, zeatin, was isolated in the 1950s (Miller et al. 1955). Most knowledge about cytokinin biosynthesis, storage, and mobilization had been gained from studies using bacterial plant pathogens (Brzobohaty et al. 1994). The proposed models about the cytokinin mode of action were based mostly on the study of mutants affected in cytokinin signaling (Kakimoto 1998). However, there was little knowledge about plant enzymes involved in cytokinin metabolism. There were different hypotheses proposing cytokinin biosynthesis. In the case of plant tissues transformed by *Agrobacterium tumefaciens*, the *isopentenyltransferase* (*ipt*) gene contained in the T-DNA is responsible for cytokinin synthesis. Earlier experiments in tobacco, maize, and Arabidopsis had suggested that plant isopentenyltransferases (IPTs) could also function in cytokinin biosynthetic pathways (Blackwell and Horgan 1994; Kakimoto 2001; Takei et al. 2001), but these had not been demonstrated to produce cytokinins in plants. Using an activation tagging strategy in petunia, Zubko and colleagues (2002) found a line showing a high cytokinin content phenotype (shooting, reduced apical dominance, and delayed senescence and flowering). This line, named *shooting* (*sho*), had a T-DNA insertion near a gene with homology to IPTs. When expressed in other plant species, this gene caused comparable cytokinin-specific effects. While the *ipt* gene from *Agrobacterium* increases zeatin content, *Sho* expression in petunia and tobacco resulted in increased concentrations of N6-(D2-isopentenyl) adenosine (2iP) derivatives. Further analysis of these plants suggested that the activity of the plant enzyme encoded by *SHO* is sufficient to produce active cytokinins in planta.

18.2.1.3 Discovery of a Rate-Limiting Enzyme in Auxin Biosynthesis

Auxin is an essential hormone related to different processes of plant growth and development, such as cell division and elongation, differentiation, tropisms, apical dominance, senescence, abscission, and flowering (Davies 1995). While information on auxin biosynthesis has been well documented, there is little knowledge about the pathways and rate-limiting steps. Many pathways had been proposed, including tryptophan-dependent and tryptophan-independent pathways (Bartel 1997). However, studying auxin biosynthesis is difficult due to the lack of auxin-deficient mutants. Recessive mutants with elevated levels of free auxin were available, but it was not known whether this was due to indirect effects of the mutation on auxin biosynthesis or conjugation. Moreover, the mutants were sterile, or showed heterogeneous phenotypes. By means of activation tagging, a mutant with elevated concentrations of free auxin was isolated and characterized (Zhao et al. 2001). The overexpressed gene *YUCCA* encodes an enzyme flavin-monooxygenase that catalyzes the hydroxylation of the amino group of the tryptamine, which is a rate-limiting step of auxin biosynthesis. *YUCCA* belongs to a YUC family of at least nine members, and some members showed overlapping functions (Marsch-Martínez et al. 2002, 2006; Cheng et al. 2006; Kim et al. 2007). This explains the difficulty in finding these genes using conventional loss-of-function mutagenesis strategies.

18.2.1.4 Unraveling the Brassinosteroid Perception Pathway

Brassinosteroids (BRs) represent another class of phytohormones with a general role in plant growth and development. Mutations that affect either biosynthesis or perception cause phenotypes in which different aspects of plant development are affected, including dwarf stature, round and curled leaves, reduced male fertility, delayed flowering, and senescence (Altmann 1998). While biosynthetic mutants can be rescued by application of BRs, perception mutants cannot be rescued in this way and are termed "brassinosteroid insensitive" (*bri*). All known *bri* mutants represent the same locus encoding a leucine-rich repeat receptor-like protein kinase (LRR-RLK; Li and Chory 1997), BRI1, for which the extracellular LRR domain has been shown to sense BR (He et al. 2000). However, whether the sensing occurs directly or through a secreted steroid-binding protein was not known (Li and Chory 1997; Schumacher and Chory 2000). There is also little knowledge about other signaling factors that, in combination with BRI1, are involved in the signaling cascade resulting in growth and development changes. To answer these questions, Li and collaborators (2001) performed gain-of-function genetic screens for suppressors of a weak *bri1* allele. *bri1-5* plants are late flowering dwarfs with short inflorescences, and have leaves with short petioles and curled lamina. Unlike for other alleles, *bri1-5* plants are fertile, allowing the use of suppressor strategy to discover two important components of BR perception and signaling. The first suppressor mutant, *bri1 suppressor-Dominant* (*brs1-1D*), recovered *bri1-5* mutant phenotypes as inflorescence length and flowering time, with larger and uncurled leaves (Li et al. 2001). The overexpressed gene *BRS1* encodes a secreted carboxy-peptidase. Interestingly, an insertional loss-of-function allele did not show a mutant phenotype, while the suppression phenotype it confers was selective for *bri1*. Moreover, it was shown to be BR-dependent, and required both a functional BRI1 protein kinase domain and BRS1 protease activity. As *brs1-1D* can suppress mutations only in the extracellular domain of BRI1, and it is a secreted protein, it is likely that BRS1 acts at early BRI1-mediated signaling when, most likely, it proteolytically processes a rate-limiting protein.

Using the same strategy, another suppressor mutant identified was *bak1-1D* (*bri1-associated receptor kinase-1-1Dominant*), which was affected in a LRR-RLK distinct to BRI (Li et al. 2002). Again, this mutation suppressed the short inflorescence and petiole, and curled lamina phenotypes of the *bri1-5* mutant. Null mutations showed semi-dwarfed phenotypes, while overexpression in wild-type plants showed enhanced growth. It was shown that BRI1 and BAK1 interact and can phosphorylate one another. The auto-phosphorylation activity of BAK1 is stimulated by BRI1. Moreover, *bak1* dominant negative mutants resembled *bri1* mutants. Based on these findings, Li et al. (2002) proposed a model in which the ligand-bound BRI1 activates BAK1. Once activated, BAK1 can phosphorylate other downstream components involved in gene expression and plant growth regulation (Li et al. 2002).

18.2.1.5   Cloning the First Regulatory Gene from Trees: the Gibberellin Catabolic Enzyme GA2ox

Gibberellins (GAs) are another group of critical enzymes involved in plant growth and development (Davies 1995). Several mechanisms control the concentrations of bioactive GAs. Changes in the GA flux can lead to modifications of plant form

(Hedden and Phillips 2000). Low concentrations of GAs can result in dwarf plants with compact and dark green leaves, while higher concentrations result in taller plants with more elongated organs. One of the several mechanisms that control bioactive GA concentrations is the transcriptional regulation of genes encoding both biosynthetic and catabolic enzymes (Olszewski et al. 2002). GA 2-oxidases (GA2-ox) are the major GA catabolic enzymes important in the regulation of GA concentrations. The gene families of GA2-ox have been identified in Arabidopsis and pea (*Pisum sativum*; Elliott et al. 2001; Schomburg et al. 2003), whereas only a single gene has been reported in rice (*Oryza sativa*; Sakamoto et al. 2001). However, GA2ox genes from trees had not been cloned, and currently there is no knowledge about their function and regulation during woody plant development (Busov et al. 2003).

Poplar has become the tree model for molecular biology studies, owing to its ease of transformation and clonal propagation, relative rapid growth, availability of genomic resources, and the recently sequenced small genome (four times the size of the Arabidopsis genome; Arabidopsis Genome Initiative 2000; Wullschleger et al. 2002; Tuskan et al. 2006). Such a model can provide insights into the development of perennial plants, which in many aspects is different from the development of annual plants.

Using activation tagging, as the first example of tagging of regulatory genes in trees, Busov et al. (2003) were able to identify a dwarf tree in which a GA2ox gene was overexpressed. This tree had a similar phenotype to transgenic poplars that overexpressed a bean (*Phaseolus coccineus*) GA2ox gene, and showed quantitative shifts in the spectrum of GAs. The cloning of the GA2ox gene from poplar showed that the strategy can be successfully applied to (transformable) woody plants (Busov et al. 2003).

18.2.2 Integration of Environmental Cues Modulating Developmental Pathways

Plant development is strongly influenced by the environment. A very important environmental factor is light. Plants use light not only for photosynthesis, but also as environmental cue for development. Moreover, light regulation of development relies on a complex network of interactions between each sensing component, acting through endogenous programs in which different hormones participate. Plants can sense the quality, quantity, direction, and duration of light through various photoreceptors. In Arabidopsis, five phytochromes (phyA–phyE) are the red/far-red sensors (Fankhauser and Chory 1997). Mutants in *phyB* show a long hypocotyl phenotype, amongst others, and *phyB-4* mis-sense mutants can still produce a pigment capable of nearly normal photo-transformation, still showing the long hypocotyl phenotype but weaker than the null *phy-B* mutation (Elich and Chory 1997).

Fig. 18.3 Activation tagging suppressor screen. An activation tagging T-DNA vector was introduced into a weak *phyB-4* mutant genotype showing a long hypocotyl phenotype. The *bas1-D* (phy*B* *a*ctivation-tagged *s*uppressor*1*-*d*ominant) suppressor phenotype was identified from the resulting activation tag population

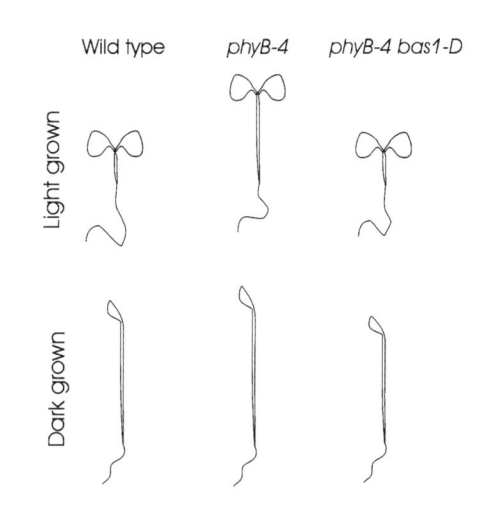

This mutant with reduced signaling activity was used to perform a screen for suppressors of the phenotype (Neff et al. 1999). One of the mutants found in the screen showed a shorter hypocotyl than the original *phy4-B* mutant. Figure 18.3 shows representations of the wild-type, mutant, and suppressor phenotypes. This mutant was designated as *bas1-D* (phy*B* *a*ctivation-tagged *s*uppressor*1*-*d*ominant). The overexpressed gene responsible for the phenotype encoded the cytochrome P450 CYP72B1. Interestingly, this gene also acts below *PHYA* and *CRY1*, showing that it is a component of other pathways. The most active brassinosteroid, i.e., brassinolide, was not detected in the mutant, but a hydroxylated form accumulated. Moreover, decreased expression of the gene results in enhanced responses to BL and reduced responses to light. Therefore, *BAS1-D* appears to play a role in connecting light to brassinosteroid signaling pathways.

18.2.2.1 Cloning Key Integrators of Flowering Signals: the Floral Inducer FT

Several genetic pathways control the transition from the vegetative to the reproductive phase in Arabidopsis. These pathways integrate both environmental and internal developmental state cues to trigger flowering (Levy and Dean 1998). Genes involved in some of these pathways had been cloned, but there was little knowledge about how flowering signals were transduced and integrated. Using activation tagging, Kardailsky and collaborators (1999) cloned a key gene in flowering signaling through identification of a dominant mutant that showed an early flowering phenotype and had terminal flowers. The activation insertion was adjacent to a gene related to a membrane-associated mammalian protein, RKIP (Raf kinase inhibitor protein), which can bind hydrophobic ligands (Schoentgen et al. 1987). RKIP regulates the activity of the RAF/MEK/ERK signal transduction pathway (Yeung et al. 1999). The insertion mapped closely to the flowering-time gene *FT*.

Therefore, the cloned gene was analyzed in *ft* mutants, and found to be mutated in different mutant alleles. Recently, FT has been shown to be a component of "florigen", the long-sought signal that induces flowering. Florigen was first thought to be signaling as an RNA molecule (Huang et al. 2005), and later confirmed to be the protein itself in Arabidopsis, rice, and cucurbits (Corbesier et al. 2007; Lin et al. 2007; Tamaki et al. 2007).

Another integrator of flowering pathways, *AGL20 or SOC1* (*Suppressor of CONSTANS*), was also cloned using activation tagging (Lee et al. 2000). In this case, it was identified as a suppressor of the *frigida* (*fri*) mutant. These late flowering mutants are affected in the vernalization pathway. The gene was shown not only to be regulated by long day-dependent and autonomous pathways, but also to be a target of *CONSTANS* (*CO*), one of the components of the long day-dependent pathway (Samach et al. 2000).

18.2.3 Meristem, Embryo, and Organ Development

18.2.3.1 Discovery of a miRNA that Targets a Family of Crucial Morphogenesis Regulators

Most organ development studies aim at understanding the factors that determine organ fate. However, very few had been directed to gaining an understanding about the molecular mechanisms guiding organ shape. One of these few studies has been shown to address the question of how a flat organ develops. Wild-type snapdragon leaves are flat, while the *cincinnata* (*cin*) mutant has crinkly leaves (Nath et al. 2003). This mutant lacks the correct regulation of cell division, which is normally differential across the leaf. The result is that *cin* leaves are not flat but show negative curvature. *CIN*, a gene belonging to the TCP family, is expressed in front of or overlapping with the mitotic arrest zone in the developing leaf (Nath et al. 2003), but it is not clear as to how the *CIN* pattern of expression is attained.

Although work on microRNAs and elucidation of its functions have been reported, direct evidence showing the regulatory role of microRNAs in defined processes of plant morphogenesis is lacking. Using activation tagging, Palatnik and colleagues (2003) discovered a mutant showing uneven leaf shape and curvature. Interestingly, the gene that was activated by the activation tagging insertion was not a regular protein-coding gene, but a microRNA. This microRNA was found to target transcripts of different TCP genes controlling leaf development for cleavage. It was shown that microRNA-mediated transcript cleavage was essential to avoid "ectopic" TCP expression, delimiting the expression of these genes to specific areas. This mechanism, in which a microRNA targets *TCP4* genes for cleavage, may be commonly present in plants for production of leaf morphogenesis, as TCP genes with microRNA target sequences have been detected in various plant species with varying leaf shapes (Palatnik et al. 2003).

18.2.3.2 Two Closely Linked Genes Involved in Leaf and Vasculature Development

let (*Lettuce*) mutants were identified during an activation tag screen as dominant mutants with leaves lacking petioles. Furthermore, the inflorescence branching and silique shape in *let* mutants were affected (van der Graaff et al. 2000). In this mutant, the *LEAFY PETIOLE* (*LEP*) gene, belonging to the AP2/EREBP gene family, was shown to be overexpressed. The lack of petioles in the mutant leaves was caused by the ectopic conversion of petioles into leaf blades. This phenotype indicated alterations in the development of the leaf proximodistal axis, and the gene was proposed to be modulating either the patterning of this axis, or cell division activity in the marginal meristem (van der Graaff et al. 2000).

Besides petiole conversion into a leaf blade, all *let* aerial organs showed vascular bundles of increased size. This phenotype was due to a greater number of xylem, phloem (pro)cambial, and pericycle cells. Most interestingly, the activation tag in the *let* mutant not only affected expression of the *LEP* gene, but also induced overexpression of the contiguous gene *VASCULAR TISSUE SIZE* (*VAS*; van der Graaff et al. 2002). This gene was found to be unique in the Arabidopsis genome, and to possess an acid secretory signal peptide, a putative acid transmembrane region, and an AAI domain shared by different protein families. *LEP* and *VAS* were arranged in tandem in the genome, and it was their combined overexpression that caused the vasculature phenotype. However, each gene contributed to specific features of the phenotype. *LEP* overexpression alone caused an increased number of xylem cells, while the increase in (pro)cambial and pericycle cells was the result of *VAS* overexpression alone. The effect of *LEP* in xylem cells further confirmed its possible role as a cell division-promoting factor.

As *LEP* and *VAS* show complementary functions, and both genes are simultaneously affected by the activation tag, their positions adjacent to each other in the Arabidopsis genome suggest that they may be coordinately regulated. This example illustrates the usefulness of the activation tagging strategy, as knockout mutations in *LEP* and *VAS* do not display mutant phenotypes in the vasculature, even when the Arabidopsis genome lacks *VAS* homologs. It is clear that expression pattern analysis in wild-type tissues and gene overexpression are the key tools to decipher the role of these genes in plant development (van der Graaff et al. 2002).

Activation tagging of *STURDY*, which encodes a patatin-like protein, also resulted in vasculature in Arabidopsis (Huang et al. 2001). Some of the features displayed in the *sturdy* dominant phenotype include stiff inflorescence stems, thicker leaves, shorter siliques, larger seeds, round-shaped flowers, and delayed growth.

18.2.3.3 Two Closely Related Genes with Functions in Meristem and Organ Development

dornröschen-D is another activation tagging mutant in which leaf development is affected (Kirch et al. 2003). In this case, a transposon was used as the tag that harbors the enhancer, and the gene affected belongs to the AP2/ERF family (Okamuro et al. 1997). This gene is expressed in a specific subdomain of meristem stem cells, in lateral organ anlagen, and in a transient way in the distal domain of organ primordia. Meristematic cells located in different domains have different activities. For example, cells at the center divide slowly and behave as pluripotent stem cells, while cells at the periphery of the meristem are the origin of new organs (Bowman and Eshed 2000). The control of these two processes lies in the interactions of genes expressed in different domains (Brand et al. 2001). In the *drn-D* mutant, meristem activity is arrested prematurely. Moreover, radialized lateral organs are formed, and the expression of meristem genes like *SHOOTMERISTEM-LESS*, *CLAVATA3*, and *WUSCHEL* is changed.

DRN is an important player in meristem and organ development. It can induce cytokinin-independent shoot formation when overexpressed in root explants (Banno et al. 2001). Moreover, *DRN* and its closest relative, *DORNRÖSCHEN-LIKE* (*DRNL*), interact with *PHAVOLUTA* and other type III HD-ZIP family members to control *Arabidopsis* embryo patterning (Chandler et al. 2007). *DRNL/SOB/ESR2/BOL* has been associated with the development of other organs (Ikeda et al. 2006; Marsch-Martínez et al. 2006; Ward et al. 2006; Nag et al. 2007). It has been shown to play a role in cotyledon development, and to regulate *CUP-SHAPED COTYLEDON 1* that is involved in meristem specification (Ikeda et al. 2006). The genes are also necessary for stamen emergence in Arabidopsis (Nag et al. 2007). Moreover, activation tagged mutants with *DRNL* overexpression exhibited altered leaf and organ formation due to changes in cell expansion and in the regulation of proliferation/differentiation pathways (Marsch-Martínez et al. 2006).

18.2.3.4 Discovery of an Extra Embryonic Function of a Well-Known Meristem Gene

Using the inducible XVE activation tagging system, Zuo et al. (2000) discovered an unknown function of the previously studied *WUSCHEL* gene (Laux et al. 1996). The new *WUSCHEL* gain-of-function phenotype was characterized for the formation of somatic embryos without exogenous supplementation of hormones (Zuo et al. 2002). The inducibility of this system allowed the particular phenotype to be studied, because constitutive *WUSCHEL* overexpression otherwise gave raise to impaired development.

18.2.3.5 Other Examples of Genes Involved in Floral Organ Growth

Using homozygous *apetala2-1* (*ap2-1*) mutants as the genetic background (Weigel et al. 2000), Ito and Meyerowitz (2000) performed an activation tagging screen to search for genes involved in cellular differentiation and growth of floral organs. One mutant, *28-5*, was shown to possess wide heart-shaped ovaries, which was due to the presence of more enlarged cells. Overexpression of a P450 gene (*CYP78A9*) in *28-5* mutant and wild-type plants resulted in alterations in silique development, promoting silique growth in the absence of fertilization, and producing larger fertilized siliques. The similarity of double mutant siliques to siliques of *Capsella bursa-pastoris*, a distant mustard relative of Arabidopsis, suggests that the gene product is involved in an evolutionarily conserved process sculpting carpel shape.

The aerial organs of higher plants are formed from cells originating at the meristems (leaves from the shoot meristem, and floral organs from the floral meristem), and differentiate into specific tissues. The correct delimitation of boundaries of meristem gene expression is required for correct organ development. Chalfun et al. (2005) identified a mutant, *ds* (*downwards siliques*), in which siliques were oriented downward. The overexpressed gene encoded a member of the AS2/LOB (Asymmetric leaves 2/Lateral Organ Boundaries) family (Lin et al. 2003). AS2 has been shown to downregulate KNOTTED1-LIKE homeobox (KNOX) genes in newly formed organs (Semiarti et al. 2001). KNOX genes are expressed in meristematic cells, and are downregulated for an organ to develop (Lincoln et al. 1994; Long et al. 1996). *BP* is a KNOX gene that causes a downward silique phenotype when mutated in Arabidopis ecotype L*er* (Douglas et al. 2002; Venglat et al. 2002). *DS* and *BP* were found to downregulate each other reciprocally in the flower. The *DS* loss-of-function phenotype was similar to the wild type, most likely due to functional redundancy with other members of the AS2 family. Double mutants for *DS* and a related member of the family *AS2* (*LBD6*, which also downregulates *BP* and other KNOX genes) showed that they act redundantly to shape Arabidopsis floral organs.

18.3 Transcription Factors Regulating Secondary Metabolic Pathways

There are also examples of genes that have been cloned from different plants as regulatory genes of secondary metabolism. Interestingly, most of these genes are transcription factors. Three examples are given below. In Madagascar periwinkle (*Catharantus roseu*s), a positive selection system was used to isolate *ORCA3* that codes for a jasmonate-responsive transcriptional regulator of primary and secondary metabolism (van der Fits and Memelink 2000). *ORCA3* is an APETALA2 (AP2)-domain transcription factor. Its overexpression resulted in the increased expression of several metabolite biosynthetic genes, and consequently the enhanced accumulation of terpenoid indole alkaloids. It was therefore proposed that this gene could link plant stress responses to metabolic changes (van der Fits and Memelink 2000).

This example illustrates the use of activation tagging to increase the accumulation of natural products in plants. This is important since many are useful bioactive molecules, but mostly present only in specific tissues and at low concentrations. Using activation tagging in Arabidopsis, Borevitz and colleagues (2000) found a mutant in which the overexpression of a *MYB* transcription factor gene resulted in overactivation of phenylpropanoid biosynthetic genes. As a result, the plant accumulated greater amounts of lignin, hydroxycinnamic acid esters, and flavonoids, including different anthocyanins. This increased accumulation of anthocyanins caused many vegetative organs of the mutant to show an intense purple pigmentation, which allowed the mutant to be identified.

In tomato, the activation tagged *ant1* mutant also showed intense purple pigmentation in vegetative tissues (Mathews et al. 2003). Interestingly, a *MYB* transcription factor gene was also overexpressed in this mutant. This gene, as for some other *MYB* factors identified in other plants (Quattrocchio et al. 1999), is a general regulator of anthocyanin pathways, since its overexpression caused the upregulation of genes involved in the early and later steps of anthocyanidin biosynthesis, anthocyanin glycosylation, and transport of anthocyanins into the vacuole. The isolation of these genes suggests that activation tagging can override the stringent genetic controls that regulate the accumulation of specific natural bioactive molecules at different stages of plant development (Borevitz et al. 2000). This provided the possibility for high-throughput strategies to discover structural and regulatory genes of metabolic pathways of interesting biomolecules.

18.4   Activation Tagging Genes that Confer Resistance to Biotic and Abiotic Stresses

18.4.1   *Genes Conferring Resistance to Pathogens*

Activation tagging has also allowed cloning of components of pathogen resistance pathways. One such component is *CDR1*. Activation tagged *cdr1* mutants expressed many pivotal defense maker genes, and were resistant to virulent Pst DC3000 (Xia et al. 2004). This gene, coding for an apoplastic aspartate protease, is necessary for the function of some R genes that are essential in triggering race- or cultivar-specific disease resistance pathways upon recognition of pathogen avirulence (avr) gene products (Dangl and Jones 2001). In another example, a screen of a T-DNA activation tagged library of 12,000 lines for resistance to *Sclerotinia sclerotiorum* resulted in selection of a mutant resistant at the germination stage in the presence of the pathogen, and also resistant at the adult plant stage (Boccongelli et al. 2003). The putative tagged gene encoded the alpha-3 subunit of 20S proteasome. In a specific targeted screen for overexpressors of the PR-1 promoter, using 30,000 T-DNA activation tagged lines, 12 lines were obtained with enhanced promoter activity (Grant et al. 2003). Characterization of one mutant, *adr1*, revealed constitutive expression of a number of key defense genes, and accumulation of SA but not of JA or ET. The *adr1* plants exhibited resistance against the biotrophic pathogens *Hyaloperonospora parasitica* and

Erysiphe cichoracearum, but not the necrotrophic fungus *Botrytis cinerea*. The *ADR1* gene encodes a protein with homology to protein kinases, a nucleotide-binding domain, and leucine-rich repeats. Although the *adr1* mutant displayed slower growth (homozygotes were severely retarded), regulated expression of *ADR1* showed disease resistance in the absence of yield penalty, supporting the use of *ADR1* in crop protection.

In a recent screen of 13,000 T-DNA activation tagged rice lines (Mori et al. 2007), a lesion-mimic mutant was obtained that showed resistance to the rice blast fungus *Magnaporthe grisea*. Interestingly, the mutant did not show normal inheritance in following generations, but partial silencing of the tagged gene. The tagged gene OsAT1 belongs to a group of acyl-CoA-dependent acyltransferases of the BAHD family, with 14 rice members that are partially redundant.

18.4.2 Genes Conferring Resistance to Drought

The *SHINE* gene was identified from Arabidopsis through a gain-of-function mutant that displayed shiny leaves with enhanced epicuticular wax deposition (Aharoni et al. 2004). The *SHN* gene is member of the AP2/ERF family of transcription factors, and was independently described as the *WIN1* gene (Broun et al. 2004). Overexpression of the SHN gene in Arabidopsis reduced stomatal density and changed other epidermal properties, increasing cuticle permeability and water loss. The *SHN* overexpressors showed drought resistance in dry-down experiments (Aharoni et al. 2004), suggested to be due to the reduced stomatal index and lower transpiration. In addition to the *SHN1* gene, the paralogs *SHN2* and *SHN3* also show the same phenotype of increased epicuticular wax on overexpression. The three *SHN* genes show very specific expression patterns in different inflorescence tissues, such as *SHN2* in anther and silique dehiscent zones. The individual *SHN* expression zones correspond to external layers involved in protection of tissue to desiccation.

The *HARDY (HRD)* gene was identified in Arabidopsis (Karaba et al. 2007) via a gain-of-function activation tag mutant with extended root branching and thickness, making the roots resistant to pulling. The *hrd-D* mutant roots showed extra cortical cell layers, and a more compact stele bearing vascular tissue. The *HRD* gene belongs to group IIIb of the AP2/ERF-like transcription factors, with four Arabidopsis and six rice group members (Nakano et al. 2006) for which no phenotypic function had been shown, different from the well-studied *DREB/CBF* group IIIc genes that provide abiotic stress tolerance. The Arabidopsis *HRD* gene was overexpressed in rice under control of the CaMV35S promoter, and displayed increased canopy growth and biomass yield, and improved water use efficiency and drought resistance. Enhanced growth was also observed in the form of increased roots and extra bundle sheath cells, compared to the wild type.

18.4.3   Activation Tagging of Genes Involved in Sensing Nutritional Status

Sugar status influences expression of different plant genes. Genes involved in source activities, like photosynthesis, are downregulated, while those with sink functions (such as storage proteins) are upregulated (Rolland et al. 2002). Using the same approach as in the earlier example, namely, activation tagging mutant screens for alterations in reporter gene expression, Masaki et al. (2005) found an activator of a subset of sugar-responsive genes under low sugar conditions. Using a sweet potato sporamin promoter fused to LUC and GUS reporters, a mutant (*asml2*) was identified with enhanced expression. The overexpressed gene codes for a protein with a CCT (CONSTANS, CONSTANS-like, TOC1) domain. The mutant showed alterations in the expression of several endogenous sugar-responsive genes, suggesting that ASML2 might be a transcriptional activator that modulates expression of a subgroup of sugar-inducible genes.

The examples described show that activation tagging strategies allow gene discovery in a broad range of studies, and activation tagging populations can be screened for visual developmental/morphological alterations, metabolic alterations, and resistance/tolerance to biotic, physical, or chemical stresses. Moreover, studies of plant growth under unusual conditions, or screens for altered responses to diverse stimuli, such as light, nutrients, other organisms, controlled environments, and signaling compounds, can also make use of this strategy. Furthermore, alterations in the regulation of gene expression such as through gene-silencing processes, or altered expression of reporter genes fused to promoters of interest, can also be screened using activation tagging. Other types of alterations in regulation at the level of the genome, transcriptome or proteome can be studied, given a suitable system to monitor changes.

18.5   Additional Considerations

Activation tagging has many advantages, but it also has limitations that should be considered. Some of the limitations are inherent to the system, and some are related to the 35S enhancer that is commonly used in this strategy.

18.5.1   Role of Enhancers

Using only enhancer elements as activators seems to have advantages over constitutive promoters, because they can promote tissue- or condition-specific upregulation (Neff et al. 1999). Constitutive overexpression (using the whole 35 S promoter) of a gene can lead to severely affected plants that are not viable (van der Graaff et al. 2000), or have phenotypes that are more difficult to interpret. In this way, the phenotypes that enhancers produce are more likely to reveal the true function of the gene than when it is constitutively overexpressed (van der Graaff et al. 2000). However, a possible limitation of activation tagging using enhancers in general might be the presence of insulators in the plant genome (Bell et al. 2001). Insulators have not been well studied in plants, but they play an important role in gene regulation, and might decrease or block the effect of an enhancer in a tag. Post-transcriptional regulation could also decrease the phenotypic effects of overexpressing a gene. Translation efficiency or regulation, as well as protein modification might modulate the level of an active protein, and therefore could "mask" the effects of gene activation by the enhancer.

18.5.2 Instability and Nonviable Phenotypes

Some researchers, using the present systems, have reported activation tag allele instability (Weigel et al. 2000). In this case, the use of transgenic plants transformed with the mutant allele can provide a stable phenotype (Tani et al. 2004). Moreover, while recessive mutations can be normally maintained in heterozygous plants showing wild-type phenotypes, and produce homozygous progeny that display the mutant phenotype, an important drawback of activation tagging as described here (using a constitutive strong enhancer) might be that dominant mutations that cause total sterility cannot be maintained. One way to find the genes involved in such mutants is to isolate DNA from the original mutant, and find the position of the insertion in the genome. Candidate genes can then be introduced as (inducible) overexpression constructs into a wild-type plant, and the range of phenotype(s) evaluated.

In more severe cases, the use of the 35S enhancer may lead to dominant mutations causing lethal or extremely weak phenotypes that are not possible to recover. Moreover, while it has a broad range of activity, this enhancer is not active in all plant tissues (Benfey et al. 1989), which can limit gene function discovery in certain tissues. The use of tissue-specific, condition-specific, or inducible activators can solve these problems by allowing the recovery of lethal or very severe dominant mutations, and target tissues that cannot be studied using the present systems, and aid in the study of gene functions of, for example, those genes that provoke early phenotypes that mask their roles at later stages when constitutively overexpressed. A few reports of chemically inducible and transactivation systems have used T-DNA as a tag. Further improvements might involve the use of different enhancers, and of transposons as tags.

The availability of the Arabidopsis genomic sequence is very helpful because it allows the position of the insertions to be identified, and gives information about the genomic context (Arabidopsis Genome Initiative 2000). However, this strategy can also be used in plants for which the genome has not been sequenced. In plants where

(T-DNA) transformation strategies have low efficiencies, activation tagging using transposons might be the best approach.

18.6 Conclusions

Classic activation tagging strategies employ enhancer sequences that can promote the overexpression of genes. The main advantage of this strategy is that it generates dominant mutations that are distinguishable directly in the primary transgenic plants, in contrast to recessive mutations. As demonstrated by the identification of many genes with previously unknown functions, it complements the limitations of loss-of-function strategies, and can help discovering the function of a broad range of genes involved in different processes. In general, the activation tagging approach has a wide variety of applications that can range from the scientific search for new gene functions to the direct generation of commercially valuable and otherwise interesting crops.

References

Aarts MGM, Dirkse WG, Stiekema WJ, Pereira A (1993) Transposon tagging of a male sterility gene in *Arabidopsis*. Nature 363:715–717

Aharoni A, Dixit S, Jetter R, Thoenes E, van Arkel G, Pereira A (2004) The SHINE clade of AP2 domain transcription factors activates wax biosynthesis, alters cuticle properties, and confers drought tolerance when overexpressed in Arabidopsis. Plant Cell 16:2463–2480

Altmann T (1998) Recent advances in brassinosteroid next term molecular genetics. Curr Opin Plant Biol 1:378–383

Arabidopsis Genome Initiative (2000) Analysis of the genome sequence of the flowering plant *Arabidopsis thaliana*. Nature 408:796–815

Ayliffe MA, Pryor AJ (2007) Activation tagging in plants - generation of novel, gain-of-function mutations. Aust J Agric Res 58:490–497

Banno H, Ikeda Y, Niu QW, Chua NH (2001) Overexpression of Arabidopsis *ESR1* induces initiation of shoot regeneration. Plant Cell 13:2609–2618

Bartel B (1997) Auxin biosynthesis. Annu Rev Plant Physiol Plant Mol Biol 48:51–66

Bell AC, West AG, Felsenfeld G (2001) Gene regulation - insulators and boundaries: versatile regulatory elements in the eukaryotic genome. Science 291:447–450

Benfey PN, Ren L, Chua NH (1989) The CaMV 35S enhancer contains at least two domains which can confer different developmental and tissue-specific expression patterns. EMBO J 8:2195–2202

Blackwell JR, Horgan R (1994) Cytokinin biosynthesis by extracts of *Zea mays*. Phytochemistry 35:339–342

Boccongelli C, Chilosi G, Magro P, Bressan RA (2003) Isolation of a new Arabidopsis mutant with enhanced disease tolerance to *Sclerotinia sclerotiorum*. J Plant Pathol 85:277 (abstract)

Borevitz JO, Xia Y, Blount J, Dixon RA, Lamb C (2000) Activation tagging identifies a conserved MYB regulator of phenylpropanoid biosynthesis. Plant Cell 12:2383–2394

Bouché N, Bouchez D (2001) Arabidopsis gene knockout: phenotypes wanted. Curr Opin Plant Biol 4:111–117

Bowman JL, Eshed Y (2000) Formation and maintenance of the shoot apical meristem. Trends Plant Sci 5:110–115

Brand U, Hobe M, Simon R (2001) Functional domains in plant shoot meristems. BioEssays 23:134–141

Broun P, Poindexter P, Osborne E, Jiang C-Z, Riechmann JL (2004) WIN1, a transcriptional activator of epidermal wax accumulation in *Arabidopsis*. Proc Natl Acad Sci USA 101: 4706–4711

Brzobohaty B, Moore I, Palme K (1994) Cytokinin metabolism: implications for regulation of plant growth and development. Plant Mol Biol 26:1483–1497

Busov VB, Meilan R, Pearce DW, Ma C, Rood SB, Strauss SH (2003) Activation tagging of a dominant gibberellin catabolism gene (*GA 2-oxidase*) from poplar that regulates tree stature. Plant Physiol 132:1283–1291

Chalfun A Jr, Franken J, Mes JJ, Marsch-Martinez N, Pereira A, Angenent GC (2005) *ASYM-METRIC LEAVES2-LIKE1* gene, a member of the AS2/LOB family, controls proximal–distal patterning in *Arabidopsis* petals. Plant Mol Biol 57:559–575

Chandler JW, Cole M, Flier A, Grewe B, Werr W (2007) The AP2 transcription factors *DORN-ROSCHEN* and *DORNROSCHEN-LIKE* redundantly control *Arabidopsis* embryo patterning via interaction with *PHAVOLUTA*. Development 134:1653–1662

Cheng Y, Dai X, Zhao Y (2006) Auxin biosynthesis by the YUCCA flavin monooxygenases controls the formation of floral organs and vascular tissues in *Arabidopsis*. Genes Dev 20:1790–1799

Coen ES, Meyerowitz EM (1991) The war of the whorls: genetic interactions controlling flower development. Nature 353:31–37

Corbesier L, Vincent C, Jang S, Fornara F, Fan Q, Searle I, Giakountis A, Farrona S, Gissot L, Turnbull C, Coupland G (2007) FT protein movement contributes to long-distance signaling in floral induction of *Arabidopsis*. Science 316:1030–1033

Dangl JL, Jones JD (2001) Plant pathogens and integrated defence responses to infection. Nature 411:826–833

Davies PJ (1995) Plant hormones: physiology, biochemistry, and molecular biology. Kluwer, London, pp 6–7

Douglas SJ, Chuck G, Dengler RE, Pelecanda L, Riggs CD (2002) *KNAT1* and *ERECTA* regulate inflorescence architecture in Arabidopsis. Plant Cell 14:547–558

Elich TD, Chory J (1997) Biochemical characterization of Arabidopsis wild-type and mutant phytochrome B holoproteins. Plant Cell 9:2271–2280

Elliott RC, Ross JJ, Smith JJ, Lester DR, Reid JB (2001) Feed-forward regulation of gibberellin deactivation in pea. J Plant Growth Regul 20:87–94

Fankhauser C, Chory J (1997) Light control of plant development. Annu Rev Cell Dev Biol 13:203–229

Ferreira FJ, Kieber JJ (2005) Cytokinin signaling. Curr Opin Plant Biol 8:518–525

Grant JJ, Chini A, Basu D, Loake GJ (2003) Targeted activation tagging of the Arabidopsis NBS-LRR gene, *ADR1*, conveys resistance to virulent pathogens. Mol Plant Microbe Interact 16:669–680

Gu Z, Steinmetz LM, Gu X, Scharfe C, Davis RW, Li W-H (2003) Role of duplicate genes in genetic robustness against null mutations. Nature 421:63–66

Haake V, Cook D, Riechmann JL, Pineda O, Thomashow MF, Zhang JZ (2002) Transcription factor CBF4 is a regulator of drought adaptation in Arabidopsis. Plant Physiol 130:639–648

Hake S, Vollbrecht E, Freeling M (1989) Cloning *Knotted*, the dominant morphological mutant in maize using *Ds2* as a transposon tag. EMBO J 8:15–22

Hardtke CS, Berleth T (1998) The *Arabidopsis* gene *MONOPTEROS* encodes a transcription factor mediating embryo axis formation and vascular development. EMBO J 17:1405–1411

He Z, Wang Z-Y, Li J, Zhu Q, Lamb C, Ronald P, Chory J (2000) Perception of brassinosteroids by the extracellular domain of the receptor kinase BRI1. Science 288:2360–2363

Hedden P, Phillips AL (2000) Manipulation of hormone biosynthetic genes in transgenic plants. Curr Opin Biotechnol 11:130–137

Heyl A, Schmülling T (2003) Cytokinin signal perception and transduction. Curr Opin Plant Biol 6:480–488

Huang S, Cerny RE, Bhat DS, Brown SM (2001) Cloning of an Arabidopsis patatin-like gene, STURDY, by activation T-DNA tagging. Plant Physiol 125:573–584

Huang T, Böhlenius H, Eriksson S, Parcy F, Nilsson O (2005) The mRNA of the Arabidopsis gene FT moves from leaf to shoot apex and induces flowering. Science 309:1694–1696

Hwang I, Sheen J (2001) Two-component circuitry in Arabidopsis cytokinin signal transduction. Nature 413:383–389

Ichikawa T, Nakazawa M, Kawashima M, Iizumi H, Kuroda H, Kondou Y, Tsuhara Y, Suzuki K, Ishikawa A, Seki M, Fujita M, Motohashi R, Nagata N, Takagi T, Shinozaki K, Matsui M (2006) The FOX hunting system: an alternative gain-of-function gene hunting technique. Plant J 48:974–985

Ikeda Y, Banno H, Niu Q-W, Howell SH, Chua N-H (2006) The ENHANCER OF SHOOT REGENERATION 2 gene in Arabidopsis regulates CUP-SHAPED COTYLEDON 1 at the transcriptional level and controls cotyledon development. Plant Cell Physiol 47:1443–1456

Imaizumi R, Sato S, Kameya N, Nakamura I, Nakamura Y, Tabata S, Ayabe S, Aoki T (2005) Activation tagging approach in a model legume, Lotus japonicus. J Plant Res 118:391–399

Ito T, Meyerowitz E (2000) Overexpression of a gene encoding a cytochrome P450, CYP78A9, induces large and seedless fruit in Arabidopsis. Plant Cell 12:1541–1551

Jeong D-H, An S, Kang H-G, Moon S, Han J-J, Park S, Lee HS, An K, An G (2002) T-DNA insertional mutagenesis for activation tagging in rice. Plant Physiol 130:1636–1644

Kakimoto T (1996) CKI1, a histidine kinase homolog implicated in cytokinin signal transduction. Science 274:982–985

Kakimoto T (1998) Cytokinin signaling. Curr Opin Plant Biol 1:399–403

Kakimoto T (2001) Identification of plant cytokinin biosynthetic enzymes as dimethylallyl diphosphate:ATP/ADP isopentenyltransferases. Plant Cell Physiol 42:677–685

Karaba A, Dixit S, Greco R, Aharoni A, Trijatmiko KR, Marsch-Martinez N, Krishnan A, Nataraja KN, Udayakumar M, Pereira A (2007) Improvement of water use efficiency in rice by expression of HARDY, an Arabidopsis drought and salt tolerance gene. Proc Natl Acad Sci USA 104:15270–15275

Kardailsky I, Shukla VK, Ahn JH, Dagenais N, Christensen SK, Nguyen JT, Chory J, Harrison MJ, Weigel D (1999) Activation tagging of the floral inducer FT. Science 286:1962–1965

Kempin SA, Savidge B, Yanofsky MF (1995) Molecular basis of the cauliflower phenotype in Arabidopsis. Science 267:522–525

Kim JI, Sharkhuu A, Jin JB, Li P, Jeong JC, Baek D, Lee SY, Blakeslee JJ, Murphy AS, Bohnert HJ, Hasegawa PM, Yun D-J, Bressan RA (2007) yucca6, a dominant mutation in Arabidopsis, affects auxin accumulation and auxin-related phenotypes. Plant Physiol 145:722–735

Kirch T, Simon R, Grünewald M, Werr W (2003) The DORNROSCHEN/ENHANCER OF SHOOT REGENERATION1 gene of Arabidopsis acts in the control of meristem cell fate and lateral organ development. Plant Cell 15:694–705

Kuromori T, Wada T, Kamiya A, Yuguchi M, Yokouchi T, Imura Y, Takabe H, Sakurai T, Akiyama K, Hirayama T, Okada K, Shinozaki K (2006) A trial of phenome analysis using 4000 Ds-insertional mutants in gene-coding regions of Arabidopsis. Plant J 47:640–651

Laux T, Mayer KF, Berger J, Jurgens G (1996) The WUSCHEL gene is required for shoot and floral meristem integrity in Arabidopsis. Development 122:87–96

Lee H, Suh S-S, Park E, Cho E, Ahn JH, Kim S-G, Lee JS, Kwon YM, Lee I (2000) The AGAMOUS-LIKE 20 MADS domain protein integrates floral inductive pathways in Arabidopsis. Genes Dev 14:2366–2376

Levy YY, Dean C (1998) The transition to flowering. Plant Cell 10:1973–1990

Lewin B (2000) Genes VII. Oxford University Press, Oxford

Li J, Chory J (1997) A putative leucine-rich repeat receptor kinase involved in brassinosteroid signal transduction. Cell 90:929–938

Li J, Lease KA, Tax FE, Walker JC (2001) BRS1, a serine carboxypeptidase, regulates BRI1 signaling in *Arabidopsis thaliana*. Proc Natl Acad Sci USA 98:5916–5921

Li J, Wen J, Lease KA, Doke JT, Tax FE, Walker JC (2002) BAK1, an *Arabidopsis* LRR receptor-like protein kinase, interacts with BRI1 and modulates brassinosteroid signaling. Cell 110: 213–222

Liljegren SJ, Ditta GS, Eshed Y, Savidge B, Bowman JL, Yanofsky MF (2000) *SHATTERPROOF* MADS-box genes control seed dispersal in *Arabidopsis*. Nature 404:766–770

Lin WC, Shuai B, Springer PS (2003) The *Arabidopsis LATERAL ORGAN BOUNDARIES-domain* gene *ASYMMETRIC LEAVES2* functions in the repression of *KNOX* gene expression and in adaxial-abaxial patterning. Plant Cell 15:2241–2252

Lin M-K, Belanger H, Lee Y-J, Varkonyi-Gasic E, Taoka K-I, Miura E, Xoconostle-Cázares B, Gendler K, Jorgensen RA, Phinney B, Lough TJ, Lucas WJ (2007) FLOWERING LOCUS T protein may act as the long-distance florigenic signal in the cucurbits. Plant Cell 19:1488–1506

Lincoln C, Long J, Yamaguchi J, Serikawa K, Hake S (1994) A *knotted1*-like homeobox gene in Arabidopsis is expressed in the vegetative meristem and dramatically alters leaf morphology when overexpressed in transgenic plants. Plant Cell 6:1859–1876

Long JA, Moan EI, Medford JI, Barton MK (1996) A member of the KNOTTED class of homeodomain proteins encoded by the *STM* gene of *Arabidopsis*. Nature 379:66–69

Marsch-Martínez N, Greco R, Van Arkel G, Herrera-Estrella L, Pereira A (2002) Activation tagging using the *En-I* maize transposon system in Arabidopsis. Plant Physiol 129:1544–1556

Marsch-Martínez N, Greco R, Becker JD, Dixit S, Bergervoet JHW, Karaba A, de Folter S, Pereira A (2006) *BOLITA*, an Arabidopsis AP2/ERF-like transcription factor that affects cell expansion and proliferation/differentiation pathways. Plant Mol Biol 62:825–843

Martienssen RA (1998) Functional genomics: probing plant gene function and expression with transposons. Proc Natl Acad Sci USA 95:2021–2026

Masaki T, Tsukagoshi H, Mitsui N, Nishii T, Hattori T, Morikami A, Nakamura K (2005) Activation tagging of a gene for a protein with novel class of CCT-domain activates expression of a subset of sugar-inducible genes in *Arabidopsis thaliana*. Plant J 43:142–152

Mathews H, Clendennen SK, Caldwell CG, Liu XL, Connors K, Matheis N, Schuster DK, Menasco DJ, Wagoner W, Lightner J, Wagner DR (2003) Activation tagging in tomato identifies a transcriptional regulator of anthocyanin biosynthesis, modification, and transport. Plant Cell 15:1689–1703

Matsuhara S, Jingu F, Takahashi T, Komeda Y (2000) Heat-shock tagging: a simple method for expression and isolation of plant genome DNA flanked by T-DNA insertions. Plant J 22:79–86

Miklos GLG, Rubin GM (1996) The role of the genome project in determining gene function: insights from model organisms. Cell 86:521–529

Miller CO, Skoog F, Von Saltza MH, Strong F (1955) Kinetin, a cell division factor from deoxyribonucleic acid. J Am Chem Soc 77:1392

Mori M, Tomita C, Sugimoto K, Hasegawa M, Hayashi N, Dubouzet JG, Ochiai H, Sekimoto H, Hirochika H, Kikuchi S (2007) Isolation and molecular characterization of a *Spotted leaf 18* mutant by modified activation-tagging in rice. Plant Mol Biol 63:847–860

Muller B, Sheen J (2007) Advances in cytokinin signaling. Science 318:68–69

Nag A, Yang Y, Jack T (2007) *DORNRÖSCHEN-LIKE*, an AP2 gene, is necessary for stamen emergence in Arabidopsis. Plant Mol Biol 65:219–232

Nakano T, Suzuki K, Fujimura T, Shinshi H (2006) Genome-wide analysis of the ERF gene family in Arabidopsis and rice. Plant Physiol 140:411–432

Nath U, Crawford BCW, Carpenter R, Coen E (2003) Genetic control of surface curvature. Science 299:1404–1407

Neff MM, Nguyen SM, Malancharuvil EJ, Fujioka S, Noguchi T, Seto H, Tsubuki M, Honda T, Takatsuto S, Yoshida S, Chory J (1999) BAS1: a gene regulating brassinosteroid levels and light responsiveness in *Arabidopsis*. Proc Natl Acad Sci USA 96:15316–15323

Odell JT, Nagy F, Chua NH (1985) Identification of DNA sequences required for activity of the cauliflower mosaic virus 35S promoter. Nature 313:810–812

Okamuro JK, Caster B, Villarroel R, Van Montagu M, Jofuku KD (1997) The AP2 domain of *APETALA2* defines a large new family of DNA binding proteins in *Arabidopsis*. Proc Natl Acad Sci USA 94:7076–7081

Olszewski N, Sun TP, Gubler F (2002) Gibberellin signaling: biosynthesis, catabolism, and response pathways. Plant Cell 14 suppl:S61–S80

Palatnik JF, Allen E, Wu X, Schommer C, Schwab R, Carrington JC, Weigel D (2003) Control of leaf morphogenesis by microRNAs. Nature 425:257–263

Putterill J, Robson F, Lee K, Simon R, Coupland G (1995) The *CONSTANS* gene of *Arabidopsis* promotes flowering and encodes a protein showing similarities to zinc finger transcription factors. Cell 80:847–857

Qu S, Desai A, Wing R, Sundaresan V (2008) A versatile transposon-based activation tag vector system for functional genomics in cereals and other monocot plants. Plant Physiol 146: 189–199

Quattrocchio F, Wing J, van der Woude K, Souer E, de Vetten N, Mol J, Koes R (1999) Molecular analysis of the *anthocyanin2* gene of petunia and its role in the evolution of flower color. Plant Cell 11:1433–1444

Ratcliffe OJ, Kumimoto RW, Wong BJ, Riechmann JL (2003) Analysis of the Arabidopsis *MADS AFFECTING FLOWERING* gene family: *MAF2* prevents vernalization by short periods of cold. Plant Cell 15:1159–1169

Riechmann JL, Ratcliffe OJ (2000) A genomic perspective on plant transcription factors. Curr Opin Plant Biol 3:423–434

Rolland F, Moore B, Sheen J (2002) Sugar sensing and signaling in plants. Plant Cell 14 suppl: S185–S205

Ross-Macdonald P, Coelho PSR, Roemer T, Agarwal S, Kumar A, Jansen R, Cheung K-H, Sheehan A, Symoniatis D, Umansky L, Heidtman M, Nelson FK, Iwasaki H, Hager K, Gerstein M, Miller P, Roeder GS, Snyder M (1999) Large-scale analysis of the yeast genome by transposon tagging and gene disruption. Nature 402:413–418

Sakamoto T, Kobayashi M, Itoh H, Tagiri A, Kayano T, Tanaka H, Iwahori S, Matsuoka M (2001) Expression of a gibberellin 2-oxidase gene around the shoot apex is related to phase transition in rice. Plant Physiol 125:1508–1516

Samach A, Onouchi H, Gold SE, Ditta GS, Schwarz-Sommer Z, Yanofsky MF, Coupland G (2000) Distinct roles of CONSTANS target genes in reproductive development of *Arabidopsis*. Science 288:1613–1616

Schneider A, Kirch T, Gigolashvili T, Mock H-P, Sonnewald U, Simon R, Flügge U-I, Werr W (2005) A transposon-based activation-tagging population in *Arabidopsis thaliana* (TAMARA) and its application in the identification of dominant developmental and metabolic mutations. FEBS Lett 579:4622–4628

Schoentgen F, Saccoccio F, Jollès J, Bernier I, Jollès P (1987) Complete amino acid sequence of a basic 21-kDa protein from bovine brain cytosol. Eur J Biochem 166:333–338

Schomburg FM, Bizzell CM, Lee DJ, Zeevaart JAD, Amasino RM (2003) Overexpression of a novel class of gibberellin 2-Oxidases decreases gibberellin levels and creates dwarf plants. Plant Cell 15:151–163

Schumacher K, Chory J (2000) Brassinosteroid signal transduction: still casting the actors. Curr Opin Plant Biol 3:79–84

Semiarti E, Ueno Y, Tsukaya H, Iwakawa H, Machida C, Machida Y (2001) The *ASYMMETRIC LEAVES2* gene of *Arabidopsis thaliana* regulates formation of a symmetric lamina, establishment of venation and repression of meristem-related homeobox genes in leaves. Development 128:1771–1783

Skoog F, Miller CO (1957) Chemical regulation of growth and organ formation in plant tissues cultured in vitro. Symp Soc Exp Biol 54:118–130

Sommer H, Beltrán JP, Huijser P, Pape H, Lönnig W-E, Saedler H, Schwarz-Sommer Z (1990) *Deficiens*, a homeotic gene involved in the control of flower morphogenesis in *Antirrhinum majus*: the protein shows homology to transcription factors. EMBO J 9:605–613

Sorokin AP, Walsh S, Baumann K, Nichols J, Bevan M, Jones JDG, Martin C, Clarke JH (2008) Induction of phenotypic variation by activation of genes harbouring a maize *Spm* element in their promoter regions using a TnpAVP16 fusion protein. Plant J 53:587–594

Suzuki Y, Uemura S, Saito Y, Murofushi N, Schmitz G, Theres K, Yamaguchi I (2001) A novel transposon tagging element for obtaining gain-of-function mutants based on a self-stabilizing *Ac* derivative. Plant Mol Biol 45:123–131

Takei K, Sakakibara H, Sugiyama T (2001) Identification of genes encoding adenylate isopentenyltransferase, a cytokinin biosynthesis enzyme, in *Arabidopsis thaliana*. J Biol Chem 276:26405–26410

Tamaki S, Matsuo S, Wong HL, Yokoi S, Shimamoto K (2007) Hd3, a protein is a mobile flowering signal in rice. Science 316:1033–1036

Tani H, Chen X, Nurmberg P, Grant JJ, SantaMaria M, Chini A, Gilroy E, Birch PRJ, Loake GJ (2004) Activation tagging in plants: a tool for gene discovery. Funct Integrat Genomics 4:258–266

Tuskan GA, DiFazio S, Jansson S, Bohlmann J, Grigoriev I, Hellsten U, Putnam N, Ralph S, Rombauts S, Salamov A, Schein J, Sterck L, Aerts A, Bhalerao RR, Bhalerao RP, Blaudez D, Boerjan W, Brun A, Brunner A, Busov V, Campbell M, Carlson J, Chalot M, Chapman J, Chen G-L, Cooper D, Coutinho PM, Couturier J, Covert S, Cronk Q, Cunningham R, Davis J, Degroeve S, Déjardin A, dePamphilis C, Detter J, Dirks B, Dubchak I, Duplessis S, Ehlting J, Ellis B, Gendler K, Goodstein D, Gribskov M, Grimwood J, Groover A, Gunter L, Hamberger B, Heinze B, Helariutta Y, Henrissat B, Holligan D, Holt R, Huang W, Islam-Faridi N, Jones S, Jones-Rhoades M, Jorgensen R, Joshi C, Kangasjärvi J, Karlsson J, Kelleher C, Kirkpatrick R, Kirst M, Kohler A, Kalluri U, Larimer F, Leebens-Mack J, Leplé J-C, Locascio P, Lou Y, Lucas S, Martin F, Montanini B, Napoli C, Nelson DR, Nelson C, Nieminen K, Nilsson O, Pereda V, Peter G, Philippe R, Pilate G, Poliakov A, Razumovskaya J, Richardson P, Rinaldi C, Ritland K, Rouzé P, Ryaboy D, Schmutz J, Schrader J, Segerman B, Shin H, Siddiqui A, Sterky F, Terry A, Tsai C-J, Uberbacher E, Unneberg P, Vahala J, Wall K, Wessler S, Yang G, Yin T, Douglas C, Marra M, Sandberg G, Van de Peer Y, Rokhsar D (2006) The genome of black cottonwood, *Populus trichocarpa* (Torr. & Gray). Science 313:1596–1604

van der Fits L, Memelink J (2000) *ORCA3*, a jasmonate-responsive transcriptional regulator of plant primary and secondary metabolism. Science 289:295–297

van der Graaff E, Dulk-Ras AD, Hooykaas PJ, Keller B (2000) Activation tagging of the *LEAFY PETIOLE* gene affects leaf petiole development in *Arabidopsis thaliana*. Development 127:4971–4980

van der Graaff E, Hooykaas PJJ, Keller B (2002) Activation tagging of the two closely linked genes *LEP* and *VAS* independently affects vascular cell number. Plant J 32:819–830

Veit B, Mathern VJ, Hake S (1990) A tandem duplication causes the *Kn1-O* allele of *Knotted*, a dominant morphological mutant of maize. Genetics 125:623–631

Venglat SP, Dumonceaux T, Rozwadowski K, Parnell L, Babic V, Keller W, Martienssen R, Selvaraj G, Datla R (2002) The homeobox gene *BREVIPEDICELLUS* is a key regulator of inflorescence architecture in *Arabidopsis*. Proc Natl Acad Sci USA 99:4730–4735

Vollbrecht E, Veit B, Sinha N, Hake S (1991) The developmental gene *Knotted-1* is a member of a maize homeobox gene family. Nature 350:241–243

Walden R, Fritze K, Hayashi H, Miklashevichs E, Harling H, Schell J (1994) Activation tagging: a means of isolating genes implicated as playing a role in plant growth and development. Plant Mol Biol 26:1521–1528

Ward JM, Smith AM, Shah PK, Galanti SE, Yi HK, Demianski AJ, van der Graaff E, Keller B, Neff MM (2006) A new role for the Arabidopsis AP2 transcription factor, *LEAFY PETIOLE*, in gibberellin-induced germination is revealed by the misexpression of a homologous gene, *SOB2/DRN-LIKE*. Plant Cell 18:29–39

Weigel D, Meyerowitz EM (1994) The ABCs of floral homeotic genes. Cell 78:203–209

Weigel D, Alvarez J, Smyth DR, Yanofsky MF, Meyerowitz E (1992) *LEAFY* controls floral meristem identity in *Arabidopsis*. Cell 69:843–859

Weigel D, Ahn JH, Blázquez MA, Borevitz JO, Christensen SK, Fankhauser C, Ferrándiz C, Kardailsky I, Malancharuvil EJ, Neff MM, Nguyen JT, Sato S, Wang Z-Y, Xia Y, Dixon RA, Harrison MJ, Lamb CJ, Yanofsky MF, Chory J (2000) Activation tagging in Arabidopsis. Plant Physiol 122:1003–1014

Wilson KLD, Swinburne K, Coupland G (1996) A Dissociation insertion causes a semidominant mutation that increases expression of *TINY*, an Arabidopsis gene related to *APETALA2*. Plant Cell 8:659–671

Wullschleger SD, Jansson S, Taylor G (2002) Genomics and forest biology: *Populus* emerges as the perennial favorite. Plant Cell 14:2651–2655

Xia Y, Suzuki H, Borevitz J, Blount J, Guo Z, Patel K, Dixon RA, Lamb C (2004) An extracellular aspartic protease functions in *Arabidopsis* disease resistance signaling. EMBO J 23:980–988

Yanofsky MF, Ma H, Bowman JL, Drews GN, Feldmann KA, Meyerowitz EM (1990) The protein encoded by the *Arabidopsis* homeotic gene *agamous* resembles transcription factors. Nature 346:35–39

Yeung K, Seitz T, Li S, Janosch P, McFerran B, Kaiser C, Fee F, Katsanakis KD, Rose DW, Mischak H, Sedivy JM, Kolch W (1999) Suppression of Raf-1 kinase activity and MAP kinase signalling by RKIP. Nature 401:173–177

Zhao Y, Christensen SK, Fankhauser C, Cashman JR, Cohen JD, Weigel D, Chory J (2001) A role for flavin monooxygenase-like enzymes in auxin biosynthesis. Science 291:306–309

Zubko E, Adams CJ, Macháèková I, Malbeck J, Scollan C, Meyer P (2002) Activation tagging identifies a gene from *Petunia hybrida* responsible for the production of active cytokinins in plants. Plant J 29:797–808

Zuo J, Niu Q-W, Chua NH (2000) An estrogen receptor-based transactivator XVE mediates highly inducible gene expression in transgenic plants. Plant J 24:265–273

Zuo J, Niu Q-W, Frugis G, Chua NH (2002) The *WUSCHEL* gene promotes vegetative-to-embryonic transition in *Arabidopsis*. Plant J 30:349–359

Chapter 19
Regulatory Mechanisms of Homologous Recombination in Higher Plants

K. Osakabe, K. Abe, M. Endo, and S. Toki

19.1 Introduction

Homologous recombination (HR) is one of two major repair systems for double-stranded DNA breaks (DSBs). HR is the process by which a strand of DNA is broken and joined to the end of a different DNA molecule (Fig. 19.1). This repair system is generally the more accurate pathway of DSB repair, ensuring repair without any loss of genetic information. However, it requires the presence of the undamaged sister chromatid or homologous chromosome as a template.

DSBs are also repaired by an alternative pathway, referred to as non-homologous end joining (NHEJ), because the broken ends are ligated directly without the need for a homologous template. As a consequence, NHEJ is a relatively inaccurate process and is frequently accompanied by insertion or deletion of DNA sequences (reviewed in Ma et al. 2005; Burma et al. 2006). HR is the primary DSB repair pathway in many prokaryotes and some lower eukaryotes (e.g. budding and fission yeasts and moss). In contrast, NHEJ is the predominant pathway in eukaryotes, including higher plants (Puchta and Hohn 1996; Kanaar et al. 1998). Although HR is not the predominant system in eukaryotes, it is an essential process in maintaining genome integrity and variability. While NHEJ is known to function throughout the cell cycle, HR is important during the late S and G2 phases of the cell cycle and during early stages of development. Thus, a deficiency in either repair pathway is highly toxic to eukaryotes (Takata et al. 1998; Essers et al. 2000). HR also forms the basis of the mechanism of programmed genetic recombination, such as meiotic recombination, thus producing genetic variation. Indeed, breeding programs depend heavily on the crossover events of meiotic HR.

K. Osakabe, K. Abe, M. Endo, and S. Toki
Division of Plant Sciences, National Institute of Agrobiological Sciences, 2-1-2 Kannondai, Tsukuba, Ibaraki 305-8602, Japan
e-mail: stoki@affrc.go.jp

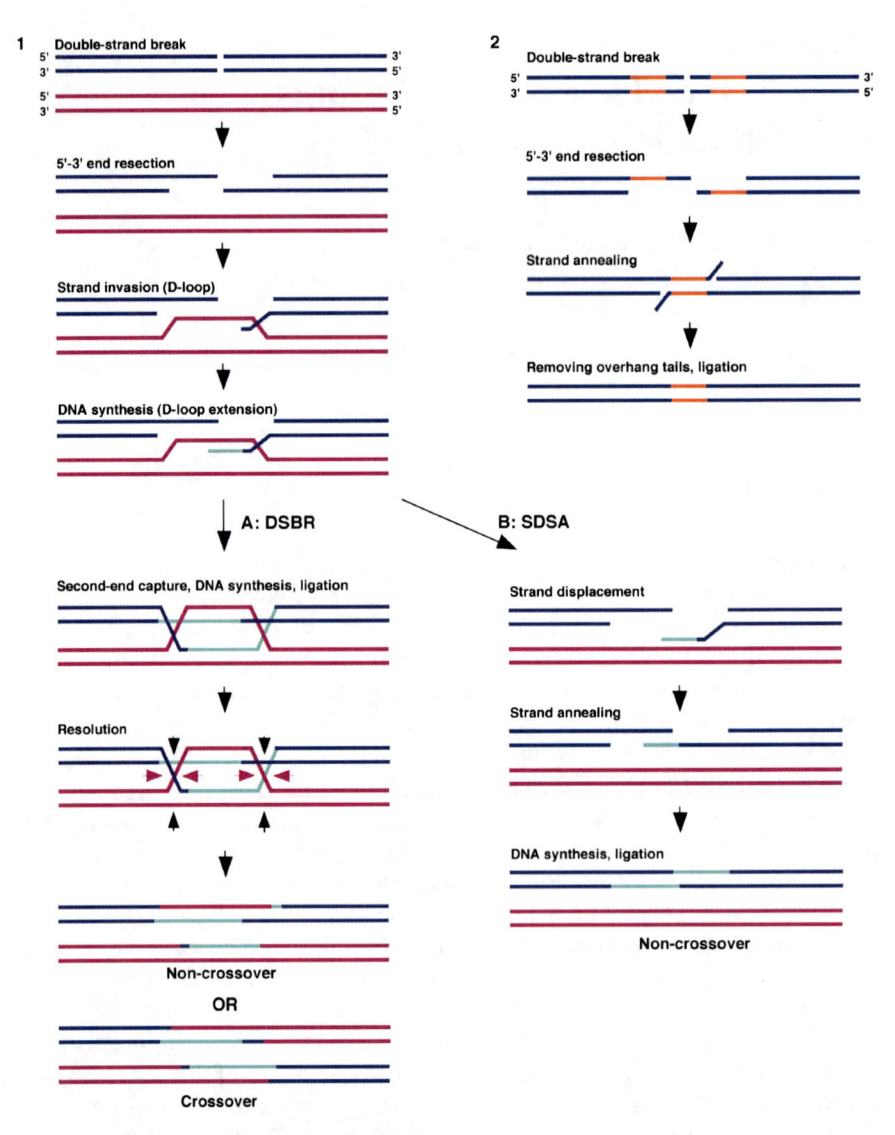

Fig. 19.1 Mechanisms of homologous recombination. The double-stranded DNA break repair (DSBR; *1A*), synthesis-dependent strand-annealing (SDSA; *1B*), and single-strand annealing (SSA; *2*) models are illustrated. In panel 1, *light blue lines* indicate newly synthesized DNA. In panel 2, *orange lines* indicate direct repeat sequences

Genetic and biochemical studies have provided evidence on the proteins involved in HR and their functions. Although studies of plant systems have contributed to a better understanding of genetics, molecular details have come from studies in yeast and mammals. The completion of *Arabidopsis* and rice genome sequences has greatly facilitated the search for genes involved in HR. In addition, the collection of T-DNA/transposon tagging mutant lines in *Arabidopsis* has helped the identification of function of some genes in plants. These include several HR repair-gene-deficient *Arabidopsis* mutants showing hypersensitivity to DSB-inducing agents. Unlike vertebrates, however, with a few exceptions these mutants do not show embryonic lethality or severe growth defects. This could represent a distinct advantage of plant systems for the study of HR.

In this chapter, recent advances in the HR mechanism in eukaryotes are discussed, including signal transduction in higher plants in relation to HR repair.

19.2 Molecular Mechanism of HR

The initial step in HR is DSB formation. DSBs can be generated by ionizing or ultraviolet (UV) irradiation, especially UV-C, or chemicals. Cellular processes such as DNA replication or repair of other types of DNA lesions also give rise to DSBs. The formation of DSBs can be induced during programmed genomic rearrangements, such as meiosis (Keeney and Neale 2006), the production of the variable regions of vertebrate immunoglobulin heavy and light chains (Lieber et al. 2004) and mating-type switching in yeast (Haber 1998). Studies of the mechanisms of genetic recombination have shown that DSB formation is the critical event for the initiation of recombination (Dudás and Chovanec 2004; Ma et al. 2005). In meiosis, DSBs are formed by the Spo11 protein, which cleaves DNA via a topoisomerase-like reaction to generate covalent protein-DNA linkages to the $5'$ DNA ends on either side (Bergerat et al. 1997).

In eukaryotes, the sequential steps of HR are carried out by members of the Rad52 epistasis group (e.g. Rad50, Rad51, Rad52, Rad54, Rad55 Rad57, Rad59, Mre11 and Xrs2), which were identified originally in a genetic screen for X-ray-sensitive mutants in budding yeast (Pâques and Haber 1999). HR can be initiated by DSBs. In yeast, the ends of the duplex at the break are processed extensively by a complex comprising Mre11, Rad50 and Xrs2 (MRX) to produce single-stranded $3'$ overhangs. The counterpart of the *XRS2* gene in vertebrates and plants is *NBS1*. Thus, the complexes formed in vertebrates and plants are referred to as Mre11/Rad50/NBS1 (MRN; Table 19.1).

The Rad51 protein binds the ends of the DNA break to form a Rad51/single-stranded DNA (ssDNA) complex. This step is facilitated by Rad52, Rad55 and Rad57. Rad55 and Rad57 are so-called Rad51 paralogues in budding yeast. Similarly, vertebrates and *Arabidopsis* possess five Rad51 paralogues, namely Rad51B, Rad51C, Rad51D, Xrcc2 and Xrcc3 (reviewed in Thacker 1999). In mammals and *Arabidopsis*, BRCA2 has been identified as an important co-factor of Rad51 (Siaud et al. 2004; Galkin et al. 2005). It is interesting to note that a counterpart gene of Rad52 has not been identified in the genome of plants, including *Arabidopsis* and rice, suggesting that other co-factor proteins, such as Rad51 paralogues, may serve as the function of Rad52 in plants.

The resulting nucleoprotein filament can invade a homologous double-stranded DNA molecule. This is followed by DNA synthesis to extend the end of the

Table 19.1 HR and NHEJ genes in budding yeast, vertebrates and *Arabidopsis*

Budding yeast	Vertebrates	*Arabidopsis*	Function
HR proteins - common			
Mre11	Mre11	AtMre11 (Hartung and Puchta 1999)	DSB end processing
Rad50	Rad50	AtRad50 (Gallego et al. 2001)	DSB end processing
Xrs2	NBS	AtNBS1 (Waterworth et al. 2007)	DSB end processing
Rad51	Rad51	AtRad51 (Doutriaux et al. 1998)	Strand invasion
Rad52	Rad52	–	Recombination mediator
Rad55	Rad51B	AtRad51B (Osakabe et al. 2002)	Recombination mediator
Rad57	Rad51C	AtRad51C (Osakabe et al. 2002)	Recombination mediator; Holliday junction resolution
	Rad51D	AtRad51D (Bleuyard et al. 2005)	Recombination mediator
	XRCC2	AtXRCC2 (Bleuyard et al. 2005)	Recombination mediator
	XRCC3	AtXRCC3 (Osakabe et al. 2002)	Holliday junction resolution
–	BRCA2	AtBRCA2a, AtBRCA2b (Siaud et al. 2004)	Recombination mediator; Rad51/ssDNA nucleofilament assembly
Meiosis			
Spo11	Spo11	AtSpo11-1 (Grelon et al. 2001), AtSpo11-2 (Stacey et al. 2006)	Meiotic recombination initiation
–	Mei1	AtPRD1 (De Muyt et al. 2007)	DSB end processing
Com1/Sae2	CtIP	AtCom1/AtSae2/AtGR1 (Uanschou et al. 2007)	DSB end processing
Dmc1	Dmc1	AtDmc1 (Klimyuk and Jones 1997)	Strand invasion
Mnd1	Mnd1	Mnd1 (Kerzendorfer et al. 2006)	Meiotic homologous pairing
Hop2	Hop2	AHP2 (Schommer et al. 2003)	Meiotic homologous pairing
Signal transduction			
Tel1	ATM	AtATM (Garcia et al. 2003)	Signalling kinase in the response to DNA damage
Mec1	ATR	AtATR (Culligan et al. 2004)	Signalling kinase in the response to DNA stalled replication fork
γ-H2AX	γ-H2AX	γ-H2AX (Friesner et al. 2005)	DNA damage recognition

invading strand. The Rad54 protein, a member of the SWI2/SNF2 family of ATP-dependent chromatin remodelling factors, plays diverse roles throughout HR processes, including stabilization of nucleoprotein filament, facilitation of D-loop formation and heteroduplex extension (Tan et al. 2003). This strand generates a co-joined molecule that contains the double Holliday junction proposed in models of meiotic crossover recombination and DSB repair (Dudás and Chovanec 2004). The interlinked molecules are then processed by branch migration, resolution of the Holliday junction and DNA ligation.

19.2.1   Current Models of HR

Currently, HR is divided into three major sub-pathways. The models of each sub-pathway have been proposed to explain the molecular mechanisms of HR

(Fig. 19.1). These are based largely on work done in yeast (reviewed in Pâques and Haber 1999; Aylon and Kupiec 2004; Krogh and Symington 2004). Regulation of usage of HR sub-pathways promotes genome stability in mitosis and genome variability in meiosis.

19.2.1.1 The Double-Stranded DNA Break-Repair Model

The first model for HR repair is the double-stranded DNA break-repair (DSBR) model (Fig. 19.1, model 1A), which was initially proposed based on observations from studies on yeast transformation using linear DNA fragments that were introduced into the yeast genome by HR (Orr-Weaver and Szostack 1983a, b). This model has since been modified from the original concept, but key features are maintained. These features are multiple enzyme-catalyzed steps (Fig. 19.1, model 1A), including (1) processing of DSB by nucleolytic resection to produce single-stranded 3' overhangs, (2) formation of a nucleoprotein filament on the single-strand DNA ends, (3) homology search and strand invasion into a homologous sequence for D-loop formation, (4) extension of DNA synthesis from the 3' end of the invading strand, (5) capture of the second 3' end of ssDNA by annealing to the extended D-loop and subsequent formation of Holliday junctions (HJs) and (6) resolution of HJs and re-ligation of broken DNA ends. The DSBR model can explain several features of meiotic recombination and linked crossover/gene conversion as different outcomes of HR. In meiosis, recombination occurs between aligned homologous chromosomes, rather than sister chromatids. This promotes genetic mixing between maternal and paternal homologous chromosomes.

19.2.1.2 The Synthesis-Dependent Strand Annealing Model

In contrast to meiotic recombination, mitotic DSB repair via HR may result in low occurrence of crossover events (Ira et al. 2006). This is contrary to the DSBR model, which predicts an equal number of crossover and noncrossover outcomes. The synthesis-dependent strand-annealing (SDSA) model was proposed to account for this phenomenon in mitotic recombination (Strathern et al. 1982; Nassif et al. 1994). The SDSA model is an important mechanism for understanding mitotic recombination. Also, it is able to participate in the formation of gene conversion products upon meiotic recombination. The DSBR and SDSA models share several sequential steps, from processing the DSB to producing single-stranded 3' overhangs, thereby forming the D-loop (Fig. 19.1, model 1B). The key distinct feature of the SDSA model is that the D-loop dissolves after DNA synthesis has taken place, and the disengaged invading strand re-anneals with the other 3' end of the ssDNA. Consequently, this pathway forms gene conversion not associated with crossover. Although SDSA can use a sister chromatid, the homologous chromosome or an ectopic region of homology in the genome as a recombination substrate, in somatic cells it is thought that mainly a sister chromatid located adjacent to the recombination-inducing lesion is used. If the template for gene conversion is located on the sister chromatid, then the repair can restore precisely the original sequence at the break.

19.2.1.3 Single-Strand Annealing Model

When a DSB occurs between the tandemly repeated sequences, repair of the lesion is highly efficient and results in the deletion of one of the repeats as well as the DNA sequence between the direct repeats (Fig. 19.1, model 2). This event, so-called single-strand annealing (SSA), was initially proposed by Lin et al. (1984). In this mechanism, the long 3′ single-stranded overhangs are produced by an exonuclease-catalyzed resection. If these overhangs possess complementary sequences, then the single-stranded DNAs can anneal with each other to form a chimeric double-stranded DNA molecule. Single-stranded tails are removed by nucleases, and the resulting gap and/or nicks are filled in by DNA synthesis and ligation. This process is accompanied by the deletion of one of the repeats and the DNA between the direct repeats. Therefore, this pathway is considered mutagenic.

19.2.1.4 HR Pathways in Plants

All the three HR sub-pathways discussed above are thought to function in higher plants. In meiosis, crossover events are a prerequisite for generating variation in progeny, which is an essence of breeding science. Thus, DSBR-type recombination must occur during meiosis in higher plants. In somatic cells of higher plants, evidence from several lines of study suggested that gene-conversion-type recombination explained by SDSA is the most common type of recombination. In the SDSA model, there is a concern of whether two single-stranded DNAs invade simultaneously (Fig. 19.2. model A), or only one single-stranded DNA participates in invasion (Fig. 19.2, model B). Puchta (1998) suggested that one single end was used efficiently for invasion, and this was called one-sided invasion (OSI). Two types of constructs for gene targeting experiments were used. While one construct showed the homology region at one end, the other contained the homology region at both ends. If the homology at both ends is required for SDSA, then the loss of one homologous end would result in a reduction in HR frequency. However, the HR frequency obtained with the one homologous-ended construct was only threefold lower than for the construct with homology at both ends. Hence, the homology at one end of the DSB is sufficient for efficient HR in plant cells.

Furthermore, the OSI model can be extended to a model accounting for the 'ectopic gene targeting' event occurring in gene targeting experiments in plants (Hanin et al. 2000; Endo et al. 2006a). The ectopic gene targeting recombinant is

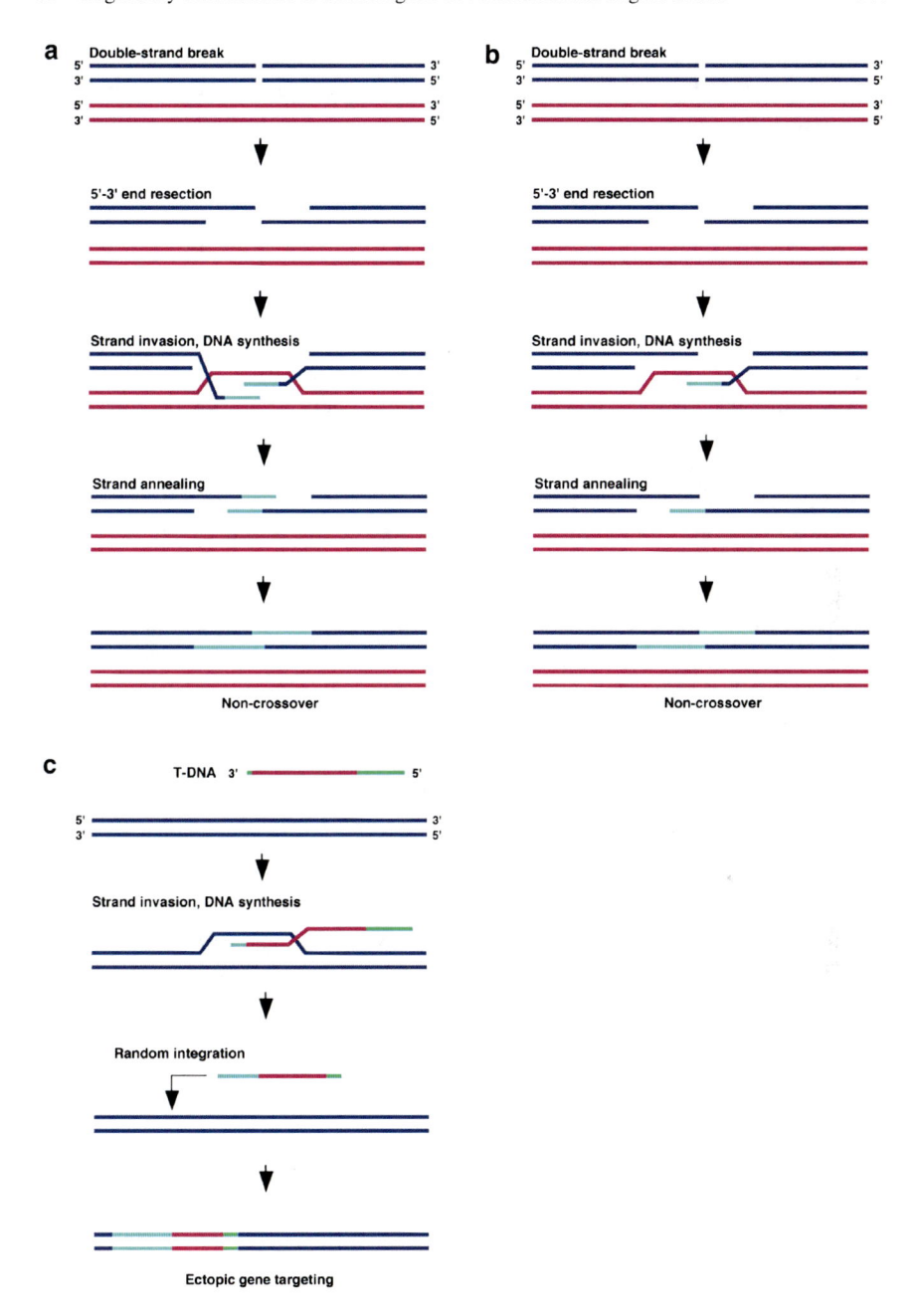

Fig. 19.2 SDSA models. (**A**) Simple SDSA model. Both 3′ ends invade the template, and initiate new DNA synthesis. (**B**) SDSA model with bubble migration. Only one end invades the template DNA. (**C**) Ectopic gene targeting model. Template DNA is copied to T-DNA. The synthesized T-DNA is integrated elsewhere in the genome, without change to the target locus

thought to be an outcome of the event in which the DNA sequence is copied from the target locus to the targeting construct, and the modified construct is integrated elsewhere in the genome. As proposed above, if one single-stranded end of the introduced targeting construct invades the target gene, then DNA synthesis of the invading single-stranded DNA of the construct occurs and, after some DNA synthesis, the modified DNA fragment can be used for an integration event elsewhere in the genome, i.e. other than the target locus (Fig. 19.2, model C). This type of recombination can be overcome by the integration of counter-selection markers in a gene targeting vector outside the homologous regions. Terada et al. (2002) reported that the usage of the diphtheria toxin gene as a counter-selection marker enriched perfect targeting recombinants in a rice gene targeting system.

As mentioned above, meiotic recombination occurs between homologous chromosomes, while recombination between homologous chromosomes occurs rarely in the somatic cells of higher plants. According to Gisler et al. (2002), the frequency of allelic recombination has been estimated to be approximately 10^{-4}. Similar frequencies have been reported in HR repair using ectopic homologous sequences (Shalev and Levy 1997; Puchta 1999).

Sister chromatids can be used as a template for HR in the late S to G2 stage of the cell cycle. This HR pathway works efficiently in mammals. Although there has been limited information regarding HR repair between sister chromatids in higher plants, due to experimental constraints, Li et al. (2008) reported recently that *Mu* transposon-induced DSBs can be efficiently repaired by SDSA using sister chromatids in maize germinal cells and the early stage of somatic cells. In contrast, other studies reported intrachromosomal DSB repair in higher plants (Athma and Peterson 1991; Chiurazzi et al. 1996; Xiao and Peterson 2000; Siebert and Puchta 2002; Orel et al. 2003). Both the SSA and SDSA pathways are thought to be involved in the repair of a break in a directly repeated sequence, resulting in deletion between direct repeats in SSA and gene conversion in SDSA (Fishman-Lobell et al. 1992). Comparison of SSA and SDSA showed that the frequency of the former event was higher than that of the latter by approximately fivefold (Orel et al. 2003). The high frequency of SSA can lead to the establishment of a method for the excision of sequences for biotechnological applications. Successive elimination of marker genes by site-specific recombinases, such as FLP and Cre recombinases, has been reported (reviewed in Hare and Chua 2002). However, the recognition sequences of recombinases are present in the genome. For this reason, a combination of different site-specific recombinases should be applied if multiple genomic changes are required. An alternative system would be DSB repair coupled with highly specific rare-cutting restriction endonucleases, such as I-SceI. It has been demonstrated that a marker gene can be efficiently eliminated from the genome via HR (Siebert and Puchta 2002). Thus, a sequence comprising a selection marker for the selection of putative transformants, a counter-selection marker for enrichment of the deletion event, and a restriction endonuclease regulated under the control of an inducible promoter, flanked by rare restriction sites, can eliminate the required sequence from the plant genome efficiently in one step.

19.3   Meiotic Recombination in Plants

The process of meiosis is highly conserved throughout the eukaryotic kingdom. It occupies a central role in the life cycles of all sexually reproducing organisms. Meiosis differs from mitosis in that one round of DNA replication precedes two sequential cell divisions, meiosis I and meiosis II. As a result, an initial diploid cell can lead to the production of four haploid cells. Prophase I has often been a target of interest, because this stage involves DSB initiation, homology search, pairing, synapsis, recombination and crossover of homologous chromosomes (Hamant et al. 2006). In this section, we discuss recent advances in the genetics of meiotic recombination in *Arabidopsis*, focusing on early events during meiosis. Other aspects of plant meiosis have been reviewed recently (Li and Ma 2006; Hamant et al. 2006; Bogdanov et al. 2007; Mezard et al. 2007).

19.3.1   Meiotic Recombination Initiation

Meiotic recombination is initiated by programmed DSBs catalyzed by the Spo11 protein (Keeney et al. 1997; Keeney 2001). In *Arabidopsis*, AtSpo11-1 and AtSpo11-2 proteins are orthologues of yeast (*Saccharomyces cerevisiae*) Spo11 and are required for meiotic recombination. *atspo11-1* and *atspo11-2* mutants show a sterile phenotype and defects in synapsis and recombination during meiosis (Grelon et al. 2001; Stacey et al. 2006). The double mutant, *atspo11-1/atspo11-2*, shows a similar phenotype to that of the corresponding *atspo11-1* and *atspo11-2* single mutants, indicating that AtSpo11-1 and AtSpo11-2 proteins are required for the same steps in meiotic recombination (Hartung et al. 2007).

19.3.2   DSB Processing

In yeast, in addition to Spo11, nine other proteins (Rad50, Mre11, NBS1/Xrs2, Rec102, Rec104, Rec114, Ski8, Mer2 and Mei4) are required for the formation of meiotic DSBs (Keeney 2001). Three of these proteins (Mre11, Rad50 and NBS1/Xrs2) appear to be widely conserved, but they are also involved in somatic DSB repair (Keeney 2001). In yeast and vertebrates, the MRX(N) complex (MRX in yeast, MRN in vertebrates) is required for the formation of Spo11-induced DSBs and the processing of these DSB ends (Smith and Nicolas 1998). The functions of Mre11, Rad50 and NBS1 have been studied in *Arabidopsis*. Both *atmre11* and *atrad50* mutants are sterile and display defects in synapsis and chromosome fragmentation during meiosis (Gallego et al. 2001; Bleuyard et al. 2004a; Puizina et al. 2004). AtMre11 and AtRad50 are physically interactive (Daoudal-Cotterell et al. 2002). On the other hand, *atnbs1* mutants show normal fertility, and AtNBS1 proteins have been shown to interact with AtMre11 (Waterworth et al. 2007). These results suggest that the function of the MRN(X) complex in meiotic DSB processing is conserved between yeast and plants.

Although *Arabidopsis* possesses a Ski8 structural homologue, this protein is not required for meiotic recombination (Jolivet et al. 2006). Rec102, Rec104, Rec114, Mer2 and Mei4 are not conserved across the kingdoms, at least at the amino acid level, and their counterparts in *Arabidopsis* have not yet been identified. Recently, new members of the family of factors required for the initiation of meiotic recombination have been characterized in *Arabidopsis*. *Arabidopsis* PRD1 has been shown to be required for meiotic DSB formation, and interacts with AtSpo11-1. It was suggested that AtPRD1 could be a functional homologue of the vertebrate Mei1 protein, but the yeast homologue has not been detected in the databases (De Muyt et al. 2007). The other protein, the Com1/Sae2 homologue, has been identified recently in many eukaryotes, including plants (Prinz et al. 1997; Uanschou et al. 2007). Interestingly, the *Arabidopsis COM1* gene was originally identified as being upregulated in response to ionizing irradiation, and was designated as *AtGR1* (*Arabidopsis* gamma response gene 1; Deveaux et al. 2000). *atcom1-1* mutants show a sterile phenotype and display defects in synapsis and chromosome fragmentation during meiosis. Chromosome fragmentation depends on AtSpo11-1 but not the AtDmc1 protein, which is a homologue of the bacterial recombinase RecA (Uanschou et al. 2007), indicating that AtCom1 acts downstream of AtSpo11-1 and upstream of AtDmc1 during meiosis.

19.3.3 Strand Invasion

Single-strand DNA tails created by Spo11 and the MRN complex search for and invade a homologous sequence. Rad51 and Dmc1, which are homologues of the bacterial recombinase RecA, catalyze this step. Rad51 functions in both mitotic and meiotic recombination, whereas Dmc1 is meiosis-specific (Bishop et al. 1992; Shinohara et al. 1992). Both proteins have been identified in *Arabidopsis*, and the corresponding mutants have been characterized (Klimyuk and Jones 1997; Couteau et al. 1999; Li et al. 2004).

The *atrad51-1* mutant shows a completely sterile phenotype and displays defects in synapsis and chromosome fragmentations during meiosis. These chromosome fragmentations observed in the *atrad51-1* mutant depend on AtSpo11-1 (Li et al. 2004). These results indicate that AtRad51 acts downstream of AtSpo11-1 during meiotic recombination. In contrast, *atdmc1* mutants show reduced fertility, but nevertheless produce a small number of seeds. In addition, *atdmc1* mutants are unable to form chiasmata (Couteau et al. 1999). These results indicate that AtDmc1 is required for bivalent formation of chromosome and chromosome segregation during meiotic recombination in *Arabidopsis*. In rice, biochemical analyses of two Dmc1 proteins (Dmc1A and Dmc1B) have been completed (Sakane et al. 2008).

Interestingly, rice Dmc1A and Dmc1B proteins could promote strand exchange in the absence of RPA with long DNA substrates containing several thousand base pairs. In contrast, the human Dmc1 protein strictly required RPA to promote strand exchange with these long DNA substrates. The DNA binding activity of the rice Dmc1A protein is greater than that of the rice Dmc1B protein. The biochemical difference between the two rice Dmc1 proteins may indicate that these proteins have functional differences during meiosis in rice (Sakane et al. 2008).

In addition to AtRad51 and AtDmc1, the five Rad51 paralogues (Rad51B, Rad51C, Rad51D, XRCC2 and XRCC3) identified in vertebrates are also present in the *Arabidopsis* genome, while yeast has only two Rad51 paralogues (Rad55 and Rad57; Osakabe et al. 2002; Bleuyard et al. 2005; Durrant et al. 2007). AtRad51C and AtXRCC3 function in both meiotic and mitotic recombination, whereas AtRad51B, AtRad51D and AtXRCC2 are mitosis-specific. *atrad51C* and *atxrcc3* mutants show a completely sterile phenotype and display chromosome fragmentation during meiosis (Bleuyard and White 2004; Abe et al. 2005; Bleuyard et al. 2005; Li et al. 2005). The chromosome fragmentation observed in *atrad51C* and *atxrcc3* mutants depends on AtSpo11-1 (Bleuyard et al. 2004b; Li et al. 2005). The meiotic phenotypes of *atrad51C* and *atxrcc3* mutants are similar to those of *atrad51-1* mutants. Moreover, AtRad51C interacts with AtXRCC3 in yeast two-hybrid assays (Osakabe et al. 2002). These results indicate that AtRad51C and AtXRCC3 proteins form a heterodimer complex and cooperate with AtRad51 downstream of AtSpo11-1 during meiotic recombination.

Several proteins also cooperate with Dmc1/Rad51 in strand invasion. Homologues of BRCA2, Mnd1 and AHP2 were identified recently as key players in meiotic recombination in *Arabidopsis* (Schommer et al. 2003; Siaud et al. 2004; Kerzendorfer et al. 2006). Despite its absence in yeast, the BRCA2 protein appears to be conserved amongst eukaryotes, including *Arabidopsis*. In humans, mutations of the *BRAC2* gene are associated with susceptibility to breast cancer. BRCA2 facilitates the loading of Rad51 on single-stranded DNA. *Arabidopsis* has two closely related BRCA2 homologues. In *Arabidopsis*, the silencing of the two *AtBRCA2* genes by RNA interference (RNAi) leads to a sterile phenotype and chromosome fragmentation. The chromosome fragmentations observed in *AtBRCA2-RNAi* plants depend on AtSpo11-1 (Siaud et al. 2004). These results indicate that AtBRCA2 acts downstream of AtSpo11-1 during meiosis. In addition, AtBRCA2 interacts with both AtDmc1 and AtRad51 in yeast two-hybrid assays (Siaud et al. 2004). This result indicates that the BRCA2 function of loading Rad51 and Dmc1 on single-stranded DNA is conserved in plants.

In yeast, Mnd1 and Hop2/AHP2 form a complex and function in the early stages of meiotic chromosome pairing and DSB repair. Both proteins have been identified in *Arabidopsis*, and the corresponding mutants have been characterized. *atmnd1* and *atahp2* mutants show a sterile phenotype and display defects in synapsis and chromosome fragmentation during meiosis (Schommer et al. 2003; Kerzendorfer et al. 2006). The chromosome fragmentation observed in *atmnd1* mutants depends on AtSpo11-1. These results indicate that AtMnd1 and AtAHP2 act downstream of AtSpo11-1 during meiotic recombination. In addition, AtMnd1 interacts with

AtAHP2 in yeast two-hybrid assays (Kerzendorfer et al. 2006). These data suggest that the function of the Mnd1-Hop2/AHP2 heterodimer is conserved from yeast to plants.

19.3.4   Crossover Pathway (Resolution of Double Holliday Junction)

After DNA strand invasion into the partner chromatid, DNA synthesis, using the partner as a template, and subsequent ligation form a double Holliday junction (dHJ). The resolution pattern of this dHJ determines the recombination products, i.e. crossovers (COs) or noncrossovers (NCOs; Whitby 2005). In most organisms, COs are distributed non-randomly, such that one CO event inhibits the occurrence of another event in a distance-dependent manner. This phenomenon is known as 'CO interference' (Hillers 2004). In yeast, there are two types of COs. The class I COs are interference sensitive and require the ZMM proteins (Zip1, Zip2, Zip3, Zip4, Mer3, Msh4 and Msh5), Mlh1 and Mlh3, while the class II COs are interference insensitive and require the Mus81 and Mms4 proteins (de los Santos et al. 2003; Whitby 2005).

Experimental evidence indicates that *Arabidopsis*, like yeast and human, also possesses two CO pathways. Recently, several members of the class I CO pathway (e.g. AtZip1, AtZip4, AtMer3, AtMsh4 and AtMsh5) have been identified in *Arabidopsis*, and some corresponding mutants (e.g. *atzip4*, *atmer3*, *atmsh4* and *atmsh5*) have been characterized (Higgins et al. 2004, 2005; Chen et al. 2005; Mercier et al. 2005; Chelysheva et al. 2007; Kuromori et al. 2008; Lu et al. 2008). *atzip4*, *atmer3*, *atmsh4* and *atmsh5* mutants showed a normal phenotype in early meiotic prophase stage, but a reduced number of chiasmata and bivalents (Higgins et al. 2004; Chen et al. 2005; Mercier et al. 2005; Chelysheva et al. 2007; Kuromori et al. 2008; Lu et al. 2008). In addition, the *atzip4* mutant showed reduced class I COs (Mercier et al. 2005; Chelysheva et al. 2007). These results indicate that AtZip4, AtMer3, AtMsh4 and AtMsh5 are involved in the class I COs in *Arabidopsis*.

Recently, the Mus81 protein has been identified in *Arabidopsis*, and the corresponding mutants have been characterized (Berchowitz et al. 2007; Higgins et al. 2008). The *Arabidopsis* genome contains two Mus81 genes, only one being functional, whereas the other is a non-functional pseudo-gene. *atmus81* mutants exhibit a reduction of class II COs (Berchowitz et al. 2007; Higgins et al. 2008), indicating that AtMus81 is involved in class II COs in *Arabidopsis*. Interestingly, two Rad51 paralogues, Rad51C and XRCC3, have been reported to possess resolvase activity for Holliday junctions in vitro, but it is not clear whether resolvase is involved in crossover in *Arabidopsis* (Liu et al. 2004). Other aspects of the crossover pathway in plant meiosis have been reviewed recently (Hamant et al. 2006; Mezard et al. 2007).

19.4   Signal Transduction from DSB to HR Repair

As discussed above, DSBs are required for the initiation of HR. However, un-repaired DSBs are highly toxic for living organisms. A single un-repaired DSB in yeast, even in a dispensable gene, can result in cell death (Bennett et al. 1993). For maintenance of genome integrity and cell survival, it is critical for cells to repair such breaks accurately and faithfully. Thus, living organisms have evolved a signalling network for sensing DNA damage and cell cycle control, as well as repair machinery.

The HR repair pathway responding to DNA damage is regulated by several major steps. Initially, sensor proteins may recognize DNA damage when it occurs, and then send a structural modification signal to mediator proteins. This modification is converted to a compatible form via signal amplification by transducer proteins. Two important protein kinases involved in sensing and signalling DNA damage in eukaryotes are *ataxia telangiectasia* mutated (ATM), and *ataxia telangiectasia* mutated and Rad3-related (ATR).

ATM activity is induced by DNA damage, particularly DSBs. The phosphory-lation targets of ATM include proteins involved in DNA repair, cell cycle control and apoptosis, histone H2AX, Artemis, MDC1, NBS1, p53, Chk2 and DNA-PKcs (reviewed in Koundrioukoff et al. 2004). In mammals, ATM defects result in IR sensitivity, prevent activation of ATM-interacting proteins (Shiloh 2003) and affect meiosis (Barlow et al. 1998). *Arabidopsis atm* mutants are hypersensitive to γ-radiation and methylmethane sulfonate, but not to UV-B light. Consistent with the radiation sensitivity, *atm* mutants failed to induce transcription of genes involved in the repair and/or detection of DNA breaks upon irradiation. Transcrip-tional induction of AtRAD51, AtPARP1, AtGR1, AtLIG4 by IR is defective in *atm* mutants (Garcia et al. 2003). ATR, an essential gene in mammals, is involved in the recognition of stalled replication forks. In mouse, ATR defects can lead to early embryonic lethality, while loss of ATR in conditional knockout embryonic stem cells causes death (Brown and Baltimore 2000; de Klein et al. 2000). *Arabidopsis atr* mutants are viable, fertile and phenotypically normal in the absence of exoge-nous DNA damaging agents (Culligan et al. 2004), but they are hypersensitive to hydroxyurea (HU), aphidicolin and UV-B light, although only slightly sensitive to γ-radiation.

One of the earliest known responses at DNA break sites is the phosphorylation of thousands of molecules of the histone variant H2AX (Rogakou et al. 1998). This phosphorylated H2AX (γ-H2AX) is associated with the retention of repair and signalling factor complexes at the sites of DNA damage. Friesner et al. (2005) investigated the ATR- and ATM-dependent formation of γ-H2AX foci in M-phase cells of *Arabidopsis atm* and *atr* mutants exposed to IR. It was observed that the majority of these foci were ATM-dependent, whereas only ~10% of IR-induced γ-H2AX foci required functional ATR. These results indicate that a distinct subset of IR-induced damage was recognized by ATR in *Arabidopsis* even in the absence of DNA replication. γ-H2AX forms in an ATM-dependent manner in response to

DSB-inducing agents such as X-rays, etoposide and bleomycin, but not UV light or methylmethane sulfonate agents that induce primarily DNA dimmers and alkyl adducts respectively (Burma et al. 2001). In contrast, H2AX is phosphorylated in an ATR-dependent manner when DNA synthesis is blocked. ATR-mediated γ-H2AX foci were detected in S-phase cells in response to hydroxyurea or UV light, but only a few non-replicating G1 cells had γ-H2AX foci under the same treatment (Ward and Chen 2001).

HR repair is proficient mainly during the late S to G2 phases of the cell cycle (Takata et al. 1998), perhaps because of availability of the template (sister chromatid) and DSB end processing regulated by CDK (Ira et al. 2004), as well as Ctp1 protein (Limbo et al. 2007; Sartori et al. 2007). Ctp1 is required for the function of Mre11 in the 5' to 3' resection of DSBs ends in *Scizosaccharomyces pombe* (Limbo et al. 2007). This protein is present only in the S/G2 phases of the cell cycle, and not in the G1 phase. Thus, this Ctp1 expression profile might account for the limitation of active HR repair in the S and G2 phase of the cell cycle. Ctp1 is a homologue of Sae2/Com1 in *S. cerevisiae*, CtIP in mammals and *Arabidopsis* gamma response-1 (AtGR1) in *Arabidopsis*. Ctp1 shares a conserved C-terminal domain with AtGR1, which was isolated as a 2.0-kb cDNA encoding an *Arabidopsis* PEST-box protein by screening for genes accumulating after ionizing radiation (Deveaux et al. 2000). It was revealed that basal AtGR1 levels were associated with the progression through mitosis, whereas elevated intracellular levels of AtGR1 seemed to induce changes between the S and M phases of the cell cycle. It was hypothesized that transient radiation-induced AtGR1 accumulation reflected DNA damage-dependent transient cell cycle arrest before mitosis, which was necessary to accomplish DNA repair prior to chromosome segregation and cytokinesis. Furthermore, results of other studies suggest the existence of a DNA damage checkpoint system, which regulates the cell cycle upon DNA damage to repair DSBs in plants. Preuss and Britt (2003) isolated the *Arabidopsis* mutants *suppressors of gamma* (*sog*) that cannot undergo ionizing radiation-induced cell cycle arrest. *sog* exhibit hypersensitivity to hydroxyurea, indicating that these mutants are not proficient in DNA damage repair. De Schutter et al. (2007) reported that the WEE1 gene is not rate-limiting for cell cycle progression under normal growth conditions, but it is a critical target for the ATR-ATM signalling cascade that inhibits the cell cycle upon activation of the DNA integrity checkpoints, coupling mitosis to DNA repair in cells that suffer DNA damage.

The ATM-ATR pathway is highly conserved, and the mechanisms of signal transduction and checkpoints appear to be shared between mammals and plants. However, cell cycle regulation in plants and mammals may be differed after the checkpoint response. In some cases, G2 arrest by DSBs does not induce cell death, but it may trigger the transition from the mitotic cycle to endocycle in plants. Endoreduplication is a variant of the cell cycle, in which cells cease dividing, although cells continue to grow and replicate DNA (D'Amato 1964; Joubès and Chevalier 2000; Sugimoto-Shirasu and Roberts 2003). As a result, the ploidy of cells increases.

In plants and mammals, different cell cycle responses have been demonstrated in mutants of chromatin assembly factor 1 (CAF-1), which is involved in nucleosome assembly following DNA replication (Smith and Stillman 1989, 1991; Shibahara and Stillman 1999; Tagami et al. 2004) and nucleotide excision repair (Ridgeway and Almouzni 2000). The CAF-1 complex consists of three subunits, these being FAS1, FAS2 and MSI1 in *Arabidopsis* (Kaya et al. 2001; Henning et al. 2003) and p150, p60 and p48 in mammals (Smith and Stillman 1989). *fas1* and *fas2* mutants of *Arabidopsis* were originally described as mutations causing stem fasciation, and abnormal phyllotaxy, leaf shape, root growth and flower organ number (Reinholz 1966; Leyser and Furner 1992). Other than alterations in postembryonic development, these mutants are viable and fertile. CAF-1-deficient *Arabidopsis* mutants showed increased levels of DNA DSBs and frequency of HR, retardation of G2 phase and increased level of ploidy (Endo et al. 2006b). G2 retardation by checkpoint may bypass the mitotic cycle to endocycle in *Arabidopsis fas* mutants. In vertebrates, CAF-1 deficiencies exhibit more severe phenotypes, while depletion of CAF-1 in human cells can lead to cell death. Ectopic expression of a dominant-negative form of the p150 subunit of CAF-1 caused severe early developmental defects in *Xenopus laevis* (Quivy et al. 2001), and induced S-phase arrest accompanied by DNA damage and S-phase checkpoint activation in human cells (Ye et al. 2003). In human cells, knockdown of the p60 subunit of CAF-1 by RNAi has resulted in death of proliferating cells, but not in quiescent cells (Nabatiyan and Krude 2004).

19.5 Conclusions

Elucidation of the HR mechanism in higher plants is important not only for understanding the fundamental mechanisms of this repair system, but also for the application of HR in molecular breeding. Currently, reliable gene targeting (GT) systems have been developed for *Arabidopsis* (Hanin et al. 2000; Endo et al. 2006a) and rice (Terada et al. 2002, 2007; Endo et al. 2007), but the technology is not applicable to other plant species. Although Shaked et al. (2005) reported that overexpression of the yeast *RAD54* gene in *Arabidopsis* resulted in an increased GT efficiency, the mechanism of GT upregulation is not clear. The homologous recombination mechanism, especially upregulation of the HR-dependent phase of the cell cycle and induction by DSBs, needs to be clarified. This knowledge may enhance the efficiency of gene targeting in plants. Future study should focus on the regulation of crossover/noncrossover and crossover interference in meiotic recombination. A better understanding of these mechanisms may facilitate the establishment of systems combining desired genetic recombination with classical breeding approaches.

Acknowledgements The authors thank colleagues for stimulating discussions, and acknowledge financial support of a PROBRAIN (Program for Promotion of Basic Research Activities for

Innovative Biosciences) grant to S.T. from the Bio-Oriented Technology Research Advancement Institution (BRAIN) of Japan, and of grants from the Ministry of Agriculture, Forestry and Fishery of Japan and budget for Nuclear Research from the Ministry of Education, Culture, Sports, and Technology of Japan. M.E. was also supported by a fellowship from Grant-in-Aid for JSPS (Japan Society for the Promotion of Science).

References

Abe K, Osakabe K, Nakayama S, Endo M, Tagiri A, Todoriki S, Ichikawa H, Toki S (2005) *Arabidopsis RAD51C* gene is important for homologous recombination in meiosis and mitosis. Plant Physiol 139:896–908

Athma P, Peterson T (1991) *Ac* induces homologous recombination at the maize *P* locus. Genetics 128:163–173

Aylon Y, Kupiec M (2004) New insights into the mechanism of homologous recombination in yeast. Mutat Res 566:231–248

Barlow C, Liyanage M, Moens PB, Tarsounas M, Nagashima K, Brown K, Rottinghaus S, Jackson SP, Tagle D, Ried T, Wynshaw-Boris A (1998) Atm deficiency results in severe meiotic disruption as early as leptonema of prophase I. Development 125:4007–4017

Bennett CB, Lewis AL, Baldwin KK, Resnick MA (1993) Lethality induced by a single site-specific double-strand break in a dispensable yeast plasmid. Proc Natl Acad Sci USA 90:5613–5617

Berchowitz LE, Francis KE, Bey AL, Copenhaver GP (2007) The role of *AtMUS81* in interference-insensitive crossovers in *A. thaliana*. PLoS Genet 3:e132

Bergerat A, de Massy B, Gadelle D, Varoutas PC, Nicolas A, Forterre P (1997) An atypical topoisomerase II from *Archaea* with implications for meiotic recombination. Nature 386: 414–417

Bishop DK, Park D, Xu L, Kleckner N (1992) DMC1: a meiosis-specific yeast homolog of *E. coli* recA required for recombination, synaptonemal complex formation, and cell cycle progression. Cell 69:439–456

Bleuyard JY, White CI (2004) The *Arabidopsis* homologue of Xrcc3 plays an essential role in meiosis. EMBO J 23:439–449

Bleuyard JY, Gallego ME, White CI (2004a) Meiotic defects in the *Arabidopsis* rad50 mutant point to conservation of the MRX complex function in early stages of meiotic recombination. Chromosoma 113:197–203

Bleuyard JY, Gallego ME, White CI (2004b) The *atspo11-1* mutation rescues *atxrcc3* meiotic chromosome fragmentation. Plant Mol Biol 56:217–224

Bleuyard JY, Gallego ME, Savigny F, White CI (2005) Differing requirements for the *Arabidopsis* Rad51 paralogs in meiosis and DNA repair. Plant J 41:533–545

Bogdanov YF, Grishaeva TM, Dadashev SY (2007) Similarity of the domain structure of proteins as a basis for the conservation of meiosis. Int Rev Cytol 257:83–142

Brown E, Baltimore D (2000) ATR disruption leads to chromosomal fragmentation and early embryonic lethality. Genes Dev 15:397–402

Burma S, Chen BP, Murphy M, Kurimasa A, Chen DJ (2001) ATM phosphorylates histone H2AX in response to DNA double-strand breaks. J Biol Chem 276:42462–42467

Burma S, Chen BP, Chen DJ (2006) Role of non-homologous end joining (NHEJ) in maintaining genomic integrity. DNA Repair 5:1042–1048

Chelysheva L, Gendrot G, Vezon D, Doutriaux MP, Mercier R, Grelon M (2007) Zip4/Spo22 is required for class I CO formation but not for synapsis completion in *Arabidopsis thaliana*. PLoS Genet 3:e83

Chen C, Zhang W, Timofejeva L, Gerardin Y, Ma H (2005) The *Arabidopsis ROCK-N-ROLLERS* gene encodes a homolog of the yeast ATP-dependent DNA helicase MER3 and is required for normal meiotic crossover formation. Plant J 43:321–334

Chiurazzi M, Ray A, Viret JF, Perera R, Wang XH, Lloyd AM, Signer ER (1996) Enhancement of somatic intrachromosomal homologous recombination in *Arabidopsis* by the HO endonuclease. Plant Cell 8:2057–2066

Couteau F, Belzile F, Horlow C, Grandjean O, Vezon D, Doutriaux MP (1999) Random chromosome segregation without meiotic arrest in both male and female meiocytes of a *dmc1* mutant of *Arabidopsis*. Plant Cell 11:1623–1634

Culligan KM, Tissier A, Britt AB (2004) ATR regulates a G2-phase cell-cycle checkpoint in *Arabidopsis thaliana*. Plant Cell 16:1091–1104

D'Amato F (1964) Endopolyploidy as a factor in plant tissue development. Caryologia 17:41–52

Daoudal-Cotterell S, Gallego ME, White CI (2002) The plant Rad50-Mre11 protein complex. FEBS Lett 516:164–166

de Klein A, Muijtjens M, van Os R, Verhoeven Y, Smit B, Carr AM, Lehmann AR, Hoeijmakers JH (2000) Targeted disruption of the cell-cycle checkpoint gene ATR leads to early embryonic lethality in mice. Curr Biol 20:188–189

de los Santos T, Hunter N, Lee C, Larkin B, Loidl J, Hollingsworth NM (2003) The Mus81/Mms4 endonuclease acts independently of double-Holliday junction resolution to promote a distinct subset of crossovers during meiosis in budding yeast. Genetics 164:81–94

De Muyt A, Vezon D, Gendrot G, Gallois JL, Stevens R, Grelon M (2007) AtPRD1 is required for meiotic double strand break formation in *Arabidopsis thaliana*. EMBO J 26:4126–4137

De Schutter K, Joubès J, Cools T, Verkest A, Corellou F, Babiychuk E, Van Der Schueren E, Beeckman T, Kushnir S, Inzé D, De Veylder L (2007) *Arabidopsis* WEE1 kinase controls cell cycle arrest in response to activation of the DNA integrity checkpoint. Plant Cell 19:211–225

Deveaux Y, Alonso B, Pierrugues O, Godon C, Kazmaier M (2000) Molecular cloning and developmental expression of AtGR1, a new growth-related *Arabidopsis* gene strongly induced by ionizing radiation. Rad Res 154:355–364

Doutriaux MP, Couteau F, Bergounioux C, White C (1998) Isolation and characterisation of the *RAD51* and *DMC1* homologs from *Arabidopsis thaliana*. Mol Gen Genet 257:283–291

Dudás A, Chovanec M (2004) DNA double-strand break repair by homologous recombination. Mut Res 566:131–167

Durrant WE, Wang S, Dong X (2007) *Arabidopsis* SNI1 and RAD51D regulate both gene transcription and DNA recombination during the defense response. Proc Natl Acad Sci USA 104:4223–4227

Endo M, Osakabe K, Ichikawa H, Toki S (2006a) Molecular characterization of true and ectopic gene targeting events at acetolactate synthase gene in *Arabidopsis*. Plant Cell Physiol 47: 372–379

Endo M, Ishikawa Y, Osakabe K, Nakayama S, Kaya H, Araki T, Shibahara K, Abe K, Ichikawa H, Valentine L, Hohn B, Toki S (2006b) Increased frequency of homologous recombination and T-DNA integration in *Arabidopsis* CAF-1 mutants. EMBO J 25:5579–5590

Endo M, Osakabe K, Ono K, Handa H, Shimizu T, Toki S (2007) Molecular breeding of a novel herbicide-tolerant rice by gene targeting. Plant J 52:157–166

Essers J, van Steeg H, de Wit J, Swagemakers SM, Vermeij M, Hoeijmakers JH, Kanaar R (2000) Homologous and non-homologous recombination differentially affect DNA damage repair in mice. EMBO J 19:1703–1710

Fishman-Lobell J, Rudin N, Haber JE (1992) Two alternative pathways of double-strand break repair that are kinetically separable and independently modulated. Mol Cell Biol 12: 1292–1303

Friesner JD, Liu B, Culligan K, Britt AB (2005) Ionizing radiation-dependent γ-H2AX focus formation requires ataxia telangiectasia mutated and ataxia telangiectasia mutated and Rad3-related. Mol Biol Cell 16:2566–2576

Galkin VE, Esashi F, Yu X, Yang S, West SC, Egelman EH (2005) BRCA2 BRC motifs bind RAD51-DNA filaments. Proc Natl Acad Sci USA 102:8537–8542

Gallego ME, Jeanneau M, Granier F, Bouchez D, Bechtold N, White CI (2001) Disruption of the *Arabidopsis RAD50* gene leads to plant sterility and MMS sensitivity. Plant J 25:31–41

Garcia V, Bruchet H, Camescasse D, Granier F, Bouchez DL, Tissier A (2003) AtATM is essential for meiosis and the somatic response to DNA damage in plants. Plant Cell 15:119–132

Gisler B, Salomon S, Puchta H (2002) The role of double-strand break-induced allelic homologous recombination in somatic plant cells. Plant J 32:277–284

Grelon M, Vezon D, Gendrot G, Pelletier G (2001) *AtSPO11-1* is necessary for efficient meiotic recombination in plants. EMBO J 20:589–600

Haber JE (1998) Mating-type switching in *Sacharomyces cerevisiae*. Annu Rev Genet 32:561–599

Hamant O, Ma H, Cande WZ (2006) Genetics of meiotic prophase I in plants. Annu Rev Plant Biol 57:267–302

Hanin M, Volrath S, Bogucki A, Briker M, Ward E, Paszkowski J (2000) Gene targeting in *Arabidopsis*. Plant J 28:671–677

Hare PD, Chua NH (2002) Excision of selectable marker genes from transgenic plants. Nature Biotechnol 20:575–580

Hartung F, Puchta H (1999) Isolation of the complete cDNA of the Mre11 homologue of *Arabidopsis* (accession No. AJ243822) indicates conservation of DNA recombination mechanisms between plants and other eukaryotes. Plant Physiol 121:312

Hartung F, Wurz-Wildersinn R, Fuchs J, Schubert I, Suer S, Puchta H (2007) The catalytically active tyrosine residues of both SPO11-1 and SPO11-2 are required for meiotic double-strand break induction in *Arabidopsis*. Plant Cell 19:3090–3099

Henning L, Taranto P, Walser M, Schönrock N, Gruissem W (2003) *Arabidopsis* MSI1 is required for epigenetic maintenance of reproductive development. Development 130:2555–2565

Higgins JD, Armstrong SJ, Franklin FC, Jones GH (2004) The *Arabidopsis MutS* homolog *AtMSH4* functions at an early step in recombination: evidence for two classes of recombination in *Arabidopsis*. Genes Dev 18:2557–2570

Higgins JD, Sanchez-Moran E, Armstrong SJ, Jones GH, Franklin FC (2005) The *Arabidopsis* synaptonemal complex protein ZYP1 is required for chromosome synapsis and normal fidelity of crossing over. Genes Dev 19:2488–2500

Higgins JD, Buckling EF, Franklin FC, Jones GH (2008) Expression and functional analysis of *AtMUS81* in *Arabidopsis* meiosis reveals a role in the second pathway of crossing-over. Plant J 54:152–162

Hillers KJ (2004) Crossover interference. Curr Biol 14:R1036–R1037

Ira G, Pellicioli A, Balijja A, Wang X, Fiorani S, Carotenuto W, Liberi G, Bressan D, Wan L, Hollingsworth NM, Haber JE, Foiani M (2004) DNA end resection, homologous recombination and DNA damage checkpoint activation require CDK1. Nature 431:1011–1017

Ira G, Satory D, Haber JE (2006) Conservative inheritance of newly synthesized DNA in double-strand break-induced gene conversion. Mol Cell Biol 26:9424–9429

Jolivet S, Vezon D, Froger N, Mercier R (2006) Non conservation of the meiotic function of the Ski8/Rec103 homolog in *Arabidopsis*. Genes Cells 11:615–622

Joubès J, Chevalier C (2000) Endoreduplication in higher plants. Plant Mol Biol 43:735–745

Kanaar R, Hoeijmakers JH, van Gent DC (1998) Molecular mechanisms of DNA double strand break repair. Trends Cell Biol 8:483–489

Kaya H, Shibahara KI, Taoka KI, Iwabuchi M, Stillman B, Araki T (2001) *FASCIATA* genes for chromatin assembly factor-1 in *Arabidopsis* maintain the cellular organization of apical meristems. Cell 104:131–142

Keeney S (2001) Mechanism and control of meiotic recombination initiation. Curr Topics Dev Biol 52:1–53

Keeney S, Neale MJ (2006) Initiation of meiotic recombination by formation of DNA double-strand breaks: mechanism and regulation. Biochem Soc Trans 34:523–525

Keeney S, Giroux CN, Kleckner N (1997) Meiosis-specific DNA double-strand breaks are catalyzed by Spo11, a member of a widely conserved protein family. Cell 88:375–384

Kerzendorfer C, Vignard J, Pedrosa-Harand A, Siwiec T, Akimcheva S, Jolivet S, Sablowski R, Armstrong S, Schweizer D, Mercier R, Schlogelhofer P (2006) The *Arabidopsis thaliana MND1* homologue plays a key role in meiotic homologous pairing, synapsis and recombination. J Cell Sci 119:2486–2496

Klimyuk VI, Jones JD (1997) *AtDMC1*, the *Arabidopsis* homologue of the yeast *DMC1* gene: characterization, transposon-induced allelic variation and meiosis-associated expression. Plant J 11:1–14

Koundrioukoff S, Polo S, Almouzni G (2004) Interplay between chromatin and cell cycle checkpoints in the context of ATR/ATM-dependent checkpoints. DNA Repair 3:969–978

Krogh BO, Symington LS (2004) Recombination proteins in yeast. Annu Rev Genet 38: 233–271

Kuromori T, Azumi Y, Hayakawa S, Kamiya A, Imura Y, Wada T, Shinozaki K (2008) Homologous chromosome pairing is completed in crossover defective *atzip4* mutant. Biochem Biophys Res Commun 370:98–103

Leyser HMO, Furner IJ (1992) Characterization of three shoot apical meristem mutants of *Arabidopsis thaliana*. Development 116:397–403

Li W, Ma H (2006) Double-stranded DNA breaks and gene functions in recombination and meiosis. Cell Res 16:402–412

Li W, Chen C, Markmann-Mulisch U, Timofejeva L, Schmelzer E, Ma H, Reiss B (2004) The *Arabidopsis AtRAD51* gene is dispensable for vegetative development but required for meiosis. Proc Natl Acad Sci USA 101:10596–10601

Li W, Yang X, Lin Z, Timofejeva L, Xiao R, Makaroff CA, Ma H (2005) The *AtRAD51C* gene is required for normal meiotic chromosome synapsis and double-stranded break repair in *Arabidopsis*. Plant Physiol 138:965–976

Li J, Wen TJ, Schnable PS (2008) Role of RAD51 in the repair of *MuDR*-induced double-strand breaks in maize (*Zea mays* L.). Genetics 178:57–66

Lieber MR, Ma Y, Pannicke U, Schwarz K (2004) The mechanism of vertebrate nonhomologous DNA end joining and its role in V(D)J recombination. DNA Repair 3:817–826

Limbo O, Chahwan C, Yamada Y, de Bruin RA, Wittenberg C, Russell P (2007) Ctp1 is a cell-cycle-regulated protein that functions with Mre11 complex to control double-strand break repair by homologous recombination. Mol Cell 28:134–146

Lin FL, Sperle K, Sternberg N (1984) Model for homologous recombination during transfer of DNA into mouse L cells: role for DNA ends in the recombination process. Mol Cell Biol 4:1020–1034

Liu Y, Masson JY, Shah R, O'Regan P, West SC (2004) RAD51C is required for Holliday junction processing in mammalian cells. Science 303:243–246

Lu X, Liu X, An L, Zhang W, Sun J, Pei H, Meng H, Fan Y, Zhang C (2008) The *Arabidopsis MutS* homolog *AtMSH5* is required for normal meiosis. Cell Res 18:589–599

Ma Y, Lu H, Schwarz K, Lieber MR (2005) Repair of double-strand DNA breaks by the human nonhomologous DNA end joining pathway: the iterative processing model. Cell Cycle 4:1193–1200

Mercier R, Jolivet S, Vezon D, Huppe E, Chelysheva L, Giovanni M, Nogue F, Doutriaux MP, Horlow C, Grelon M, Mezard C (2005) Two meiotic crossover classes cohabit in *Arabidopsis*: one is dependent on *MER3*, whereas the other one is not. Curr Biol 15:692–701

Mezard C, Vignard J, Drouaud J, Mercier R (2007) The road to crossovers: plants have their say. Trends Genet 23:91–99

Nabatiyan A, Krude T (2004) Silencing of chromatin assembly factor 1 in human cells leads to cell death and loss of chromatin assembly during DNA synthesis. Mol Cell Biol 24: 2853–2862

Nassif N, Penney J, Pal S, Engels WR, Gloor GB (1994) Efficient copying of nonhomologous sequences from ectopic sites via P-element-induced gap repair. Mol Cell Biol 14:1613–1625

Orel N, Kyryk A, Puchta H (2003) Different pathways of homologous recombination are used for the repair of double-strand breaks within tandemly arranged sequences in the plant genome. Plant J 35:604–612

Orr-Weaver TL, Szostack JW (1983a) Yeast recombination: the association between double-strand gap repair and crossing-over. Proc Natl Acad Sci USA 80:4417–4421

Orr-Weaver TL, Szostack JW (1983b) Multiple, tandem plasmid integration in *Saccharomyces cerevisiae*. Mol Cell Biol 3:747–749

Osakabe K, Yoshioka T, Ichikawa H, Toki S (2002) Molecular cloning and characterization of RAD51-like genes from *Arabidopsis thaliana*. Plant Mol Biol 50:71–81

Pâques F, Haber JE (1999) Multiple pathways of recombination induced by double-strand breaks in *Saccharomyces cerevisiae*. Microbiol Mol Biol Rev 63:349–404

Preuss SB, Britt AB (2003) A DNA-damage-induced cell cycle checkpoint in *Arabidopsis*. Genetics 164:323–324

Prinz S, Amon A, Klein F (1997) Isolation of *COM1*, a new gene required to complete meiotic double-strand break-induced recombination in *Saccharomyces cerevisiae*. Genetics 146: 781–795

Puchta H (1998) Repair of genomic double-strand breaks in somatic plant cells by one-sided invasion of homologous sequences. Plant J 13:331–339

Puchta H (1999) Double-strand break-induced recombination between ectopic homologous sequences in somatic plant cells. Genetics 152:1173–1181

Puchta H, Hohn B (1996) From centiMorgans to base pairs: homologous recombination in plants. Trends Plant Sci 1:340–348

Puizina J, Siroky J, Mokros P, Schweizer D, Riha K (2004) Mre11 deficiency in *Arabidopsis* is associated with chromosomal instability in somatic cells and Spo11-dependent genome fragmentation during meiosis. Plant Cell 16:1968–1978

Quivy JP, Grandi P, Almouzni G (2001) Dimerization of the largest subunit of chromatin assembly factor 1: importance in vitro and during *Xenopus* early development. EMBO J 20:2015–2027

Reinholz E (1966) Radiation induced mutants showing changed inflorescence characteristics. Arabid Inf Serv 3:19–20

Ridgeway P, Almouzni G (2000) CAF-1 and the inheritance of chromatin states: at the cross roads of DNA replication and repair. J Cell Sci 113:2647–2658

Rogakou E, Pich D, Orr A, Ivanova V, Bonner W (1998) DNA double stranded breaks induce histone H2AX phosphorylation on serine 139. J Biol Chem 273:5858–5868

Sakane I, Kamataki C, Takizawa Y, Nakashima M, Toki S, Ichikawa H, Ikawa S, Shibata T, Kurumizaka H (2008) Filament formation and robust strand exchange activities of the rice DMC1A and DMC1B proteins. Nucleic Acids Res 36:4266–4276

Sartori AA, Lukas C, Coates J, Mistrik M, Fu S, Bartek J, Baer R, Lukas J, Jackson SP (2007) Human CtIP promotes DNA end resection. Nature 450:509–514

Schommer C, Beven A, Lawrenson T, Shaw P, Sablowski R (2003) *AHP2* is required for bivalent formation and for segregation of homologous chromosomes in *Arabidopsis* meiosis. Plant J 36:1–11

Shaked H, Melamed-Bessudo C, Levy AA (2005) High-frequency gene targeting in *Arabidopsis* plants expressing the yeast RAD54 gene. Proc Natl Acad Sci USA 102:12265–12269

Shalev G, Levy AA (1997) The maize transposable element Ac induces recombination between the donor site and a homologous ectopic sequence. Genetics 146:1143–1151

Shibahara K, Stillman B (1999) Replication-dependent marking of DNA by PCNA facilitates CAF-1-coupled inheritance of chromatin. Cell 96:575–585

Shiloh Y (2003) ATM and related protein kinases: safeguarding genome integrity. Nature Rev Cancer 3:155–168

Shinohara A, Ogawa H, Ogawa T (1992) Rad51 protein involved in repair and recombination in *S. cerevisiae* is a RecA-like protein. Cell 69:457–470

Siaud N, Dray E, Gy I, Gerard E, Takvorian N, Doutriaux MP (2004) Brca2 is involved in meiosis in *Arabidopsis thaliana* as suggested by its interaction with Dmc1. EMBO J 23:1392–1401

Siebert R, Puchta H (2002) Efficient repair of genomic double-strand breaks by homologous recombination between directly repeated sequences in the plant genome. Plant Cell 14: 1121–1131

Smith KN, Nicolas A (1998) Recombination at work for meiosis. Curr Opin Genet Dev 8:200–211

Smith S, Stillman B (1989) Purification and characterization of CAF-1, a human cell factor required for chromatin assembly during DNA replication *in vivo*. Cell 58:15–25

Smith S, Stillman B (1991) Stepwise assembly of chromatin during DNA replication *in vitro*. EMBO J 10:971–980

Stacey NJ, Kuromori T, Azumi Y, Roberts G, Breuer C, Wada T, Maxwell A, Roberts K, Sugimoto-Shirasu K (2006) *Arabidopsis* SPO11-2 functions with SPO11-1 in meiotic recombination. Plant J 48:206–216

Strathern JN, Klar AJS, Hicks JB, Abraham JA, Ivy JM, Nasmyth KA, McGill C (1982) Homothallic switching of yeast mating type cassettes is initiated by a double-stranded cut in the *MAT* locus. Cell 31:183–192

Sugimoto-Shirasu K, Roberts K (2003) "Big it up": endoreduplication and cell-size control in plants. Curr Opin Plant Biol 6:544–553

Tagami H, Ray-Gallet D, Almouzni G, Nakatani Y (2004) Histone H3.1 and H3.3 complexes mediate nucleosome assembly pathways dependent or independent of DNA synthesis. Cell 116:51–61

Takata M, Sasaki MS, Sonoda E, Morrison C, Hashimoto M, Utsumi H, Yamaguchi-Iwai Y, Shinohara A, Takeda S (1998) Homologous recombination and non-homologous end-joining pathways of DNA double-strand break repair have overlapping roles in the maintenance of chromosomal integrity in vertebrate cells. EMBO J 17:5497–5508

Tan TL, Kanaar R, Wyman C (2003) Rad54, a Jack of all trades in homologous recombination. DNA Repair 2:787–794

Terada R, Urawa H, Inagaki Y, Tsugane K, Iida S (2002) Efficient gene targeting by homologous recombination in rice. Nature Biotechnol 20:1030–1034

Terada R, Johzuka-Hisatomi Y, Saitoh M, Asao H, Iida S (2007) Gene targeting by homologous recombination as a biotechnological tool for rice functional genomics. Plant Physiol 144: 846–856

Thacker J (1999) A surfeit of *RAD51*-like genes? Trends Genet 15:166–168

Uanschou C, Siwiec T, Pedrosa-Harand A, Kerzendorfer C, Sanchez-Moran E, Novatchkova M, Akimcheva S, Woglar A, Klein F, Schlogelhofer P (2007) A novel plant gene essential for meiosis is related to the human *CtIP* and the yeast *COM1/SAE2* gene. EMBO J 26:5061–5070

Ward IM, Chen J (2001) Histone H2AX is phosphorylated in an ATR-dependent manner in response to replicational stress. J Biol Chem 276:47759–47762

Waterworth WM, Altun C, Armstrong SJ, Roberts N, Dean PJ, Young K, Weil CF, Bray CM, West CE (2007) NBS1 is involved in DNA repair and plays a synergistic role with ATM in mediating meiotic homologous recombination in plants. Plant J 52:41–52

Whitby MC (2005) Making crossovers during meiosis. Biochem Soc Trans 33:1451–1455

Xiao YL, Peterson T (2000) Intrachromosomal homologous recombination in *Arabidopsis* induced by a maize transposon. Mol Gen Genet 263:22–29

Ye X, Franco AA, Santos H, Nelson DM, Kaufman PD, Adams PD (2003) Defective S phase chromatin assembly causes DNA damage, activation of the S phase checkpoint, and S phase arrest. Mol Cell 11:341–351

Chapter 20
Synthetic Promoter Engineering

M. Venter and F.C. Botha

20.1 Introduction

Continuous deciphering of the regulatory complexities that govern transcriptional control has allowed for more refined plant genetic engineering strategies. In recent years, advances in plant genomics, high-throughput platforms and computational assistance, combined with the emergence of technologies such as virus induced gene silencing (VIGS) and RNA interference (RNAi), have greatly accelerated novel experimental design strategies. In the midst of these new developments, however, targeted control of gene activity in transgenic research remains a challenge. It is progressively becoming more evident that the accurate dissection and functional interpretation of *cis*-regulatory context within the promoter could facilitate possible prediction of gene expression and might offer clues on the *trans*-acting factors that regulate the sites and levels of gene activity. With special emphasis on the promoter, molecular fine-tuning of *cis*-motif architecture has set the stage for synthetic and predictive biotechnology applications and holds much promise towards enhancing the molecular engineering toolbox for plant geneticists. Numerous other factors and compositional properties affect plant inducible transgene activity. In this chapter, we will focus primarily on, and highlight key examples of, synthetic promoters designed to overcome drawbacks unable to be addressed by conventional wild-type promoters. In addition, the concept of 'bottom-up' experimental design and systematic design strategies for the future development of synthetic promoters, underscoring the importance of in silico assistance, are highlighted.

M. Venter

Department of Genetics, Stellenbosch University, Private Bag X1, Matieland 7602, South Africa
e-mail: mauritz@sun.ac.za

F.C. Botha
BSES, P.O. Box 68, Indooroopilly, 4068 Qld, Australia

Institute for Plant Biotechnology, Stellenbosch University, Private Bag X1, Matieland 7602, South Africa

E-C. Pua and M.R. Davey (eds.),
Plant Developmental Biology – Biotechnological Perspectives: Volume 2,
DOI 10.1007/978-3-642-04670-4_20, © Springer-Verlag Berlin Heidelberg 2010

20.2 Promoters: Biotechnology Tools Combining Molecular 'Switch' and 'Sensor' Capabilities

New developments in recombinant DNA technologies have led to the rapid diversification of plant genetic engineering applications, ranging from conventional characterisation of gene function in basic research to high-level expression and 'fine-tuning' of transgene expression for agricultural, environmental and biopharmaceutical purposes. The following are the major elements that play an important part in experimental design: (1) the specific transgene of interest, (2) the transformation procedure, (3) selectable markers, (4) cellular targeting, (5) the promoter, (6) enhancers and (7) gene silencing. For high-level constitutive expression, or precise control of transgene activity in response to a specific cue, promoters are the key for successful genetic engineering strategies (Yoshida and Shinmyo 2000; Lessard et al. 2002; Müller and Wassenegger 2004; Gurr and Rushton 2005). Numerous other elements such as introns, $5'$- and $3'$- untranslated leader sequences, ribosomal DNA and polyadenylation signals within the *cis*-functional context influence transgene inducibility and levels of expression. This discussion will be limited to the current understanding of the factors that control the function of the core promoter and the role of the $5'$-upstream *cis*-regulatory architecture. Both these issues are important for accurate and refined synthetic promoter design.

20.2.1 The Promoter

The eukaryotic promoter is a unique stretch of DNA sequence consisting of a core, proximal and distal region (Fig. 20.1). The core promoter usually comprises distinct elements, such as the TATA-box (consensus TATAAA sequence), Initiator (Inr) and/or a downstream promoter element (DPE). Transcription is initiated by RNA polymerase II and an orchestrated assembly of general TFs (TFIIA, TFIIB, TFIID,

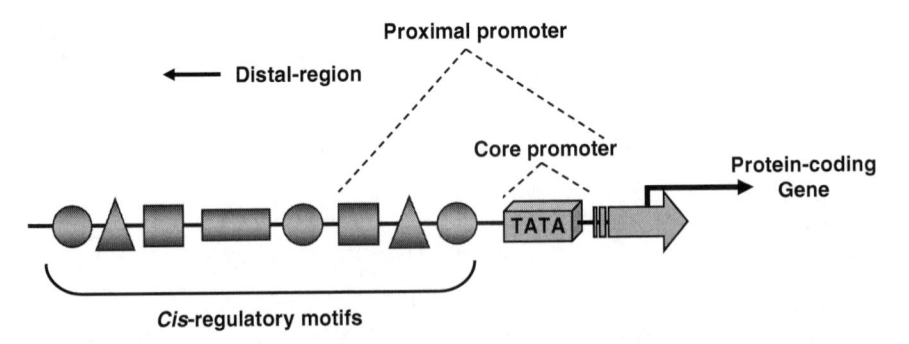

Fig. 20.1 Simplified diagrammatic representation of the eukaryotic promoter containing TATA-box and $5'$-upstream-specific configuration of *cis*-motifs

TFIIE, TFIIF and TFIIH) to form the so-called pre-initiation complex (PIC) around the transcriptional start site. Stepwise formation of the PIC of TATA-box containing promoters is facilitated by initial interaction of TFIID containing a so-called TATA-box-binding protein (TBP; Novina and Roy 1996; Smale and Kadonaga 2003). Conversely, a new core promoter element, XCPE1 (X gene core promoter element 1), has recently been discovered in TATA-less promoters, directing TBP-dependent, but TFIID-independent, RNA polymerase II transcriptional initiation (Tokusumi et al. 2007). The proximal and distal promoter regions include a compact arrangement or combination of *cis*-regulatory sequences (motifs or TF-binding sites; Fig. 20.1) that, when bound to specific *trans*-acting transcription factor (TF) proteins, can modulate the regulation and activity of the adjacent protein-coding gene (reviewed in Novina and Roy 1996; Butler and Kadonaga 2002; Smale and Kadonaga 2003; Heintzman and Ren 2007). The simplistic view of a 'universal' eukaryotic core-promoter structure has been challenged, and several new evidence-based developments are continuously shedding light on the diversity of core-promoter architecture and selectivity (reviewed in Müller et al. 2007). The core promoter is not only the key determinant for basal transcription, but it also contributes significantly to transcriptional regulation and tissue-specificity (Smale 2001; Hochheimer and Tjan 2003). Although several other transcriptional co-regulators (i.e. co-activators, co-repressors, mediator and chromatin modulators) contribute to the complexity of transcriptional initiation and regulation, any wild-type or modified promoters of plant, bacterial, mammalian or viral origin need to conform to the unified 'rules' of eukaryotic gene regulation (Orphanides and Reinberg 2002) in a plant genetic engineering strategy.

20.2.2   Spatial and Temporal Control of Transgene Activity in Plants

Transcriptional regulation in plants is a combination and synergistic interplay of *cis*- and *trans*-acting factors responsible for a diverse array of gene expression events within a specific tissue or organ, during a particular condition (e.g. development, differentiation) or in response to environmental stimuli (Singh 1998; Schwechheimer and Bevan 1998). This well-established view on gene regulation and characterisation of promoter architecture has allowed elucidation of *cis*-regulatory sequences or modules (specific arrangement of *cis*-motifs) that regulate the site and level of transgene activity in response to plant and non-plant signals (Guilfoyle 1997; Reynolds 1999; Müller and Wassenegger 2004). In recent years, several genetic engineering strategies have incorporated a wide variety of promoters of plant, viral and bacterial origin to confer constitutive, tissue-specific and/or inducible transgene expression in both dicotyledons and monocotyledons (Yoshida and Shinmyo 2000; Lessard et al. 2002; Müller and Wassenegger 2004). There are two major types of promoter, namely wild-type (unmodified/native or truncated) and synthetic (combining wild-type or newly synthesized core-promoter structure with multiple repeats, specific combinations or rearrangements of plant or non-plant *cis*-regulatory elements). Promoters are categorized as inducible, constitutive, tissue- and developmental-stage specific. Inducible promoters can be regulated by (1) hormones, i.e. gibberellin, auxin, ethylene and methyl jasmonate, (2) biotic and abiotic stresses, i.e. drought, salinity, wounding, insects, microbes and pathogens, (3) environmental stimuli, i.e. temperature and light, and (4) chemicals, i.e. copper, ethanol, tetracycline and dexamethasone. Depending on the biotechnological application, both high-level constitutive and targeted transgene-inducible expression systems are of considerable importance and are therefore major focus areas for future development of synthetic promoters.

20.2.3   Cauliflower Mosaic Virus 35S: the 'Workhorse' Promoter in Plant Biotechnology

Currently, the choice and availability of promoter is still relatively limited. In recent years, only a few promoters, and recombinant derivatives thereof, have been used in biotechnological applications to control transgene expression. Modification made to these promoters includes incorporation of compositional elements such as introns, untranslated leader sequences and *cis*-regulatory enhancers to ensure either constitutive or inducible expression in response to various environmental, physical or chemical cues (Yoshida and Shinmyo 2000; Lessard et al. 2002; Müller and Wassenegger 2004). Some of the characterised promoters that have been used in a plethora of fundamental plant studies and biotechnological applications include the cauliflower mosaic virus (CaMV 35S; Odell et al. 1985), maize alcohol dehydrogenase-1 (Adh-1; Ellis et al. 1987), maize ubiquitin (Ubi-1; Christensen et al. 1992), phenylalanine ammonia-lyase (PAL; Ohl et al. 1990), chalcone synthase (CHS8; Schmid et al. 1990), patatin (Liu et al. 1990) and nopaline synthase (nos; An et al. 1990) promoters, constituting specific, or a combination of, expression features in dicot- and monocotyledon plant cells. Among them, the CaMV 35S promoter has proven its staying power. The CaMV 35S promoter confers high-level constitutive or near-constitutive transgene activity and is one of the most widely used promoters in plant molecular biological research. Since its original description and characterisation in the early 1980s (Franck et al. 1980; Guilley et al. 1982; Odell et al. 1985), several extensive investigations have reported on functional subdomain dissection and modification, comparative promoter analysis and transgenic risk assessment of the CaMV 35S promoter (Kay et al. 1987; Fang et al. 1989; Pietrzak et al. 1989; Benfey and Chua 1990; Lam 1994; Holtorf et al. 1995; Mitsuhara et al. 1996; Hull et al. 2002). The CaMV 35S promoter region is divided into so-called subdomains, of which the first comprises a core or minimal promoter of −46 bp relative to the transcriptional start site, followed by a set of six subdomains (designated A1, B1, B2, B3, B4 and B5) upstream from −46 bp to −343 bp, which confer a variety of gene expression patterns when modified in combinatorial gain-of-function studies (Benfey and Chua 1990; Benfey et al. 1990). More recently, results from two studies re-evaluating the combinatorial and functional subdomain organization of the CaMV 35S promoter revealed that modification or rearrangement (swapping) of 35S domains could aid in the future development of a variety of synthetic CaMV 35S promoters with minimum sequence homology, yet able to constitutively drive transgene activity equivalent to the unmodified wild-type CaMV 35S promoter (Bhullar et al. 2003, 2007). The CaMV 35S promoter serves as a valuable molecular model for functional *cis*-engineering in plant synthetic promoters.

20.2.4 Hurdles that Necessitate Promoter Modification

The need for development of a wide range of modified and synthetic promoters is spurred on by the inability of current conventional strategies using wild-type promoters to address particular challenges in specific biotechnological applications. Strong constitutive transgene expression will always remain a desirable trait and efforts for isolation or modification of alternative promoters, comparable or even better than currently available promoters such as CaMV 35S, are continuing (Maiti and Shepard 1998; Foster et al. 1999; Schenk et al. 2001; Bhattacharyya et al. 2002; Schünmann et al. 2003; Cazzonelli et al. 2005; Xiao et al. 2005). In addition, with special focus on elucidation of *cis*-regulatory function and inducible transgene expression, strategies concerning the precise spatial and temporal control of transgene activity have attracted much attention and several elegant systems have been developed (see Sect. 20.3). Contributing to the complexity of promoter characterisation and inducibility in transgenic research, however, is the well-known fact that spatial and temporal activity of transgene expression may vary due to positional integration into the chromatin (also known as 'position effect'), which leads to variable transgene regulation among independent transgenic lines and individual transformants (Dean et al. 1988; Van Leeuwen et al. 2001). Chromatin integration plays an important part in plant epigenetics, and several chromatin remodelling mechanisms and DNA methylation may interfere with the transcriptional machinery, leading to unreliable transgene expression and, ultimately, transgene silencing (Matzke and Matzke 1998a, b; Meyer 2000, 2001; Richards and Elgin 2002). In contrast to the advances and advantageous applications of gene silencing in plant genetics, the greatest hurdle with regard to high-level transgene expression and/or targeted inducibility for industrial, biopharmaceutical and metabolic engineering purposes is homology-dependent gene silencing (HDGS), which occurs on a transcriptional and posttranscriptional level (reviewed in Meyer and Saedler 1996; Kooter et al. 1999; Fagard and Vaucheret 2000; De Wilde et al. 2000). HDGS is thought to be caused by (1) a transgene or promoter sharing homology with an endogenous gene or promoter, (2) two homologous transgenes and/or (3) the repeated use of the same promoter (Kooter et al. 1999; De Wilde et al. 2000).

HDGS combined with other factors, such as specific and/or combinatorial inducibility of plant transgenes (e.g. in response to environmental stimuli or pathogen attack) and multiple-transgene expression (also known as stacking or pyramiding), which are of considerable interest and one of the key priorities in modern plant biotechnology (François et al. 2002; Halpin 2005), are major challenges currently facing plant genetic engineering strategies. To overcome the limitations of conventional wild-type promoter usage and predictive transgene expression, there are four major objectives for the development of synthetic promoters (uni-or bidirectional) that are tailor-made for particular biotechnological applications: firstly, to increase promoter availability, secondly, to control the expression of multiple transgenes, thirdly, to prevent homology-dependent gene silencing (HDGS) and fourthly, to ensure more refined control of transgene expression in a tissue- and environment-specific manner.

20.3 Synthetic Promoters: Refinement of *cis*-Regulatory Architecture Leads to Targeted Inducibility and High-Level Expression of Single or Multiple Transgenes

To gain insight in the complexity and rearrangement of *cis*-motif regulatory logic and *trans*-interaction in response to specific exogenous or endogenous signals, the majority of synthetic promoter systems have been developed and evaluated implementing reporter genes such as luciferase (LUC), green fluorescent protein (GFP), β-glucuronidase (GUS) and/or chloramphenicol acetyltransferase (CAT). An optimized synthetic system or module, able to control single or multiple transgene activity in a quantitative, spatial and temporal mode, needs to conform to specific requirements such as versatility, inducer specificity, minimal basal (background) transcriptional activity from the core promoter, minimal depletion of endogenous TFs (also known as TF 'sequestration'), rapid induction response, increased promoter strength and transgene induction without pleiotropic effect on plants (Wang et al. 2003; Gurr and Rushton 2005; Tavva et al. 2006). Depending on the application and synthetic *cis*-engineered design strategy, several systems may not satisfy all these criteria. Nevertheless, synthetic promoters are powerful molecular tools and, given the flexibility of transgene control combined with an extensive range of optional design strategies, it is increasingly evident that synthetic promoter engineering may contribute significantly to future biotechnology applications and elucidation of gene function in basic research (Moore et al. 2006; Venter 2007). Combined with strong constitutive transgene expression, several applications in plant transgenic research require regulatory systems conferring inducible control of transgene activity with a high degree of specificity and/or versatility. In this context, we have highlighted specific investigations with special emphasis on *cis*-motif modification and two-component transactivation, which serve as key examples for synthetic promoter strategies.

20.3.1 cis-*Motif Context Modified: the Centre for Synthetic Promoter Engineering*

The primary component of a synthetic promoter is the region known as a regulatory module, fused upstream to the core promoter. The *cis*-motif context (i.e. motif position, spacing, orientation, copy number and specific combination) of a synthetic module serves as the principal target for promoter modification strategies. Furthermore, modification of *cis*-regulatory architecture does not only allow 'fine-tuning' and/or enhancement of transgene transcriptional activity and inducibility, but sheds light on the functional role of defined *cis*-motif sequences binding to particular TFs in response to a specific stimulus (Fig. 20.2a). Synthetic promoters have successfully been used in a number of plant studies to either elucidate the role of *cis*-regulatory elements or to modulate targeted inducibility (using prior knowledge of *cis*-motif function), independently and/or within a specific *cis*-motif arrangement. Examples include *cis*-motifs that are strongly associated with heat shock, light, development, tissue-specificity, wounding, pathogen attack, sugar sensing and reactive oxygen species, and cold stress (Pietrzak et al. 1989; Comai et al. 1990; Gilmartin et al. 1990; Ni et al. 1995, 1996; Mitsuhara et al. 1996; Puente et al. 1996; Rushton et al. 2002; Geisler et al. 2006; Mazarei et al. 2008; Zhu et al. 2008). In addition, synthetic promoters may play a more prominent role in future biofuel strategies utilizing transgenic plants (Taylor et al. 2008).

20.3.1.1 CaMV 35S *cis*-Motif Context Re-evaluated

Modification of *cis*-regulatory context can be performed by two basic 'cut-and-paste' strategies: firstly, by rearrangement (also known as 'swapping' or 'shuffling') of defined modules, each containing a number of *cis*-motifs within native context, and secondly, by construction of novel synthetic modules by *cis*-motif insertion (of plant or non-plant origin). Both strategies can be used to develop synthetic promoters with additional features and/or enhanced transcriptional activity (or at least functionally equivalent) when compared to wild-type counterparts, but with minimum sequence homology in an attempt to reduce HDGS, as discussed above in Section 20.2.4. Bhullar et al. (2003) evaluated both strategies (domain swapping versus novel *cis*-regulatory context) using domain A (core promoter plus subdomain A1) of the CaMV 35S promoter as model (see Sect. 20.2.3) compared to expression levels of the wild-type 35S promoter in transient and stably transformed tobacco plants. Results from this study showed that synthetic promoters constructed by domain swapping conferred lower GUS reporter gene activity than the wild-type 35S promoter and, on the other hand, synthetic promoters comprising novel *cis*-regulatory context revealed GUS expression patterns comparable to wild-type 35S promoter (Bhullar et al. 2003). More recently, Bhullar et al. (2007) further improved the strategy to construct synthetic 35S promoters with minimum sequence homology by re-evaluating regulatory information content and selective contribution to

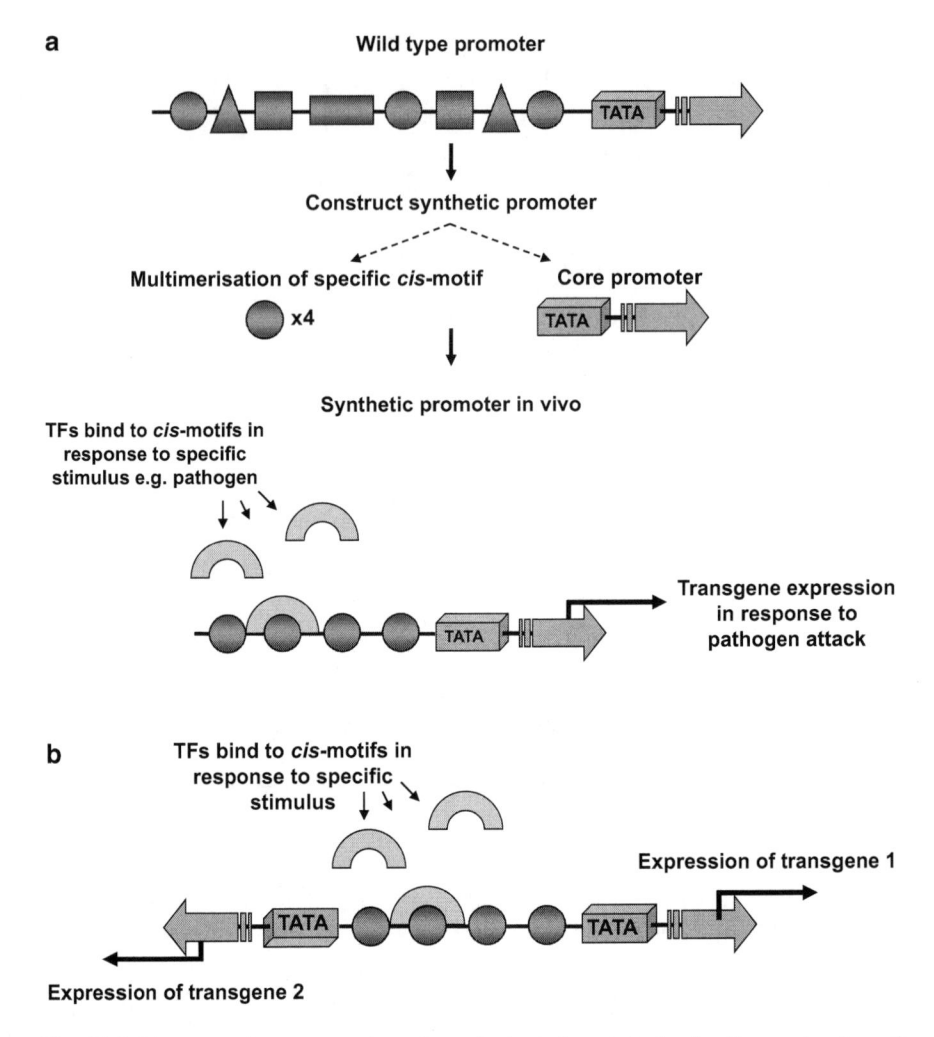

Fig. 20.2 Diagrammatic representation of two basic design strategies for the construction of synthetic promoters. (**a**) Unidirectional synthetic promoter design. The core-promoter region (used as vehicle to drive transgene expression) is fused to a synthetic stretch of DNA comprising multiple repeats of a specific *cis*-motif (multimerisation). Prior knowledge of *cis*-motif specifically associated with induction to a specific stimulus (e.g. pathogen attack) allows for construction of synthetic promoter to confer transgene expression in response to (in our example) pathogen attack. (**b**) Bidirectional synthetic promoter design. Fusion of an additional core promoter on the opposite side allows regulation of two transgenes using the same *cis*-regulatory architecture

CaMV 35S promoter activity of specific 35S subdomains B5, B4 and B2 (see Sect. 20.2.3). Specific results highlighted in this study suggest that the subdomain B2 may not comprise *cis*-motifs of significant function, but rather plays a vital role in maintaining appropriate distance between proximal and distal 35S promoter regions, thus underscoring the importance of *cis*-regulatory 'spacing' combined with *cis*-motif functional content. Functional dissection of 35S *cis*-regulatory architecture suggests that, although domain swapping might be a less efficient strategy compared to the development of synthetic modules comprising novel *cis*-regulatory architecture (Bhullar et al. 2003), both these strategies could certainly be employed to develop functionally equivalent synthetic promoters with decreased sequence homology to limit the level of promoter HDGS.

20.3.1.2 Gaining Insight from Synthetic *cis*-Regulatory Complexity

Extrapolating valuable information for future synthetic promoter design, Rushton et al. (2002) showed in vivo how variations of *cis*-motif copy number, specific order and spacing relative to each other and to the TATA-box of the CaMV 35S core promoter could dramatically influence promoter strength and inducibility in response to pathogen-induced signalling. Optimization of synthetic promoters revealed that multimerisation of upstream *cis*-motifs progressively improved promoter strength, and single base pair differences in the core *cis*-motif sequences had a dramatic effect on promoter inducibility (Rushton et al. 2002). In addition, this study revealed that (1) an increase in motif copy number did not necessarily enhance promoter inducibility in response to pathogen attack, (2) a more refined selection of different *cis*-motif combinations allowed higher inducibility with lower background expression and (3) the specific *cis*-motifs shared combined inducibility in response to pathogen and wounding and/or (4) across different plant species (Rushton et al. 2002). Sawant et al. (2001) implemented a synthetic promoter design strategy by incorporating computational database assistance and prior knowledge of gene expression in plants (Sawant et al. 1999), to construct a 450-bp synthetic regulatory module known as *Pcec* (complete expression cassette) comprising two units known as *Pmec* (138-bp minimum expression cassette with TATA-box) and a 312-bp 'transcription activation module'. Initial results revealed that *Pcec*, comprising core sequences from eight types of *cis*-motifs and their variants most commonly found and specifically arranged (as statistically represented) in the promoters of highly expressed plant genes, conferred a higher GUS expression level in stably transformed tobacco plants than the wild-type CaMV 35S promoter (Sawant et al. 2001). The *Pcec* synthetic module served in its purpose to confer high-level transgene expression for biotechnological applications. However, by structural dissection and modification of the *Pcec* module in a later study, Sawant et al. (2005) extrapolated fundamental information on the complexity of *trans*-interactions and combinatorial effect of *cis*-motifs (individually and in combination), highlighting several key factors to be considered for future synthetic design strategies. In this study, the effects of individual *cis*-motifs (each of the eight *cis*-motifs in *Pcec*), in multimeric form and/or in combination fused upstream to the minimal expression cassette (*Pmec*), were evaluated on a transient level measuring GUS reporter gene activity 48 h after particle bombardment onto tobacco leaves. Results from this investigation (Sawant et al. 2005) revealed that, although each *cis*-motif contributed to enhance basal transcriptional activity (ranging from two- to eightfold), the complete expression cassette (*Pcec*), comprising a combination of all eight *cis*-motifs, enhanced basal transcriptional GUS activity by 110-fold. This observation led Sawant et al. (2005) to suggest that the combined presence of all the *cis*-motifs (individually able to enhance transcriptional activity) functions synergistically and that in vivo titration with any one of the *cis*-motifs (by co-bombardment of double-stranded oligonucleotides comprising 18 copies of the individual motif) caused nearly complete inactivity of the synthetic *Pcec* promoter. In addition, these results highlighted one of the major challenges currently facing synthetic promoter design strategies, by showing that an excess of synthetic *cis*-motifs could deplete endogenous TFs necessary for cellular and/or housekeeping function. In contrast, results showed that complete occupancy of the synthetic TF-binding sites substantially contributes to the stability of the PIC assembly on the TATA-box, thus suggesting that hierarchical TF-assembly to a combination of multiple *cis*-motifs within a synthetic module is an orchestrated event necessary for efficient functioning of a synthetic promoter (Sawant et al. 2005). Measuring reporter gene activity in transformed tobacco plants, Cazzonelli and Velten (2008) evaluated a combination of synthetic promoters, comprising different arrangements (multiple repeats and/or combinations) of short directly repeated (DR) enhancer elements of gemini-, nano-, badna- and caulimoviral origin upstream of a CaMV 35S core promoter. In an attempt to elucidate the effect of individual *cis*-motif interactions on overall promoter function, Cazzonelli and Velten (2008) showed that multimerisation of the more weakly *cis*-enhancer elements most often produced a linear additive effect, conferring increasing reporter gene activity in direct proportion to *cis*-enhancer copy number. Similarly, Cazzonelli and Velten (2008) showed that synthetic promoters comprising specific combinations of *cis*-enhancer elements frequently conferred enhanced reporter gene expression markedly greater than the combined effect of individual enhancer activities. From the examples highlighted in this subsection, it is evident that (1) multimerisation remains a viable technique for synthetic promoter design, (2) there is a fine balance between *cis*-motif/*trans*-factor binding and synthetic promoter efficiency, and (3) prior knowledge of *cis*-motif function (from TF-databases and literature) followed by in vivo evaluation of 'cut-and-paste' *cis*-motif multimerisation and/or combination could allow a more accurate assembly of *cis*-motif 'building blocks' for predictable inducibility and enhanced transgene activity.

20.3.1.3 Bidirectionalisation Improves Transcriptional Activity and Overall Versatility

A unidirectional synthetic promoter can be bidirectionalised, by fusing an additional core promoter in the opposite direction on the 5′-end of a regulatory module and thus allowing simultaneous expression of two transgenes (Fig. 20.2b; Barfield and Pua 1991; Xie et al. 2001). Compared to other more conventional technologies used for the introduction and expression of multiple transgene stacking in plants (i.e. crossing, and sequential or co-transformation; reviewed in François et al. 2002, Halpin 2005), a synthetically engineered bidirectional promoter may offer additional advantages with regard to enhanced transcriptional activity and inducibility. Li et al. (2004) have shown that bidirectional promoters comprising two divergently arranged core promoters, both from either CaMV 35S or cassava vein mosaic virus (CsVMV) per cassette and separated by duplicated enhancer repeats, could dramatically enhance GUS and modified-GFP reporter gene activity in tobacco and grape, compared to an equivalent unidirectional module of the same promoter. Although duplicated enhancer elements increased unidirectional reporter gene activity, a bidirectional configuration of the doubled enhanced unidirectional version permits a more stable and/or optimal conformation of the transcriptional machinery, functioning more efficiently when compared to a promoter in unidirectional context (Li et al. 2004). Stability of the general transcriptional assembly was similarly emphasized by Chaturvedi et al. (2006), showing that bidirectionalisation of the *Pcec* module (see Sect. 20.3.1.2) allowed for efficient transcriptional initiation in both directions. In addition, combining high-level transgene expression with bidirectional inducibility, Xie et al. (2001) and Chaturvedi et al. (2006) showed that bidirectional configuration of a single multifactorial module comprising specific inducible *cis*-motifs allows simultaneous induction and enhanced expression of transgenes in both directions, in response to a specific stimulus (i.e. salicylic acid) and/or condition (i.e. leaf senescence). Although the improved versatility of a synthetic bidirectional promoter may seem apparent, it should be noted that the functional properties of individual *cis*-regulatory elements within such a central module may differ due to changes in the orientation and relative distance, thus leading to diverse transgene activity and inducibility in opposite directions. This orientation-dependent characteristic (known as vectorial activity) of specific *cis*-regulatory elements has previously been demonstrated in plants using synthetic constructs to examine the bidirectional *cis*-regulatory nature of the mannopine synthase (MAS) promoter (Guevara-Garcia et al. 1999). The T-DNA MAS promoter modulates bidirectional activity of two genes (*mas1'* and *mas2'*), from one regulatory module that can be divided into different subdomains. In this study by Guevara-Garcia et al. (1999), combinations and/or multiple copies of these subdomains were evaluated in diverse orientations, fused to the CaMV 35S core promoter driving GUS reporter gene activity in transgenic tobacco plants. The initial purpose of this study was to gain insight on the regulatory properties of divergently transcribed *mas1'* and *mas2'*, by using synthetic promoters. In terms of biotechnological applications, however, the major outcome we wish to highlight from this study was that a promoter fragment (or subdomain) comprising several *cis*-motifs can confer a diverse transgene expression level and tissue-specific pattern, acting as a transcriptional silencer in one orientation and an enhancer in the opposite direction (Guevara-Garcia et al. 1999). The vectorial behaviour of *cis*-elements, first described in plants by Guevara-Garcia et al. (1999), certainly contributes to the complexity of *cis*-regulatory fine-tuning in the design and evaluation of bidirectional synthetic modules. As a consequence of the results of these studies, combined with (1) the advantage of multiple transgene stacking and (2) the benefits offered by synthetic promoter design (see above), it appears that bidirectionalisation allows for overall enhancement of transcriptional activity. Further interpretation of this characteristic and elucidation of the fundamental rules governing uni- and/or bidirectional transcriptional control are likely to have a major impact on the refinement and predictive design principles for future development of stable synthetic modules.

20.3.2 Two-Component Transactivated Gene Switches: Promising Systems for Flexible Transgene Expression

Two-component gene switches are ideal molecular systems to achieve tightly regulated transgene expression in a tissue-specific mode and/or in response to a specific inducer. Generally, the two-component transactivation system comprises two transcriptional units, namely the TF expressed by a strong constitutive or tissue-specific promoter, and the inducible (target) regulatory unit consisting of a minimal promoter region and multimeric upstream *cis*-motif-repeats specific for binding to the TF expressed by the first unit (Fig. 20.3). As can be expected, synthetic 'cut-and-paste' modifications can be made to both the promoter and TF (specific construction of chimeric TF by selective grouping of DNA and/or chemical-binding domains) of the 1st unit and the promoter of the 2nd unit, allowing improved versatility and refined transgene inducibility. Typically, the two-component gene switch is based on the principle that the second transcriptional unit alone is functionally inactive and will become active only when 'transactivated' by the TF product of the first unit.

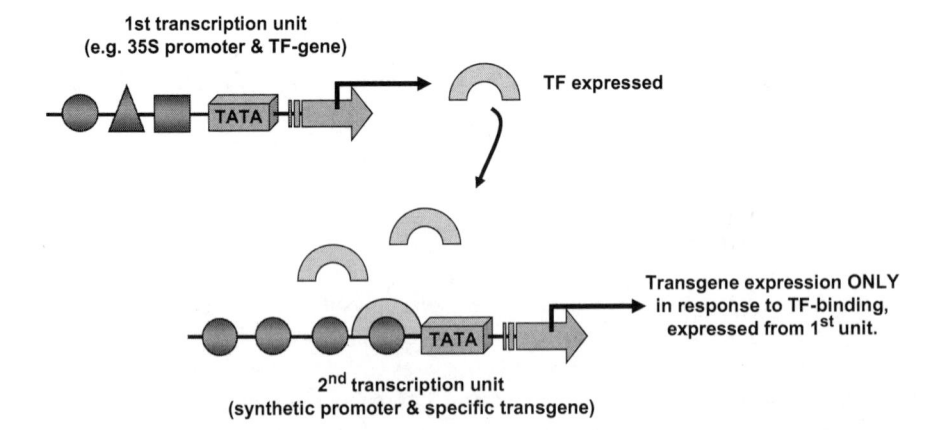

Fig. 20.3 Simplified diagrammatic representation of a two-component transactivation system

20.3.2.1   Tight Control by Targeted *cis-trans* Interaction

In the last few years, numerous studies (see references in Sect. 20.3.2.2) have reported on the design and use of chimeric two-component systems integrating combinations of core promoter and multimeric repeats or specific combinations of upstream heterologous *cis*-motifs of prokaryotic, insect and/or mammalian origin. One of the most widely used examples is the pOp/LhG4 transactivation system (Moore et al. 1998, 2006). This system comprises a synthetic pOp-promoter (consisting of two regulatory *E.coli lac* operator elements upstream of a CaMV 35S core promoter) known as the 'reporter' unit, and a transcription 'activator' unit constitutively or spatially expressing a chimeric LhG4 TF, which is a fusion between a mutant *lac*-repressor and the transcription-activation domain II from the *Saccharomyces cerevisiae* Gal4 protein (Moore et al. 1998). The pOp-promoter is transcriptionally inactive when introduced alone in reporter plant lines devoid of the chimeric LhG4 TF. However, transgene expression driven by the pOp-promoter can be activated in the F1-progeny when reporter lines are crossed with activator lines expressing LhG4. Initially described using tobacco plants, this system has subsequently been used to investigate developmental features in *Arabidopsis*, maize and tobacco (Moore et al. 2006). It has been shown that the reporter gene expression profile is accurately reflected by the expression pattern of the activator, thus allowing transcriptional control of the transgene only in the progeny plant cells conferring regulated expression of the LhG4 TF depending on the selected promoter of the activator unit. Although the pOp/LhG4 system is particularly useful to direct spatial transgene transcriptional activity without external interference or stimuli (Moore et al. 1998), it serves as a complementary platform for chemical induction, thus offering additional temporal control, as reported more recently (Craft et al. 2005; Samalova et al. 2005). An even more recent investigation by Chaturvedi et al. (2007) elegantly accentuated the two-component transactivation strategy by deploying a reporter unit expressing GUS from a strong seed-specific chickpea legumin promoter of which the TATA-box had been mutated to TGTA, and an activator unit using the *Pcec* synthetic constitutive promoter (see Sect. 20.3.1.2) to express a mutated TATA-box-binding protein (TBP) general TF (designated as TBP_{m3}), designed to recognize only the mutated TGTA core-promoter region for transcriptional activation (Chaturvedi et al. 2007). This system was evaluated using combinations of transactivation units—e.g. unit one seed-specific *cis*-regulatory region plus core promoter containing TATA or TGTA combined with unit two expressing TBP or TBP_{m3}—and revealed tightly regulated tissue-specific transgene expression in stably transformed transgenic tobacco plants. Reporter gene expression results of this study confirmed the basic desired operational properties of a well-designed two-component system, where abolishment of transcriptional activity of the reporter unit containing the mutated TGTA region could be completely restored by co-expression of the mutated TBP_{m3} from unit two. Focusing on synthetic promoter design and utilization principles, Chaturvedi et al. (2007) have shown that using a 'cut-and-paste' strategy on the GUS-reporter unit, i.e. replacement of the native legumin core promoter with a synthetic *Pmec* core promoter (see Sect. 20.3.1.2), did not alter the tissue-specific regulatory properties of the legumin promoter, thus confirming the flexibility of promoter dynamics in *cis*-context.

20.3.2.2 Inducible Fine-Tuning by a Chemical Trigger

The applied specificity and targeted inducibility of a two-component transactivation system is largely increased when binding of the TF to the multimeric *cis*-regulatory complex is dependent on effective ligand conformation conferred by interaction of a specific inducing compound (usually a chemical substance). More specifically, inducible systems enabling transgene activation (or repression) in response to non-plant inducing compounds, such as chemicals or specific hormones, have gained much momentum allowing flexible control of transgene activity in both dicots and monocots without influencing the status quo with regard to normal plant processes, growth and development. The desired properties, comparisons, modifications, strengths, limitations and potential applications of the most widely used and well established transactivated and chemically inducible systems have been discussed in several studies (Ni et al. 1995; Gatz 1997; Gatz and Lenk 1998; Reynolds 1999; Guilfoyle and Hagen 1999; Zuo and Chua 2000; Böhner and Gatz 2001; Padidam 2003; Wang et al. 2003; Tang et al. 2004; Moore et al. 2006). Although a number of chemical-inducible systems are available, modification and improved versions of the original systems (e.g. allowing dual control of one system by two different inducers) are continuously being developed and tested in plants. Most of these systems are originally derived from non-plant sources and the different compounds used for induction include the following: tetracycline, pristinamycin, dexamethasone, tebufenozide (commercial name: RH5992), methoxyfenozide (commercial name: Intrepid-F2), ethanol, copper and β-estradiol (Zuo and Chua 2000; Padidam 2003; Wang et al. 2003; Tang et al. 2004; Moore et al. 2006). The majority of these systems implement a multimeric *cis*-motif engineering strategy, where the synthetic TF/ligand-target promoter contains a compact array of multiple *cis*-motif repeats (usually ranging from two to ten) eventually permitting enhanced binding of the TF/ligand complex and expression of the transgene of interest. With regard to conventional synthetic promoter design, although an increase of *cis*-motif copy number may not necessarily enhance transcriptional inducibility (Rushton et al. 2002), a two-component, non-plant inducible system could allow continuous *cis*-motif/trans-factor interaction (initiated by chemical induction) without major effect on endogenous *cis*-motif/TF-occupancy. Depending on the application, not all systems exhibit the desired properties in terms of inducibility, specificity and non-toxicity. For example, the glucocorticoid (dexamethasone)-inducible system is detrimental in *Arabidopsis* (Kang et al. 1999), *L. japonicus* (Andersen et al. 2003) and tobacco (Amirsadeghi et al. 2007). Nevertheless, due to the highly conserved nature of gene regulation between species and continuous efforts for deciphering of *cis*-regulatory complexity, possibilities for innovative system designs and the potential benefits offered by these systems for future plant biotechnological applications are exciting. Pioneering work by Kodama et al. (2007) demonstrated how utilization of a mammalian xenobiotic biosensor regulatory system, combining a synthetic promoter design and transactivation strategy, could lead to the development of transgenic plants acting as bio-monitors for environmental pollutants. The major building blocks for this synthetic system comprised a mouse aryl hydrocarbon receptor (AhR) and AhR-nuclear translocator (Arnt), which in mammals mediates expression of certain genes (including drug metabolizing genes) in response to exposure to dioxins, pollutants and other related xenobiotic compounds. In response to dioxins, the ligand-dependent AhR translocates to the nucleus after heterodimerization with the Arnt, and binds to *cis*-regulatory promoter regions containing xenobiotic response elements (XREs; Barouki et al. 2007). The synthetic reporter unit of the plant transactivation system contained six repeated XRE sequences fused to a minimal CaMV 35S promoter. GUS reporter gene activity, measured in transgenic tobacco plants, was markedly enhanced in response to treatments with dioxin compounds such as 3-methylcholanthrene (MC), indigo and β-naphthoflavone (βNF; Kodama et al. 2007). Furthermore, modification of the wild-type AhR (by replacing the AhR activation domain with that of the *Herpes simplex* virus protein VP16) allowed improved inducibility of GUS expression, specifically in response to MC, in a time- and dose-dependent mode (Kodama et al. 2007). So far, most inducible gene expression studies have focused primarily on the establishment, evaluation and modification of novel systems in plants, the examples highlighted in Section 20.3.2 clearly demonstrating that a refined *cis*-engineering strategy, combining synthetic promoter design principles with two-component transactivation systems, could greatly advance the product pipeline for future plant biotechnological tools.

20.4 The Way Forward: Systematic Engineering and Integration Leads to Accurate Design

The integration of biological data extrapolated from (1) whole-genome sequence assemblies, (2) high-throughput technologies (i.e. microarrays and chromatin immunoprecipitation (ChIP) assays) and (3) systematic computational analyses of promoter architecture and regulatory networks have allowed for a holistic deciphering of highly conserved processes governing transcriptional control (Pilpel et al. 2001; Orphanides and Reinberg 2002; Wasserman and Sandelin 2004; Blais and Dynlacht 2005; Tompa et al. 2005; Hoheisel 2006; Barrera and Ren 2006; Elnitski et al. 2006; Nguyen and D'haeseleer 2006; Heintzman and Ren 2007; Kolchanov et al. 2007). More specifically, implementation of a systems approach, combining multiple datasets to decipher transcriptional control, has set the stage for synthetic engineering and predictive biology. The goal of such a combined interaction (between systems biology and synthetic engineering) is to obtain good agreement between experimental observation and reconstruction (or mimicking) of regulatory codes within a biological system

(Andrianantoandro et al. 2006; Barrett et al. 2006; Drubin et al. 2007). The promoter serves as one of the key molecular components of such a biological system, which can be modified and used as a 'device' with a defined practical purpose to accurately tune transgene expression in an inducible or constitutive mode. In addition to deciphering gene function within a genetic circuit integrating information from TF-activity and - networks in the cellular environment, re-designing of the regulatory code in *cis*-configuration and the use of synthetic promoter libraries have been well documented in the fields of molecular medicine, as well as biopharmaceutical and metabolic engineering (Jensen and Hammer 1998; Tornøe et al. 2002; Yew 2005; Martinelli and De Simone 2005; Alper et al. 2005; Mijakovic et al. 2005; Miksch et al. 2005; Weber and Fusenegger 2006; Hammer et al. 2006; Rud et al. 2006; Sørensen et al. 2006; Greber and Fussenegger 2007; Tyo et al. 2007). In plant systems, however, synthetic promoter design principles are still in their infancy and applications in plant biotechnology programs are relatively limited. Therefore, using established design concepts employed in other organisms could diversify and/or advance synthetic engineering strategies in plants. Experimental efforts conducted in model organisms (i.e. *Escherichia coli*, *S. cerevisiae* and *Drosophila melanogaster*) have been devoted to the deciphering of promoter architecture and reconstruction of *cis*- and *trans*-regulatory logic, subsequently demonstrating the 'programming' of gene expression and modelling of synthetic gene networks using a so-called bottom-up approach (Pilpel et al. 2001; Buchler et al. 2003; Werner et al. 2003; Beer and Tavazoie 2004; Zhou and Wong 2004; Istrail and Davidson 2005; Guido et al. 2006; Janssens et al. 2006; Mayo et al. 2006; Cox et al. 2007; Murphy et al. 2007). From these examples, it is evident that complex mathematical modelling, systematic computational integration and high-throughput datasets all contribute to gain a more holistic view of the specific interaction between promoter architecture, TF-binding specificity and gene expression. Among a plethora of computational tools available for plant-omics analyses (Vision and McLysaght 2004; Rhee et al. 2006), synthetic promoter engineering relies heavily on computational (in silico) assistance, integrating specifically two major in silico components—(1) *cis*-motif/module detection algorithm and (2) TF-binding site database assistance (Hehl and Wingender2001; Tompa et al. 2005)—that should play an integral part in future design concepts. Although the basic strategy for successful synthetic promoter design relies on defining the *cis*-regulatory logic (cf. detect statistical overrepresentation of 'true' *cis*-motifs), it is likely that combining computational and experimental biology in plants will expedite the progress in redesigning a unique and refined synthetic regulatory code. As a result, functionality of overrepresented *cis*-motifs, within a modified (synthetic) context, can be evaluated (e.g. by selective inducibility). Furthermore, accurate reconstruction of *cis*-regulatory logic can be used in a bottom-up strategy to predict transgene expression conferred in response to specific, or a variety of, conditions (Rushton et al. 2002; Geisler et al. 2006; Venter 2007; Cazzonelli and Velten 2008). Within the complexity of different regulatory mechanisms, networks and synergistic *cis*/*trans*-interactions that govern transcriptional control, much information on distinct gene expression patterns can be extrapolated from deciphering the *cis*-regulatory code within the promoter region. While gaps in our understanding of eukaryotic gene regulation still remain, it is evident that employing an integrative approach (combining multiple datasets to decipher biological processes in the cellular environment) could greatly accelerate synthetic promoter design.

References

Alper H, Fischer C, Nevoigt E, Stephanopoulos G (2005) Tuning genetic control through promoter engineering. Proc Natl Acad Sci USA 102:12678–12683

Amirsadeghi S, McDonald AE, Vanlerberghe GC (2007) A glucocorticoid-inducible gene expression system can cause growth defects in tobacco. Planta 226:453–463

An G, Costa MA, Ha SB (1990) Nopaline synthase promoter is wound inducible and auxin inducible. Plant Cell 2:225–233

Andersen SU, Cvitanich C, Hougaard BK, Roussis A, Grønlund M, Jensen DB, Frøkjaer LA, Jensen EO (2003) The glucocorticoid-inducible GVG system causes severe growth defects in both root and shoot of the model legume *Lotus japonicus*. Mol Plant Microbe Interact 16: 1069–1076

Andrianantoandro, E. Basu S, Karig DK, Weiss R (2006) Synthetic biology: new engineering rules for an emerging discipline. Mol Systems Biol 2:2006.0028. www.molecularsystemsbiology.com

Barfield DG, Pua EC (1991) Gene transfer in plants of *Brassica juncea* using *Agrobacterium tumefaciens*-mediated transformation. Plant Cell Rep 10:308–314

Barouki R, Coumoul X, Fernandez-Salguero PM (2007) The aryl hydrocarbon receptor, more than a xenobiotic-interacting protein. FEBS Lett 581:3608–3615

Barrera LO, Ren B (2006) The transcriptional regulatory code of eukaryotic cells – insights from genome-wide analysis of chromatin organization and transcription factor binding. Curr Opin Cell Biol 18:291–298

Barrett CL, Kim TY, Kim HU, Palsson BØ, Lee SY (2006) Systems biology as a foundation for genome-scale synthetic biology. Curr Opin Biotechnol 17:488–492

Beer MA, Tavazoie S (2004) Predicting gene expression from sequence. Cell 117:185–198

Benfey PN, Chua N-H (1990) The cauliflower mosaic virus 35S promoter: combinatorial regulation of transcription in plants. Science 250:959–966

Benfey PN, Ren L, Chua N-H (1990) Combinatorial and synergistic properties of CaMV 35S enhancer subdomains. EMBO J 9:1685–1696

Bhattacharyya S, Dey N, Maiti IB (2002) Analysis of cis-sequence of subgenomic transcript promoter from the Figwort mosaic virus and comparison of promoter activity with the cauliflower mosaic virus promoters in monocot and dicot cells. Virus Res 90:47–62

Bhullar S, Chakravarthy S, Advani S, Datta S, Pental D, Kumar Burma P (2003) Strategies for development of functionally equivalent promoters with minimum sequence homology for transgene expression in plants: cis-elements in a novel DNA context versus domain swapping. Plant Physiol 132:988–998

Bhullar S, Datta S, Advani S, Chakravarthy S, Gautam T, Pental D, Kumar Burma P (2007) Functional analysis of the cauliflower mosaic virus 35S promoter: re-evaluation of the role of subdomains B5, B4, and B2 in promoter activity. Plant Biotechnol J 5:696–708

Blais A, Dynlacht BD (2005) Constructing transcriptional regulatory networks. Genes Dev 19:1499–1511

Böhner S, Gatz C (2001) Characterization of novel target promoters for the dexamethasone-inducible/tetracycline-repressible regulator TGV using luciferase and isopentenyl transferase as sensitive reporter genes. Mol Gen Genet 264:860–870

Buchler NE, Gerland U, Hwa T (2003) On schemes of combinatorial transcription logic. Proc Natl Acad Sci USA 100:5136–5141

Butler JEF, Kadonaga JT (2002) The RNA polymerase II core promoter: a key component in the regulation of gene expression. Genes Dev 16:2583–2592

Cazzonelli CI, Velten J (2008) In vivo characterization of plant promoter element interaction using synthetic promoters. Transgenic Res 17:437–457

Cazzonelli CI, McCallum EJ, Lee R, Ramón Botella J (2005) Characterization of a strong, constitutive mung bean (*Vigna radiate* L.) promoter with a complex mode of regulation in planta. Transgenic Res 14:941–967

Chaturvedi CP, Sawant SV, Kiran K, Mehrotra R, Lodhi N, Ansari SA, Tuli R (2006) Analysis of polarity in the expression from a multifactorial bidirectional promoter designed for high-level expression of transgenes in plants. J Biotechnol 123:1–12

Chaturvedi CP, Lodhi N, Ansari SA, Tiwari S, Srivastava R, Sawant SV, Tuli R (2007) Mutated TATA-box/TATA binding protein complementation system for regulated transgene expression in tobacco. Plant J 50:917–925

Christensen AH, Sharrock RA, Quail PH (1992) Maize polyubiquitin genes: structure, thermal perturbation of expression and transcript splicing, and promoter activity following transfer to protoplasts by electroporation. Plant Mol Biol 18:675–689

Comai L, Moran P, Maslyar D (1990) Novel and useful properties of a chimeric plant promoter containing CaMV 35S and MAS elements. Plant Mol Biol 15:373–381

Cox RS III, Surette MG, Elowitz MB (2007) Programming gene expression with combinatorial promoters. Mol Systems Biol 3:145

Craft J, Samalova M, Baroux C, Townley H, Martinez A, Jepson I, Tsiantis M, Moore I (2005) New pOp/LhG4 vectors for stringent glucocorticoid-dependent transgene expression in *Arabidopsis*. Plant J 41:899–918

Dean C, Jones J, Favreau M, Dunsmuir P, Bedbrook J (1988) Influence of flanking sequences on variability of expression levels of an introduced gene in transgenic tobacco plants. Nucleic Acids Res 16:9267–9283

De Wilde C, Van Houdt H, De Buck S, Angenon G, De Jaeger G, Depicker A (2000) Plants as bioreactors for protein production: avoiding the problem of transgene silencing. Plant Mol Biol 43:347–359

Drubin DA, Way JC, Silver PA (2007) Designing biological systems. Genes Dev 21:242–254

Ellis JG, Llewellyn DJ, Dennis ES, Peacock WJ (1987) Maize *Adh1* promoter sequences control anaerobic regulation: addition of upstream promoter elements from constitutive genes is necessary for expression in tobacco EMBO J 6:11–16

Elnitski L, Jin VX, Farnham PJ, Jones SJM (2006) Locating mammalian transcription factor binding sites: a survey of computational and experimental techniques. Genome Res 16:1455–1464

Fagard M, Vaucheret H (2000) (Trans) gene silencing in plants: how many mechanisms? Annu Rev Plant Physiol Plant Mol Biol 51:167–194

Fang R-X, Nagy F, Sivasubramaniam S, Chua N-H (1989) Multiple *cis*-regulatory elements for maximal expression of the cauliflower mosaic virus 35S promoter in transgenic plants. Plant Cell 1:141–150

Foster E, Hattori H, Labbe H, Ouellet T, Fobert PR, James LE, Iyer VN, Miki BL (1999) A tobacco cryptic constitutive promoter, *tCUP*, revealed by T-DNA tagging. Plant Mol Biol 41:45–55

Franck A, Guilley H, Jonard G, Richards K, Hirth L (1980) Nucleotide-sequence of cauliflower mosaic-virus DNA. Nucleic Acids Res 21:285–294

François IEJA, Broekaert WF, Cammue BPA (2002) Different approaches for multi-transgene-stacking in plants. Plant Sci 163:281–295

Gatz C (1997) Chemical control of gene expression. Annu Rev Plant Physiol Plant Mol Biol 48:89–108

Gatz C, Lenk I (1998) Promoters that respond to chemical inducers. Trends Plant Sci 9:352–358

Geisler M, Kleczkowski LA, Karpinski S (2006) A universal algorithm for genome-wide *in silico* identification of biologically significant gene promoter putative *cis*-regulatory-elements; identification of new elements for reactive oxygen species and sucrose signaling in *Arabidopsis*. Plant J 45:384–398

Gilmartin PM, Sarokin L, Memelink J, Chua N-H (1990) Molecular light switches for plant genes. Plant Cell 2:369–378

Greber D, Fussenegger M (2007) Mammalian synthetic biology: engineering of sophisticated gene networks. J Biotechnol 130:329–345

Guerva-Garcia A, López-Bucio J, Herrera-Estrella L (1999) The mannopine synthase promoter contains vectorial *cis*-regulatory elements that act as enhancers and silencers. Mol Gen Genet 262:608–617

Guido NJ, Wang X, Adalsteinsson D, McMillen D, Hasty J, Cantor CR, Elston TC, Collins JJ (2006) A bottom-up approach to gene regulation. Nature 439:856–860

Guilfoyle TJ (1997) The structure of plant gene promoters. In: Setlow JK (ed) Genetic engineering. Plenum Press, New York, pp 15–47

Guilfoyle TJ, Hagen G (1999) Potential use of hormone-responsive elements to control gene expression in plants. In: Reynolds PHS (ed) Inducible gene expression in plants. CABI Publishing, Wallingford, pp 219–236

Guilley H, Dudley RK, Jonand G, Balazs E, Richards KE (1982) Transcription of cauliflower mosaic virus DNA: detection of promoter sequences and characterization of transcripts. Cell 30:763–773

Gurr SJ, Rushton PJ (2005) Engineering plants with increased disease resistance: how are we going to express it? Trends Biotechnol 23:283–290

Halpin C (2005) Gene stacking in transgenic plants – the challenge for 21st century plant biotechnology. Plant Biotechnol J 3:141–155

Hammer K, Mijakovic I, Jensen PR (2006) Synthetic promoter libraries - tuning of gene expression. Trends Biotechnol 24:53–55

Hehl R, Wingender E (2001) Database-assisted promoter analysis. Trends Plant Sci 6:251–255

Heintzman ND, Ren B (2007) The gateway to transcription: identifying, characterizing and understanding promoters in the eukaryotic genome. Cell Mol Life Sci 64:386–400

Hochheimer A, Tjan R (2003) Diversified transcription initiation complexes expand promoter selectivity and tissue-specific gene expression. Genes Dev 17:1309–1320

Hoheisel JD (2006) Microarray technology: beyond transcript profiling and genotype analysis. Nature Rev Genet 7:200–210

Holtorf S, Apel K, Bohlmann H (1995) Comparison of different constitutive and inducible promoters for the transgenes in *Arabidopsis thaliana*. Plant Mol Biol 29:637–646

Hull R, Covey SN, Dale P (2002) Genetically modified plants and the 35S promoter: assessing the risk and enhancing the debate. Microb Ecol Health Dis 12:1–5

Istrail S, Davidson EH (2005) Logic functions of the genomic cis-regulatory code. Proc Natl Acad Sci USA 102:4954–4959

Janssens H, Hou S, Jaeger J, Kim A-R, Myasnikova E, Sharp D, Reinitz J (2006) Quantitative and predictive model of transcriptional control of the *Drosophila melanogaster even skipped* gene. Nature Genet 38:1159–1165

Jensen PR, Hammer K (1998) Artificial promoters for metabolic optimization. Biotechnol Bioeng 58:191–195

Kang H-G, Fang Y, Singh KB (1999) A glucocorticoid-inducible transcription system causes severe growth defects in *Arabidopsis* and induces defense-related genes. Plant J 20:127–133

Kay R, Chan A, Daly M, McPherson J (1987) Duplication of CaMV 35S promoter sequences creates a strong enhancer for plant genes. Science 236:1299–1302

Kodama S, Okada K, Inui H, Ohkawa H (2007) Aryl hydrocarbon receptor (AhR)-mediated reporter gene expression systems in transgenic tobacco plants. Planta 227:37–45

Kolchanov NA, Merkulova TI, Ignatieva EV, Ananko EA, Oshchepkov DY, Levitsky VG, Vasiliev GV, Klimova NV, Merkulov VM, Hodgman TC (2007) Combined experimental and computational approaches to study the regulatory elements in eukaryotic genes. Brief Bioinform 8:266–274

Kooter JM, Matzke MA, Meyer P (1999) Listening to the silent genes: transgene silencing, gene regulation and pathogen control. Trends Plant Sci 4:340–345

Lam E (1994) Analysis of the tissue-specific elements in the CaMV 35S promoter. In: Nover L (ed) Results and Problems in Cell Differentiation, Plant Promoters and Transcription Factors, vol 20. Springer, Berlin Heidelberg, pp 181–196

Lessard PA, Kulaveerasingam H, York GM, Strong A, Sinskey AJ (2002) Manipulating gene expression for the metabolic engineering of plants. Metab Eng 4:67–79

Li ZT, Jayasankar S, Gray DJ (2004) Bi-directional duplex promoters with duplicated enhancers significantly increase transgene expression in grape and tobacco. Transgenic Res 13:143–154

Liu XJ, Prat S, Willmitzer L, Frommer WB (1990) Cis regulatory elements directing tuber-specific and sucrose-inducible expression of a chimeric class I patatin promoter/GUS–gene fusion. Mol Gen Genet 223:401–406

Maiti IB, Shepard RJ (1998) Isolation and expression analysis of peanut chlorotic streak caulimovirus (PCISV) full-length transcript (FLt) promoter in transgenic plants. Biochem Biophys Res Comm 244:440–444

Martinelli R, De Simone V (2005) Short and highly efficient synthetic promoters for melanoma-specific gene expression. FEBS Lett 579:153–156

Matzke MA, Matzke AJM (1998a) Position effects and epigenetic silencing of plant transgenes. Curr Opin Plant Biol 1:142–148

Matzke MA, Matzke AJM (1998b) Epigenetic silencing of plant transgenes as a consequence of diverse cellular defence responses. Cell Mol Life Sci 54:94–103

Mayo AE, Setty Y, Shavit S, Zaslaver A, Alon U (2006) Plasticity of the *cis*-regulatory input function of a gene. PLoS Biol 4:e45

Mazarei M, Teplova I, Hajimorad MR, Stewart CN (2008) Pathogen phytosensing: plants to report plant pathogens. *Sensors* 8(4):2628–2641

Meyer P (2000) Transcriptional transgene silencing and chromatin components. Plant Mol Biol 43:221–234

Meyer P (2001) Chromatin remodelling. Curr Opin Plant Biol 4:457–462

Meyer P, Saedler H (1996) Homology dependent gene silencing in plants. Annu Rev Plant Physiol Plant Mol Biol 47:23–48

Mijakovic I, Petranovic D, Jensen PR (2005) Tunable promoters in systems biology. Curr Opin Biotechnol 16:329–335

Miksch G, Bettenworth F, Friehs K, Flaschel E, Saalbach A, Twellmann T, Nattkemper TW (2005) Libraries of synthetic-phase and stress promoters as a tool for fine-tuning of recombinant proteins in *Escherichia coli*. J Biotechnol 120:25–37

Mitsuhara I, Ugaki M, Hirochika H, Ohshima M, Murakami T, Gotoh Y, Katayose Y, Nakamura S, Honkura R, Nishimiya S, Ueno K, Mochizuki A, Tanimoto H, Tsugawa H, Otsuki Y, Ohashi Y (1996) Efficient promoter cassettes for enhanced expression of foreign genes in dicotyledonous and monocotyledonous plants. Plant Cell Physiol 37:49–59

Moore I, Gälweiler L, Grosskopf D, Schell J, Palme K (1998) A transcription activation system for regulated gene expression in transgenic plants. Proc Natl Acad Sci USA 95:376–381

Moore I, Samalova M, Kurup S (2006) Transactivated and chemically inducible gene expression in plants. Plant J 45:651–683

Müller AE, Wassenegger M (2004) Control and silencing of transgene expression. In: Christou P, Klee H (eds) Handbook of Plant Biotechnology, vol 1. Wiley, Hoboken, NJ, pp 291–330

Müller F, Demény MA, Tora L (2007) New problems in RNA polymerase II transcription initiation: matching the diversity of core promoters with a variety of promoter recognition factors. J Biol Chem 282:14685–14689

Murphy KF, Balazsi G, Collins JJ (2007) Combinatorial promoter design for engineering noisy gene expression. Proc Natl Acad Sci USA 104:12726–12731

Nguyen DH, D'haeseleer P (2006) Deciphering principles of transcription regulation in eukaryotic genomes. Mol Systems Biol 2:2006.0012

Ni M, Cui D, Einstein J, Narasimhulu S, Vergara CE, Gelvin SB (1995) Strength and tissue specificity of chimeric promoters derived from octopine and manopine synthase genes. Plant J 7:661–676

Ni M, Cui D, Gelvin SB (1996) Sequence-specific interactions of wound-inducible nuclear factors with mannopine synthase 2' promoter wound-responsive elements. Plant Mol Biol 30:77–96

Novina CD, Roy AL (1996) Core promoters and transcriptional control. Trends Genet 12:351–355

Odell JT, Nagy F, Chua N-H (1985) Identification of DNA sequences required for activity of the cauliflower mosaic virus 35S promoter. Nature 313:810–812

Ohl S, Hedrick SA, Chory J, Lamb CJ (1990) Functional properties of a phenylalanine ammonia-lyase promoter from Arabidopsis. Plant Cell 2:95–106

Orphanides G, Reinberg D (2002) A unified theory of gene expression. Cell 108:439–451

Padidam M (2003) Chemically regulated gene expression in plants. Curr Opin Plant Biol 6:169–177

Pietrzak M, Burri M, Herrero J-J, Mosbach K (1989) Transcriptional activity is inducible in the cauliflower mosaic virus 35S promoter engineered with the heat shock consensus sequence. FEBS Lett 249:311–315

Pilpel Y, Sudarsanam P, Church GM (2001) Identifying regulatory networks by combinatorial analysis of promoter elements. Nature Genet 29:153–159

Puente P, Wei N, Deng XW (1996) Combinatorial interplay of promoter elements constitutes the minimal determinants for light and developmental control of gene expression in *Arabidopsis*. EMBO J 15:3732–3743

Reynolds PHS (1999) Inducible control of gene expression: an overview. In: Reynolds PHS (ed) Inducible gene expression in plants. CABI Publishing, Wallingford, pp 1–9

Rhee SY, Dickerson J, Xu D (2006) Bioinformatics and its applications in plant biology. Annu Rev Plant Biol 57:335–360

Richards EJ, Elgin SCR (2002) Epigenetic codes for heterochromatin formation and silencing: rounding up the usual suspects. Cell 108:489–500

Rud I, Jensen PR, Naterstad K, Axelsson L (2006) A synthetic promoter library for constitutive gene expression in *Lactobacillus plantarum*. Microbiology 152:1011–1019

Rushton PJ, Reinstadler A, Lipka V, Lippok B, Somssich IE (2002) Synthetic plant promoters containing defined regulatory elements provide novel insights into pathogen- and wound-induced signaling. Plant Cell 14:749–762

Samalova M, Brzobohaty B, Moore I (2005) pOp6/LhGR: a stringently regulated and highly responsive dexamethasone-inducible gene expression system for tobacco. Plant J 41:919–935

Sawant S, Singh PK, Gupta SK, Madanala R, Tuli R (1999) Conserved nucleotide sequences in highly expressed genes in plants. J Genet 78:123–131

Sawant S, Singh PK, Madanala R, Tuli R (2001) Designing of an artificial expression cassette for the high-level expression of transgenes in plants. Theor Appl Genet 102:635–644

Sawant SV, Kiran K, Mehrotra R, Chaturvedi CP, Ansari SA, Singh P, Lodhi N, Tuli R (2005) A variety of synergistic and antagonistic interactions mediated by *cis*-acting DNA motifs regulate gene expression in plant cells and modulate stability of the transcription complex formed on a basal promoter. J Exp Bot 56:2345–2353

Schenk PM, Remans T, Sági L, Elliott AR, Dietzgen RG, Swennen R, Ebert PR, Grof CPL, Manners JM (2001) Promoters for pregenomic RNA of banana streak badnavirus are active for transgene expression in monocot and dicot plants. Plant Mol Biol 47:399–412

Schmid J, Doerner PW, Clouse SD, Dixon RA, Lanb CJ (1990) Developmental and environmental regulation of a bean chalcone synthase promoter in transgenic tobacco. Plant Cell 2:619–631

Schünmann PHD, Llewellyn DJ, Surin B, Boevink P, De Feyter RC, Waterhouse PM (2003) A suite of novel promoters and terminators for plant biotechnology. Funct Plant Biol 30:443–452

Schwechheimer C, Bevan M (1998) The regulation of transcription factor activity in plants. Trends Plant Sci 10:378–383

Singh KB (1998) Transcriptional regulation in plants: the importance of combinatorial control. Plant Physiol 118:1111–1120

Smale ST (2001) Core promoters: active contributors to combinatorial gene regulation. Genes Dev 15:2503–2508

Smale ST, Kadonaga JT (2003) The RNA polymerase II core promoter. Annu Rev Biochem 72:449–479

Sørensen SJ, Burmølle M, Hansen LH (2006) Making bio-sense of toxicity: new developments in whole-cell biosensors. Curr Opin Biotechnol 17:11–16

Tang W, Luo X, Sameuls V (2004) Regulated gene expression with promoters responding to inducers. Plant Sci 166:827–834

Tavva VS, Dinkins RD, Palli SR, Collins GB (2006) Development of a methoxyfenozide-responsive gene switch for applications in plants. Plant J 45:457–469

Taylor LE II, Dai Z, Decker SR, Brunecky R, Adney WS, Ding S-Y, Himmel ME (2008) Heterologous expression of glycosyl hydrolases in planta: a new departure for biofuels. *Trends Biotechnol* 26:413–424

Tokusumi Y, Ma Y, Song X, Jacobson RH, Takada S (2007) The new core promoter element XCPE1 (X Core Promoter Element 1) directs activator-, mediator-, and TATA-binding protein-dependant but TFIID-independent RNA polymerase II transcription from TATA-less promoters. Mol Cell Biol 27:1844–1858

Tompa M, Li N, Bailey TL, Church GM, De Moor B, Eskin E, Favorov AV, Frith MC, Fu Y, Kent WJ, Makeev VJ, Mironov AA, Noble WS, Pavesi G, Pesole G, Régnier M, Simonis N, Sinha S, Thijs G, Van Helden J, Vandenbogaert M, Weng Z, Workman C, Ye C, Zhu Z (2005) Assessing computational tools for the discovery of transcription factor binding sites. Nature Biotechnol 23:137–144

Tornøe J, Kusk P, Johansen TE, Jensen PR (2002) Generation of a synthetic mammalian promoter library by modification of sequences spacing transcription factor binding sites. Gene 297:21–32

Tyo KE, Alper HS, Stephanopoulos GN (2007) Expanding the metabolic engineering toolbox: more options to engineer cells. Trends Biotechnol 25:132–137

Van Leeuwen W, Ruttink T, Borst-Vrenssen AWM, Van der Plas LHW, Van der Krol AR (2001) Characterization of position-induced spatial and temporal regulation of transgene promoter activity in plants. J Exp Bot 52:949–959

Venter M (2007) Synthetic promoters: genetic control through *cis* engineering. Trends Plant Sci 12:118–124

Vision TJ, McLysaght A (2004) Computational tools and resources in plant genome informatics. In: Christou P, Klee H (eds) Handbook of Plant Biotechnology, vol 1. Wiley, Hoboken, NJ, pp 201–228

Wang R, Zhou X, Wang X (2003) Chemically regulated expression systems and their applications in transgenic plants. Transgenic Res 12:529–540

Wasserman WW, Sandelin A (2004) Applied bioinformatics for the identification of regulatory elements. Nature Rev Genet 5:276–287

Weber W, Fussenegger M (2006) Pharmacologic transgene control systems for gene therapy. J Gene Med 8:535–556

Werner T, Fessele S, Maier H, Nelson PJ (2003) Computer modelling of promoter organisation as a tool to study transcriptional coregulation. FASEB J 17:1228–1237

Xiao K, Zhang C, Harrison M, Wang Z-Y (2005) Isolation and characterization of a novel plant promoter that directs strong constitutive expression of transgenes in plants. Mol Breed 15:221–231

Xie M, He Y, Gan S (2001) Bidirectionalization of polar promoters in plants. Nature Biotechnol 19:677–679

Yew NS (2005) Controlling the kinetics of transgene expression by plasmid design. Adv Drug Deliver Rev 57:769–780

Yoshida K, Shinmyo A (2000) Transgene expression systems in plant, a natural bioreactor. J Biosci Bioeng 90:353–362

Zhou Q, Wong WH (2004) CisModule: *de novo* discovery of *cis*-regulatory modules by hierarchical mixture modelling. Proc Natl Acad Sci USA 101:12114–12119

Zhu Q, Song B, Zhang C, Ou Y, Xie C, Liu J (2008) Construction and functional characteristics of tuber-specific and cold-inducible chimeric promoters in potato. Plant Cell Rep 27:47–55

Zuo J, Chua N-H (2000) Chemical-inducible systems for regulated expression of plant genes. Curr Opin Biotechnol 11:146–151

Index